MW01479491

AVIONICS: SYS
AND TROUBLESHOOTING

A practical guide to non-traditional avionics

Thomas K. Eismin

First Edition
Copyright 2002

Illustrations and page design by:
Luke Collier and Stacey Strickler

ISBN Number 0-9708109-1-1

Published in the USA by
AVOTEK, a Division of Select Aerospace Industries, Inc.
P.O. Box 219 Weyers Cave Virginia 24486

ORDER NUMBER
T-AVSAT0101

1-800-828-6839
(540) 432-0180
FAX (540) 432-0193

To my parents Nick and Mary Lou. You showed me how to live, work hard, respect others, and enjoy life. My mother taught me how to love and be loved. My father taught me about technology and electronics. Thank you for all you have done.

PREFACE
AVIONICS: SYSTEMS AND TROUBLESHOOTING

For decades aircraft have been dependant on electronic systems for various navigation, communication, and flight control functions. These electronic systems have been termed Avionics (short for aviation electronics). Today the term avionics reaches far beyond the traditional systems. Modern aircraft employ avionics for electrical power management, engine thrust control, regulation of cabin pressure, and even passenger entertainment. In general, today's high performance aircraft use avionics to control virtually every system.

Contemporary avionics are quite different from those found on older aircraft. Digital electronics and computer- based systems have replaced analog devices. Modern avionics no longer employ electro-mechanical instruments; they have been replaced with solid-state displays. The most recent aircraft utilize integrated systems that share a variety of components.

This trend toward the increased use of electronic systems has created a shift in the way aircraft are maintained. Aircraft technicians (A&P mechanics) are expected to perform routine preventative maintenance and line replacement of many avionics systems. To aid in system diagnostics, central maintenance computers are used to monitor almost every function of the aircraft. Line technicians can now isolate faults from the flight deck and correct defects that previously required avionics specialists. Avionics specialists, or R&E (radio and electronics) technicians troubleshoot and repair only a small fraction of electronic systems onboard the aircraft.

In today's environment, it is extremely important for avionics technicians to understand the technology of aircraft systems. It is equally important for A&P technicians to possess a thorough understanding of avionics. Avionics: Systems and Troubleshooting provides the reader with a study of integrated electronic systems as they apply to flight line maintenance.

Avionics: Systems and Troubleshooting begins with information common to a variety of avionics systems. Aircraft manuals, wiring diagrams, and schematics are each studied and their common features discussed. Since all avionics systems require electrical power, technicians must have a good working knowledge of power systems. Power distribution systems for a variety of corporate and transport category aircraft are studied. Common digital data formats, such as, ARINC 429 and 629 are presented in great detail along with their related troubleshooting techniques. Complex avionics systems common to most modern aircraft are also presented in this text. Integrated displays, autopilot/autoflight, flight management, GPS, ACARS, passenger entertainment, and central diagnostic systems are all discussed. Complex systems are presented from the perspective of the flight line technician. System architecture, operation, and troubleshooting techniques are all discussed for a variety of common avionics systems.

Since this text presents functional concepts of avionics systems, its information is valuable to a variety of individuals. Avionics technicians can utilize Avionics: Systems and Troubleshooting to gain a better understanding of line troubleshooting and computer controlled aircraft systems. A&P mechanics should study this text for insight into system evaluation of modern complex aircraft. Avionics engineers and bench technicians can also gain a better perspective of the entire aircraft through the study of this text.

For educators, this text is designed as a "ready to use" course on advanced electronics/avionics. The materials presented in this book will bring your students well ahead of any FAA advanced electronics requirements and satisfy the needs for both corporate and air carrier technicians. A student study guide is available for this text that provides a variety of questions presented in assignment format. Each page is designed so the assignments can be turned in individually. The instructor can choose from a variety of questions and find all answers on the instructor's CD-ROM. The CD-ROM also contains each figure presented in the text. Over 500 illustrations can easily be made into handouts, lecture notes, or copied into unit exams. This text along with the CD-ROM is truly instructor friendly.

ACKNOWLEDGMENTS

The author would like to thank the following individuals, companies, and corporations for their generous contributions to making this text possible.

Air Transport Association, Inc.
Airbus Industrie
Airbus Service Company, Inc
AlliedSignal Bendix King
Beech Aircraft Corporation
Boeing Commercial Airplane Group
Collins Commercial Avionics
Garmin International Inc.
Honeywell, Inc.
IFR Systems, Inc.
Northwest Airlines Inc.
Raytheon, Inc.
Rockwell International Corporation

Special thanks to:
Luke Collier, Gale Littlewood, Melinda Pickering, Lorie Schrimshaw, Justin Shafer, Stacey Strickler and Marsha Stultz.

TABLE OF CONTENTS

CHAPTER 1
AVIONICS TECHNICAL DATA

INTRODUCTION

Modern avionics equipment performs a variety of functions and can be found on virtually any type of aircraft. With this type of diversity it is often difficult to keep track of all the paper work associated with aircraft electronic systems. Troubleshooting and repair of these systems require access to different manuals and often require considerable time to simply find the defective component or wire on the aircraft.

To help technicians in their quest to keep avionics systems airworthy several standards have been developed. The most commonly used standard for aircraft reference materials has been established by the **ATA (Air Transport Association)**. The ATA represents airlines, aircraft manufacturers, and various system manufactures in an effort to ensure uniformity in various facets of the aviation industry. Although the ATA has its largest impact in transport category aircraft, many of the standards also apply to general aviation type aircraft.

The ATA has helped to gain standardization for reference materials and for component locations. As aircraft grow in size and complexity, it was soon recognized that the ATA should set standards for locating various components, parts, and sections of the aircraft. A series of stations and zones soon became recognized as a means to identify locations on the aircraft. The following chapter will introduce the reader to the basic organization of aircraft technical manuals and component location systems.

PRESENTATION OF TECHNICAL DATA

One of the biggest challenges for today's technician is to keep up with an overwhelming amount of paper work. As aircraft have become increasingly dependent on electronics for basic flight and navigation, the documentation for these systems has increased significantly. Major manufacturers, such as the Boeing Corporation, print literally tons of manuals covering inspection, maintenance, and repair of their avionics systems.

To help reduce the paperwork needed to keep manuals current most airlines and many general aviation shops rely on micro film to store technical data. Any revisions to the technical publications are made simply by replacing the appropriate micro film with an updated version. Micro film is commonly available in two formats, known as **Microfiche** and **Microfilm** strips. Microfiche consist of a four by six inch film card which is capable of storing up to 288 pages of data. A special reader is used to view the cards.

Microfilm strips, often referred to as *film*, are used mainly by airlines and those companies maintaining transport category aircraft. A typical film reader is shown in figure 1-1. Each film strip is enclosed by an individual cartridge which are categorized by aircraft and manual type. To access a particular section of a manual simply install the film cartridge into the reader and advance the film to the desired area.

Figure 1-1. A typical micro film reader is used to access a particular section of a maintenance manual.

Although the maintenance manual micro film format has many advantages, microfilm is not easily viewed on the flight line. To alleviate this problem, most maintenance facilities use microfilm reader/printers. Once the appropriate pages have been located on the film, the machine is used to print paper copies of any needed documents.

To further simplify the use of technical data some manufacturers have introduced their manuals on a **CD ROM** format. The major advantage of this format is that it allows near instant access to various sections of the manuals through the use of a personal computer. The speed of the CD ROM format is exceptionally beneficial when performing maintenance tasks which require information from multiple sections of the aircraft manuals.

ATA Specifications

The **Air Transport Association (ATA)** has developed a standard for the organization of technical data. The standard, called **ATA Specification 100**, (also called *ATA 100*) applies to virtually all aircraft technical publications. Maintenance manuals, minimum equipment lists, parts lists, wiring diagrams, and even computer based diagnostics are all organized using ATA 100. In general, the ATA specification 100 was designed to provide consistency between various manufacturers. Technicians who work on a variety of aircraft, such as, the B-737, the A-340, or the MD-11 need only become familiar with one system for organizing publications.

The Specification 100 system consists of three two-digit numbers. As shown in figure 1-2, the first two digits of the number designate the chapter, or specific system, of the aircraft. For example, the general number 23-00-00 represents the entire communications chapter. The second two digits represent a section within the given chapter. The number 23-20-00 represents materials covered in the communications chapter (23), VHF/UHF section (20). The first three digits of the code are specified by the ATA standard and cannot be changed by the airline or manufacturer.

The last three digits of the code are typically assigned by the manufacturer and bring further detail to the chapter. In this example, (figure 1-2) the complete number 23-

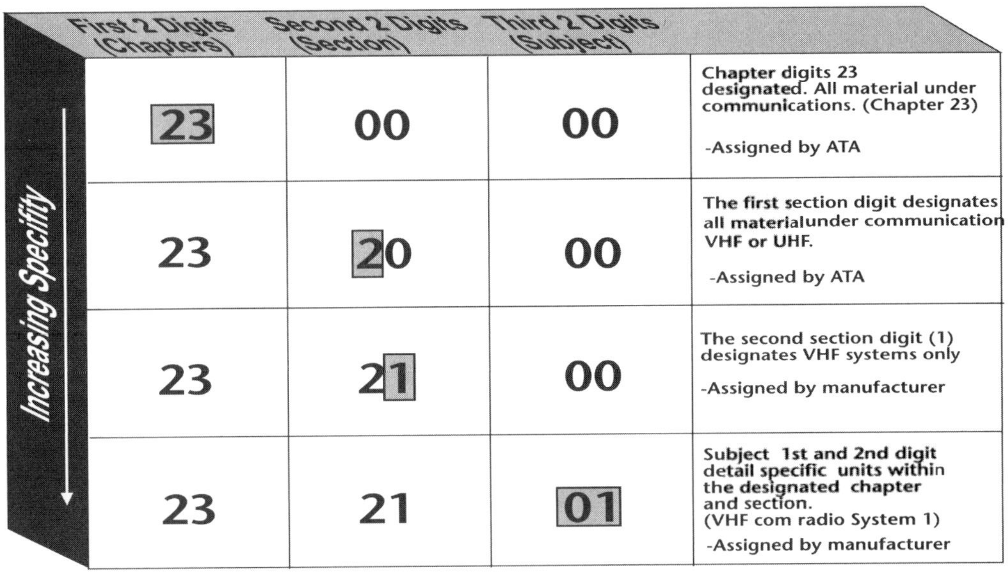

First 2 Digits (Chapters)	Second 2 Digits (Section)	Third 2 Digits (Subject)	
23	00	00	**Chapter digits 23 designated.** All material under communications. (Chapter 23) -Assigned by ATA
23	20	00	**The first section digit designates all material under communication VHF or UHF.** -Assigned by ATA
23	21	00	**The second section digit (1) designates VHF systems only** -Assigned by manufacturer
23	21	01	**Subject 1st and 2nd digit detail specific units within the designated chapter and section.** (VHF com radio System 1) -Assigned by manufacturer

Increasing Specificity

Figure 1-2 The ATA specification 100. The first two digits of the number designate the chapter, the second two digits represent a section.

21-01 designates the communications, VHF radio, system number 1. It should be noted that all numbers of a given chapter may not be used by a given manufacturer. In fact the specification 100 does not use all possible combinations for the first three digits of the code. This allows for flexibility in the standard and for future aircraft systems.

It should be noted ATA Specification 100 has been revised periodically. In May 2000, a new version of the spec was introduced that combined ATA 100 and ATA 2100 (digital data standards). This combined version of the standard is called **ATA 2200**. ATA 2200 still maintains a list of chapter, section, and subject codes for organization of aviation technical data. Since ATA 2200 is relatively new most technicians and technical publications will still refer to ATA 100. **ATA 100 has simply been superseded by ATA 2200.** The remainder of this text we will refer to the chapter, section, and subject codes as part of the ATA Spec 100 since most of the industry is familiar with that term. It should also be noted that periodic revisions to ATA 100 (and 2200) create slight differences in the layout of technical manuals. The current chapter and section numbers of the ATA specification 100 (2200) are included in appendix A of this text.

The chapter format just discussed can be categorized into five main sections, **Customized Information, General, Airframe Systems, Structures**, and **Powerplant** (figure 1-3). Airline specified data is contained in Chapters 1-4 (**customized information**). Chapters 5-12, are **General** items, which includes taxiing, towing and servicing. Chapters 20-49 are the **Airframe Systems** items, including electrical, mechanical, pneumatic, and hydraulic systems. Chapters 51-57 contains **Structures** information, such as, wings, doors, fuselage, etc. Chapters 70-80 are the **Powerplant** chapters, including all engine related items.

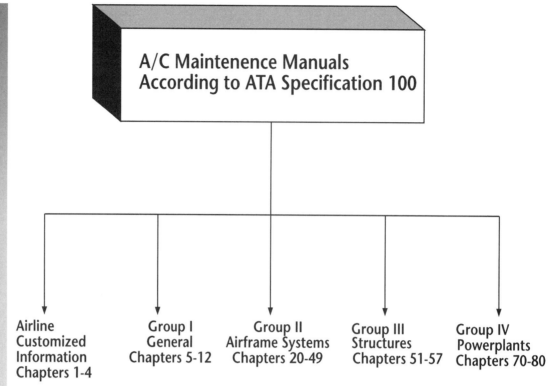

Fig. 1-3 **The 5 main sections of a maintenance manual are Customized Information, General Information, Airframe Systems, Structures and Powerplants.**

As an electronic/avionics technician you will most likely use group 2 of the manuals most often. This group contains chapter 22 (auto flight), chapter 23 (communications), chapter 24 (electrical power), chapter 31 (indicating and recording systems), chapter 34 (navigation) and chapter 45 (central maintenance systems). The general group may also be of particular interest, since in many cases the aircraft must be towed, taxied or set into a particular configuration prior to maintenance. Chapter 20 (standard practices airframe) is another chapter often used by avionics technicians. Chapter 20 contains important information, such as, wire splicing techniques which is common to various areas of the aircraft.

GAMA Specification 2

The General Aviation Manufacturers Association (GAMA) has developed a standard for the organization of aircraft publications called **GAMA Specification 2**. This specification is very similar to the ATA 100 spec, however the GAMA version is more flexible to allow for the relatively small documents used by light single-engine aircraft. The GAMA specification was designed to be compatible with the ATA spec 100.

The manufacturers of general aviation aircraft may choose between the GAMA or ATA specification. Many manufacturers of corporate type aircraft, such as the Lear 25 or Falcon 20, utilize the ATA specification even though they are actually general aviation aircraft. No matter which system is chosen for organization of technical publications, if the technician becomes familiar with the standard it will greatly reduce the time needed to find important technical information.

Aircraft Manuals

Various manuals have been designed to communicate information regarding the operation and maintenance of aircraft. The manuals most commonly used during avionics/electronics troubleshooting and repair are the **maintenance manuals, wiring diagram manuals, and parts manuals.** These manuals can either be supplied by the aircraft manufacturer or by the vendor of a given system. The **installation manual** supplied by a specific vendor is often used by avionics technicians during initial system installation. In some cases, the pilot's operation guide or **operator's manual** is used by technicians during system inspection and/or troubleshooting.

The complexity of the manual system is usually a function of the type of aircraft. Light aircraft may contain only one, relatively simple, maintenance manual which also contains all wiring diagrams. Light aircraft typically use a separate parts manual. Large and/or complex aircraft typically employ a series of manuals. Large aircraft manuals are typically organized according to ATA specification 100.

Maintenance Manuals

Aircraft **maintenance manuals** provide specific information for flight line and hangar maintenance activities. Data for both scheduled and unscheduled activities are included in most manuals. Maintenance manuals are often used by avionics technicians during troubleshooting of various systems on the aircraft. Maintenance manuals provide information on installation, and location of components; as well as troubleshooting, testing and inspection procedures. The description and operation section of a manual is often extremely helpful as an overview of the entire system under study. When in question about any system, start by reading the description and operation. Typically, the description and operation is the first subject for each section of a chapter (subject designator 00).

The correct section of the maintenance manual can easily be found through the following procedures: 1) Identify the correct ATA chapter (example; chapter 22, Auto Flight). If you are not familiar with the specific ATA chapter titles, a list is typically provided at the beginning of each maintenance manual. 2) Locate the table of contents for the chapter desired. The table of contents should be the first page(s) in that chapter. 3) Simply follow through the table of contents to find the specific system/procedure desired. It should be noted that *sections* are arranged numerically within the table of contents, and *subjects* are arranged alphabetically. For example, Autoflight - General (22-00-00) would be listed before Autoflight - Flight Director System (22-12-00). 4) If the correct chapter/section/subject number is already known, it is not necessary to reference the chapter table of contents. Simply go directly to the desired page within the manual.

An example of the chapter 22 table of contents is shown in figure 1-4. Here it can be seen that the top of the page shows the major heading (Chapter 22 - AUTO FLIGHT) and the headings for each column. Column 1 (*subject*) contains a description of the system. Column 2 (*chapter/section/subject*) lists the ATA code. This ATA number is used to find the materials in the maintenance manual. Column 3 (*page*) and column 4 (*effectivity*) contain more information which will further breakdown the contents of the chapter.

PAGE BLOCK TOPIC DESCRIPTION	
PAGE BLOCK	**INCLUDED TOPIC(S)**
001-099	Description and Operation (D/O)
101-199	Troubleshooting (T) Note: includes Electrical Schematics
201-299	Maintenance Practices (MP)
301-399	Servicing (S)
401-499	Removal and Installation (R/I)
501-599	Adjustment and Test (A/T)
601-699	Inspection and Check (I/C)
701-799	Cleaning and Painting (C/P)
801-899	Approved Repairs (AR)

Table 1-1   Chapter 22 (Auto flight) Table of Contents is used to find the materials in the maintenance manual.

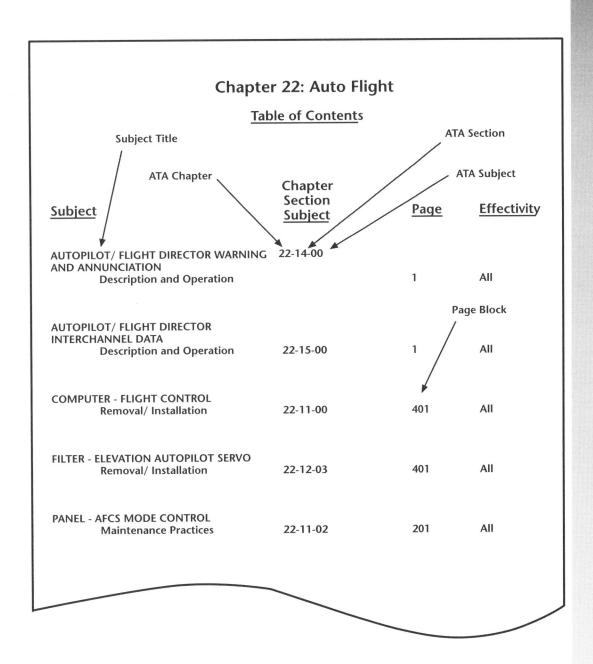

Fig. 1-4   Chapter 22 (Auto flight) Table of Contents is used to find the materials in the maintenance manual.

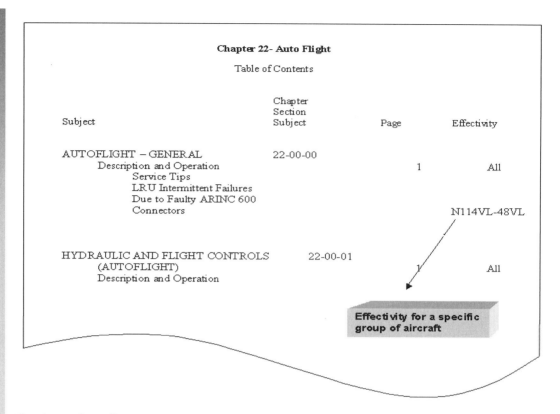

Fig. 1-5 The Effectivity column of a typical maintenance manual chapter index for a specific group of aircraft.

Page blocks

Page blocks are used to further categorize a given chapter-section-subject of the maintenance manual. The term page block refers to a set of pages, such as 401 - 499. Each page block covers a given set of topics which is consistent throughout the manuals. The topics covered by the various page blocks are shown in Table 1-1. The actual page number for any given page in the maintenance manual is comprised of a three digit number, such as 401. The page block number is located in the Table of Contents at the start of each chapter of the maintenance manual (see figure 1-4).

The first page within a given block is always page _ _ 1. For example, the first page under the topic Inspection/Check would be 601; the second page would be 602, etc. If there are more than 99 pages in any given block the next page will be numbered beginning with the letter *A*. For example, a sequence might be 601, 602, ... 698, 699, A600, A601, A602, etc.

Effectivity

The **effectivity** of a given maintenance page is used to determine if your aircraft is covered by the information stated in that section (page) of the manual. The table of contents at the beginning of each chapter of the maintenance manual contains an effectivity column (see figure 1-5). If the word **ALL** is listed under effectivity, it designates that the corresponding pages apply to all the aircraft covered by that manual. A list of "ALL" aircraft (by serial or registration number) is included in the beginning of each manual.

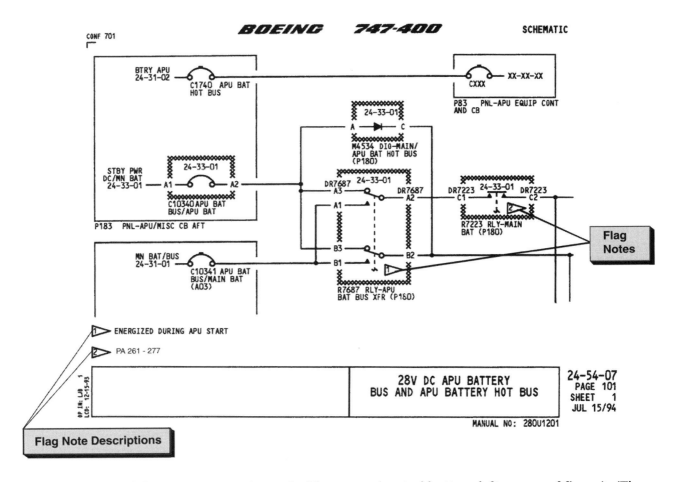

Fig. 1-6 Typical use of flag notes on a schematic (Flag notes located bottom left corner of figure). (The Boeing Commercial Airplane Company). For Training Purposes Only.

Effectivity can also be stated according to specific type of equipment installed on the aircraft, or specific aircraft serial or registration numbers. In figure 1-5, the Description and Operation subsection *Service tip: LRU intermittent failures due to faulty ARINC 600 connectors* applies only to aircraft with registration numbers N141VL to N148VL. The general information under the section *Description and Operation* applies to all aircraft covered by this maintenance manual.

Effectivity is also located in the bottom left corner on all pages of the maintenance manual. In some cases, the effectivity is stated in the verbiage of the maintenance manual or designated by a flag note. A **flag note** is used to bring attention to a specific effectivity on a diagram or schematic. As shown in figure 1-6, the flag is placed on the diagram in the effected area and the note is located in the lower left section of the page. In this example flag number two denotes this configuration is for aircraft with serial numbers PA 261 to 277.

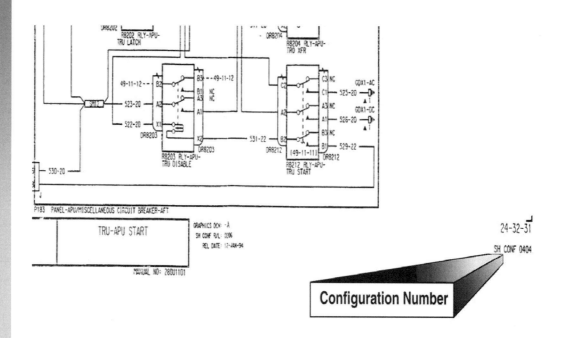

Fig 1-7 A typical schematic labeled for a specific configuration, configuration
number designates a page of the same diagram that has been modified.
(Configuration numbers located in the lower right corner of figure.) (The Boeing
Commercial Airplane Company). For Training Purposes Only.

If the effectivity change is too large to contain in a simple note, the diagram or
schematic will be labeled with a configuration number in the lower right corner of
the page (figure 1-7). A **configuration number** designates a first, second, third,
etc. page of the same diagram which has been modified for a given configuration of
aircraft. This type of effectivity change occurs only in limited situations.

Cautions and warnings notes

Whenever using any manual pay particular attention to all cautions and warnings. A **warning** calls attention to any methods, materials, or procedures that must be followed to avoid
injury or death. A **caution** calls attention to any methods, materials, or procedures that
must be followed to avoid damage to equipment on the aircraft. A **note** is used to draw
your attention to a particular procedure that will make the task easier to perform. Many of
the notes within a manual will refer the technician to chapter 20 (standard practices).
Standard practices, will be discussed later in this chapter.

Revisions

Virtually all manuals are revised periodically. For transport category aircraft, regular revisions are made every four months. At that time micro films are replaced if they are effected
by a revision and corrections are inserted into paper manuals. Temporary revisions are
used for items which require change before the regularly scheduled revisions. Temporary
revisions are filed next to the effected page for paper manuals and kept in a separate
binder for micro film. Temporary revisions are printed on yellow paper.

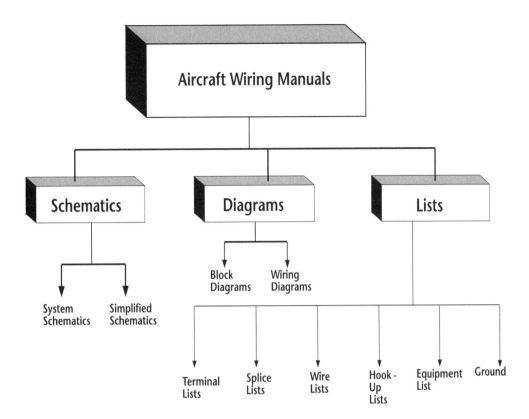

Fig. 1-8   The general classifications of wiring manuals typically used by transport category aircraft.

Wiring Manuals

A series of **wiring manuals** have been developed which contain various diagrams, charts, lists and schematics needed to maintain the various electrical/electronic systems. For light aircraft, wiring diagrams and schematics may be contained in the maintenance manual; however, all complex aircraft wiring information is contained in a separate manual of one or more volumes.

Virtually all transport category and many general aviation aircraft follow ATA Specification 100 in the design of their wiring manuals. Figure 1-8 shows the general classifications of wiring manuals which are typically used by transport category aircraft. There are three broad categories of wiring manuals, **schematics, diagrams, and lists**. The various manuals may be produced as stand alone publications or contained as part of other manuals.

Diagrams

Electrical diagrams are designed to provide an understanding of a system through the use of line drawings and/or pictures of the various system components. There are two basic types of electrical diagrams used in modern aircraft manuals, block diagrams and wiring diagrams. **Block diagrams** define electrical circuits by separating a system or subsystem into functional blocks. Block diagrams typically show the various connections between subsystems divided into categories, such as, input and output connections, or analog and digital circuits. This type of configuration makes block diagrams especially useful in gaining an initial understanding of a given system. A typical block diagram is shown in figure 1-9. **Wiring diagrams**, on the other hand, are typically very specific with details on wires, connectors, and pin numbers for a given system. Since wiring diagrams are

so specific, they typically have a limited scope. In many cases it requires several pages of wiring diagrams to detail an entire system or subsystem. Wiring diagrams are typically used during aircraft assembly, installation of systems and detailed troubleshooting.

Schematics

Schematics are typically more general than wiring diagrams. Schematics are often used for identification of basic system components during troubleshooting. In some cases, there may be two levels of schematics provided by the aircraft manufacturer, simplified schematics and system schematics. **Simplified schematics** are wiring illustrations with intermediate depth and scope. These schematics typically show all LRUs but

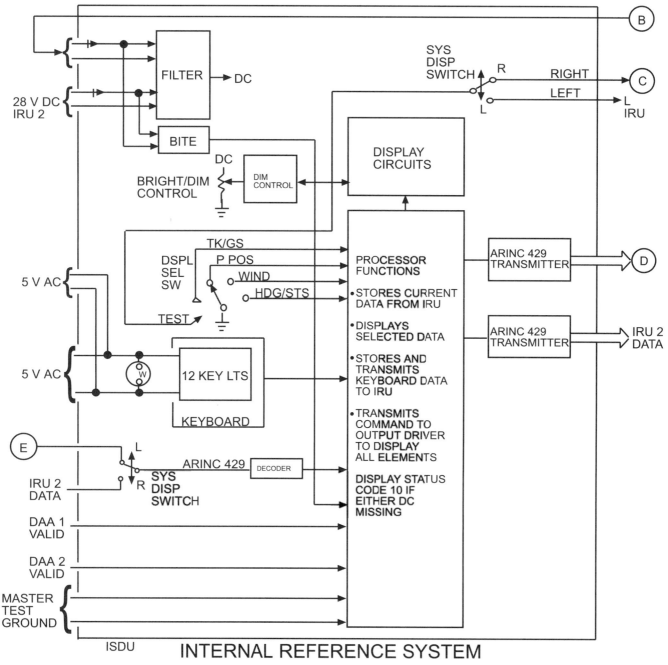

Fig. 1-9 A typical system block diagram. (The Boeing Commercial Airplane Company) For Training Purposes Only.

do not detail all wiring. Simplified schematics are often used for elementary trouble-shooting. System schematics typically contain more depth than simplified schematics however, they are still less specific than wiring diagrams. **System schematics** typically show all related components, connections and wiring within a system. In many cases, system schematics will not show specific wire or connector numbers; that information is contained in the wiring diagrams.

It should be noted, the terms *schematic* and *wiring diagram* are often used interchangeably when referring to electrical/electronic systems. Technicians often use the term wiring diagram to refer to a schematic, and visa versa. Some manufacturers even seem to confuse the terms. For example, it is easy to find an aircraft system schematic in a wiring diagram manual. It is therefore important to become familiar with the publications for the aircraft you maintain and the various types of wiring information they provide.

Wiring Manual Numbering

All schematics and diagrams are identified with a code number according to ATA specification 100. The first four digits of the code are used to identify block diagrams and simplified system schematics. If a schematic has only four digits, it is covered again in more detail on another schematic. A fifth digit is added to identify all third level schematics. If the complete system is covered in one schematic, a *-0* is used for the fifth digit. If the system requires 2, 3 or more schematics, the fifth digit is *-1, -2*, etc. For example, a simplified system schematic could be *31-22*, a system schematic could be *31-22-01*.

In many cases a given system or subsystem is too complex to include on one single page (sheet) of the manual. The **sheet number** is used to designate the number of pages required for a given schematic or diagram. The number *33-22-00, Sheet 2* indicates the second page for the third level schematic where the entire system is shown on one schematic which requires two pages (sheets).

In some cases, schematics are numbered corresponding to the maintenance manual page blocks. Schematics will always be in page block 101-199 (troubleshooting). Wiring diagrams are labeled with page numbers 1, 2, 3, etc. according to the number of pages (not sheets) for a given system or subsystem.

Lists and charts

The Aircraft Wiring Lists (AWL) contain very specific information for a given circuit. The **Equipment list, Hook-up list, and Master Wire lists** are all contained in the aircraft wiring lists. The Equipment list contains information on various electrical/electronic components and references the appropriate aircraft schematic or diagram. The Hook-up list contains information on specific wire connections to plugs and receptacles, terminal blocks, splices, and ground points. The Master Wire list contains detailed information on wire size, types and lengths. In some cases a separate splice list, ground list, and terminal list is included in the manuals to show further detail of the system.

A master wire list from an Airbus A-320 is shown in figure 1-10. The heading of the wire list identifies the meaning of each column of the list. The three major categories are *WIRE IDENTIFICATION, FROM TERMINATION A*, and *FROM TERMINATION B*.

A320 AIRCRAFT WIRING LIST											
WIRE IDENTIFICATION					FROM TERMINATION A			FROM TERMINATION B			WIRING DIAGRAM MANUAL REFERENCE	EFFECT
NUMBER	COLOR IF NOT WHITE	TYPE/ GAUGE	LENGTH IN CM	ROUTE	LOCATION ZONE	FUNCTIONAL IDENTIFICATION NUMBER	CONNECTOR PIN NUMBER	LOCATION ZONE	FUNCTIONAL IDENTIFICATION NUMBER	CONNECTOR PIN NUMBER		
2373-1705		CF 20	215	1M	223	340RH1	5	223	2749VT	20G	23-73-52	ALL
2373-1706		CF 24	310	1M	223	300RH1	4	223	340RH1	4	23-73-52	ALL
2373-1707		CF 24	310	1M	223	300RH1	3	223	340RH1	3	23-73-52	ALL
2373-1708		CF 24	310	1M	223	300RH1	2	223	340RH1	2	23-73-52	ALL
2373-1709		CF 24	310	1M	223	300RH1	1	223	340RH1	1	23-73-52	ALL
2373-1710	B	PF 24		1M	223	300RH1	50	221	320RH1	2	23-73-54	ALL
2373-1710	R	PF 24		1M	223	300RH1	49	221	320RH1	1	23-73-54	ALL
2373-1712		CF 20	115	1M	221	320RH1	3		6681VN	B1	23-73-54	ALL
2373-1713		CF 24	15	1M			20	223	300RH1	20	23-73-54	ALL
2373-1714	B	PF 24		1M	223	300RH1	5	221	2791VC	P	23-73-53	ALL
2373-1714	R	CF 24		1M	223	300RH1	4	221	2791VC	N	23-73-53	ALL
2373-1716		CF 24	15	1M			3	223	300RH1	3	23-73-53	ALL
2373-1717	B	PF 24		1M	223	300RH1	7	221	2791VC	V	23-73-53	ALL
2373-1717	R	PF 24		1M	223	300RH1	6	221	2791VC	U	23-73-53	ALL
2373-1722		CF 24	150	1M	223	300RH1	45	223	2749VT	5E	23-73-50	ALL
2373-1723		CF 24	150	1M	223	300RH1	11	223	2749VT	16K	23-73-50	ALL

Fig. 1-10   A typical master wire list. (Airbus Industrie) For Training Purposes Only.

The wire identification includes information about the wire number, color, manufacturer, length, and routing. The *From/To* termination identifies the location of each end of the wire, electrical component (LRU) to which the wire is connected, the specific connector, and the pin for that wire. The left most columns show the wire diagram ATA number and effectivity for each wire. For the A-320, the wire number identifies the ATA chapter and section of the system for that wire. In figure 1-10, notice the first four digits of the wire number (left hand column) are identical to the ATA chapter and section (right hand column). Each of the wires on this list begins with 2373; indicating ATA chapter 23 (communications), section 73 (cabin inter-communications system).

Charts are used to depict specific component or wire locations, in the aircraft or on a subsystem, such as pins or sockets of a connector plug. An example of a chart depicting a glare shield connector is shown in figure 1-11. Charts are contained in Chapter 91 of the wiring manuals.

Standard practices

The procedures and practices used repeatedly during aircraft maintenance, troubleshooting, and repair are contained in the **standard practices** chapter of each manual. According to the ATA code, standard practices are always in chapter 20. **Chapter 20** of the maintenance manual contains procedures and practices that are common to a variety of maintenance operations. **Chapter 20** of the wiring manual contains practices that are common to a variety of wiring, electrical and electronic systems.

The first section of each standard practices chapter is **safety practices**. Other pertinent information includes: electrical bonding practices, soldering of electrical connectors, installation of undersize wire insulation sleeves, proper torquing of electrical connectors, and cleaning of electrical connectors. Figure 1-12 shows an excerpt from a wiring diagram manual, standard practices chapter. Remember chapter 20 (standard practices) contains common information. Whenever a procedure cannot be found in other sections of the manual, check chapter 20. The procedure you are looking for may be a *standard practice*.

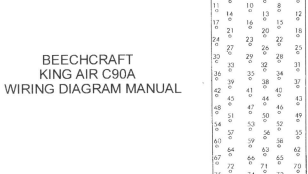

BEECHCRAFT
KING AIR C90A
WIRING DIAGRAM MANUAL

VIEW FROM BACK OF CONNECTOR

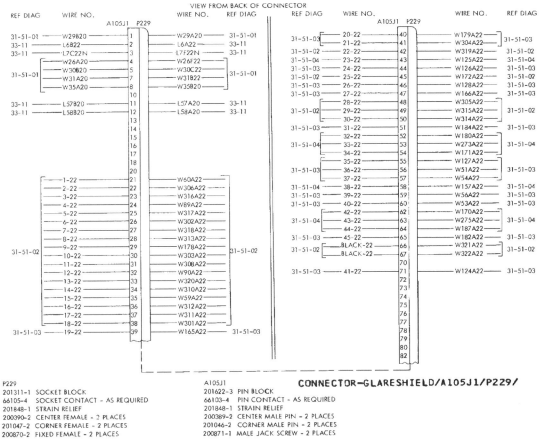

P229
201311-1 SOCKET BLOCK
66105-4   SOCKET CONTACT - AS REQUIRED
201848-1 STRAIN RELIEF
200390-2 CENTER FEMALE - 2 PLACES
201047-2 CORNER FEMALE - 2 PLACES
200870-2 FIXED FEMALE - 2 PLACES

A105J1
201622-3 PIN BLOCK
66103-4   PIN CONTACT - AS REQUIRED
201848-1 STRAIN RELIEF
200389-2 CENTER MALE PIN - 2 PLACES
201046-2 CORNER MALE PIN - 2 PLACES
200871-1 MALE JACK SCREW - 2 PLACES

CONNECTOR-GLARESHIELD/A105J1/P229/

Fig.1-11   An example of a chart depicting a glare shield connector from a wiring diagram manual. (Raytheon Aircraft Co.) For Training Purposes Only.

2. <u>Dead—end Shielding of Coaxial Cable, Single or Multiconductor Shielded Cables</u>

   A.   Remove the outer jacket over the shield braid. Use extreme caution to avoid damage to shielding and inner conductors.

   <u>NOTE</u>: A Reon R—720 cable jacket removal tool can be used on BMS 13—51 cable.

   B.   Fold the shielding back over the cable jacket 1/4 to 3/8 inch. See Figure 3.

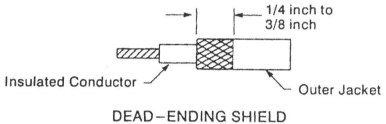

DEAD—ENDING SHIELD

Figure 3

   C.   Insulate the shield dead—end with Raychem RT—876 heat shrinkable sleeve. Select proper size sleeve to fit diameter of wire. See Subject 20—10—14.

   As an option, insulate the shield dead—end with tape, using tape wrap procedures of Subject 20—30—12.

   20—10—15   Page 4
   MAR 01/94

Fig. 1-12   Portion of a standard practices chapter of a wiring diagram manual. (The Boeing Commercial Airplane Company). For Training Purposes Only.

Using Wiring Manuals

Many light aircraft wiring diagrams and schematics are often contained in the aircraft's maintenance manual. Larger aircraft wiring diagrams and schematics are contained in separate manuals which are divided by ATA chapter. A wiring diagram from a Beechcraft King Air is shown in figure 1-13. It should be noted that adjacent to each wiring diagram is the associated parts list. This system allows for quick identification of defective components. The King Air Wiring Diagrams Manual also contains an **Alpha-Numeric Index** at the beginning of the Wiring Diagrams Manual. This index allows the technician to identify a given wiring diagram using a specific part number. To find a given wiring diagram for this aircraft, the systems can be referenced using an appropriate part number or ATA code.

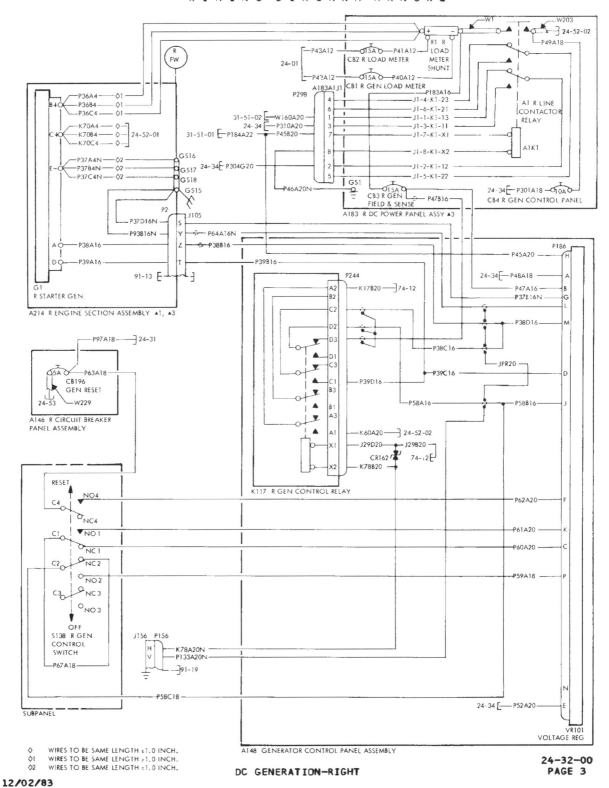

BEECHCRAFT
KING AIR C90A
WIRING DIAGRAM MANUAL

Fig. 1-13(a) Wiring diagram of DC generator from a Beechcraft King Air. For Training Purposes Only.

12/02/83

DC GENERATION-RIGHT

24-32-00
PAGE 3

B E E C H C R A F T
KING AIR C90A
W I R I N G   D I A G R A M   M A N U A L

GAMA/ATA CODE & REF DES	PART NO.	DESCRIPTION 1 2 3 4 5 6 7	UNITS PER ASSY	INSTL ZONE	USABLE ON CODE
32-00-A146		PANEL ASSY,RH CIRCUIT BREAKER/SEE CHAPTER 24-53	1	246	
-CB196	7277-2-5	CIRCUIT BREAKER,GENERATOR RESET	1	246	
-A146W229		. BUS BAR/SEE CHAPTER 24-53/.	1	246	
-A148	109-364017-13	PANEL ASSY,GENERATOR CONTROL.	1	153	
-CB162	100-361045-1	. TRANSZORB	1	153	
-A148K117	K-D4A	. RELAY,RH GENERATOR CONTROL.	1	153	
-A148P186	MS27473T20A16S	. PLUG,RH VOLTAGE REGULATOR	1	153	
-A148P186	MS27506A20-2	. CLAMP,STRAIN RELIEF	1	153	
-A148P244	SO1048-8308	. PLUG,RH GENERATOR CONTROL RELAY	1	153	
-VR101	101-364270-5	. REGULATOR,RH VOLTAGE.	1	153	
-A183	90-364151-1	PANEL ASSY,RH DC POWER.	1	621	
-A183A1	109-361030-9	. RELAY ASSY,RH LINE CONTACTOR.	1	621	
-A183A1J1	211068-1	. . RECEPTACLE,RH LINE CONTACTOR RELAY.	1	621	
-A183A1J1	66099-4	. . PIN,CONTACT	AR	621	
-A183A1K1	K-D4A	. . RELAY,LH LINE CONTACTOR	1	621	
-A183CB1	7277-2-15	. CIRCUIT BREAKER,RH GENERATOR LOAD METER . . .	1	621	
-A183CB2	7277-2-15	. CIRCUIT BREAKER,RH LOAD METER	1	621	
-A183CB3	7277-2-15	. CIRCUIT BREAKER,RH GENERATOR FIELD AND SENSE.	1	621	
-A183CB4	7277-2-10	. CIRCUIT BREAKER,RH GENERATOR CONTROL PANEL. .	1	621	
-A183GS1	131270-3	. GROUND STUD	1	621	
-A183GS1	AN960-10L	. WASHER.	1	621	
-A183GS1	MS35338-43	. LOCKWASHER.	1	621	
-A183GS1	MS21042L3	. NUT	1	621	
-A183P298	208678-1	. PLUG,RH LINE CONTACTOR RELAY.	1	621	
-A183P298	66105-4	. SOCKET,CONTACT.	AR	621	
-A183P298	MS3187-16-2	. PLUG,SEALING.	AR	621	
-A183R1	100-380007-3	. SHUNT,RH LOAD METER	1	621	
-A183W1	101-364046-56	. BUS BAR	1	621	
-A183W203	101-364046-111	. BUS BAR	1	621	
-A214		ENGINE SECTION ASSY,RH.	1	420	
-A214G1	50-369122-21	GENERATOR ASSY,RH STARTER	1	420	
-A214G1	50-369122-13	GENERATOR ASSY,RH STARTER	1	420	1
-A214GS15		GROUND STUD/CONSISTS OF THE FOLLOWING/.	1	420	
-A214GS15	MS35207-264	SCREW	1	420	
-A214GS15	MS35338-43	LOCKWASHER.	2	420	
-A214GS15	MS25082-3	NUT	1	420	
-A214GS15	AN960-10L	WASHER.	4	420	
-A214GS15	MS21042L3	NUT	1	420	
-A214GS16		GROUND STUD/TYPICAL A214GS16,A214GS17 AND . . . A214GS18/CONSISTS OF THE FOLLOWING/	3	420	
-A214GS16	AN5-7A	BOLT.	1	420	
-A214GS16	MS35338-45	LOCKWASHER.	1	420	
-A214GS16	AN960-516	WASHER.	1	420	
-A214GS16	MS21042L5	NUT	1	420	
-A214P2		PLUG,RH FIREWALL NO 2/SEE CHAPTER 91/	1	420	
-J105		RECEPTACLE,RH FIREWALL NO 2/SEE CHAPTER 91/ . .	1	621	
-J/P156		CONNECTOR,RH CABIN GROUNDING/SEE CHAPTER 91/. .	1		
-S138	13AT10S	SWITCH,RH GENERATOR CONTROL	1	245	

    1 UNITED KINGDOM ONLY   CODES OF EFFECTIVITY

Fig. 1-13(b) Associated parts list; Beechcraft King Air (Raytheon Aircraft Co.) For Training Purposes Only.

For transport category aircraft, the correct wiring diagram or schematic is often found using a given ATA code and chapter index. In some cases, specific schematics or diagrams are referenced by the maintenance manual. Wire numbers and part numbers can also be used to find their related schematic. The master wire list or equipment list could be used to find the appropriate diagram. Figure 1-14(a) shows a portion of the wire list for a Boeing 747-400. As shown in the list, wire W880-M555-18 is located on diagram 38-32-11.

BOEING 747
WIRING DIAGRAM MANUAL

```
R  M   BUNDLE PART NUMBER AND TITLE
E  O   ----------------------------- - LENGTH
V  D   BUNDLE/WIRE/GA/CO   TY FA FT  /IN  DIAGRAM   EQUIP    TERM  TT SP EQUIP      TERM TT SP  EFFECTIVITY
       ------------------------------------------------------------------------------------------------
       61B40880-WIRE BUNDLE-AFT LAVATORIES M AND N
       W880-M 0555-18      UA  010 00  38-32-11  DM0529    15        SM00003    S          ALL
          -M  0556-18      UA  010 00  38-32-11  DM0529    32        SM00002    S          ALL
          -M  0557-18      UA  010 00  38-32-11  DM0529    31        SM00001    S          ALL
          -M  1021-22      UA  011 00  38-12-11  DM0528     4        SP04496    S          ALL
          -M  1022-22      UA  011 00  37-12-11  DM0528    18        SP04498    S          ALL
          -M  1023-22      UA  011 00  32-12-11  DM0528    19        SP04500    S          ALL
          -M  1038-20      UA  012 00  38-12-11  DM0528     8        GD00616    A..E       ALL
       ------------------------------------------------------------------------------------------------
```

W880- M
0555- 18

Fig. 1-14(a) Example of a typical wire list for a Boeing 747-400. (The Boeing Commercial Airplane Company) For Training Purposes Only.

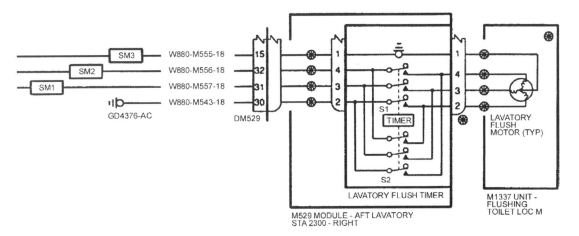

Fig. 1-14(b)   Shows a portion of the wiring diagram #38-32-11. In this figure one can easily find wire W880-M555-18 and it's related equipment. (Boeing Commercial Airplane Company) For Training Purposes Only.

The wiring diagram manuals for Airbus Industries aircraft are divided into three categories: **Aircraft Schematic Manuals (ASM), Aircraft Wiring Manuals (AWM), and Aircraft Wiring Lists (AWL)**. The ASM contains simplified and moderately complex illustrations. The AWM contains the detailed electrical/electronic diagrams. By knowing a systems function the correct AWM or ASM can be easily located using the alphabetical index. This is represented by the top left block (*Access by Function*) of figure 1-15. The top right block, (*Access by Wire Number*), can be used if the correct wire number is known. In this approach the correct AWM or ASM can be located through the master wire list (chapter 91). As shown in the center block of figure 1-15, the FIN (described in the next paragraph) can be used to reference the correct AWM or ASM page(s). This is typically the fastest way to find a given wiring diagram.

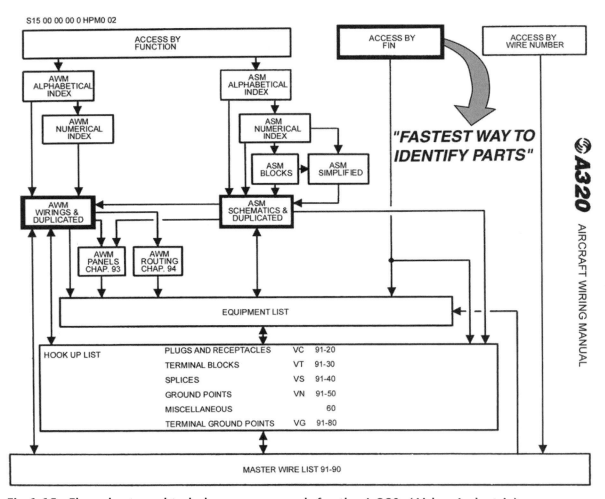

Fig 1-15   Flow chart used to help access manuals for the A-320. (Airbus Industrie)

Functional Item Numbers

The **Functional Item Number (FIN)** is a unique number given to each line replaceable unit on Airbus aircraft. The FIN is located on a placard next to each component as seen in figure 1-16. The FIN consists of a two letter code which defines the system or circuit of a specific Line Replaceable Unit (LRU). A numerical prefix or suffix is added to the code which gives each LRU a specific Functional Item Number. A typical FIN could be 12RC2.

There are 18 major categories of FIN alphabetical identifiers as shown in table 1-2. Figure 1-17 shows a break down of a typical FIN. If the FIN represents a connector plug the suffix becomes a letter. The letter *A* designates the first plug in that system the letter *B* the second plug and so on (Figure 1-17).

3006 GM

Fig. 1-16 Example of a Functional Identification Number (FIN).

ALPHABETICAL CODE FOR AIRBUS FUNCTIONAL ITEM NUMBERS	
SYSTEM IDENTIFICATION CODE (MAJOR CATEGORY)	SYSTEM FUNCTION
C.	FLIGHT CONTROL SYSTEMS
D.	DE-ICING
E.	ENGINE MONITORING
F.	FLIGHT INSTRUMENTS
G.	LANDING GEAR HYDRAULIC
H.	AIR CONDITIONING
J.	IGNITION
K.	ENGINE CONTROL AND STARTING
L.	LIGHTING
M..	INTERIOR ARRANGEMENT
P..	DC POWER SUPPLY DISTRIBUTION
Q..	FUEL
RÖ	RADIO (NAVIGATION AND COMMUNICATION)
S.Ö	RADAR NAVIGATION
TÖ...	SPECIAL ELECTRONICS
VÖ	FICTITIOUS CIRCUITS
W..	FIRE PROTECTION AND WARNING
XÖ..	AC GENERATION AND DISTRIBUTION

Table 1-2 Alphabetical code for Airbus functional item numbers.

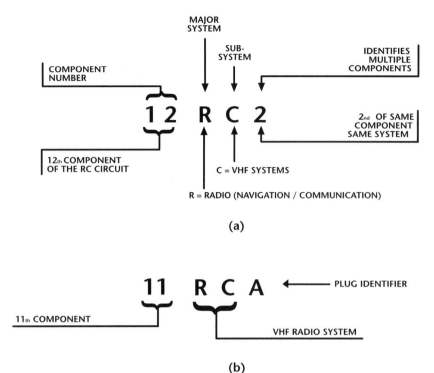

Fig 1-17 *Descri*ption of a typical FIN (a) a VHF radio component, (b) a connector plug for the VHF radio system.

The Airbus Functional Item Number is used to identify components through out the entire series of aircraft manuals. In most cases, the FIN is the fastest way to identify a given manual page or to locate a given part number of an LRU. As seen in figure 1-18, the FIN appears just below each LRU on a wiring diagram or schematic. In the center portion of this diagram, it can be seen that the flight control/ ELAC 2 push button switch has a FIN of 6CE2.

Wire Identification Systems

All aircraft wires three inches or longer must be labeled with a **wire identification number**. This number is typically comprised of an alphanumeric code based on the military standard MIL-W-5088. Figure 1-19 shows a typical wire code system. The wire code consists of a wire bundle number, circuit function letter(s), a wire number, segment letter, wire gauge, and color. It should be noted that wire bundle numbers are typically used on large aircraft only; segment letters are typically found on light aircraft. On some aircraft, portions of the wire number corresponds to the ATA chapter and section number for the system where the wire is installed. Each manufacturer has their own wire code system. Be sure to read the introduction of the wiring manuals to determine the specific code for your aircraft.

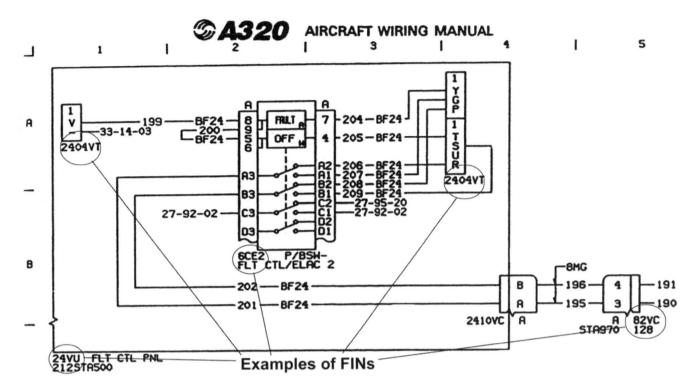

Fig 1-18   Examples of various FIN (Functional Item Numbers) For Training Purposes Only.

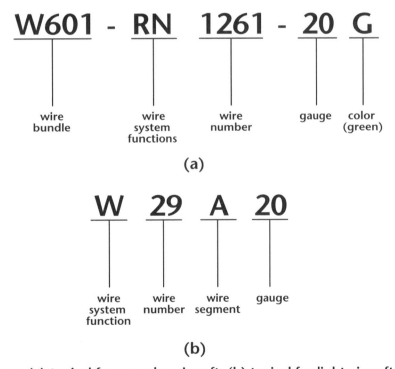

Fig 1-19   Wire code systems. (a) typical for complex aircraft; (b) typical for light aircraft.

Sensitive wires are considered critical to flight safety and must not be modified without specific manufacturers approval. Sensitive wires are typically found on fly by wire type aircraft used for primary flight controls and other crucial circuits. Sensitive wires are always identified in the wire code and often by a distinct color marking. On the A-320 for example, an *S* at the end of the wire number (2792-1568-CF22-S) designates a sensitive wire. On this aircraft, there is a pink band located at every end of a sensitive wire. It should be noted, any work performed on sensitive wires (including a simple disconnect-reconnect) typically requires an inspector's signature.

There are several different wire types used on modern complex aircraft. Two common specialty wires include: **shielded cable** and **data bus cable**. Wires also vary with different insulation or conductor types. Be sure to choose the correct type of wire whenever installing a replacement. Typically wire type is defined in the **wire lists** section of the aircraft wiring diagrams manuals. Some companies identify the wire type and gauge on the wiring diagrams.

Home Diagrams

Most aircraft components interact with other systems or subsystems throughout the aircraft. In many cases, a diagram or schematic must show these interrelations to explain the component in question. The **home diagram** of a component is the diagram where the component is shown in full detail. The home diagram will show all electrical connections and wires for that unit. A component is considered *home* only when it is shown in the ATA chapter, section and subject which corresponds to that component. Components in there home diagram are enclosed in a solid bold line. In figure 1-20, the 6XA relay, main galley supply control, (located on the right side of the schematic) is in its home diagram.

In most cases, a component which is not in it's home diagram will not be shown in detail. Components not in their home diagram are enclosed by a "crisscross" (*xxxxxxxxxx*) line. Adjacent to all components not in their home diagram is the ATA chapter where they can be found in detail (i.e. their home diagram). In figure 1-20, the LGCIU-2 (located on the left side of the schematic) is not in its home diagram. The corresponding home diagram for the LGCIU-2 is located on schematic 32-31-00, page 101 (figure 1-20 bottom left).

Individual wires may also be assigned a home diagram or schematic. If a wire is represented as a dotted line it is not in its home diagram. If a wire is not in its home diagram it will likely contain an (*H*) preceding the ATA chapter of the wire's home diagram. Figure 1-21 shows a relay where the wires from T1, T2, T3, are not located in their home diagram. Diagram number 24-21-51 must be referenced to find more detail concerning the wires to T1, T2, & T3.

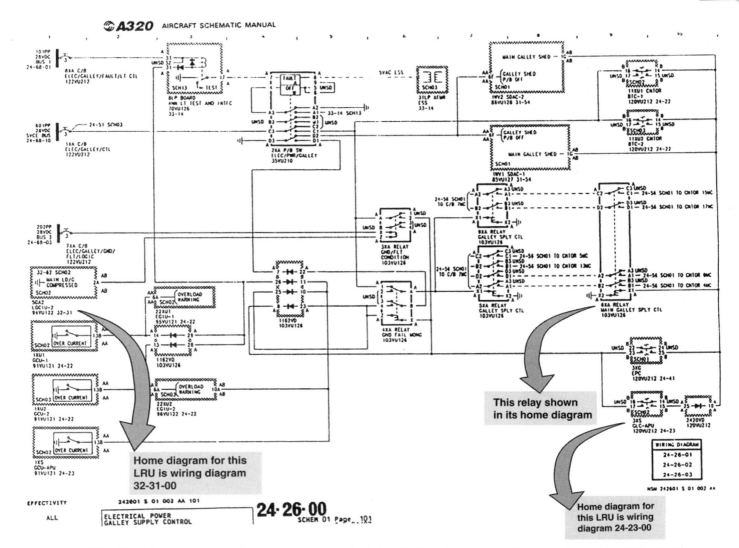

Fig. 1-20 Electrical power galley supply control. Note: 1) This is a home diagram for relay 6XA. 2) The home diagram for 3XS is located in figure 24-23. (Airbus Industrie) For Training Purposes Only.

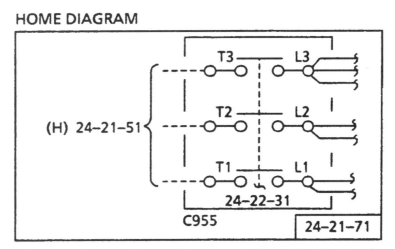

Fig. 1-21 Example of relay diagram showing home diagram of wires. (The Boeing Commercial Airplane Company) For Training Purposes Only.

Other Manuals

Most manufacturers of transport category aircraft offer a variety of specific manuals designed to be used for certain maintenance situations. The **Ramp Maintenance Manual (RMM)** is an abbreviated maintenance manual designed specifically for line maintenance and minor troubleshooting. The RMM contains most of the information a technician would need for minor repairs between flights. If additional information is needed, the RMM contains a cross references to other "more specific" manuals for that aircraft. The RMM often contains simplified electrical schematics and a reference to all wiring diagrams for each system.

Aircraft **Troubleshooting Manuals (TSM)** are designed specifically for system troubleshooting. These manuals contain block diagrams, electrical schematics, as well as, functional diagrams of mechanical, hydraulic, and pneumatic systems. The TSM is an excellent place to start when troubleshooting systems. The **Fault Isolation Manual** (FIM) is the aircraft maintenance manual which contains various repair strategies for given faults. The FIM is used during troubleshooting on many Boeing aircraft.

STATIONS AND ZONES

Each aircraft manufacture has a **component location system** that is used to identify the location of various assemblies, subassemblies, and components on the aircraft. The component location system is designed to provide a reference that can be used during engineering, manufacturing, and maintenance. The location system provides an easy means to find individual components that can be located virtually anywhere on the aircraft. The component location system is often used during avionics troubleshooting to help locate a specific wire, component or connector.

Typically, light single-engine aircraft require only a modest component location system. While the component location system for transport category aircraft is quite complex due to the number of various components and size of the aircraft. All aircraft have a vertical reference called the datum. The **datum** is an imaginary vertical plane typically located near the front of the aircraft. All horizontal measurements are taken from the datum, thus giving a consistent point of reference. The **water line** is an imaginary horizontal plane which runs the length of the aircraft from nose to tail. All vertical measurements are taken from the water line. Figure 1-22 shows two variations of a datum and water line. In one case, (figure 1-22a) the reference planes are located so all measurements are a positive value. In (figure 1-22b), the reference planes are located so that both positive and negative numbers are used. It should be noted that all measurement in front of the datum and below the water line are considered negative. Dimensions are typically given in inches; however, some foreign made aircraft use centimeter.

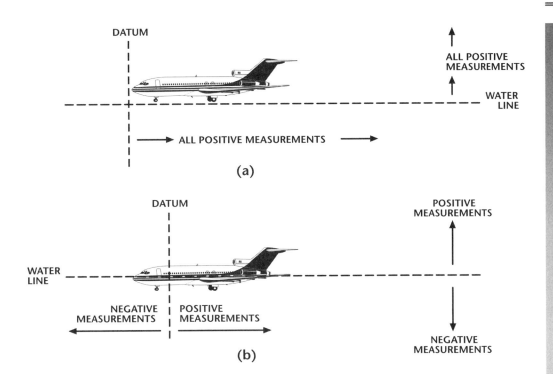

Fig. 1-22 Examples of datum and water lines. (a) all positive measurements; (b) both positive and negative measurements.

Stations

Fuselage stations (FS) are used to indicate locations longitudinally along the aircraft fuselage. Fuselage stations are typically measured from the datum, with the datum at FS 00.00. Various fuselage stations for a typical corporate type aircraft are shown in figure 1-23.

Wing stations (WS) are used to indicate locations longitudinally along the wings of the aircraft. Wing station 00.00 is located in the center of the fuselage at the butt line. The **butt line (BL)** is a vertical plane which divides the aircraft front to rear through the center of the fuselage (figure 1-24). On this aircraft the butt line numbers are used from station 00.00 to station 99.60. Outboard of station 99.60 the measurements are referred to as wing stations. On most aircraft WS numbers are used in wing areas and BL numbers indicate fuselage stations left and right of station 00.00. On all aircraft both WS and BL measurements have the same reference point, BL 00.00.

Other stations such as **nacelle stations (NS)** or **elevator stations (ES)** are often used on larger aircraft to help technicians find components in specific areas of the aircraft. It should be noted that for any station on the aircraft the 00.00 reference will always be taken from one of the three main reference planes, the datum line, the water line or the butt line.

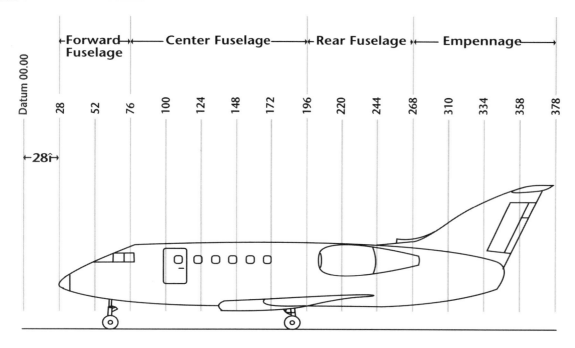

Fig. 1-23   Various fuselage stations for a corporate jet type aircraft.

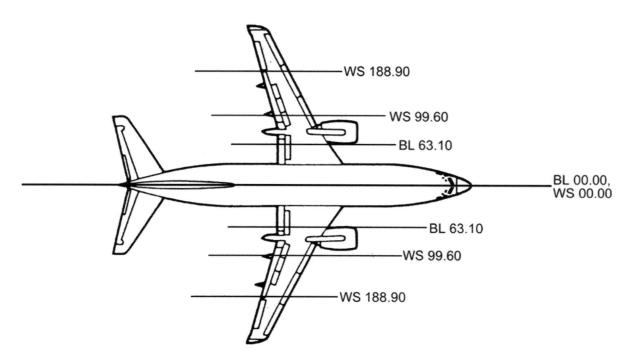

Fig. 1-24   Butt Lines and Wing Stations on a typical aircraft.

AVIONICS
TECHNICAL DATA **1**

Zones

A system of zoning has been designed by the Air Transport Association (ATA) to further identify the location of components on large aircraft. Aircraft **zones** are assigned to various regions of the aircraft specifically to aid in component location. Figure 1-25 shows an example of the zone system used on a Beechcraft King Air C90A. Here it can be seen that most zone numbers on the left side of the aircraft

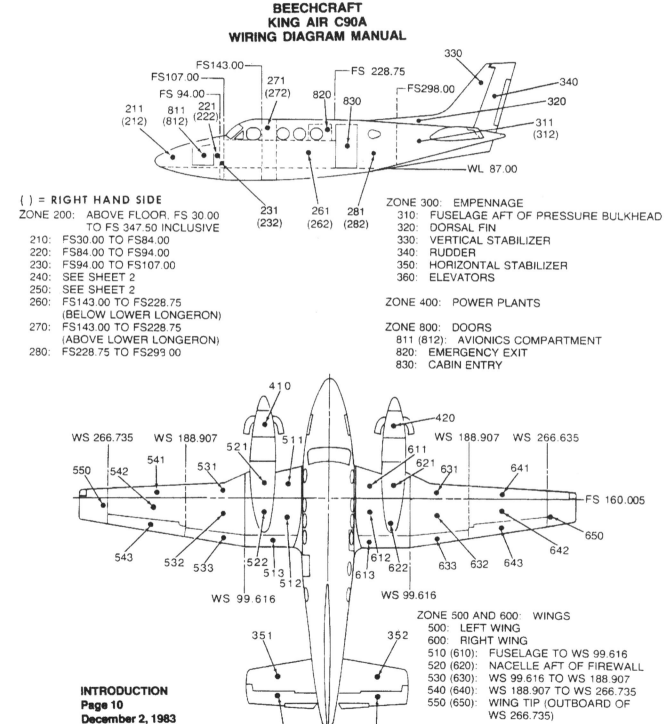

Fig. 1-25 Zone locations for a typical aircraft. (Raytheon Aircraft Co.) For Training Purposes Only.

begin with odd numbers and all zones on the right begin with even. The right/left side is determined when sitting inside the aircraft facing forward, ie. the pilot sits in the left seat, the copilot sits on the right side.

The zoning system is based on a three digit code, the first digit is the major zone designator. The second nonzero digit is the subzone, and the third nonzero digit is a specific area within the subzone.

The major zone areas are:

100   *lower half of the fuselage to the rear of the pressure bulkhead (below the main cabin)*
200   *Upper half of fuselage to the rear of the pressure bulkhead*
300   *Empennage, including aft of the rear pressure bulkhead*
400   *Powerplants and struts or pylons*
500   *Left wing*
600   *Right wing*
700   *Landing gear and landing gear doors*
800   *Doors*
900   *Reserved for uncommon differences between aircraft types not covered in zones 100-800*

Subzones (the second digit of the code) are used to designate specific stations or given areas within a major zone. Figure 1-25 shows zones 530 and 630 as between wing stations 99.616 and 188.907. Figure 1-26 shows a breakdown of the crew compartment of the King Air C90A. In this situation all three digits of the zone code are used to specify location.

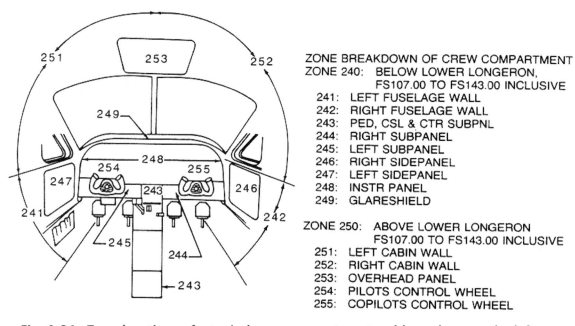

ZONE BREAKDOWN OF CREW COMPARTMENT
ZONE 240:   BELOW LOWER LONGERON,
                 FS107.00 TO FS143.00 INCLUSIVE
     241:   LEFT FUSELAGE WALL
     242:   RIGHT FUSELAGE WALL
     243:   PED, CSL & CTR SUBPNL
     244:   RIGHT SUBPANEL
     245:   LEFT SUBPANEL
     246:   RIGHT SIDEPANEL
     247:   LEFT SIDEPANEL
     248:   INSTR PANEL
     249:   GLARESHIELD

ZONE 250:   ABOVE LOWER LONGERON
                 FS107.00 TO FS143.00 INCLUSIVE
     251:   LEFT CABIN WALL
     252:   RIGHT CABIN WALL
     253:   OVERHEAD PANEL
     254:   PILOTS CONTROL WHEEL
     255:   COPILOTS CONTROL WHEEL

Fig. 1-26   Zone locations of a typical crew compartment; odd numbers on the left, even numbers on the right. (Raytheon Aircraft Co.)

To better explain component location systems used during maintenance and troubleshooting, study the following examples. Three different aircraft that represent a cross section of the industry will be examined (the Beechcraft King Air, Airbus A-320, and the Boeing 747).

IDENTIFYING COMPONENT LOCATIONS ON THE KING AIR C90A

For electronics or avionics technicians, locating aircraft components will often start at the wiring diagram. Let us assume that our task is to troubleshoot an inverter power system on a King Air C90A. The first step would be to find the wiring schematic in the wiring diagrams manual under ATA chapter 24 (electrical power). As seen in figure 1-27, each Beechcraft wiring diagram is comprised of two pages, a component description page and a system schematic page.

Through study of the schematic and operation of the system, it is determined that two circuit breakers should be tested, CB236 and CB246. To locate these circuit breakers on the aircraft, simply find the circuit breaker number in the left hand column of the component description page (figure 1-27a). Follow the component description line to the right and find the installation location in the *Installation Zone* column. Circuit breaker 236 is located in zone 246, CB246 is located in zone 612. Referring to the zone location charts (figures 1-25 and 1-26), it is determined that CB236 is installed in the right side panel of the flight deck. CB246 is located in the right wing root, center section.

IDENTIFYING COMPONENT LOCATIONS ON THE AIRBUS A-320

On transport category aircraft, locating components is often a difficult task due to the sheer size of the aircraft. On the A-320 there are over 3,000 panels and racks containing equipment, switches, and circuit breakers. Switches and circuit breakers are typically mounted on a panel. Line replaceable units (LRUs), such as the VOR receiver, autoflight computers, or EFIS display management computers are typically mounted on an equipment rack. Airbus designates all panels and racks with a VU number. For example, the electrical power distribution control switches are located on the 35VU panel on the lower-overhead section of the flight deck. The Autoflight system computers are located in the aft electronics racks 84VU and 83VU. It should be noted that a panel or rack will always have removable components. For example, if an item contains several switches and those switches are not individually replaced (ie. the complete item must be replaced as a unit) that item is considered an LRU not a panel or rack.

BEECHCRAFT
KING AIR C90A
WIRING DIAGRAM MANUAL

GAMA/ATA CODE & REF DES	PART NO.	DESCRIPTION 1 2 3 4 5 6 7	UNITS PER ASSY	INSTL ZONE	USABLE ON CODE
21-00-A146		PANEL ASSY,RH CIRCUIT BREAKER/SEE CHAPTER 24/ .	1	246	
−CB188	7277-2-2	. CIRCUIT BREAKER,LH TORQUE METER	1	246	
−CB189	7277-2-2	. CIRCUIT BREAKER,RH TORQUE METER	1	246	
−CB234	7277-2-5	. CIRCUIT BREAKER,NO.1 INVERTER POWER SELECT. .	1	246	
−CB236	7277-2-5	. CIRCUIT BREAKER,NO.2 INVERTER POWER SELECT. .	1	246	
−CB212	7277-2-5	CIRCUIT BREAKER,INVERTER NO.1,26 VAC POWER. . .	1	512	
−CB213	7277-2-10	CIRCUIT BREAKER,INVERTER NO.1,115 VAC POWER . .	1	512	
−CB214	7277-2-5	CIRCUIT BREAKER,INVERTER NO.2,26 VAC POWER. . .	1	612	
−CB215	7277-2-10	CIRCUIT BREAKER,INVERTER NO.2,115 VAC POWER . .	1	612	
−CB241	7277-2-5	CIRCUIT BREAKER,NO.2 INVERTER CONTROL	1	612	
−CB243	7277-2-5	CIRCUIT BREAKER,NO.1 INVERTER CONTROL	1	512	
−CB245	7277-2-7	CIRCUIT BREAKER,NO.1 INVERTER POWER SELECT. . .	1	512	
−CB246	7277-2-7	CIRCUIT BREAKER,NO.2 INVERTER POWER SELECT. . .	1	612	
−CR148	100-361045-1	TRANSZORB,NO.1 INVERTER POWER RELAY	1	512	
−CR149	100-361045-1	TRANSZORB,NO.2 INVERTER POWER RELAY	1	612	
−CR176	100-361045-1	TRANSZORB,NO.1 INVERTER POWER SELECT RELAY. . .	1	512	
−CR177	100-361045-1	TRANSZORB,NO.2 INVERTER POWER SELECT RELAY. . .	1	612	
−CR207	1N4005	DIODE,INVERTER SELECT RELAY	1	222	
−F114	MDA5	FUSE/TYPICAL F114 AND F115/	2	222	
−XF114	4532	FUSEHOLDER.	1	222	
−GS160		GROUND STUD/TYPICAL GS160,GS161 & GS162/MADE. . UP FROM THE FOLLOWING/	3	222	
−GS160	131270-3	GROUND STUD/BLIND/.	3	222	
−GS160	MS21042L3	NUT	3	222	
−GS160	MS35338-43	LOCKWASHER.	3	222	
−GS160	AN960-10L	WASHER.	3	222	
−J186	1506-105	TEST JACK,115 VOLTS AC/BLUE/.	1	245	
−K106	50-380048-5	RELAY,AC INVERTER OUT WARNING LIGHT	1	222	
−K144	109-381002-1	RELAY,NO.1 INVERTER POWER SELECT.	1	512	
−K145	109-381002-1	RELAY,NO.2 INVERTER POWER SELECT.	1	612	
−K153	50-380048-1	RELAY,INVERTER SELECT	1	222	
−K154	109-381002-1	RELAY,NO.1 INVERTER POWER	1	512	
−K155	109-381002-1	RELAY,NO.2 INVERTER POWER	1	612	
−S136	MS24659-21A	SWITCH,INVERTER SELECT.	1	245	
	M81714/5-1	RAIL,MODULE	1	222	
−TB107	M81714/3AB1	BLOCK,NO 1 INVERTER POWER TERMINAL.	1	222	
−TB108	M81714/3AB1	BLOCK,NO 2 INVERTER POWER TERMINAL.	1	222	
−W300		HARNESS ASSY,INVERTER/SEE CHAPTER 24-22/. . . .	1	521/621	

Fig. 1-27(a) Typical Beechcraft component description page (Raytheon Aircraft Co.). For Training Purposes Only.

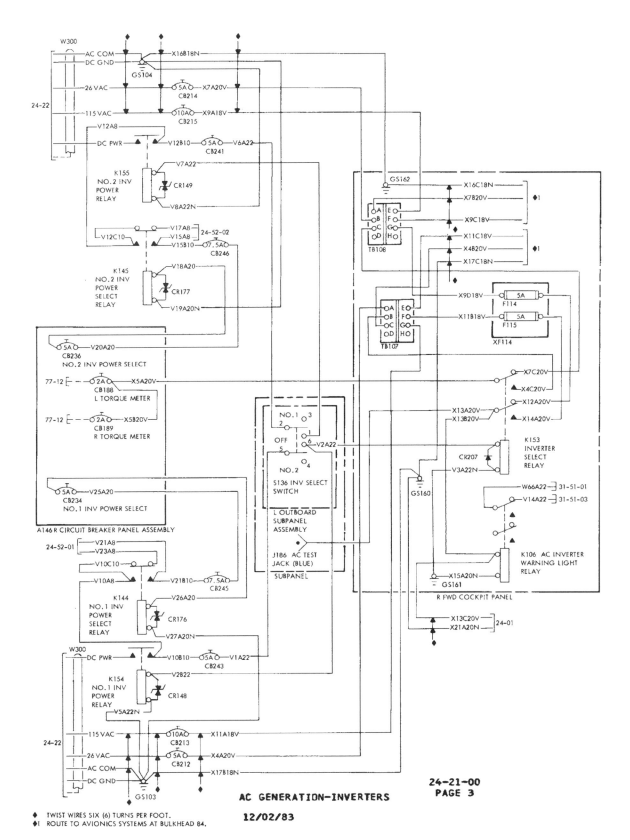

BEECHCRAFT
KING AIR C90A
WIRING DIAGRAM MANUAL

AC GENERATION—INVERTERS

24-21-00
PAGE 3

12/02/83

◆ TWIST WIRES SIX (6) TURNS PER FOOT.
◆1 ROUTE TO AVIONICS SYSTEMS AT BULKHEAD 84.

Fig. 1-27(b) Typical Beechcraft wiring diagram. (Raytheon Aircraft Co.) For Training Purposes *Only*.

Most of the avionics and computerized LRUs are located in the forward or aft equipment bays on the 90VU, 108VU, and 80VU electronic racks (see figure 1-28a). These racks are further divided into individual shelves as seen in figure 1-28b that shows the location of the Autoflight computers.

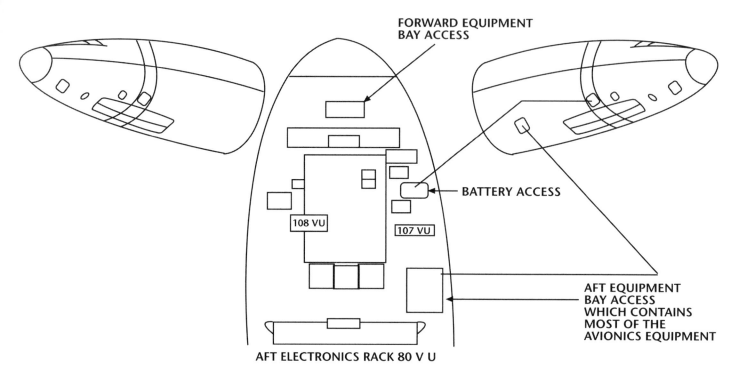

Fig. 1-28(a)   Avionics equipment locations: equipment bay access.

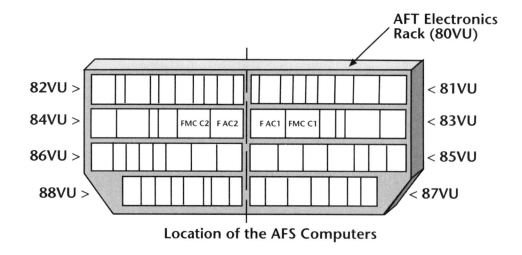

Location of the AFS Computers

Fig. 1-28(b)   A typical Avionics equipment rack showing avionics equipment locations. For Training Purposes Only.

The A-320 Aircraft Schematics Manuals (ASM) and Aircraft Wiring Manuals (AWM) identify the location of various electronics components using the **VU number**, the aircraft station number and/or the zone number. As shown in figure 1-29, the radio management panel- communication/navigation frequency selector number 1 (RMP-COM/NAV FREQ SEL.1) is located on the 11VU panel. The schematic states that 11VU is located on the center pedestal at zone 210 and fuselage station 460. It should be noted that all station numbers are measured in centimeters. Zone and station information also shows adjacent connector plugs and ground connections (see figure 1-29).

The **wire routing charts** for the A-320 will help the technician to locate wire bundles which run through the aircraft. The AWM, chapter 24 contains a wire zoning and routing diagram as seen in figure 1-30. This diagram can be extremely helpful when troubleshooting a defective wire.

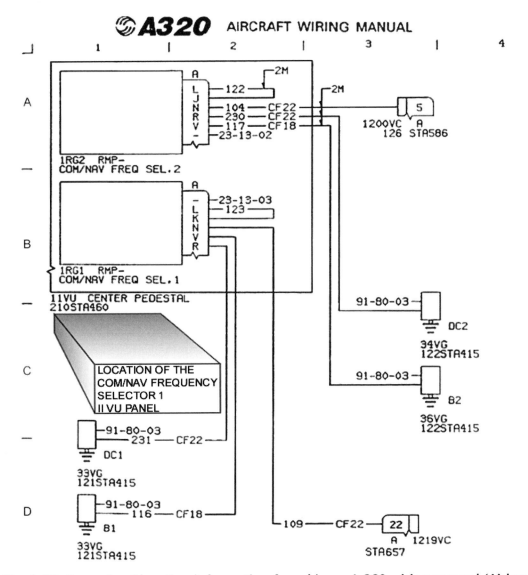

Fig. 1-29 Example of location information found in an A-320 wiring manual (Airbus Industrial). **For Training Purposes Only.**

IDENTIFYING COMPONENT LOCATIONS ON THE BOEING 747

Because the Boeing wiring diagrams do not contain equipment location numbers, their component location system is somewhat more complicated than that of the Airbus system. In the Boeing system component locations for specific equipment are

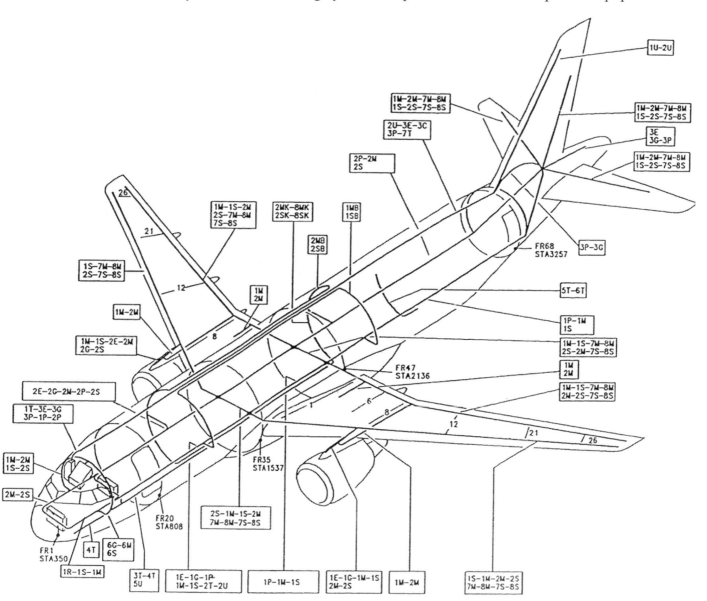

Fig. 1-30(a) A-320 zone and main wiring routing diagram. (Airbus Industrie) For Training Purposes Only.

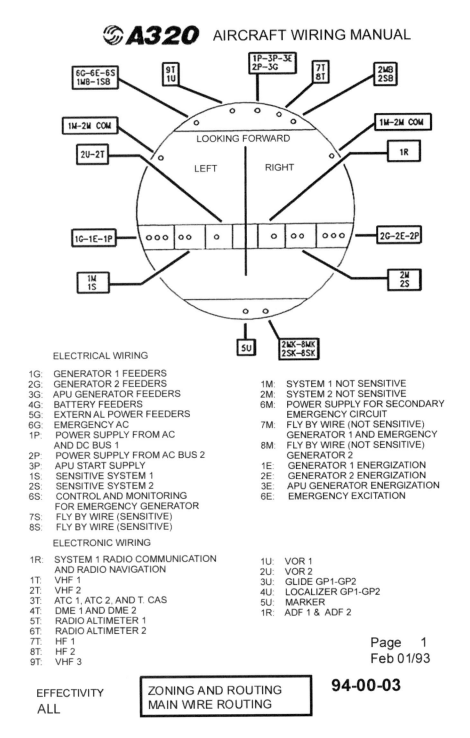

ELECTRICAL WIRING

1G: GENERATOR 1 FEEDERS
2G: GENERATOR 2 FEEDERS
3G: APU GENERATOR FEEDERS
4G: BATTERY FEEDERS
5G: EXTERN AL POWER FEEDERS
6G: EMERGENCY AC
1P: POWER SUPPLY FROM AC
      AND DC BUS 1
2P: POWER SUPPLY FROM AC BUS 2
3P: APU START SUPPLY
1S: SENSITIVE SYSTEM 1
2S: SENSITIVE SYSTEM 2
6S: CONTROL AND MONITORING
      FOR EMERGENCY GENERATOR
7S: FLY BY WIRE (SENSITIVE)
8S: FLY BY WIRE (SENSITIVE)

1M: SYSTEM 1 NOT SENSITIVE
2M: SYSTEM 2 NOT SENSITIVE
6M: POWER SUPPLY FOR SECONDARY
      EMERGENCY CIRCUIT
7M: FLY BY WIRE (NOT SENSITIVE)
      GENERATOR 1 AND EMERGENCY
8M: FLY BY WIRE (NOT SENSITIVE)
      GENERATOR 2
1E: GENERATOR 1 ENERGIZATION
2E: GENERATOR 2 ENERGIZATION
3E: APU GENERATOR ENERGIZATION
6E: EMERGENCY EXCITATION

ELECTRONIC WIRING

1R: SYSTEM 1 RADIO COMMUNICATION
      AND RADIO NAVIGATION
1T: VHF 1
2T: VHF 2
3T: ATC 1, ATC 2, AND T. CAS
4T: DME 1 AND DME 2
5T: RADIO ALTIMETER 1
6T: RADIO ALTIMETER 2
7T: HF 1
8T: HF 2
9T: VHF 3

1U: VOR 1
2U: VOR 2
3U: GLIDE GP1-GP2
4U: LOCALIZER GP1-GP2
5U: MARKER
1R: ADF 1 & ADF 2

Page   1
Feb 01/93

94-00-03

EFFECTIVITY
ALL

ZONING AND ROUTING
MAIN WIRE ROUTING

Fig. 1-30(b) A-320 zone and main wiring routing diagram. (Airbus Industrie) For Training Purposes Only.

identified on the **Equipment List**. The **Wiring List** is used to find specific wires; the **Splice List** for splices; the **Terminal List** for terminal strips; and the **Hook-up List** for connectors or pins within a connector. The wiring, splice, terminal, and hook-up lists are located in chapter 91 of the maintenance publications. The equipment list is a separate document subdivided by ATA chapters. As seen in figure 1-31, the maintenance manuals for the Boeing aircraft do a good job of identifying component locations. In most cases however, troubleshooting and repair of a given system will require the use of several different manuals.

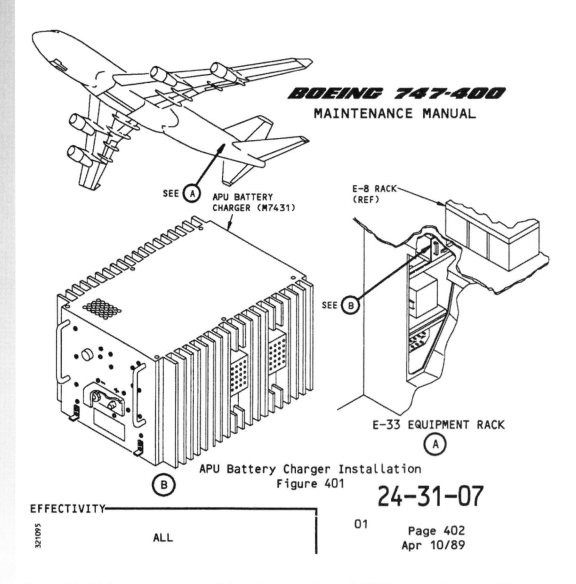

BOEING 747-400
MAINTENANCE MANUAL

SEE (A)   APU BATTERY
CHARGER (M7431)

E-8 RACK
(REF)

SEE (B)

E-33 EQUIPMENT RACK
(A)

APU Battery Charger Installation
Figure 401

24-31-07

(B)

EFFECTIVITY

321095

ALL

01   Page 402
Apr 10/89

Fig. 1-31 Maintenance manual showing locations of APU battery charger. (The Boeing Commercial Airplane Company) For Training Purposes Only.

Figure 1-32 shows a excerpt from the B-747 equipment list. In this example, the equipment (listed under the EQUIP column) includes two different diode assemblies, and two information signs for the passenger compartment upper deck. In this case, the number used to identify each piece of equipment begins with an *M*. The *M* indicates the equipment is considered an electrical system component that is not listed in another specific category. Other equipment designators are listed in Table 1-3.

Please reference figure 1-32 while reading the following paragraph. The equipment list contains component part numbers, descriptions, a vendor number, and the quantity installed on the aircraft. The diagram number and aircraft effectivity are also shown on the equipment list. The equipment location is shown just below the vendor information. The location of M01562 is shown as fuselage station 530, water line 370 and butt line R (right) 50. All dimensions are given in inches. For items located on a specific panel, the panel number is substituted for the station, water line and butt line locations. In our example notice that Panel 14 is listed for the location of both diode assemblies.

When troubleshooting from an electrical schematic, it may not be necessary to refer to the equipment list to find the location of a component. Once a technician is familiar with the aircraft, many of the panels and equipment center locations become second nature. For example, when studying the electrical power standby system control and distribution schematic, figure 1-33, it was determined that circuit breaker C80 should be checked. C80, the main battery charger circuit breaker, is found in the upper left corner of the schematic. The breaker is located on the panel (P6) as indicated on the schematic. A technician who is familiar with the B-747 would likely know that P6 is the main power circuit breaker panel located at the right side rear of the flight deck.

If needed, chapter 91 (charts) of the wiring diagrams manual can be used to find the P6 panel (see figure 1-34). This chart gives the locations of all flight deck panels.

REV	MOD	EQUIP	OPT	PART NUMBER PART DESCRIPTION	USED ON	DWG	VENDOR STATION– WL–BL	QTY	DIAGRAM	EFFECTIVITY
		M01559	1	YHLZD–9 DIODE ASSY R965			V09922 PANEL–14–	1	21–25–12	ALL
		M01559	2	69B40800–3 DIODE ASSY R965			V09922 PANEL–14–	1	21–25–12	ALL
		M01562		60B50192 INFO SIGN–PASS UPR DK			V81205 530-370-R 50	1	33–24–21	ALL
		M01564		60B50192 INFO SIGN–PASS UPR DK			V81205 550-370-R 50	1	33–24–21	ALL

MODEL	747	REV DATE	MANUAL D6–XXXXX	EQUIPMENT LIST	SECTION M01500
CUSTOMER	XXX	MAY 02/85		VOLUME–1	PAGE 1

Fig. 1-32 Example of equipment list for a Boeing 747. (The Boeing Commercial Airplane Company) For Training Purposes Only.

A SUMMARY OF THE EQUIPMENT DESIGNATORS FOR THE BOEING 747 EQUIPMENT LIST AND WIRING DIAGRAMS. (NOTE: SIMILAR DESIGNATORS ARE USED ON OTHER TRANSPORT CATEGORY FOR SPECIFIC DESIGNATORS BE SURE TO SEE THE AIRCRAFT MANUALS)	
EQUIPMENT DESIGNATOR	**EQUIPMENT CATEGORY**
A	ANTI-ICING/DE-ICING
B	ELECTRONICS EQUIPMENT: ACCESSORIES UNITS, AMPLIFIERS, ANTENNAS, CAPACITORS, COMPUTERS, CONTROL UNITS, DIRECTIONAL GYROS, FILTERS, FLUX VALVES, ILS RACKS, MUSIC REPRODUCERS, RECEIVERS, SELCAL, TRANSMITTERS, VERTICAL GYROS, AND OTHER ELECTRONICS NOT LISTED IN A SPECIFIC CATEGORY
C	CIRCUIT BREAKERS, CURRENT LIMITERS, AND MISCELLANEOUS PROTECTIVE DEVICES
D	CONNECTORS
E	EQUIPMENT RACKS
F	FUEL SYSTEM COMPONENTS
G	GENERATORS AND RELATED COMPONENTS
GD	GROUNDS (AIRFRAME)
J	JUNCTION BOXES
K	COMMAND POST MISSION SYSTEM ELECTRONIC MODUALS
L	LIGHTS, LAMPS, AND RELATED ASSEMBLIES
M	ELECTRICAL SYSTEMS: BATTERIES, BALLAST ASSEMBLIES, BELLS, CHIMES, CONTROL UNITS, HEATERS, HORNS, LAVORATORY ASSEMBLIES, MOTORS, PHASE ADAPTORS, POWER UNITS, PUMPS, STRATO LIGHT ASSEMBLIES

Table 1-3 A summary of the equipment designators for the Boeing 747 equipment list and wiring diagrams.

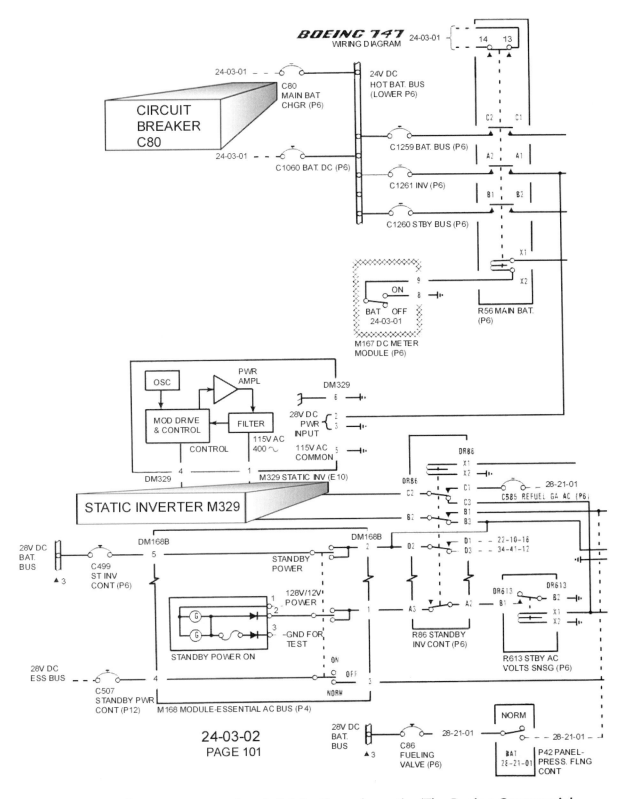

Fig. 1-33 Portion of a standby system control and distribution schematic. (The Boeing Commercial Airplane Company) For Training Purposes Only.

The exact circuit breaker (C80) can be identified by the label on the front side of the P6 panel (MAIN BATT CHGR), or circuit breaker C80 can be identified using chapter 91. For example, figure 1-35 shows the rear side of P6 to further identify the location of each circuit breaker. It should be noted, this figure shows only a portion of P6; the entire panel requires 8 pages of charts.

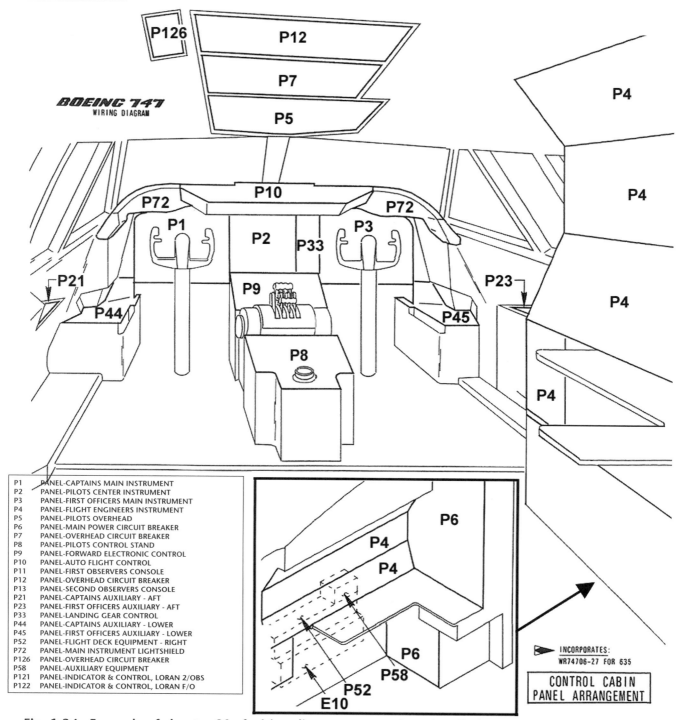

P1	PANEL-CAPTAINS MAIN INSTRUMENT
P2	PANEL-PILOTS CENTER INSTRUMENT
P3	PANEL-FIRST OFFICERS MAIN INSTRUMENT
P4	PANEL-FLIGHT ENGINEERS INSTRUMENT
P5	PANEL-PILOTS OVERHEAD
P6	PANEL-MAIN POWER CIRCUIT BREAKER
P7	PANEL-OVERHEAD CIRCUIT BREAKER
P8	PANEL-PILOTS CONTROL STAND
P9	PANEL-FORWARD ELECTRONIC CONTROL
P10	PANEL-AUTO FLIGHT CONTROL
P11	PANEL-FIRST OBSERVERS CONSOLE
P12	PANEL-OVERHEAD CIRCUIT BREAKER
P13	PANEL-SECOND OBSERVERS CONSOLE
P21	PANEL-CAPTAINS AUXILIARY - AFT
P23	PANEL-FIRST OFFICERS AUXILIARY - AFT
P33	PANEL-LANDING GEAR CONTROL
P44	PANEL-CAPTAINS AUXILIARY - LOWER
P45	PANEL-FIRST OFFICERS AUXILIARY - LOWER
P52	PANEL-FLIGHT DECK EQUIPMENT - RIGHT
P72	PANEL-MAIN INSTRUMENT LIGHTSHIELD
P126	PANEL-OVERHEAD CIRCUIT BREAKER
P58	PANEL-AUXILIARY EQUIPMENT
P121	PANEL-INDICATOR & CONTROL, LORAN 2/OBS
P122	PANEL-INDICATOR & CONTROL, LORAN F/O

INCORPORATES:
WR74706-27 FOR 635

CONTROL CABIN
PANEL ARRANGEMENT

Fig. 1-34 Example of chapter 91 of wiring diagrams manual used to find control panel locations. (The Boeing Commercial Airplane Company) For Training Purposes Only.

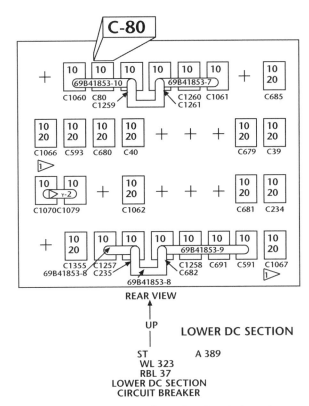

Fig. 1-35   Diagram showing the rear of the P6 panel, main circuit breaker power. (The Boeing Commercial Airplane Company) For Training Purposes Only.

As with most transport category aircraft, the majority of the B-747 electronic/avionics equipment is located in various equipment racks found throughout the aircraft. On the schematic for a given system, the equipment rack identifier is located just below the component. All equipment racks are designated by an *E* number and all panels by a *P* number. To indicate a shelf on a rack a dash and the shelf number is added, such as, E003-2 (rack 003, shelf 2).

Figure 1-33 shows a static inverter on the left side of the diagram. Shown just below the bold outlines of the inverter is the identifier *M329 STATIC INV (E10)*. *M329* can be used to find the static inverter on the aircraft main equipment list. *E10* identifies the inverter is located in the equipment rack *E10*. As seen from the chart in figure 1-36, *E10* is located on the copilot's side towards the rear of the flight deck. Chapter 91 also contains other charts showing the locations of various items, such as junction boxes, service interphone jacks, antennas, and coaxial cables (see figure 1-37). Remember, chapter 91 of the wiring diagram manual contains panel and rack locations.

Using The Boeing 747 Component Locator Guide

The Component Locator Guide for the 747-400 aircraft is a 5 x 8 inch manual of approximately 450 pages. The manual is designed for 747-400 line technicians and provides quick access to the locations of various components. The guide also provides a list of ATA chapters for maintenance or wiring manuals. There are three sections to the manual: the *Index, Access/Area*, and *Recognition/Identification*. Each section of the manual is color coded for quick reference. The example below will help to explain the use of the Component Locator Guide.

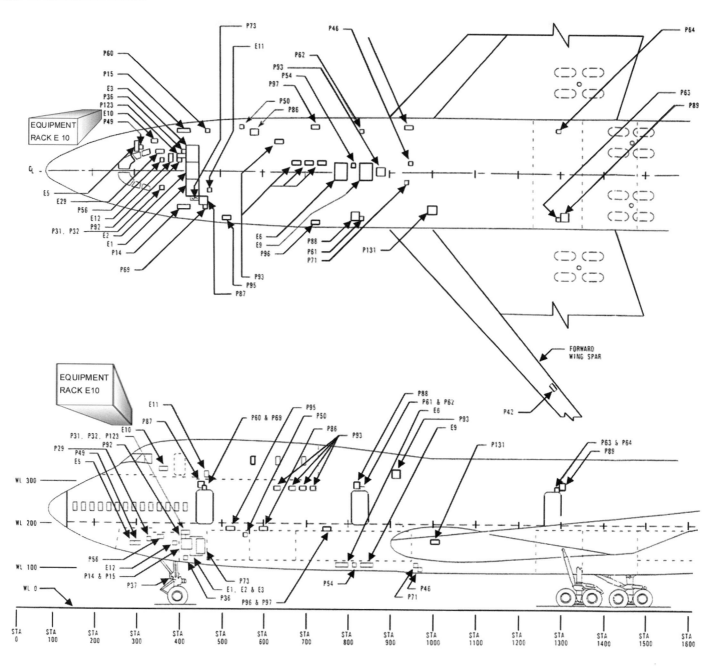

Fig. 1-36   Panel and equipment locations diagram. (The Boeing Commercial Airplane Company) For Training Purposes Only. (CONTINUED ON NEXT PAGE)

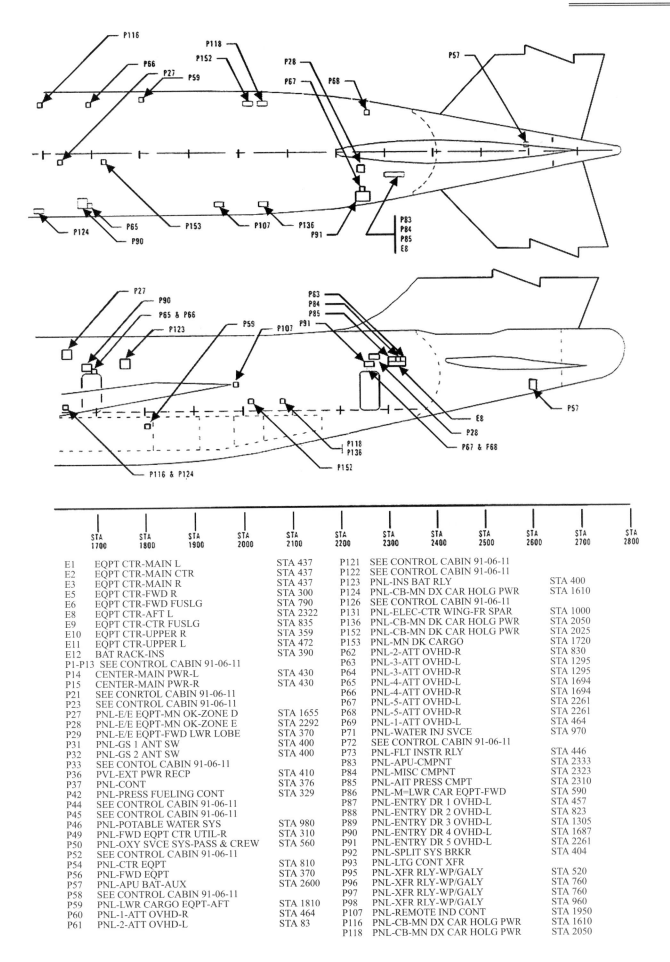

| | | | STA 2100 | STA 2200 | STA 2300 | STA 2400 | STA 2500 | STA 2600 | STA 2700 | STA 2800 |

E1	EQPT CTR-MAIN L	STA 437	
E2	EQPT CTR-MAIN CTR	STA 437	
E3	EQPT CTR-MAIN R	STA 437	
E5	EQPT CTR-FWD R	STA 300	
E6	EQPT CTR-FWD FUSLG	STA 790	
E8	EQPT CTR-AFT L	STA 2322	
E9	EQPT CTR-CTR FUSLG	STA 835	
E10	EQPT CTR-UPPER R	STA 359	
E11	EQPT CTR-UPPER L	STA 472	
E12	BAT RACK-INS	STA 390	
P1-P13	SEE CONTROL CABIN 91-06-11		
P14	CENTER-MAIN PWR-L	STA 430	
P15	CENTER-MAIN PWR-R	STA 430	
P21	SEE CONRTOL CABIN 91-06-11		
P23	SEE CONTROL CABIN 91-06-11		
P27	PNL-E/E EQPT-MN OK-ZONE D	STA 1655	
P28	PNL-E/E EQPT-MN OK-ZONE E	STA 2292	
P29	PNL-E/E EQPT-FWD LWR LOBE	STA 370	
P31	PNL-GS 1 ANT SW	STA 400	
P32	PNL-GS 2 ANT SW	STA 400	
P33	SEE CONTOL CABIN 91-06-11		
P36	PVL-EXT PWR RECP	STA 410	
P37	PNL-CONT	STA 376	
P42	PNL-PRESS FUELING CONT	STA 329	
P44	SEE CONTROL CABIN 91-06-11		
P45	SEE CONTROL CABIN 91-06-11		
P46	PNL-POTABLE WATER SYS	STA 980	
P49	PNL-FWD EQPT CTR UTIL-R	STA 310	
P50	PNL-OXY SVCE SYS-PASS & CREW	STA 560	
P52	SEE CONTROL CABIN 91-06-11		
P54	PNL-CTR EQPT	STA 810	
P56	PNL-FWD EQPT	STA 370	
P57	PNL-APU BAT-AUX	STA 2600	
P58	SEE CONTROL CABIN 91-06-11		
P59	PNL-LWR CARGO EQPT-AFT	STA 1810	
P60	PNL-1-ATT OVHD-R	STA 464	
P61	PNL-2-ATT OVHD-L	STA 83	
P121	SEE CONTROL CABIN 91-06-11		
P122	SEE CONTROL CABIN 91-06-11		
P123	PNL-INS BAT RLY	STA 400	
P124	PNL-CB-MN DX CAR HOLG PWR	STA 1610	
P126	SEE CONTROL CABIN 91-06-11		
P131	PNL-ELEC-CTR WING-FR SPAR	STA 1000	
P136	PNL-CB-MN DK CARGO HOLG PWR	STA 2050	
P152	PNL-CB-MN DK CAR HOLG PWR	STA 2025	
P153	PNL-MN DK CARGO	STA 1720	
P62	PNL-2-ATT OVHD-R	STA 830	
P63	PNL-3-ATT OVHD-L	STA 1295	
P64	PNL-3-ATT OVHD-R	STA 1295	
P65	PNL-4-ATT OVHD-L	STA 1694	
P66	PNL-4-ATT OVHD-R	STA 1694	
P67	PNL-5-ATT OVHD-L	STA 2261	
P68	PNL-5-ATT OVHD-R	STA 2261	
P69	PNL-1-ATT OVHD-L	STA 464	
P71	PNL-WATER INJ SVCE	STA 970	
P72	SEE CONTROL CABIN 91-06-11		
P73	PNL-FLT INSTR RLY	STA 446	
P83	PNL-APU-CMPNT	STA 2333	
P84	PNL-MISC CMPNT	STA 2323	
P85	PNL-AIT PRESS CMPT	STA 2310	
P86	PNL-M=LWR CAR EQPT-FWD	STA 590	
P87	PNL-ENTRY DR 1 OVHD-L	STA 457	
P88	PNL-ENTRY DR 2 OVHD-L	STA 823	
P89	PNL-ENTRY DR 3 OVHD-L	STA 1305	
P90	PNL-ENTRY DR 4 OVHD-L	STA 1687	
P91	PNL-ENTRY DR 5 OVHD-L	STA 2261	
P92	PNL-SPLIT SYS BRKR	STA 404	
P93	PNL-LTG CONT XFR		
P95	PNL-XFR RLY-WP/GALY	STA 520	
P96	PNL-XFR RLY-WP/GALY	STA 760	
P97	PNL-XFR RLY-WP/GALY	STA 760	
P98	PNL-XFR RLY-WP/GALY	STA 960	
P107	PNL-REMOTE IND CONT	STA 1950	
P116	PNL-CB-MN DX CAR HOLG PWR	STA 1610	
P118	PNL-CB-MN DX CAR HOLG PWR	STA 2050	

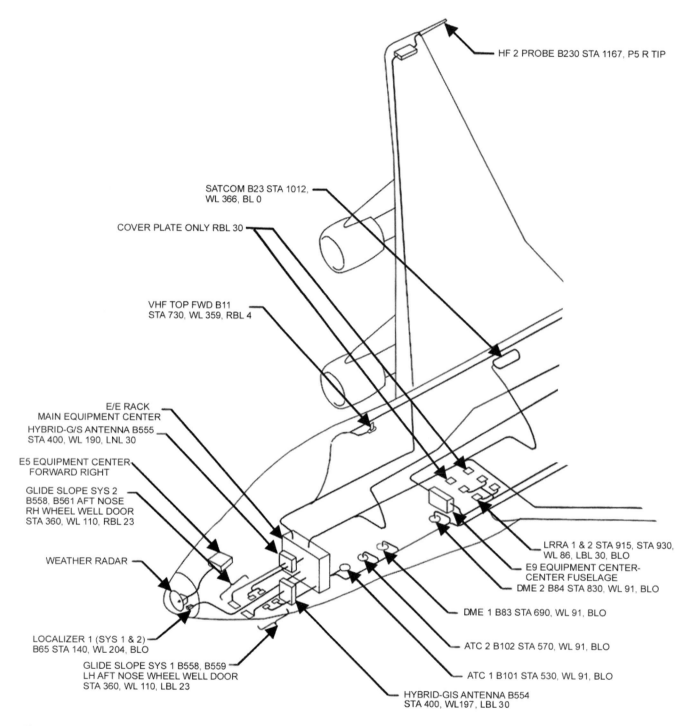

HF 2 PROBE B230 STA 1167, P5 R TIP

SATCOM B23 STA 1012,
WL 366, BL 0

COVER PLATE ONLY RBL 30

VHF TOP FWD B11
STA 730, WL 359, RBL 4

E/E RACK
MAIN EQUIPMENT CENTER

HYBRID-G/S ANTENNA B555
STA 400, WL 190, LNL 30

E5 EQUIPMENT CENTER
FORWARD RIGHT

GLIDE SLOPE SYS 2
B558, B561 AFT NOSE
RH WHEEL WELL DOOR
STA 360, WL 110, RBL 23

WEATHER RADAR

LRRA 1 & 2 STA 915, STA 930,
WL 86, LBL 30, BLO

E9 EQUIPMENT CENTER-
CENTER FUSELAGE

DME 2 B84 STA 830, WL 91, BLO

DME 1 B83 STA 690, WL 91, BLO

LOCALIZER 1 (SYS 1 & 2)
B65 STA 140, WL 204, BLO

ATC 2 B102 STA 570, WL 91, BLO

GLIDE SLOPE SYS 1 B558, B559
LH AFT NOSE WHEEL WELL DOOR
STA 360, WL 110, LBL 23

ATC 1 B101 STA 530, WL 91, BLO

HYBRID-GIS ANTENNA B554
STA 400, WL197, LBL 30

Fig. 1-37 Typical antenna and coaxial cable arrangement diagram. (The Boeing Commercial Airplane
Company) For Training Purposes Only. (CONTINUED ON NEXT PAGE)

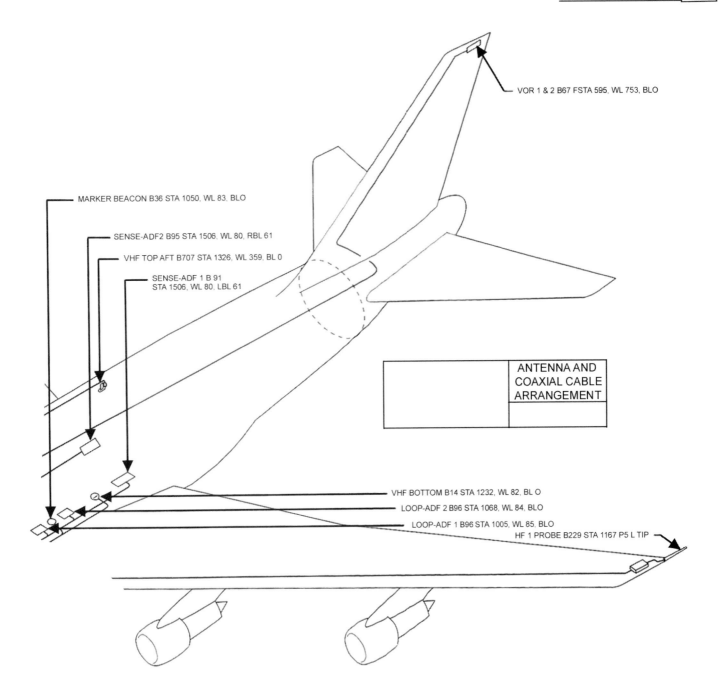

VOR 1 & 2 B67 FSTA 595, WL 753, BLO

MARKER BEACON B36 STA 1050, WL 83, BLO

SENSE-ADF2 B95 STA 1506, WL 80, RBL 61

VHF TOP AFT B707 STA 1326, WL 359, BL 0

SENSE-ADF 1 B 91
STA 1506, WL 80, LBL 61

ANTENNA AND
COAXIAL CABLE
ARRANGEMENT

VHF BOTTOM B14 STA 1232, WL 82, BL O

LOOP-ADF 2 B96 STA 1068, WL 84, BLO

LOOP-ADF 1 B96 STA 1005, WL 85, BLO

HF 1 PROBE B229 STA 1167 P5 L TIP

Assume the technician wishes to find the center auto pilot elevator servo. To locate the component, proceed as follows:

1) Determine the correct ATA chapter for the auto pilot servo (chapter 22, Autoflight). A list of ATA chapters is included in the locator guide introduction.

2) In the yellow index pages find ATA 22. Next find the correct component in that chapter (see figure 1-38).

3) Read the information in each column of the servo - elevator A/P, center to determine: quantity (1), access/area (STA2600, WL320, LBL20), page showing the access diagram (28), page showing the recognition diagram (49), and ATA chapter (22-12-01).

ATA 22	AUTOFLIGHT			ACCESS	RECOG-NITION	ATA 22
COMPONENT		QTY	ACCESS/AREA	(pink)	(blue)	CH/SEC
ACCELEROMETER - AFT MODAL SUPPRESSION, LEFT		1	BELOW MAIN DECK - STA 2260	8	46	22-21-03
ACCELEROMETER - AFT MODAL SUPPRESSION, RIGHT		1	BELOW MAIN DECK - STA 2260	8	46	22-21-03
ACCELEROMETER - FWD MODAL SUPPRESSION, LEFT		1	ABOVE NOSE GEAR WHEEL WELL	8	46	22-21-03
ACCELEROMETER - FWD MODAL SUPPRESSION, RIGHT		1	ABOVE NOSE GEAR WHEEL WELL	8	46	22-21-03
COMPUTER - FLIGHT CONTROL, CENTER		1	MEC - E1-3	22,42	8	22-11-01
COMPUTER - FLIGHT CONTROL, LEFT		1	MEC - E1-1	22,42	8	22-11-01
COMPUTER - FLIGHT CONTROL, RIGHT		1	MEC - E1-4	22,42	8	22-11-01
PACKAGE - ROLLOUT POWER CONTROL, CENTER		1	FR, FL, ZONE 324	31	47	22-13-04
PACKAGE - CNTRL LATERAL CONTROL, LEFT		1	WING LANDING GEAR WELL	25	48	27-11-08
PACKAGE - ROLLOUT POWER CONTROL, LEFT		1	FR, FL, ZONE 324	31	47	22-13-04
PACKAGE - CNTRL LATERAL CONTROL, RIGHT		1	WING LANDING GEAR WELL	25	48	27-11-08
PACKAGE - ROLLOUT POWER CONTROL, RIGHT		1	FR, FL, ZONE 324	31	47	22-13-04
PANEL - PASSENGER OXYGEN AND YAW DAMPER		1	FLIGHT DECK - P5	41	3	22-21-01
PANEL - AFCS MODE CONTROL		1	FLIGHT DECK - P10	41	2	22-11-02
SERVO - ELEVATOR A/P, CENTER		1	STA 2600, WL320, LBL 20	28	49	22-12-01
SERVO - ELEVATOR A/P, LEFT		1	STA 2600, WL320, RBL 12	28	49	22-12-01
SERVO - ELEVATOR A/P, RIGHT		1	STA 2600, WL320, RBL 22	28	49	22-12-01
SERVO - LATERAL A/P		1	WING LANDING GEAR WELL	25	48	22-13-01
SWITCH - AFCS GO-AROUND		4	FLIGHT DECK - PILOT'S CONTROL STAND	41	45	22-11-04
SWITCH - CAPTAIN A/P DISENGAGE, S282		1	FLIGHT DECK - CONTROL WHEEL	41	45	22-11-03
SWITCH - F/O A/P DISENGAGE, S283		1	FLIGHT DECK - CONTROL WHEEL	41	45	22-11-03

Fig. 1-38 Chapter 22 of Boeing 747 component locators guide. (The Boeing Commercial Airplane Company) For Training Purposes Only.

4) Turn to access page 28 (access pages are pink). This figure shows the various access panels in the horizontal stabilizer (figure 1-39a). The recognition page will show which panel provides access to the servo.

5) Turn to recognition page 49 (recognition pages blue). This will show that the center elevator servo is located behind panel 315 A (figure 1-39b). This diagram also gives a basic drawing of the servo to help the technician recognize the unit.

6) Find chapter 22-12-01 in the maintenance and/or wiring diagram manuals for the information needed to finish the needed repair.

The Component Locator Guide has provided the technician all the necessary information to locate the center elevator servo. Other LRUs can be found in a similar manner.

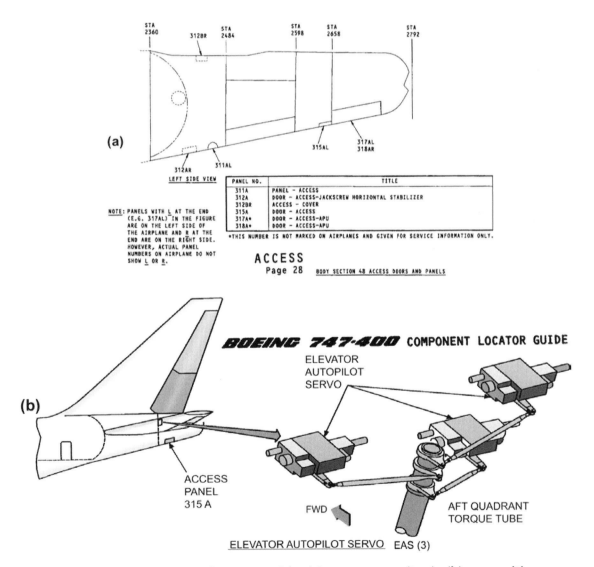

Fig. 1-39   Example of Boeing 747 component locators guide. (a) access page (top); (b) recognition page.(bottom) (The Boeing Commercial Airplane Company) For Training Purposes Only.

CHAPTER 2
POWER DISTRIBUTION, DATA BUS
AND DIGITAL SYSTEMS

INTRODUCTION

In many respects all aircraft and their avionics systems are virtually the same. This chapter will present topics that are common to virtually every modern aircraft. In order to fully understand the operation of an avionics systems, the technician must first become familiar with the aircraft's electrical power distribution. On transport category aircraft the power distribution system is very complex involving several computers for operation and control. The following pages will present the reader with information on power distribution for both simple and complex aircraft.

Modern avionics systems operate using a variety of computers and microprocessor circuits. These computer based systems rely on digital data for communication between components. There are currently several formats which are used for transmission of digital data. The materials presented in this chapter will help the avionics technician become familiar with commonly used data bus systems.

Many of the computer based systems also employ electronics which are sensitive to the discharge of static electricity. Standard techniques for handling these components will be presented here as well as multiplexing/demultiplexing techniques and analog to digital converters.

POWER DISTRIBUTION SYSTEMS

An aircraft's **power distribution system** is comprised of one or more electrical distribution points. The distribution points are often made of a solid copper bar referred to as a **bus** or **bus bar**. In some cases, the distribution point is a simple terminal stud where several wires connect. On light, single-engine aircraft, there is often only one bus which is used to distribute the positive voltage of the electrical power. The negative voltage is connected to the various loads through the metal structure of the airframe. Any aircraft which uses the airframe to distribute negative voltage is considered to have a **Single-Wire Electrical System.** A single-wire electrical system may also be referred to as a **Negative Ground System.** As aircraft become more complex so do their power distribution systems. Virtually all business type and transport category aircraft contain several power distribution busses. It is very important for electronics technicians to understand the characteristics of a bus system in order to effectively troubleshoot various avionic systems.

Bus Hierarchy

Since aircraft are designed with safety as a primary consideration, it stands to reason that electrical systems require special attention. Light aircraft which fly during night or poor weather conditions rely heavily on electronic equipment to provide a safe and comfortable flight. Modern transport category aircraft rely on electrical

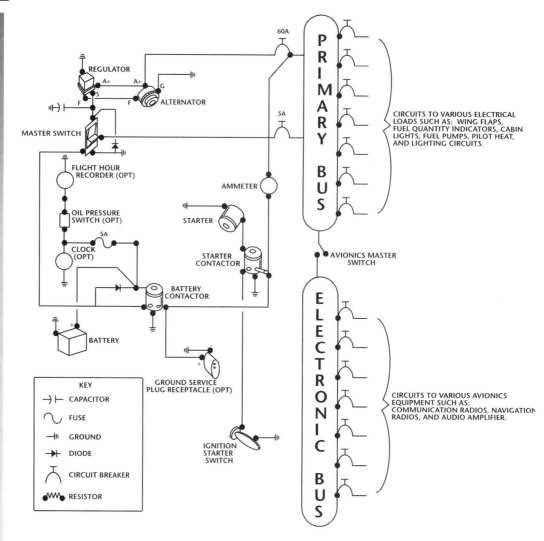

Fig. 2-1   A simple power distribution system. For Training Purposes Only.

systems to power computers which operate virtually every system on the aircraft. On aircraft with more than one distribution bus, the busses are arranged into a hierarchy ranging from least to most critical. The **bus hierarchy** is designed to ensure that the most critical electrical systems are the least likely to fail.

Light Aircraft Power Distribution

Figure 2-1 illustrates a simple power distribution system which has only a two stage hierarchy. That is, the electrical systems may be powered by the engine-driven generator or by the aircraft's battery. Figure 2-2 shows the electrical load distribution of a Beechcraft King Air. It should be noted that Beechcraft is currently owned by Raytheon Corporation and the King Air is often referred to as the *Raytheon King Air*. This text will use the term *Beechcraft King Air* or *King Air*. The King Air system is comprised of: 1) a main battery bus, 2) an isolation bus, 3) a left and right generator bus, 4) four dual fed busses, 5) a 26 VAC bus, 6) two subpanel busses and 7) a hot battery bus.

The King Air has both a DC and AC power system. Direct current is used to power the majority of the aircraft systems. AC is supplied through one of two static inverters, labeled *INV NO. 1* and *INV NO. 2* on the schematic. As seen in the upper left

ELECTRICAL LOAD DISTRIBUTION

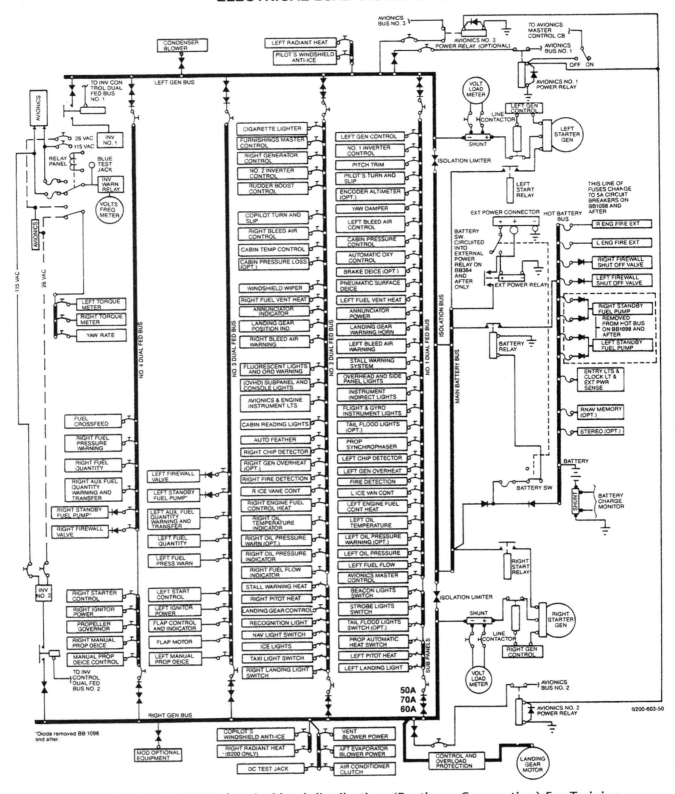

Fig. 2-2 Beechcraft King Air C90A electrical load distribution. (Raytheon Corporation) For Training Purposes Only.

portion of the diagram, 26V and 115V AC power is used by various avionics systems. During maintenance of these systems, be sure AC power is available from one of these static inverters.

The diagram for AC power distribution is shown in figure 2-3. Here it can be seen that one or more components of the AC system can fail without loss of AC electrical power. For example, if the left power relay should fail the left inverter would not receive power. However, the right inverter system would still be operational and inverter power would be routed though the inverter select relay to the 26V and 115V AC busses.

The King Air system is designed to allow for more than one failure before critical systems are lost. The most critical bus on this aircraft is the hot battery bus (figure 2-2). The hot battery bus receives DC power whenever a charged battery is installed in the aircraft. Notice that critical systems, such as, fire protection and fuel pumps are connected to the hot battery bus. The clock, RNAV memory, and stereo are also connected to the hot battery bus. It is advantageous that these systems remain powered even if the battery relay is turned off.

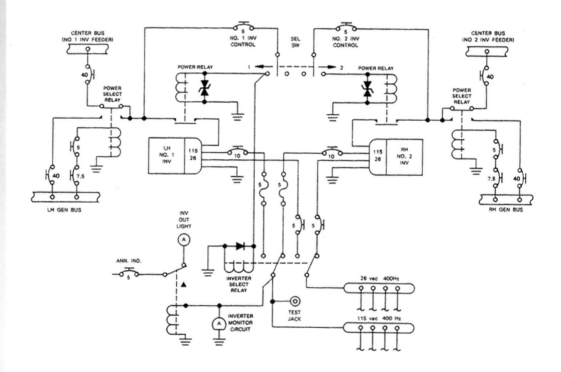

Fig. 2-3 Beechcraft King Air C90A AC power distribution system. (Raytheon Corporation) For Training Purposes Only.

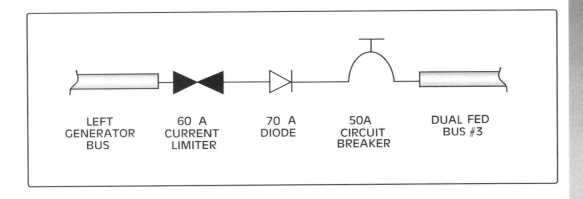

LEFT
GENERATOR
BUS   60 A
CURRENT
LIMITER   70 A
DIODE   50A
CIRCUIT
BREAKER   DUAL FED
BUS #3

Fig. 2-4 A typical circuit connecting two distribution busses.

As their names imply, the left generator bus receives power from the left generator; the right generator bus receives power from the right generator (figure 2-2). The isolation bus is used to parallel the system during normal operation. The isolation bus is connected to the left and right busses through a 325 amp isolation limiter (figure 2-2). An isolation limiter is basically a large fuse which is used to protect the system. Figure 2-4 shows a detail of the Right Generator Bus and the #3 Dual Fed Bus connection, figure 2-2 shows an expanded view of this circuit. This type of power distribution system contains a series circuit with a 60A current limiter, a 70A diode, and a 50A circuit breaker. These three components are used to isolate each Dual Fed Bus and protect the electrical system in the event of a catastrophic failure. Any given bus could short to ground loosing the electrical components connected to that bus, however, the remainder of the electrical system will operate normally. For this protection to operate correctly the current limiter must be rated at a higher value than the circuit breaker (60A-current limiter, 50A-circuit breaker). The current limiter is designed to open only if the circuit breaker fails to protect the circuit. (i.e. The current limiter is simply a backup for the circuit breaker.) The diode in the circuit is used to prevent any current from leaving the Dual Fed Bus to power the Left Generator Bus. Virtually all modern corporate and commuter type aircraft employ a split bus power distribution system similar to the one described for the Beechcraft King Air.

Transport Category Aircraft Power Distribution

On transport category aircraft, such as the Boeing 727, 737, 747 and 777 the Airbus A-300, A-320 and A-340, and the McDonnell Douglas (Boeing) MD-80 or MD-11, the power distribution systems are quite complex. There are several busses and three or more AC generators which can supply electrical power during flight. Direct current is supplied by the generators through **transformer rectifier (TR)** units. The generators and distribution systems are computer controlled and automatically monitored for malfunctions and defects.

The electrical system hierarchy for transport aircraft can be broken into two segments: 1) the hierarchy for the generators supplying power, and 2) the hierarchy for the distribution busses. The typical generator hierarchy might be arranged as shown in figure 2-5. Here it can be seen that the two engine driven generators are least critical and the static inverter would be considered the most critical AC power source.

A typical bus hierarchy is shown in figure 2-6. The diagram is divided vertically with AC shown on the left and DC on the right. The most critical busses are located on the bottom of the chart. Loss of these critical power distribution busses would most likely be catastrophic; therefore, they are typically fed by redundant power sources.

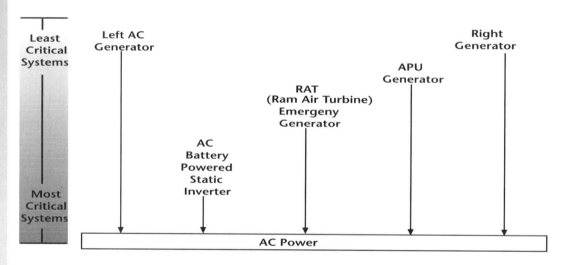

Fig. 2-5   Generator hierarchy.

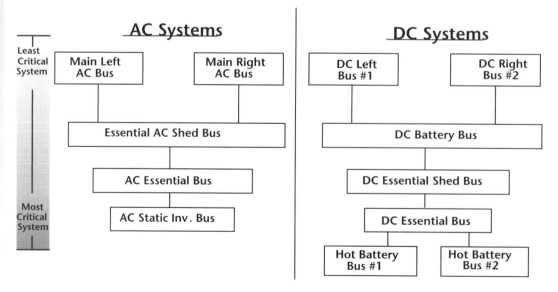

Fig. 2-6   Bus hierarchy.

There are two basic types of power distribution systems found on transport category aircraft, the **split bus system** and the **parallel system.** The split bus system is used on most twin engine commercial aircraft such as the Boeing 737, 757, 767 and 777, the McDonnell Douglas MD 80, and the Airbus Industries A-320 and A-310. Split bus power distribution systems are found exclusively on twin engine aircraft. In a split bus system each engine driven generator supplies power to a specific bus (left of right). The left and right busses operate independently and must never be connected while both generators are on line.

In a parallel electrical system, the entire electrical load is equally shared by all of the working generators. Parallel AC power distribution systems are typically found on commercial aircraft containing three or more engines such as the Boeing 727, early 747's, the Lockheed L-1011, and the McDonnell Douglas MD-11 (DC-10). A **split parallel** power distribution system is used on some modern four engine aircraft such as the Boeing 747-400. A split parallel system allows all generators to operate in parallel during normal conditions. In the event of a system malfunction, one or more generators may operate independently.

Split bus systems

The Airbus A-320 employs a state-of-the-art split bus power distribution system (figure 2-7). Since this aircraft is typical of most twin-engine transport category aircraft, it will be presented here for discussion on split bus power distribution. The

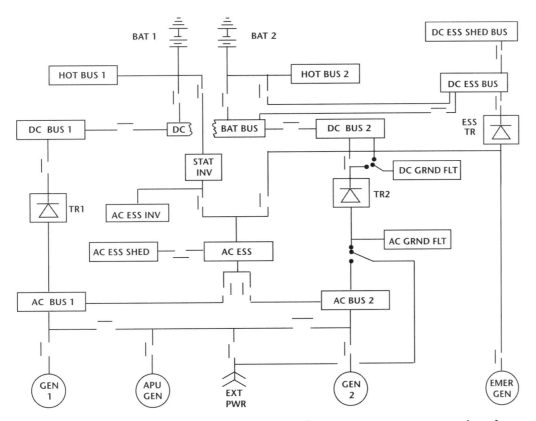

Fig. 2-7   A split bus power distribution system for a transport category aircraft. (Airbus Industrie) For Training Purposes Only.

A-320 system consists of four AC generators; two powered by the main engines, one powered by the auxiliary power unit (APU) and one emergency generator powered by a ram air turbine or RAT (described below). The two main (engine-driven) generators are least critical and a failure of one generator would cause only momentary loss of galley power. Assuming the left generator failed, the flight crew would start the APU generator to assume all loads from the failed main generator. If the right main generator should also fail, the APU generator would supply all electrical power to the aircraft. All transport category aircraft with a split bus electrical system utilize the APU to power a backup generator. Therefore, the APU must be capable of operation during normal flight. If the APU generator should fail, the Ram Air Turbine (RAT) would be used to drive an emergency power generator. If that system should fail the aircraft would still have AC power available through the two static inverters. The static inverters receive their power from the ship's DC batteries.

Since the static inverter system is the most critical AC power source on the A-320, the system must be checked each night during routine maintenance. The static inverter system is tested using the aircraft's computer diagnostic system. If the test shows an inverter system failure, it must be repaired prior to the next flight.

The **Ram Air Turbine (RAT)** is a device used on many twin-engine transport category aircraft to supply emergency hydraulic and electrical power. The RAT consists of a propeller driven hydraulic pump which is deployed from the fuselage in the event of a catastrophic hydraulic or electrical system failure (figure 2-8). As long as the aircraft has an airspeed greater than 80 knots, the RAT propeller will

Fig. 2-8 A Ram Air Turbine shown extended from the aircraft fuselage.

"wind mill" as it passes through the air. The turning propeller powers a hydraulic pump. The pressurized hydraulic fluid is routed to a hydraulic motor that is used to turn the emergency AC generator.

The emergency generator has a maximum output of 5 KVA at 115 VAC, 400 Hz. The main and APU generators have a maximum output of 90 KVA at 115 VAC, 400 Hz. Comparing these two values, it is easy to see that if the aircraft is operating on emergency generator power, the electrical system is extremely limited and the aircraft should land as soon as practical.

Parallel bus systems

The **Boeing 727** is a three engine aircraft which employs a parallel power distribution system. As seen in figure 2-9, the system contains three main engine-driven generators and one APU-driven generator which is ground operable only. Each generator is connected to a paralleling, or synchronizing bus (sync bus), through its respective generator breaker (GB) and bus tie breaker (BTB). During normal flight, generators 1, 2 & 3 are connected to the sync bus; however, if a generator fails it is isolated by opening it's GB. If AC bus number 1, 2 or 3 should short to ground, the related BTB and GB would open to isolate that bus.

On the B-727 the least critical AC loads would be powered by AC bus number 1, 2 or 3. The least critical DC loads are powered by DC bus number 1 or 2. The next most critical systems would be powered by the **essential (ESS)** AC and DC busses shown on the left side of the diagram (figure 2-9). The essential AC bus can receive power from any AC generator. The essential DC bus can be powered by either DC bus 1 or 2, or the essential TR unit. The most critical electrical loads on the aircraft are powered by the standby busses located at the top of the diagram. The standby busses will receive power from the battery even if all three engine-driven generators fail.

Split parallel system

A **split parallel** electrical power distribution system allows for flexibility in load distribution and yet maintains isolation between systems when needed. The **split system breaker (SSB)** can be seen in the simplified diagram of a split parallel system (figure 2-10). When closed, the split system breaker connects all four main-engine generators, thus paralleling the system. When open, the split system breaker isolates the right and left hand systems.

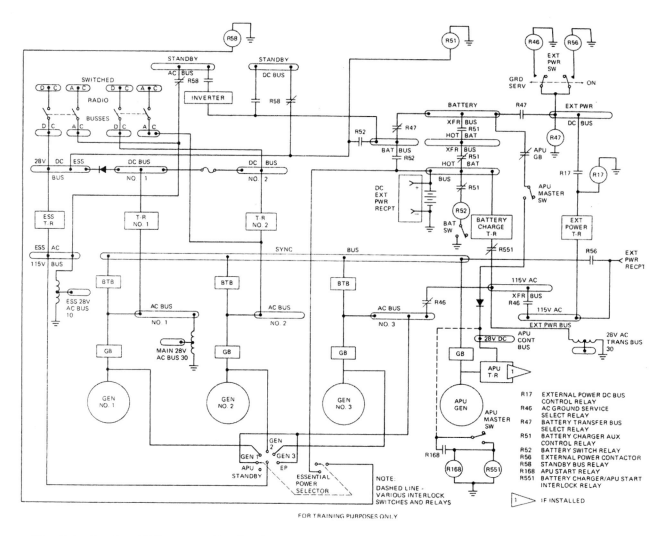

Fig. 2-9 Boeing 727 power distribution system. (Boeing Commercial Airplane Company) For Training Purposes Only.

A split parallel system is used on the **Boeing 747-400** aircraft. As seen in figure 2-11, this system employs four engine-driven integrated drive generators (IDGs), two auxiliary power unit (APU) generators, and can accept two separate external power sources (EXT 1 & EXT 2). As seen in the system schematic, the four IDGs are connected to their respective AC busses through generator control breakers (GCBs). The AC busses are paralleled through the bus tie breakers (BTBs) and the split system breaker (SSB). When the SSB is open, the right system operates independent of the left. With this system, any generator can supply power to any load bus or any combination of the IDGs can operate in parallel. It should be noted that the term *Integrated Drive Generator (IDG)* is used here to represent the AC generators. An IDG is simply an AC generator built into (integrated with) the same housing as the constant speed drive unit. A constant speed drive is a type of hydraulic transmission that turns the generator at a constant speed regardless of the engine RPM. The AC generator must rotate at a constant speed in order to maintain a constant output frequency of 400 CPS. Modern transport category aircraft employ integrated drive generators on all main engines.

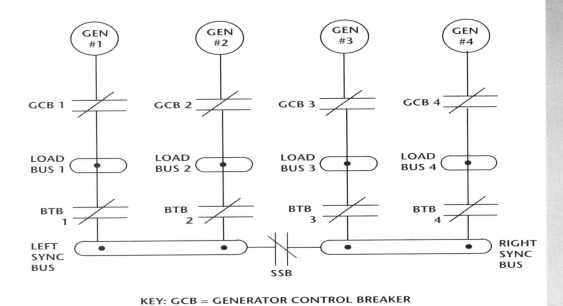

KEY: GCB = GENERATOR CONTROL BREAKER
BTB = BUS TIE BREAKER
SSB = SPLIT SYSTEM BREAKER

Fig. 2-10   A split parallel power distribution system for a transport category aircraft.
For Training Purposes Only.

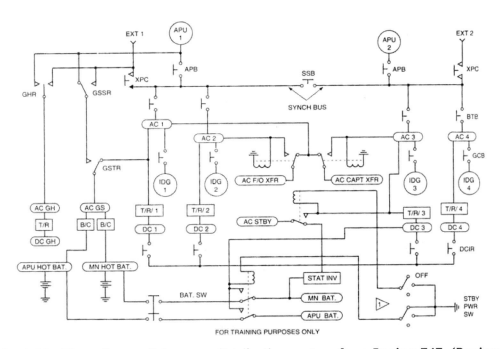

FOR TRAINING PURPOSES ONLY

Fig. 2-11   The split parallel power distribution system for a Boeing 747. (Boeing
Commercial Airplane Company) For Training Purposes Only.

Ground handling and ground service busses

Ground handling (GH) busses are used to power lighting and miscellaneous equipment for cargo loading, fueling, and cleaning the aircraft. The AC ground handling (GH) busses are powered by closing the ground handling relay (GHR) to either the APU or EXT power. The DC ground handling busses receive power from the transformer rectifier (TR) shown in figure 2-11 (lower left side of the diagram). The GH busses are not powered during normal flight.

On the Boeing 747-400 the **ground service (GS)** busses are used to light the aircraft interior, power the main battery charger and other miscellaneous systems required for maintenance and initial start-up of the aircraft. The ground service busses are controlled from the flight attendant station located at the number two left door of the aircraft. This switch energizes the ground service relay (GSR) which connects the GS bus to which ever is currently on line, the APU or EXT power.

No break power transfer

The 747-400 uses an automated power distribution control system which features a **No Break Power Transfer (NBPT).** A no break power transfer means that the automated system can change the AC power source without a momentary interruption of electrical power. For example, when external power is being used and the aircraft is preparing to depart, the engines are started and the main generators are brought on line. During a NBPT, the generator control units monitor the power source currently on line (the external power) and the power source requested by the flight crew (an engine driven generator). If the power requested is within specifications, both power supplies are paralleled for a split second and no power interruption occurs. If the requested power is out of limits the CGU tries to adjust the system and then connects the requested power to the busses. If the power systems cannot be adjusted to the correct tolerance for paralleling, the requested power source will be rejected or there will be a momentary power interruption.

During the power transfer, both sources are paralleled for a fraction of a second. This can only happen if the individual power sources are within extremely tight voltage, frequency and phase limits. For example the 747-400 requires any difference between generators to be within 10 V, 6 Hz, and 90° maximum phase difference to complete a NBPT. If the NBPT fails it is most likely that one of the power sources is out of limits. This is particularly true when switching from ground power to aircraft power. Many ground power units are poorly regulated and cannot meet the specifications for the NBPT.

Powering The Aircraft For Maintenance Purposes

Virtually all aircraft have some means to provide electrical power to various systems without starting the main or APU engines. On light aircraft, the ship's battery can be connected to the bus and used as a temporary power source for troubleshooting or maintenance of various electronic or avionics systems. Alternating current (if used) would be available through the aircraft's static inverter.

Caution: the battery is being discharged during this configuration. Always limit maintenance activities using battery power, and recharge the battery if needed.

All corporate and transport category aircraft will accept external power for prolonged ground operations of electrical systems. Many smaller general aviation aircraft also have external power receptacles. Whenever practical use external power (not battery power) to supply electrical power to the aircraft. It is important to become familiar with your particular aircraft before connecting external power. Always observe all safety precautions and be sure to warn any other technicians which might be working on the aircraft before you connect external power. With the aircraft in certain configurations, application of external power can activate systems which could injure fellow technicians and/or damage the aircraft.

Whenever connecting external power to any aircraft, always refer to the operations manual or service manual if you are unfamiliar with the procedures. The following procedures would be used to connect external power to the Beechcraft King Air. First, ensure the aircraft is in a safe configuration. Be sure no wiring or other systems are disassembled which may cause damage if power is connected, and inform all technicians working on the aircraft that external power will soon be connected. Second, place the battery switch in the on position (the ship's battery must be at least 20V). Third, turn the avionics master off. Lastly, connect the external power plug to the aircraft being sure the plug is seated firmly into the socket. Figure 2-12 shows the external power socket and related connections for a Beechcraft King Air. Like most aircraft, the polarity of the external power plug must be correct for the external power relay to close.

Fig. 2-12   Installation of a typical external power connection.

The King Air maintenance manual states several precautions which must be observed when connecting external power, briefly stated they are: 1) If unknown, determine the polarity of the external power supply plug. 2) Turn off all radio/avionics equipment and both engine generators before connecting external power. Typically the avionics master switch can be used to disconnect all avionics simultaneously. 3) A battery of at least 20 volts must be in the aircraft and the battery master switch **must be ON** prior to connecting external power. This allows the normally closed avionics master relay to energize and disconnect the avionics bus. 4) The external power supply must be capable of the correct amperage and voltage when used for engine starting purposes. For the Super King Air-200, a continuous 300 amperes at 24 to 30 volts, and 1,000 amperes for 1 second must be available. 5) Proper ground connections are needed to ensure voltage stability at high amperage. 6) The aircraft battery maybe damaged if the voltage applied is in excess of 30 volts. Most of the above suggestions apply not only to the King Air but to all general aviation aircraft.

On transport category aircraft external power typically supplies 115 V, 400 Hz alternating current. The 28 V DC is made available through one of the aircraft's transformer-rectifier units which converts the incoming 115 V AC to 28 V DC. Figure 2-13 shows the electrical system control panel used to activate electrical power on a Boeing 757 aircraft. This is the P5 panel which is located on the center over head panel of the flight deck.

The P5 panel controls all power distribution functions for the aircraft using alternate action and momentary contact type switches. The alternate action ("in or out")

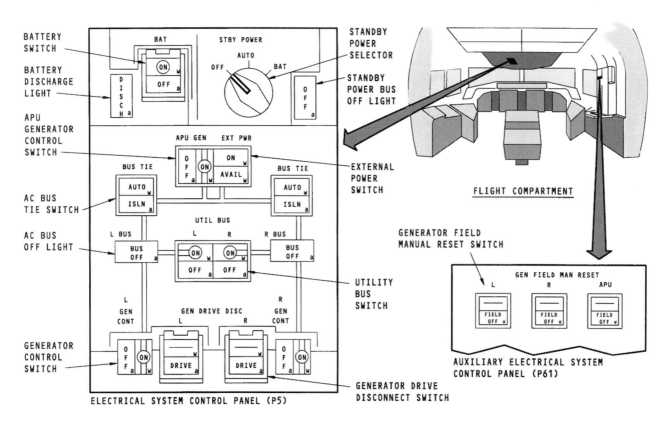

Fig. 2-13 Electrical control panels for a Boeing 757. (Boeing Commercial Airplane Company)

switches are latched to the last operated position. A mechanical flag located inside the switch indicates the switch position, such as, ON or AVAIL.

Connecting external power

To connect external power to this aircraft the following steps must be taken.

1) **Use caution**. Be sure it is safe to connect electrical power to the aircraft. If other maintenance is being performed, certain systems may be powered inadvertently when applying external power. Or, if portions of the electrical system are disconnected for maintenance, connecting external power may create a short to ground and damage the electrical system and/or the aircraft structure.

2) The external power cord must be plugged into the aircraft external power panel located near the aircraft's nose wheel (figure 2-14). If the connected power is within specifications, the AC CONNECTED light on the external power panel and the external power switch available (AVAIL) light on the P5 panel will illuminate. Also, the ground handling bus will automatically receive power. The ground handling bus will supply power to necessary lighting for minor maintenance and cleaning of the aircraft.

3) Locate the forward flight attendant panel (P21) just inside the aircraft entrance (see figure 2-15). By depressing the ground service switch on P21, the ground service bus will be powered by external power. This bus

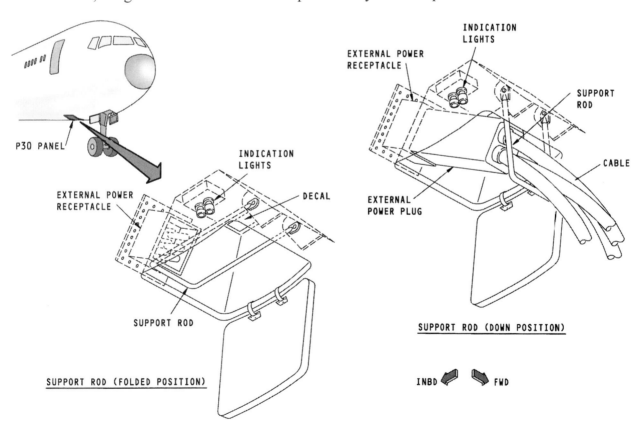

Fig. 2-14 Typical external power panel for a transport category aircraft. (Boeing Commercial Airplane Company)

supplies equipment which must be powered in both air and the ground modes. Figure 2-16 shows that the ground service bus is connected to the right AC bus during flight and to external power or APU (if selected) during ground operations.

4) If it is necessary to power the left or right AC busses, depress the external power switch on the P5 panel, figure 2-13. This will connect external power to the aircraft bus ties. If the bus tie switches are latched in the AUTO position, the electrical power will automatically connect to the AC busses. If the bus ties are in the isolation mode, depress the appropriate bus tie switch to activate its associated bus.

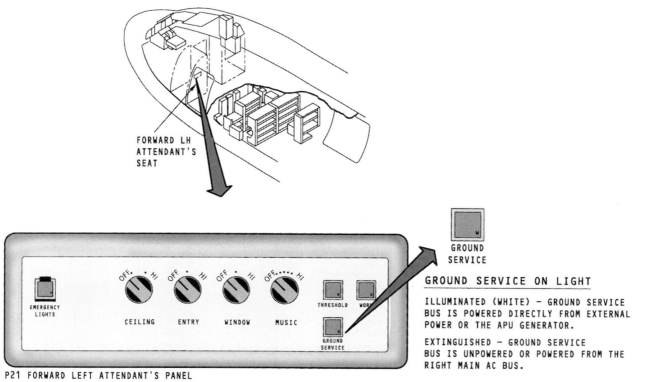

GROUND SERVICE

GROUND SERVICE ON LIGHT

ILLUMINATED (WHITE) – GROUND SERVICE BUS IS POWERED DIRECTLY FROM EXTERNAL POWER OR THE APU GENERATOR.

EXTINGUISHED – GROUND SERVICE BUS IS UNPOWERED OR POWERED FROM THE RIGHT MAIN AC BUS.

P21 FORWARD LEFT ATTENDANT'S PANEL

Fig. 2-15   Forward flight attendant's panel and ground service switch. (Boeing Commercial Airplane Company)

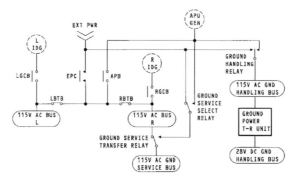

Fig. 2-16   AC external power distribution. (Boeing Commercial Airplane Company) For Training Purposes Only.

ARINC STANDARDS

ARINC Incorporated is a global corporation consisting of various U.S. and international airlines and aircraft operators and their subsidiaries. The company provides services related to a variety of aviation communication and navigation systems. ARINC technical standards have been accepted by both U.S. and international aircraft and component manufacturers. ARINC also provides standardization guidelines for engineering and development of both software and hardware systems for a variety of military and civilian aviation electronic systems.

One of the most important facets of ARINC, from a technicians point of view, would have to be the standards set for various avionics systems. Several years ago as the avionics industry was beginning to emerge, it became quite clear that some type of commonality was needed to ensure systems were compatible. As avionics became even more interactive, the need for commonality grew. ARINC has documented characteristics and specifications (standards) for a multitude of communication and navigation systems. Most of these standards are focused on transport category avionics; however, many of the general aviation avionics systems also adhere to ARINC specifications.

Individual system specifications and their related interfaces are addressed by ARINC. ARINC equipment specifications include: (#579) airborne VOR receivers, (#578) airborne ILS receivers, (#594) ground proximity warning systems, (#566) VHF communication transceivers, (#542) digital flight data recorders, (#741) aviation satellite communications, (#738) air data and inertial reference systems, (#728) analog and discrete data converter systems, (#725) electronic flight instrument (EFI), (#724) aircraft communication and reporting systems, and (#708) airborne distance measuring equipment. As discussed below, ARINC is also responsible for two of the most common data bus standards used on transport category aircraft (ARINC 429 and 629).

DIGITAL DATA BUS SYSTEMS

Many aircraft currently employ avionics systems that operate using digital electronics. Although these systems range in complexity from a simple radio receiver/transmitter to a complete autoflight system, they all adhere to some type of data bus standard. Adherence to an established standard is extremely important to ensure commonality of data transfer. A **data bus standard** can be thought of as a common language that allows several different units to communicate. Without a data bus standard, communication between various avionics systems would be impossible. It would be similar to speaking French to someone who understands only English. Even though both languages use common symbols (the alphabet, letters A-Z), without a "standard" language, communication is impossible.

There are several digital data bus standards currently in common use on modern aircraft. Agencies such as ARINC Incorporated and the General Aviation Manufacturers Association (GAMA) set standards for various data bus systems. The two ARINC data bus standards are: **ARINC 429** and **ARINC 629**. In industry, ARINC 429 and 429 (or ARINC 629 and 629) are used interchangeably, and this text will also use both of these designations. ARINC 429 is a one-way data bus also known as

Mark 33 Digital Information Transfer Systems (DITS). ARINC 629 is a faster, two-way, multi-transmitter data bus system. ARINC 429 or 629 data bus systems are found on virtually all transport category aircraft that employ digital systems. ARINC 429 bus systems are also found on many corporate and commuter aircraft.

The **ASCB (Avionics Standard Communication Bus)** and the **CSDB (Commercial Standard Digital Bus)** are two data bus systems used extensively on corporate-type general aviation aircraft, such as, the Beechcraft King Air, Cessna Citation, or Falcon 20. The ASCB format is used solely on Sperry or Honeywell avionics equipment. (It should be noted that Sperry and Honeywell have merged and are currently referred to solely as Honeywell.) The CSDB format is used on aircraft that employ systems manufactured by Collins Avionics. Collins is a subsidiary of Rockwell International and manufactures avionics for both general aviation (mainly corporate and commuter aircraft) and transport category aircraft. **Manchester** data is also common to many corporate and commercial type aircraft and is adaptable to a variety of formats. In many cases two or more data bus standards may be used on the same aircraft. It is extremely important that all avionics technicians become familiar with these data bus specifications if they intend to repair and maintain digital systems.

A point of clarification: a data bus is typically thought of as the wires which transmit digital signals between two LRUs (Line Replaceable Units), not the wiring inside the LRU. The data bus standard applies only to the data transferred on the external data bus, not to the data manipulated inside the LRU. For example, a system may employ the ARINC 429 data bus standards for communication between the Flight Management Computer (FMC) and the Thrust Management Computer (TMC). However, inside the FMC and TMC the data may be manipulated in a completely different format as specified by the LRU software. It should also be noted, each data bus is dedicated to a specific data format. For example, a technician will never find a data bus that transmits ARINC 429 and then changes format to transmit ARINC 629. These two different formats would require two different data bus cables.

A common digital bus is a pair of solid copper wires, approximately 24 gauge. Each wire is coated with a thin plastic insulation, twisted together, wrapped in a foil shield, and covered again with insulation. It is important that the shield be grounded at each end of the cable. Many data bus systems are polarity sensitive and it is therefore very important that the two conductors never be switched. That is, wire A must always be connected to terminal A; wire B to terminal B (see figure 2-17). The digital signals can be measured between wires A and B using an oscilloscope or data bus analyzer.

Manchester II Coding
Both general aviation and transport category aircraft often incorporate a Manchester data bus. The **Manchester II code** (often simply referred to as Manchester) is a serial digital data format that incorporates a voltage change in each data bit. This means: a) logic 1 begins with a positive voltage and changes to zero (or negative) voltage halfway through the bit time period, and b) logic 0 begins with a zero (or negative) voltage and transitions to a positive voltage half way through the bit time

period. This type of format allows the Manchester code to be self-clocking. A self-clocking data bus carries a timing signal with each data bit. A self-clocking data bus helps to ensure the correct timing of all digital systems connected to that bus. Figure 2-18 shows a typical Manchester code with a 10 Kbits/sec transmission rate. As can be seen in this diagram, binary 1 equals +12V and binary zero equals 0V. The time period for each bit is 100 υsec and each bit is divided in half (50 υsec). The first half of the bit contains the data and the second half of the bit contains a clock pulse. A Manchester code can be adapted to a variety of different word configurations, voltage levels and transmission speeds. In some cases, Manchester is used by the processing LRU and the data is converted to a different format for transmission on the data bus. For transmission between two LRUs, Manchester will always be transmitted on a dedicated Manchester data bus

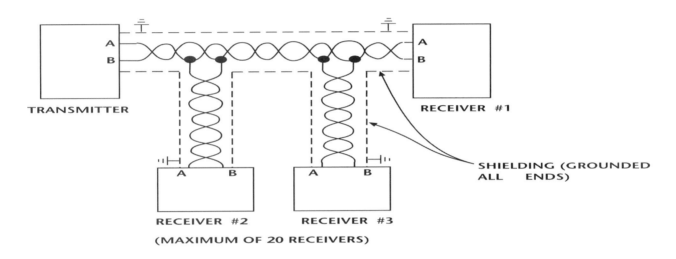

Fig. 2-17   Typical data bus cable connections for a polarity sensitive bus.

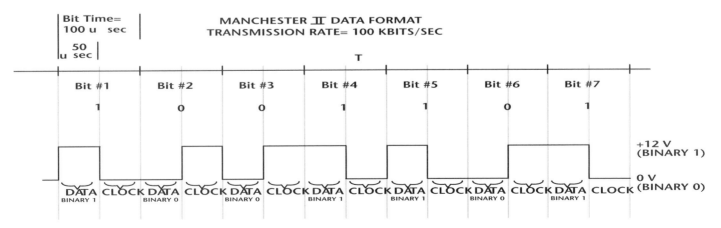

Fig. 2-18   A typical Manchester digital signal format.

Avionics Standard Communication Bus (ASCB)

The **ASCB (avionics standard communication bus)** is a bidirectional data bus operating at 0.667 MHz. This system uses multiple bus controllers that coordinate bus transmission activities. Figure 2-19 shows the structure of a typical ASCB system. In this example, there are three bus controllers and two parallel busses; this provides the redundancy needed to ensure flight safety. During operation only one bus controller is active at any given time. The other two controllers are ready in active standby.

The subsystems transmit data only by command of the bus controller. The bus controller contains the software necessary to "call" various subsystems and coordinate data transmission in the correct sequence. The bus controller sends a request for messages on both busses. This request will have the specific address of the subsystem that the controller wishes. The subsystem will then respond to the request with its own message.

Data format

All ASCB data is transmitted at **0.667 MHz** in a standard digital format using positive logic. ASCB uses a **non-return to zero (**NRZ) format for each data bit. That is, a signal of +5V is used for a binary one, 0 V for binary zero, and the data fills the

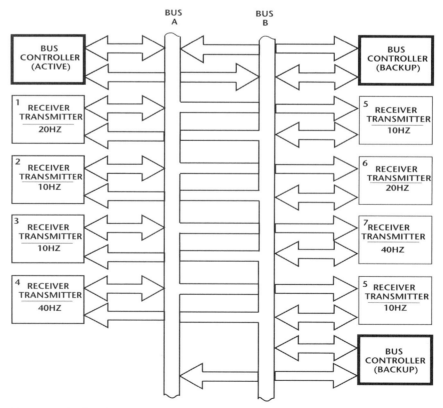

Fig. 2-19   A typical Avionics Standard Communication Bus (ASCB) bus structure.

entire bit time. There is no change in voltage level during each bit period for timing purposes. A signal loss during transmission of approximately 0.5V (allowing +4.5V to reach the receiver) is typically acceptable. A greater signal loss will most likely cause an error in data transmission (see figure 2-20). An error in data transmission can easily be seen using an oscilloscope or data bus analyzer.

The data from any subsystem will be transmitted using a specified format programmed into the transmitting LRU. The software of the receiving LRU will decode the data accordingly. All ASCB data transmissions will contain an address at the beginning of the message. Any receiver on the bus can choose to accept or ignore a given

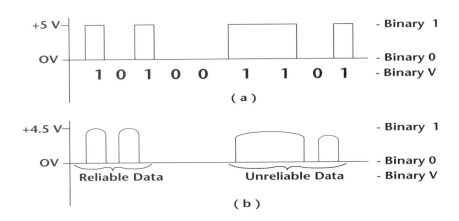

Fig. 2-20   ASCB data transmission. (a) accurate data transmission, + 5V = Binary 1; (b) inaccurate data transmission, data signal less than + 4.5 V.

message through analysis of the address. Figure 2-21 shows the format for a typical bus controller request and the transmission of information by a subsystem. Notice the information content transmitted by the subsystem is shown in the box *IRS Data*. The information transmitted before and after the actual data is used for error checking and bus control functions.

Sequencing data

The request for data from the bus controller will be transmitted in a given sequence and at specific time intervals according to the controller software. Not all subsystems will be accessed during each controller request. Critical systems will be accessed more frequently than less critical systems. For example, the bus controller may send a request for data 40 times a second (40Hz). This means that requests to a specific subsystem can be made at rates such as 40, 20, 10, 5, etc. times per second. The ASCB request/answer time periods are known as **bus frames.** The total time allowed for a given frame must be less than the update period (40 times a second or 25 milliseconds for the previous example).

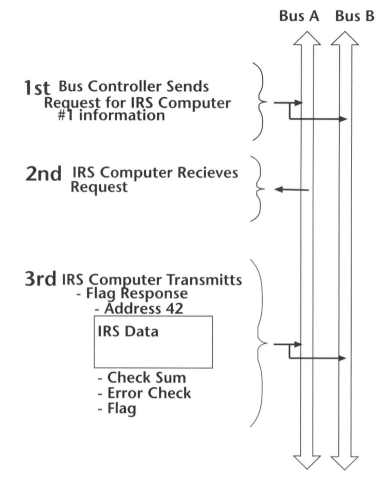

Bus A   Bus B

1st Bus Controller Sends
Request for IRS Computer
#1 information

2nd IRS Computer Recieves
Request

3rd IRS Computer Transmitts
- Flag Response
- Address 42

IRS Data

- Check Sum
- Error Check
- Flag

Fig. 2-21 Typical ASCB data bus controller configuration.

At the beginning of each bus frame the bus controller transmits a frame-start message and a control/test message. The frame start message is a "wake-up" call telling all subsystems that a new frame is starting. After the control test message, the controller requests specific subsystems to transmit data. Each subsystem is requested to respond individually as shown in figure 2-22. At the end of the bus frame, the controller waits for the specific time period and initiates the next bus frame sequence. With this configuration, the bus controller coordinates all bus activities. The failure of any single transmitter will not effect the operation of others on the bus.

Data bus cable
The typical data bus cable used for ASCB transmission is two wire, 24 gage, shielded cable. Figure 2-23 shows some of the ASCB bus specifications as discussed in this paragraph. The maximum bus cable impedance is 125Ω +/- 2Ω, and maximum capacitance is 12 +/- 2 picofarads. The end of each wire pair is terminated using a 127Ω, +/-1/4 Ω, nonconductive metal film resistor. The avionics equipment is connected to the bus through a stub bus and a transformer/coupler assembly. The maximum bus length is 150 feet, and the maximum stub length (between bus and avionics equipment) is 36 inches. All bus cable shielding must be terminated to ground at each end of the cable. A schematic of a typical bus coupler is shown in figure 2-24.

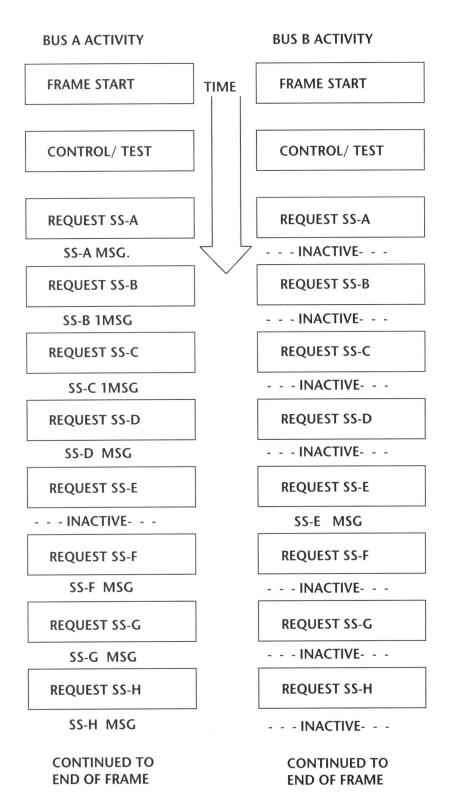

SS= Subsystem

MSG= Message from subsystem

INACTIVE= No message on that bus at this time

(i.e.- that subsystem transmits on opposite bus only)

Fig. 2-22   Example of ASCB bus activity.

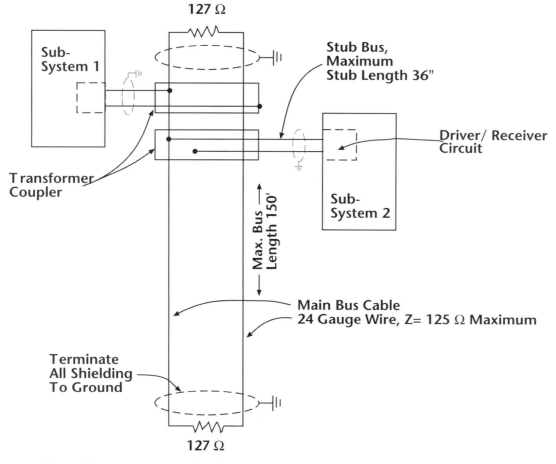

Fig. 2-23 ASCB data bus specifications. For Training Purposes Only.

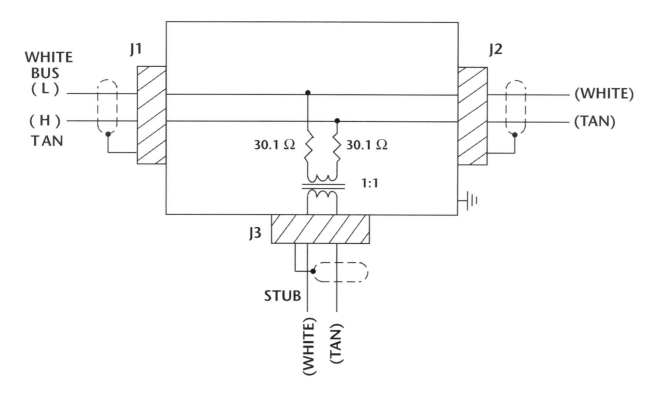

Fig. 2-24 ASCB bus coupler.

Commercial Standard Digital Bus (CSDB)

The **CSDB (commercial standard digital bus)** is a one-way data bus system between one transmitter and a maximum of 10 receivers. CSDB is used in Collins general aviation electronics equipment and is certified on a large number of aircraft systems. CSDB adheres to a specification standard RS-422A established by the Electronics Industries Association, and has also been recognized as a standard aircraft data bus by the General Aviation Manufacturers Association (GAMA). The CSDB data bus cable is a twisted pair of shielded wire which can be up to 150 feet long.

Data formats

The CSDB standard allows for operation in one of three forms: **continuous repetition, noncontinuous,** and **burst** transmissions. During continuous repetition transmission, the data is consistently updated at given intervals. Continuous repetition data is typically used for primary data, such as aircraft attitude. For noncontinuous data transmission the information is updated at a given rate only when that data is available. Noncontinuous data is typically used for secondary data, such as test data. Burst data is information that is meant for a "one time" transmission intended to announce a specific action, such as capture of the glide slope signal. Reception and transmission of the three different data forms is a function of the software within a given system. All three types of data may be transmitted on the same bus. CSDB data is transmitted on a NRZ (non-return to zero) format where logic state 1 exists when bus line A is positive with respect to line B. Logic state 0 exists when bus line B is positive with respect to line A. The CSDB system operates at a high speed of **50 Kbits/sec** or **12.5 Kbits/sec** for low speed.

Data frames

The CSDB data transmission is divided into **frames.** Each frame is a fixed time interval for a given bus. The length of each frame is a function of the maximum update rate required for that system. Each frame begins at the start of a synchronization (sync) block and ends at the start of the next sync block. The sync block con-

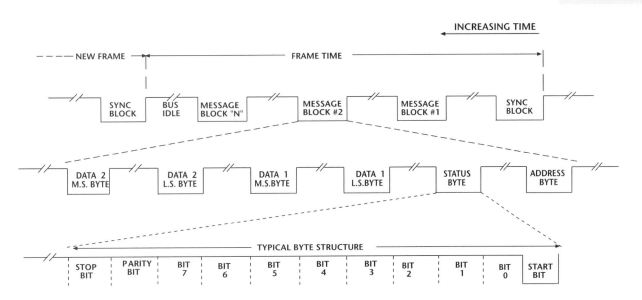

Fig. 2-25   The CSDB data format consists of bus frames, message blocks, bytes, and bits.

sists of a fixed number of bytes assigned to that bus. A typical sync block byte is the hexadecimal number "A5". As can be seen in figure 2-25, each frame contains one or more message blocks.

Message blocks

The CSDB message block is a specific serial message, consisting of a defined number of bytes. Each message is divided into an address byte, a status byte, and one or more data bytes (see figure 2-25). The data bytes are transmitted with the least significant (L.S.) byte first and the most significant (M.S.) byte last. Each byte consists of 8 bits of data along with a start bit, a parity bit, and a stop bit. Odd parity is used for CSDB transmissions.

ARINC 429

An **ARINC 429** data bus is a one way communication link between a single transmitter and one or more receivers. The 429 data bus standard is by far the most common data bus format in use today on civilian aircraft. The ARINC 429 system provides for the transmission of **32** bits of information in each data word. (A data word is sometimes referred to as a byte.) Each word is separated by a four bits of bus silence known as a four bit null. A null is a signal which is equal to neither binary 1 nor binary 0. For an ARINC 429 transmitter the nominal voltage values are as follows: a digital 1 requires a signal of **+10 +/- 1.0 V**; a digital 0 requires a **-10 +/-1.0 V** signal. The null signal is equal to **0.0 +/- 0.5 V**. Figure 2-26 shows the basic digital signal format for an ARINC 429 transmitter. As seen in the expanded view of this diagram, each bit actually consists of two voltage values, the data signal (+10 Volts or -10 Volts) and the null signal (0 Volts). The **null signal** is always transmitted immediately after the data and makes up one half of the bit.

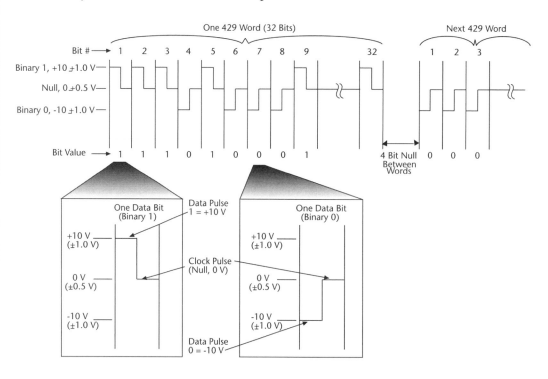

Fig. 2-26   Digital signal for an ARINC 429 data bus.

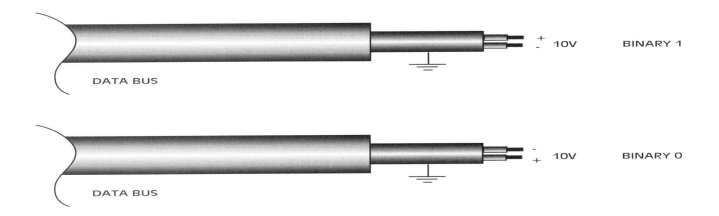

Fig. 2-27   Typical ARINC 429 data bus cable. Note the signal polarity reverses for binary 1 and binary 0.

The null signal is used for timing, or synchronizing, the data bus. Each 429 transmitter contains an internal clock; the timing signal is then transmitted through the data bus to each receiver via the null voltages. This type of data transmission is referred to as **RZ (return to zero).** Using RZ format allows ARINC 429 to be "self clocking".

The ARINC 429 signal is also a **Bipolar** format. Bipolar means the data signal actually reverses polarity (two-polarity) when it changes from binary 1 to binary 0. Remember binary 1 is equal to **+10** volts, binary 0 equals **–10** volts. Since the data bus consists of two shielded wires, reversing polarity of these wires changes the data signal from binary 1 to binary 0 (See figure 2-27).

The ARINC 429 signal is a Bipolar, Return-to-zero format. This means the bus voltage must reverse polarity to change from binary 1 to binary 0 and each bit must return to zero to provide the synchronizing pulse.

There can be a maximum of 20 receivers for any one ARINC 429 transmitter. Each receiver must have an input impedance of 12,000 Ω or greater to keep signal loss to a minimum. The transmitter impedance should be 75 +/-5 Ω. This matches the impedance of the data bus cable which should also be approximately 75 Ω.

An ARINC 429 receiver will recognize a binary 1 as any value **+6.5 to +13 V**; and a binary 0 as **-6.5 to -13 V**. The null voltage value must be **+2.5 to -2.5 V** to be accepted by the receiver. Values outside of these limits are often caused by poor electrical connections or faulty LRUs. It should be noted the receiver has a greater voltage tolerance than the transmitter (transmitter tolerance: binary 1 = +10V +/- 1V, binary 0 = -10V +/- 1V, null = 0V +/- 0.5V). The receiver must have this greater tolerance to allow for any signal loss, noise, or other forms of distortion that may occur between the transmitter and the receiver. The use of an oscilloscope or data bus analyzer can be used to detect excessive signal distortion. Oscilloscopes and data bus analyzers will be discussed later in this chapter.

The ARINC 429 standard allows for the transmission of digital data in two speeds. The majority of the digital systems transmit ARINC 429 data at **low-speed (12.0 to 14.5 Kbits/sec)**. The high-speed bus is typically reserved for systems requiring large quantities of data, which is frequently updated. For example, the flight management computer would operate at high speed. The **high-speed** bus transmits at a frequency of **100 Kbits/sec**. At high-speed, each bit lasts only 10 usec, at low speed each bit lasts 70 to 83 usec. To eliminate timing problems, the ARINC 429 system never mixes high-speed and low-speed data on the same bus. It should be noted low-speed has a range (12.5-14.5 Kbits/sec) the exact speed is determined by the transmission software design.

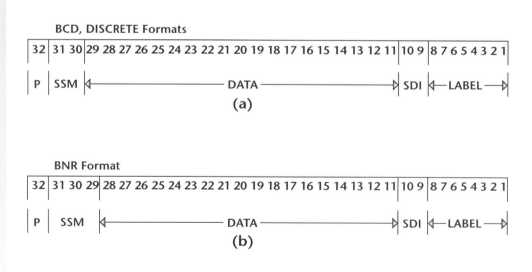

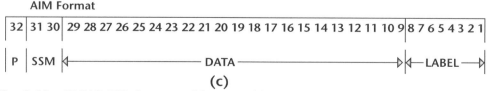

Fig. 2-28   ARINC 429 data word format. (a) BCD, Discrete; (b) Binary; (c) AIM.

429 Data word formats

One of four data word formats must be used to conform to the ARINC 429 standards: **Binary Data (BNR), Binary-Coded Decimal (BCD), Acknowledgment ISO Alphabet Maintenance (AIM),** or **Discrete**. As illustrated by figure 2-28, the ARINC 429 standard assigns the first eight digits of a byte as the **word label**, and bit number 32 is a **parity bit**. The **data field** is comprised of digits 11 through 29 for BCD, and Discrete word formats. The data field is comprised of digits 11 through 28 for the BNR word format. The data field is bits 9 through 29 for the AIM format. The **sign-status matrix (SSM)** includes bit numbers 30 and 31 for the BCD, AIM, and Discrete word formats. The SSM includes bit numbers 29, 30 and 31 for the BNR word format. Digits 9 and 10 are a **source-destination identifier (SDI)** for all formats except AIM. The AIM format does not use a SDI (bits 9 and 10 are part of the data).

To eliminate confusion, remember each of the four ARINC 429 formats (BNR, BCD, AIM & discrete) are comprised of a 32 bit word. Each word has the same basic structure as shown in figure 2-28. However, the SSM may vary in size and the SDI may be eliminated for certain word formats.

The **SDI** serves as the address for the 32-bit word. That is, the SDI identifies the source or destination of the word. All information sent to a common serial bus is received by any receiver connected to that bus. Each receiver accepts only that information labeled with the correct SDI. An SDI set to binary 00 represents an All Call situation. In other words, if the SDI is set to 00 all receivers connected to the bus will "listen" to the transmitted data. If the SDI is set to 01, 10, or 11 only receiver(s) assigned that specific code (SDI) will "listen" to the transmitted data.

The information, or data, of an ARINC 429 transmission must be contained within bits number 11 through 28 for BNR; bits 11 through 29 for BCD and Discrete; bits 9 through 29 for AIM. This data is the actual message which is to be transmitted. For example, an airspeed indicator may transmit the binary message 0110101001. Translated to decimal, this could mean 425, or an airspeed of 425 knots. It should be noted that in some cases, all bits of the data field are not needed. **Pad Bits** "fill in" any portion of the data field which are not used. Pad Bits are always set to binary 0. The letter P is often used to represent Pad Bits.

The **sign-status matrix (SSM)** provides information which might be common to several peripherals. Typically bit number 29 is used for *common* information, such as north, south, plus, minus, right, left, etc. Bits 30 and 31 are used to transmit information concerning system status. For example, if bit 30 and 31 are set to 00 it could mean failure warning. Indicating a malfunction in the transmitter system. The specific meaning of the SSM digits may vary between word formats. The ARINC 429 specification provides details as to the SSM meanings.

The ARINC 429 **parity bit** allows the receiver to detect errors that may have occurred **during data transmission**. The parity bit does not check the accuracy of the data itself, it only verifies accurate transmission on the data bus. ARINC 429 uses **odd parity**. This means that properly transmitted data will always contain an odd

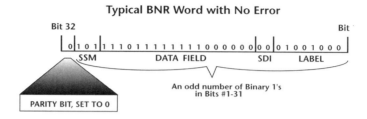

Fig. 2-29 The ARINC 429 parity bit is bit number 32. The parity bit is always set to allow for an odd number of binary 1s in the complete ARINC 429 data word (bits 1-32).

number of binary 1s in data bits 1-32 (figure 2-29). Just prior to transmitting data onto the bus, the LRU's transmitter circuitry "counts" the number of binary 1s in data bits 1-31. If there is an even number, the parity bit is set to 1; if there is an odd number the parity bit is set to 0. This ensures the data leaving the transmitter **always** has an odd number of 1s in bits 1-32. When the data is received by an LRU, the circuitry "counts" the binary 1s. The receiver will only consider the data valid if there is an odd number of binary 1s in data bits 1-32. Therefore, if one or more data bits should be lost during transmission the receiver will most likely detect a fault due to inaccurate parity.

The receiver also performs a **reasonableness check** designed to detect any unreasonable information. For example, if the airspeed data changed from 205 knots to 12 knots in a fraction of a second, the receiver would ignore that information since it is unreasonable. The reasonableness check along with the parity bit ensures that all defects in data transmission will be ignored by the receiver. If the defect continues, a system fault flag will be displayed on the instrument panel to inform the pilot. The defect may also be recorded in the system's nonvolatile memory to be used later for maintenance purposes.

Word labels and equipment identifiers

Once a receiver accepts a word from the bus it must determine how to decode the data field. To do this, the receiver "reads" the word label. Knowing the label allows the receiver's software to access the correct program to decode the data. As shown in figure 2-28, the word label is contained in bits 1-8 of each data word. There are 256 possible combinations of word labels in the ARINC 429 code. The ARINC 429 standard assigns specific names to each of the possible 256 labels. Each label is assigned a specific word format and it cannot change. As mentioned earlier, the four available formats are BNR, BCD, AIM, and Discrete.

It should be noted that ARINC 429 was first developed in the late 1970s. At that time there was a limited use of digital systems onboard aircraft and 256 label combinations was more than adequate. In the following years, a digital revolution took place and a multitude of systems were developed. Soon, 256 label combinations were not enough to cover all the digital systems. At that time, equipment identifiers were added to most of the labels. The **equipment identifier** is a subcategory of the ARINC 429 label providing a greater variety of label combinations. The equipment

LABEL	EQPT . ID (HEX)	PARAMETER NAME	UNITS	RANGE (SCALE)	SIG. DIG.	POSITIVE SENSE	RESOLUTION	MINIMUM TRANSIT INTERVAL (msec) 2	MAXIMUM TRANSIT INTERVAL (msec) 2
1 1 2	0 0 2	Runway Length	Feet	20480	11		10	250	500
	0 A 1	Selected EPR		4	12		0.001	100	200
	0 A 1	Selected N1	RPM	4096	12		1	100	200
	0 B B	Flap Lever Position - Left	Deg/180	+/-180	18		0.000687	80	160
1 1 4	0 0 2	Desired Track	Deg/180	+/-180	12		0.05	100	200
	0 2 9	Brake Temperature (Left Inner L/G)	Deg C	2048	11		1	100	200
	0 2 F	Ambient Pressure	PSIA	32	14		0.002	100	200
	0 B B	Flap Lever Position - Right	Deg/180	+/-180	18		0.000687	80	160
	0 C C	Wheel Torque Output	Lb./ft.	16384	12		4	50	100
	1 0 A	Selected Ambient Static Pressure	PSIA	1.5-20.0	11		0.016	100	500
	1 0 B	Selected Ambient Static Pressure	PSIA	1.5-20.0	11		0.016	100	500

Table 2-1   ARINC 429 binary data word definitions.

identifier is a hexadecimal code assigned by ARINC. Table 2-1 shows a portion of the BNR data standards for labels 112 and 114. Here it can be seen that label 112 has 4 different equipment identifiers, and each identifier employs different units, range, significant digits, resolution, and transmission intervals.

The equipment identifier is not transmitted as part of the 429 data word. Since ARINC 429 employs a one-way data bus with only one transmitter, the software in each receiver "knows" what equipment identifier is assigned to each label. This situation is valid only if the transmitter and the receiver are both programmed for identical label/identifier combinations. Only one equipment identifier can be used for any given label on any given 429 bus. During aircraft operations the label is enough to identify the specific word configuration because the system software "knows" the assigned equipment identifier. However, when using a data bus analyzer (discussed later in this chapter), the equipment identifier may become critical. Since the data bus analyzer can read all combinations of a given ARINC 429 label, the user must "tell" the analyzer what equipment identifier is used when receiving or transmitting a given label. Once programmed with the correct label/identifier combination the analyzer can transmit or receive data in the correct format. Of course the equipment identifier is also important during system engineering.

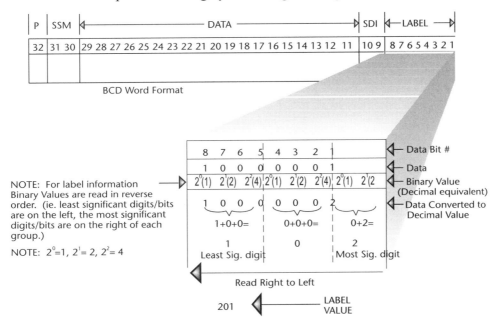

Fig. 2-30   Decoding an ARINC 429 word label.

Decoding the word label

The label data (bits 1-8) is comprised of two three-bit groups and one two-bit group (see figure 2-30). Bits 1 & 2 comprise the most significant digit of the word label, and bits 6, 7 & 8 comprise the least significant digit of the label. To determine the decimal value of the label, simply convert each binary group into it's decimal equivalent. Remember, the binary values are set in reverse order. For example, bit number $8 = 2^0$, bit number $7 = 2^1$, and bit number $6 = 2^2$ (figure 2-30). In this example, bit number 8 is set to binary 1, bit 7 is set to 0, and bit 6 is set to 0; this group is therefore equal to a decimal value of 1. This is the least significant group. Bit numbers 3, 4 and 5 are each set to 0 and therefore the decimal value of this group is 0. Bit number 2 is set to 0 and bit 1 is set to 1 therefore the most significant group is set to a decimal value of 2. In this example, the word label is 201. According to the ARINC 429 specifications, 201 is the label for DME distance.

Binary data

Refer to figure 2-31 during the following discussion on binary data. The octal notation code of the first eight bits is read as described above to achieve the word label. The bits 011/000/01 represent 206 (602 reversed). According to the ARINC 429 code, 206 represents computed airspeed (see figure 2-31a). The SDI label of 00 indicates transmission of these data to all receivers connected to the serial bus. The

Label #	Parameter name	Units	Range	Significant Digits	Resolution
206	Computed Airspeed	Knots	1024	14	0.0625
207	Max. Allowable Airspeed	Knots	1024	12	0.25

(a)

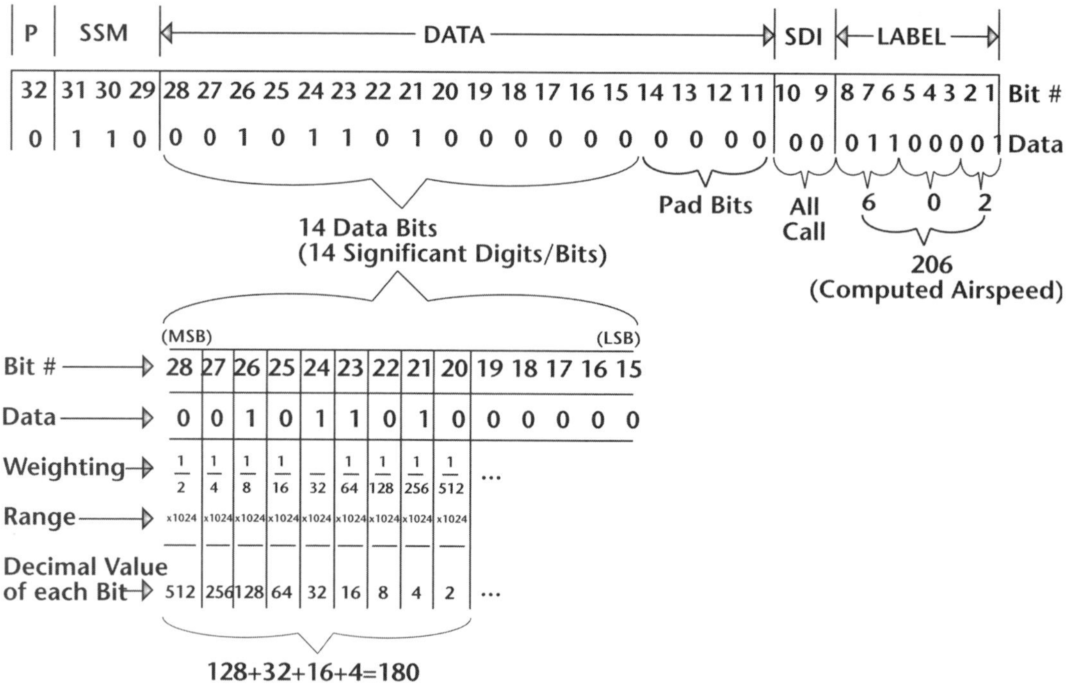

(b)

Fig. 2-31 Decoding an ARINC 429 Binary word data field. (a) The ARINC 429 specification table for Binary words 206 and 207; (b) analysis of the data field.

data field for label 206 is measured in units of knots, has a range (scale) of 1024, has 14 significant digits, and has a resolution of 0.0625. The **range** determines the maximum and minimum values for a given label. In this case the minimum computed airspeed is 0 knots and the maximum would be 1024. (Although we know no transport category aircraft can travel at 1024 knots, this value represents the maximum value that can be transmitted using label 206.) Since a minimum range value is not stated, it is assigned a value of zero. The number of **significant digits** determines how many bits will be used for the data field. For label 206 there are 14 bits of data. Since the BNR format allows for 18 bits in the data field, there will be 4 pad bits (bit numbers 11-14). As discussed earlier, all pad bits are set to zero. The **resolution** for a given label defines the smallest unit of data that can be transmitted. The resolution for label 206 is 0.0625 knots.

To decode the data segment of a BNR word a **weighting factor** must be determined for each bit in the data field. This is accomplished by assigning the most significant bit (MSB) a value of 1/2, the next digit 1/4th, the next digit 1/8th, the next 1/16th, and so on down to the least significant bit (LSB) as seen in see figure 2-31b. Each weighting factor is then multiplied by the maximum value of the range. In this example, the weighting factor is multiplied by 1024. Multiplying the weighting factor by the range determines the decimal value for each bit. To determine the value of the data field, simply add the decimal values for all data bits assigned digital 1. In this example, bit numbers 26, 24, 23, and 21 are binary 1, all other data bits are binary 0. Therefore, the value of the data field is 128 (decimal value of bit #26) + 32 (decimal value of bit #24) + 16 (decimal value of bit #23) + 4 (decimal value of bit #21) = 180. The computed airspeed for the example in figure 2-31 is 180 knots.

In this example (figure 2-31), only a portion of the data field was used. The receiver "knew" the extent of the data field by the label code through internal software programming. Once the label is analyzed by the receiver, the computer recalls information about that data and how it will be presented. This allows the computer software to make a determination as to how the data and SSM should be read.

In figure 2-31 the SSM digits 29, 30, and 31 are binary 0, 1, and 1, respectively. According to the 429 specification, a SSM of 011 for label 206 represents a normal operation of plus value data. The parity bit (number 32) is a 0, which denotes an odd number of binary 1s in the transmitted word. No error is present according to the parity bit.

Binary coded decimal

The **binary coded decimal (BCD)** word format uses bits 11-29 for the data field and allows only bits 30 & 31 for the SSM. The data field is divided into four, four-bit groups and one three-bit group as shown in figure 2-32a. The least significant digit is found in the four-bit group on the far right of the data field. In this example, DME is transmitting a distance of 150.71 NM. Once again, the label was used by the receiver's software to determine how the data would be analyzed. In figure 2-32b, it can be seen that the ARINC 429 standard defines the unit's range or scale, significant digits, direction, and resolution for each label. In this example, the units are

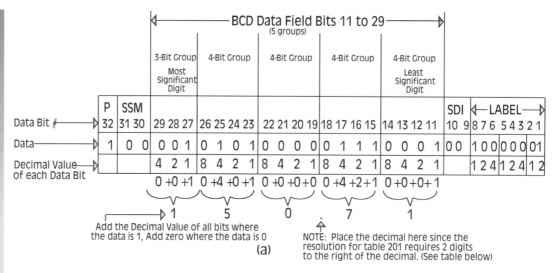

(a)

Label	Equip ID	Parameter Name	Units	Range	Significant Digits	Resolution
200	0 4	Drift Angle	DEG	± 180	4	0.1
201	0 9	DME Distance	NM	-1 to 399.99	5	0.01
230	0 6	True Airspeed	Knots	100 to 599	3	1.0

(b)

Fig. 2-32 Decoding an ARINC 429 Binary Coded Decimal (BCD) data field. (a) example of transmitted data; (b) specifications for label 200, 201, & 230.

nautical miles (NM), the range is -1 to 399.99, there are 5 significant digits, and a resolution of 0.01. The resolution is used to determine where to place the decimal point after the data field has been decoded.

Studying figure 2-32a it can be seen that bits 11-14 create the least significant digit of the data. In this example, bit number 11 is set to binary 1 and bits 12-14 are binary 0. This is decoded to be decimal value 1. Remember, bits 11-14 form a four-bit group to create the least significant digit of the data. Bits 15-18 form the next four-bit group (data bits 0111 are decoded as decimal value 7). This process is repeated for each four-bit group, 19-22 & 23-26. The most significant digit is formed by three bits of the data field (bits 27-29). In this example, the data field is decoded to be the digits 1,5,0,7,and 1 (15071). To locate the decimal point in the data, simple reference the 429 specification. In this case, the specification shows a resolution of 0.01. Therefore, the decoded data must contain two digits to the right of the decimal point. The decoded data is therefore equal to 150.71 nautical miles.

AIM word format

The AIM **(acknowledgment, ISO-alphabet, maintenance)** format is used for systems that require large amounts of data transfer. ISO is an acronym for the International Standards Organization. This group has identified a matrix that provides graphic, numerical, and alphabetic symbols for a group of binary digits. The ISO-alphabet used for ARINC 429 data uses a 7 bit code for each character. The values of bits 1-4 determine the matrix **row**, bits 5-7 determine the matrix **column**. For example see figure 2-33, here the number five is represented by the binary bits 0110101. (NOTE: b_1 is shown on the right; b_7 is shown on the left.)

STANDARD CODE

(column) Bit #5,6&7

b7 b6 b5 Bits					$0\,0\,0$	$0\,0\,1$	$0\,1\,0$	$0\,1\,1$	$1\,0\,0$	$1\,0\,1$	$1\,1\,0$	$1\,1\,1$
b4	b3	b2	b1	COL→ ROW↓	0	1	2	3	4	5	6	7
0	0	0	0	0	NUL	DLE	SP	0	@	P	`	p
0	0	0	1	1	SOH	DC1	!	1	A	Q	a	q
0	0	1	0	2	STX	DC2	"	2	B	R	b	r
0	0	1	1	3	ETX	DC3	#	3	C	S	c	s
0	1	0	0	4	EOT	DC4	$	4	D	T	d	t
0	1	0	1	5	ENQ	NAK	%	5	E	U	e	u
0	1	1	0	6	ACK	SYN	&	6	F	V	f	v
0	1	1	1	7	BEL	ETB	'	7	G	W	g	w
1	0	0	0	8	BS	CAN	(	8	H	X	h	x
1	0	0	1	9	HT	EM	)	9	I	Y	i	y
1	0	1	0	10	LF	SUB	*	:	J	Z	j	z
1	0	1	1	11	VT	ESC	+	;	K	[	k	{
1	1	0	0	12	FF	FS	,	<	L	\	l	/
1	1	0	1	13	CR	GS	–	=	M	]	m	}
1	1	1	0	14	SO	RS	.	>	N	^	n	~
1	1	1	1	15	SI	US	/	?	O	—	o	DEL

(row) Bit # 1,2,3&4

Fig. 2-33   ARINC 429 ISO alphabet.

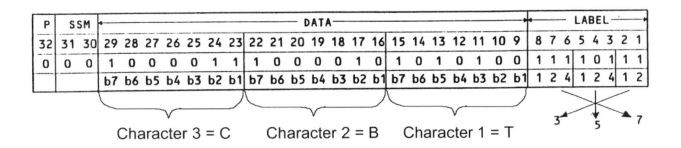

Fig. 2-34   A typical ARINC 429 AIM data word format.

As seen in figure 2-34, the data field for an AIM word is comprised of three, seven-bit groups. Each of these seven-bit groups can be decoded using the ISO-alphabet matrix (figure 2-33). Since this format does not provide much data on each 32-bit word, a complete message may take several words. The ARINC 429 specification allows for a maximum of 127 individual words.

The ARINC 429 AIM word format is comprised of three types of data files: **initial word, intermediate word(s),** and **final word**. As the name implies the initial word is always the first in the series. The initial word defines the total number of words in the message to be transmitted. A maximum of 126 words can follow the initial word. Initial words are identified by the SSM being set to 01. Intermediate word(s) contain the data to be transmitted (SSM=00). The final word (SSM=10) is used to signal the end of the data words. After the final word a 4 bit null would be transmitted and a new message would begin. The exact formats for the initial, intermediate, and final word all vary slightly and are identified in the ARINC 429 specification.

Discrete words

The **discrete** word format is used to transmit the status of several individual components or systems in a 32-bit word. These discretes are not defined by the ARINC specification. The manufacturer can determine what discrete signals will be represented by the various bits of the data field. For example, the manufacturer may assign bit 11 as engine 1, fire loop 1 - fail, and bit 12 as engine 1, fire loop 2 - fail. In this example, if bits 11 or 12 are binary 1 it would indicate a failure in that specific fire loop. If bits 11 or 12 are binary 0 it would indicate the respective fire loop has not failed. Discrete data word formats use bits 11 through 29 for the data field. Any discrete word bits that are not used are referred to as pad bits and set to binary 0.

For a typical aircraft using the ARINC 429 data bus system most of the information is transmitted using the binary or binary coded decimal word formats. On the Boeing 767 for example, approximately 65% of all 429 data is transmitted in either the BNR or BCD formats.

ARINC 629

ARINC 629 is a new digital data bus format that offers more flexibility and greater speed than the 429 system. ARINC 629 permits up to 120 receiver/transmitters to share a **bidirectional serial data bus**. The bus can be either a twisted wire pair or a fiber optic cable up to 100 meters long. ARINC 629 was introduced to the commercial aviation industry on the Boeing 777. A military version of 629 can also be found on some of the newer military aircraft. As aircraft are developed, they will most likely adopt ARINC 629 as their digital data bus standard.

ARINC 629 has several major improvements over other data bus systems. First, there is a substantial weight saving when compared to the ARINC 429 system. The older 429 specification is a one-way data bus requiring a separate bus for each transmitter; 629 is bidirectional and can accommodate up to 120 receiver/transmitters on each bus. Figure 2-35 shows a simplified diagram of the 429 and 629 bus structures. Here it can be seen that the bidirectional 629 system requires much less data cable than the older 429 system. Second, the 629 bus operates at a speed of 2 megabits/second, this is much faster than many older bus formats. (ARINC 429 operates at a maximum of 100 kilobits/second.) Third, the 629 requires no separate bus control unit since each LRU connected to the bus and is designed to independently coordinate transmission activities. Fourth, the 629 bus system employs an inductive coupling unit to connect each LRU to the data bus. The inductive coupler permits easy connection to the bus without physical interruption of the bus wires; hence improving bus reliability (i.e. fewer loose connections).

Signal characteristics

The ARINC 629 data signal actually changes format during the transmission process. The data initially produced by the transmitting LRU is a simple digital signal in a Manchester format. (The Manchester format was discussed earlier in this chapter.) This current signal is then converted to a signal known as a *doublet* for transmission on the 629 data bus. A **doublet** signal emits a short pulse whenever the bit value changes from binary 1 to 0 or vise versa. The doublet signal will also transmit a short pulse for any timing signals transmitted on the bus (i.e. a doublet pulse

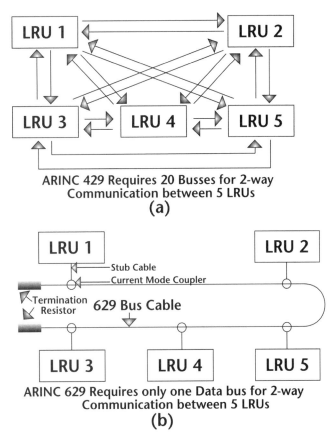

Fig. 2-35 Typical ARINC data bus structures. (a) One-way ARINC 429; (b) two-way ARINC 629.

occurs whenever there is a change in logic state). The receiving LRU converts the doublet signals back into a Manchester format for use by the receiver's processor circuitry.

In order to better understand the ARINC 629 signal characteristics, a short review of digital data signals will be presented here. As seen in figure 2-36a, a simple digital signal requires a +10 V for binary 1 and 0V for binary 0. (Note: The voltage values may vary for different systems.) A very weak current flow occurs for binary 1 and no current flows for binary 0. The voltage never changes polarity since this is a single polarity format. A bipolar return-to-zero type signal is shown in figure 2-36b. Here the polarity on the bus is such that a +10 V equals binary 1. The bus polarity reverses to –10V for Binary 0. This is a bipolar format used by ARINC 429 systems. The signal also contains a clock pulse with each data bit. When voltage equals zero, the clock (null) signal is produced. There is no current flow during the clock pulse. Figure 2-36c shows a typical Manchester II format. This Manchester signal is bipolar and contains a clock pulse with each data bit. A +5V produces a binary 1 and –5V produces binary 0.

A **Doublet Signal** produces a short positive and negative pulse (spike) on the bus whenever the data value changes from binary 1 to binary 0 (see figure 2-37). Notice that with a doublet signal the bus is active only for a short period of time during each change of the binary data. The short spike alerts the receiver that the data bit has changed. From this *short spike* signal, the processor can conclude the value of the binary data (1 or 0).

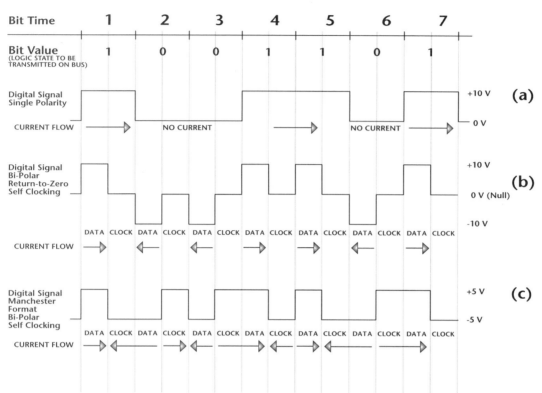

Fig. 2-36   Digital signal characteristics. a) single polarity signal; b) bipolar signal typical of ARINC 429 data; c) bipolar Manchester signal.

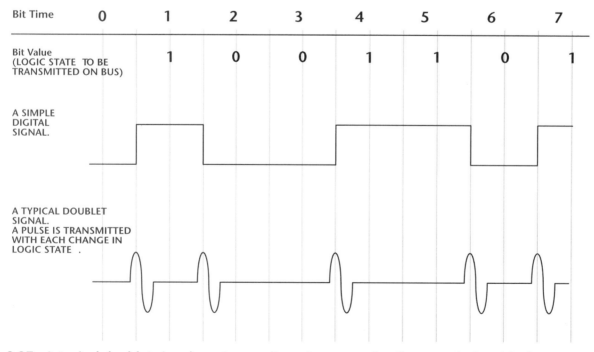

Fig. 2-37   A typical doublet signal creates a voltage (or current) spike at each data bit change.

A doublet can be either a voltage specific or current specific signal. That is, a voltage doublet will have limits set to the voltage parameters of the signal. A current doublet will have limits set to the current parameters of the signal. Figure 2-38 shows an example of typical voltage and current doublets.

Figure 2-39 shows the various forms of data required for transmission and reception by an ARINC 629 system. Study this figure during the following discussion. ARINC 629 data goes though several steps between transmitter and receiver. Step 1: the Manchester signal is generated in the transmitting LRU. Step 2: the Manchester signal is transformed into a voltage doublet signal. The voltage doublet leaving the transmitter will have a maximum and minimum differential voltage. (In this case, the voltage values are +4.5V and –4.5V.) Step 3: The voltage doublet is sent to the bus via the transmitting stub cable. (Stub cables will be discussed in the next section of this chapter.) Step 4: the signal is changed into a current doublet for transmission on the 629 data bus. Step 5: the current doublet is transmitted on the 629 data bus. (The current doublet has a maximum value of + 50 mA and a minimum value of –50 mA.) Step 6: the data is changed from a current doublet back into a voltage doublet at the receiving end of the data bus. (The receiver's voltage doublet parameters are +2.5V and –2.5V; slightly less than the transmission doublet parameters.) Step 7: the data is sent to the receiving LRU on the receiver's stub cable. Step 8: the data is once again converted to a Manchester format that can be processed by the receiving LRU. It should be noted the voltage and current parameters displayed in figure 2-39 are not fixed values. The ARINC 629 specification allows the manufacture substantial leeway in the design of their specific systems. This flexibility allows for new

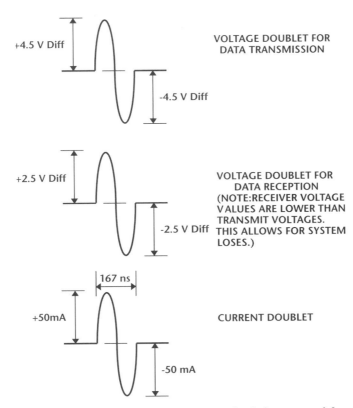

Fig. 2-38   Voltage and current doublets typical of those used for ARINC 629 data transmission.

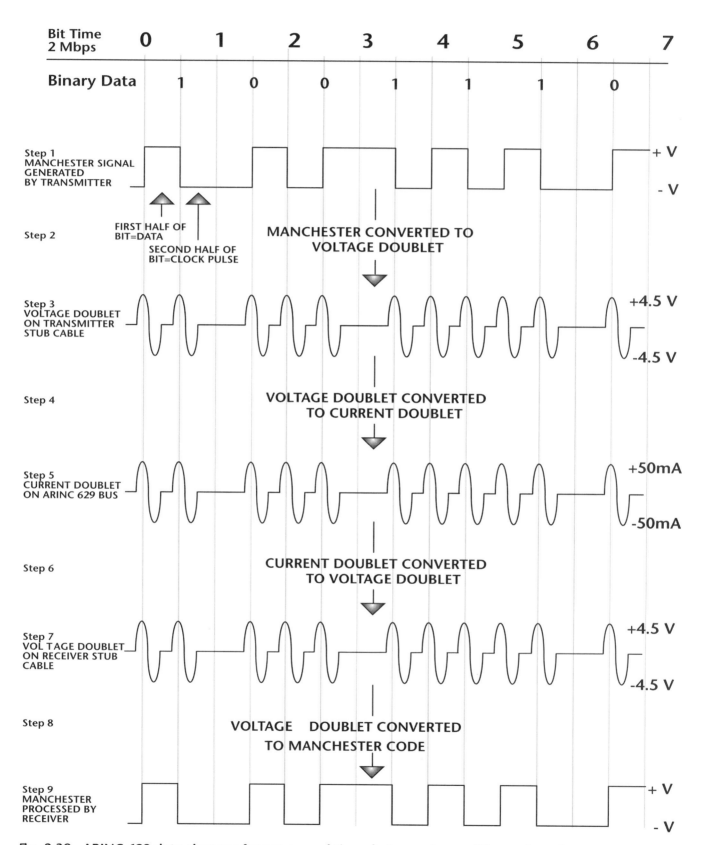

Fig. 2-39 ARINC 629 data changes format several times between transmitter and receiver.

designs to be developed without limiting engineers; hence, different aircraft may have slightly different system parameters. Be sure to check the aircraft specifications for exact details.

Data bus physical properties

ARINC 629 signals may be transmitted on a traditional copper data bus or a fiber optic network. The fiber optic network is used to transmit an optical (light) signal that can be adapted to a variety of formats. Fiber optics have the advantage of being immune to electromagnetic interference and lightning transients; however, fiber optics are still extremely limited on civilian aircraft and are beyond the scope of this text. The traditional copper wire data bus may be up to 100 meters in length with an impedance of 130 Ω +/-2%. The data bus consists of two insulated conductors of #20 AWG stranded wire. The conductor pair must be twisted together and may, or may not, be shielded. The insulation consists of a foam core with a Teflon outer shell. During maintenance be sure not to damage the insulation since water may penetrate the cable and change the impedance of the bus. Special tools are typically required for installing and removing components on the data bus. Be sure to follow approved maintenance procedures. The bus must have a termination resistor of 130 Ω +/-4% at each end. It is recommended the outer insulation of all ARINC 629 data bus cable be a unique color that is easy to distinguish from other wires in the aircraft.

The information transmitted using ARINC 629 must travel through several subsystems to enter or leave the data bus. Figure 2-40 shows the relationship of the various LRUs, and their related subsystems. The **Terminal Controller (TC)** is an LRU subsystem that moves data to and from the LRU memory. The terminal controller is the first interface between the transmitting LRU and the data bus. The TC provides various control functions for data transmission. From the terminal controller the data is sent to the **Serial Interface Module (SIM).** The SIM connects the TC to the data bus stub cable. The SIM changes the Manchester current signal from the LRU into an analog voltage doublet signal. The SIM also provides a fault monitoring function for the various signals received from the data bus. The SIM examines each data word for correct format and informs the TC in the event of a system failure.

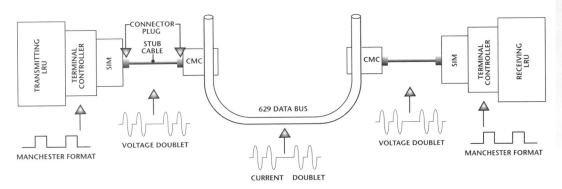

Fig. 2-40 ARINC 629 bus structure and related signal formats.

As seen in figure 2-40, the SIM is connected to the actual data bus through a stub cable and the Current Mode Coupler. The **stub cable** is a 4-wire cable used to connect the TC to the current mode coupler, and can be a maximum of 75 feet long. The stub cable uses two wires for receive mode and two wires for transmit mode. The stub cable also supplies power from the LRU to the current mode coupler.

The **Current Mode Coupler (CMC)** is an inductive coupling device that connects the LRU to the 629 data bus. The CMC contains a coupler transformer which receives induced signals from the data bus cable (figure 2-41). The CMC sends and receives electrical signals from the data bus without being physically connected to the bus. This helps to ensure integrity of the main data bus cable. The CMC actually contains two inductive couplers in one unit; one operates in active mode and one in standby. The SIM selects which coupler is operated in the active mode. As the CMC couples to the data bus the data signal changes format. During transmission the CMC changes the voltage signal (doublet) from the stub cable into a current signal for transmission on the data bus. In the receive mode the CMC changes the current signal received from the data bus into a voltage doublet sent to the stub cable and onto the receiving LRU.

Data word format
Reference figure 2-42 during the following discussions on the ARINC 629 data word format. ARINC 629 transmits a series of messages separated by various timing signals. Each message has up to 31 word strings. A 4-bit sync gap identifies the beginning of each word string. Each word string may contain up to 256 data words.

Each message begins with a **label word**. The label words are followed by a sequence of data words. The **label word** consists of 20 bits, containing a 12-bit label field, a 4-bit label extension field, a single parity bit, and a 3-bit sync pulse. The data word also contains 20 bits. The **data word** is divided into a 16-bit data field, a single parity bit, and a 3-bit sync pulse.

Before the start of each label word a special pulse is added just prior to the 3-bit sync pulse. As seen in figure 2-42 the pre-sync sync pulse (PSSP) and the pre-pre-sync sync pulse (PPSSP) distinguish the start of a label word from the start of a data word. The PSSP and PPSSP are bits that transmit for approximately one-half the time of a standard bit. These special bits are used to inform all receivers on the data bus that a new label word is being transmitted.

Transmission protocol
The ARINC 629 system can be thought of as a party line for the various electronics systems on the aircraft. Any particular unit can transmit on the bus or "listen" for information. At any given time, only one user can transmit and one or more units can receive data. The 629 system allows each transmitter to be completely independent of each other, and there is no bus controller used to coordinate transmission activities. The software of each transmitter must determine when that transmitter can talk on the bus. This "open bus" scenario creates some interesting challenges for the 629 system. The transmitter software must ensure that no one transmitter dominates the use of the bus and ensure that the higher priority systems have a chance to talk first.

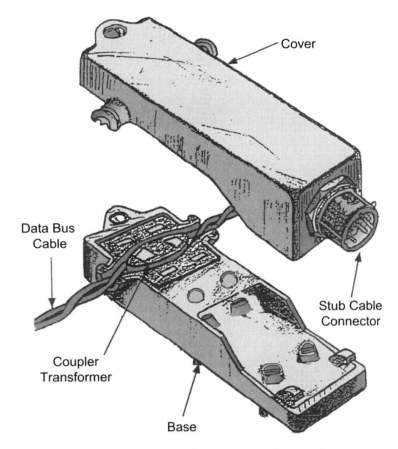

Cover

Data Bus
Cable

Stub Cable
Connector

Coupler
Transformer

Base

Fig. 2-41   A typical current mode coupler.

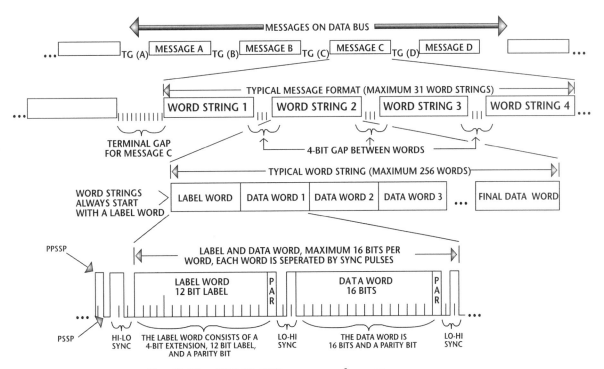

MESSAGES ON DATA BUS

... | TG (A) | MESSAGE A | TG (B) | MESSAGE B | TG (C) | MESSAGE C | TG (D) | MESSAGE D | ...

TYPICAL MESSAGE FORMAT (MAXIMUM 31 WORD STRINGS)

WORD STRING 1 | WORD STRING 2 | WORD STRING 3 | WORD STRING 4 ...

TERMINAL GAP
FOR MESSAGE C

4-BIT GAP BETWEEN WORDS

TYPICAL WORD STRING (MAXIMUM 256 WORDS)

WORD STRINGS
ALWAYS START
WITH A LABEL WORD

LABEL WORD | DATA WORD 1 | DATA WORD 2 | DATA WORD 3 | ... | FINAL DATA WORD

LABEL AND DATA WORD, MAXIMUM 16 BITS PER
WORD, EACH WORD IS SEPERATED BY SYNC PULSES

PPSSP

LABEL WORD
12 BIT LABEL | PAR | DATA WORD
16 BITS | PAR

PSSP

HI-LO
SYNC

THE LABEL WORD CONSISTS OF A
4-BIT EXTENSION, 12 BIT LABEL,
AND A PARITY BIT

LO-HI
SYNC

THE DATA WORD IS
16 BITS AND A PARITY BIT

LO-HI
SYNC

Fig. 2-42   ARINC 629 message format.

ARINC 629 is often referred to as a **Periodic, Aperiodic Multi-Transmitter Bus**. Multi-transmitter simply means the bus will support more than one transmitter; ARINC 629 allows up to 120 receiver/transmitters. According to the dictionary, *Periodic* defines something that occurs at *regular intervals*; *aperiodic* means *not occurring periodically*, or something that occurs at irregular intervals. ARINC 629 allows certain systems to transmit at regular intervals (i.e. in the periodic mode). Periodic transmissions would include information that requires updating at regular intervals, such as aircraft attitude or airspeed. Information transmitted aperiodically would be data such as flap position, system failure data, or fire warning information. This type of signal is only transmitted when the event occurs (i.e. not at regular intervals).

Transmission priority

ARINC 629 uses a series of timing signals to ensure each LRU transmits in the correct sequence. The most important LRU must have the option to transmit first, and only one transmitter can occupy the bus at any given time. There are three timing signals used to sequence the transmission of 629 data, the **Terminal Gap (TG)**, the **Synchronization Gap (SG)**, and the **Transmit Interval (TI)**. The TG is a unique time period for each transmitter, the SG is always longer than the longest TG and is common to all transmitters. The TI is the longest time interval and is also common to all transmitters on that bus. These three timing signals are stored in the memory of the LRU software. Each LRU is responsible for it's own timing. The LRUs monitor the bus for activity and keep track of all transmissions. When the bus becomes idle and the correct timing occurs, the LRU software transmits information onto the data bus.

Terminal gap

The **Terminal Gap (TG)** is a time period unique for each transmitter connected to the bus. The terminal gap determines the transmission priority for each LRU. LRUs with a high priority have a short TG. LRUs with a low priority have a long TG. (In other words, the most important information will be transmitted from an LRU with a short TG; least critical information will be transmitted from an LRU with a long TG.) No two LRUs connected to the same bus can ever have the same terminal gap. The TG priority is flexible and can be determined through software changes in the receiver/transmitters; however, all LRUs on that bus must have a different TG. Terminal Gap time periods range from 3.68 to 127.68 microseconds.

Synchronization gap

The **Synchronization Gap (SG)** is a time period common to all transmitters on the bus. The synchronization gap can be thought of as the reset signal for the bus. Once reset, the bus is open to all transmitters. Each LRU can transmit only once until after the next SG. The LRUs always transmit in order of their priority determined by their respective terminal gaps. Remember, the synchronization gap is longer than any terminal gap, and will only occur on the bus after each user has had a chance to transmit. It should be noted, the terminal gaps and synchronization gaps are simply time periods measured by the software of each LRU. The software starts a timer whenever the bus is idle. (To be idle means there is no data transmission on that bus.) If that LRU has information to transmit, the transmission will begin at the end of that LRUs TG. If the LRU does not wish to transmit, the timer continues until another LRU transmits or the SG is reached. Once the SG is complete, all LRUs reset their TG and SG timers.

Transmit interval

The **transmit interval (TI)** is a common time period for all transmitters on the bus. The TI for a specific LRU begins immediately when that LRU starts transmission. The TI inhibits any transmission from that same user until after the TI time period. Even if the TG and the SG time periods are satisfied, the LRU will not transmit until after the TI time has expired. The TI therefore creates a relatively long waiting period between two transmissions from the same LRU. In effect, the TI keeps any one user from dominating the bus. The transmit interval ranges from 0.5 to 64 milliseconds and are determined by the software engineers developing the system.

Timing description

Each LRU on an ARINC 629 data bus must determine when that LRU can transmit. The TG, SG, and TI are all used to determine the LRU transmission timing. To understand this system, study the example in figure 2-43 during the following discussion on transmission timing. Each LRU can use the bus if it meets a certain set of conditions. First, each transmitter is inactive until the **terminal gap** time for that transmitter is complete. Second, each LRU can only make one transmission then must wait until after the synchronization gap occurs before a second transmission. Third, the LRU can only make one transmission per each **transmit interval.**

Think of figure 2-43 as a snapshot of the data being transmitted on a 629 bus. The snapshot begins at the SG in the upper left corner of diagram (time progresses as we move to the right of the diagram). Remember the SG is the "reset signal" for all LRUs. At the end of the SG all TGs begin their timers. The first TG to be satisfied is TG(A). TG(A) is the shortest since LRU(A) has the highest transmission priority. At completion of TG(A), LRU(A) begins transmission. At that time the TI timer for LRU(A) begins. LRU(A) cannot transmit again until TI(A) is complete. When message A is complete the bus goes idle and the TG timers start again. LRU(B) has the next highest transmission priority. The next TG to be satisfied is TG(B). When TG(B) is complete LRU(B) begins transmission. At that time the TI timer for LRU(B) begins. LRU(B) cannot transmit again until TI(B) is complete. When message B is complete the bus goes idle and the TG timers start again. LRU(C) has the next highest transmission priority. The next TG to be satisfied is TG(C). This sequence continues until all LRUs on the bus have had an opportunity to transmit. Once all LRUs have had an opportunity to transmit, the SG will repeat and the bus will once again be open to all users. It should be noted, each LRU is not required to transmit after it's given TG. At that time the LRU may not have information to "share" with other LRUs on the bus. In this case, the LRU makes no transmission.

Periodic/aperiodic mode

As mentioned earlier, ARINC 629 operates in a **Periodic Mode** when all users complete their desired transmission prior to completion of the TI. Systems designed for operation in the periodic mode will transmit messages that have a consistent length. This ensures the update rate remains relatively constant and all transmitters have a

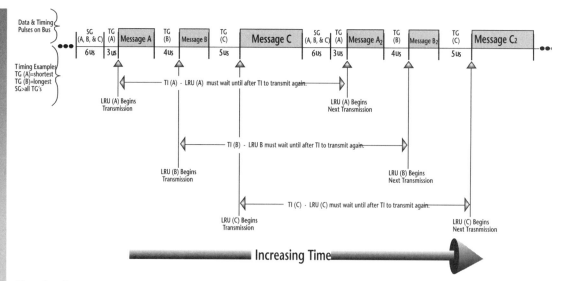

Fig. 2-43   ARINC 629 typical timing sequence for various LRU transmissions.

chance to transmit within a stable time period (i.e. periodically). If an LRU transmits a longer than average message, the TI is exceeded and the bus is operating in the **Aperiodic Mode**. Systems designed to operate in the aperiodic mode transmit messages that vary in length. These systems include data to control various flight operations, such as flap position and data for system status. Some systems are designed to operate in both the periodic and aperiodic modes depending in the specific data being transmitted.

DIGITAL BUS SYSTEM TROUBLESHOOTING AND REPAIR

In general, the digital data bus is a relatively trouble free system. Many aircraft operate for years without ever experiencing a data bus problem. Keep in mind, the actual data bus is nothing more than a pair of twisted wires. Most of the data bus problems occur at the bus connections to the various LRUs (line replaceable units). As aircraft gain flight hours, many of the LRUs are periodically replaced. Virtually all connector pins and sockets "wear out" through repetitive LRU replacement or due to the vibrations encountered during normal aircraft operations. Since data bus signals are very low power and rapidly changing, a slight increase in connector impedance can easily create a faulty bus (bus connection).

All data bus systems have a given limit as to the impedance of the bus/connectors. If a connector becomes worn or even dirty, the transmitted signal will experience loss or even stop completely. A minimal loss is acceptable; however if the bus loss exceeds given parameters, the receiver will not recognize the transmitted data. Adding to troubleshooting difficulties, a change in data bus impedance is difficult to measure. So, in most cases the actual transmitted signal is monitored and if excess loss is detected, the data bus (or most likely a connector) must be repaired.

Testing The Data Bus

There are several ways to measure the transmission characteristics of a digital data bus system (the "system" includes the connectors and/or bus couplers). The simplest test is a visual inspection of the bus and connectors. Obviously, if a bus cable has been physically broken the defect has been found. A visual inspection of the connector pins and sockets may also reveal problems. On the connectors plugs, look for worn, bent, or dirty contacts. A magnifying glass may help to identify connector problems. Connectors which are worn beyond limits may be slightly discolored due to a loss of the protective plating on the pins or sockets. Most pins and sockets are made of copper and plated with silver, gold, or other metal to prevent corrosion and enhance continuity. Dirty connectors may be cleaned using approved solvents and nonabrasive swabs. Be sure not to damage the connector during cleaning. If a connector socket is forced open by a cleaning swab, it will most likely bend out of shape and permanently damage the socket. Also never clean the contacts with an abrasive, such as sand paper. This will damage the contact plating and allow for corrosion.

An ohmmeter can also be used to detect an open in a bus wire or the bus shielding. Be careful however when using an ohmmeter; the bus cable may have low resistance to a direct current signal (ohmmeters use direct current to measure resistance) and yet have high impedance to a high frequency digital signal. An ohmmeter will only detect a catastrophic bus fault. A bus may pass an ohmmeter test and still fail to properly transmit digital data due to high impedance.

Data signal faults

There are three basic ways in which transmitted digital data can be effected by the data bus system. As shown in figure 2-44, the digital signal can experience **drop out(s)**, the signal can be **attenuated**, or the signal can be overridden by **interference.**

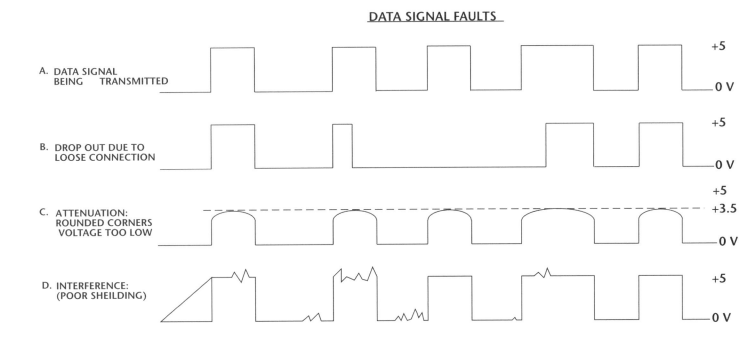

DATA SIGNAL FAULTS

A. DATA SIGNAL BEING TRANSMITTED

B. DROP OUT DUE TO LOOSE CONNECTION

C. ATTENUATION: ROUNDED CORNERS VOLTAGE TOO LOW

D. INTERFERENCE: (POOR SHEILDING)

Fig. 2-44 Digital data transmission faults. (a) proper data signal; (b) drop outs; (c) attenuation; (d) interference.

Drop outs

Drop outs occur when a portion or portions of the data signal are completely lost during transmission. Drop outs are typically caused by loose connections at a data bus connector plug. The loose connection causes an intermittent signal transmission and partial loss of the data. Intermittent connections often occur only during certain flight conditions, such as, turbulent air, large bank angle, or temperature extremes. This phenomenon often makes it difficult to find intermittent connections. In some cases, the technician can simulate turbulent air by simply wiggling the LRU or wiring harness. In many situations, the circuit must be analyzed under actual flight conditions to detect the fault.

Attenuation

Attenuation occurs when a digital data signal becomes weak or the transmitted voltage becomes too low. The attenuated signal shown in figure 2-44c is weak in voltage and the square wave form has been rounded somewhat. It should be noted that all data transmission experiences some attenuation; however, if excessive signal loss occurs, the digital data receiver will not recognize the signal. Excessive attenuation is caused by a change in data bus impedance. This can be caused by dirty connection pins, poorly soldered or crimped connections, pinched data bus cable, or excess moisture which has penetrated the data bus insulation.

Attenuation can also be caused by an excessive load placed on the data bus by one of the units connected to the bus. If this occurs, the problem can be isolated by systematically removing items which are connected to the bus. Keep in mind this type problem only occurs for LRUs with a direct connection to the data bus. LRUs which use an inductive coupling device to connect to the data bus are immune to this problem. Let us assume that receiver #2 in figure 2-45 has an internal short which places excess load on the data bus. To isolate the problem, operate the system normally then remove receiver #1. In this example the problem still remains. Next, disconnect receiver #2 from the bus. In this case, the bus operations return to normal. The problem must have been caused by receiver #2. Excess attenuation can also be caused by the transmitter on the bus. To solve this problem, the output signal of the transmitter must be checked using an oscilloscope or data bus analyzer.

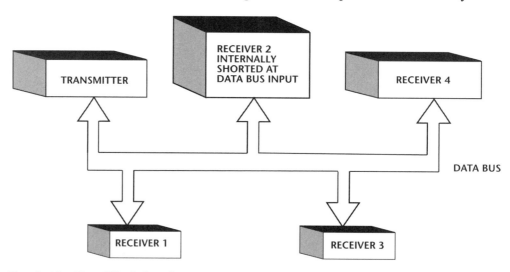

Fig. 2-45   Simplified data bus structure showing a shorted LRU.

Interference

Data bus **interference** occurs when an unwanted signal is induced into the bus wiring, hence, inhibiting data transmission. Interference is often caused by improper shielding of the data bus cable. If a bus cable is not shielded properly and it is exposed to a changing magnetic field, the digital signal will become distorted as in figure 2-44d. An excessively strong electromagnetic field near a data bus termination point (connector plug) can be another common cause of interference. In this case, it is likely that a component independent of the data bus, but located near the bus, is actually creating "too much" interference. This excessively strong interference signal induces unwanted current onto the bus and degrades the original signal beyond limits. The bus cable may actually have no faults and yet the interference "finds" its way onto the bus through "leaks" in the bus shielding at the connector plug. The source of the interference must be eliminated. Keep in mind, if the interference only occurs during certain flight conditions, such as operation of the flap motor, the technician should suspect either the data bus cable shielding or the flap motor brushes or shielding. One of the best ways to detect interference is to view the digital signal using an oscilloscope. To view the original digital signal (and the interference), connect the oscilloscope in parallel with the bus connections.

There are several other types of faults that can cause loss of digital data transmission; however, the aforementioned faults are those most likely to occur due to data bus problems. Other faults can be caused by failure of the data transmitter or receiver circuitry. Whenever analyzing the transmission of digital data, it is often wise to check the output signal at the data transmitter. The signal at this point must be within specifications and not subject to drop out, attenuation, or interference. If one of these problems is present at the transmitter (with the data bus disconnected), the fault lies within the transmitter not the data bus.

Timing tolerance faults

A **timing tolerance fault** occurs when the rise or fall of the digital signal responds too slowly to be within specifications of the data bus standard. This type of fault is typically caused by a defective transmitter. The ARINC 429 output signal timing tolerances are shown in figure 2-46. As can be seen in this diagram, total bit time is designated by the letter Y. For high-speed data transmission Y=10 usec. For low-speed, Y=Z; where Z is determined by Z=1/R. Remember, the bit rate (R) for low speed data is between 12-14.5 Kbps; for high-speed data the bit rate is 100Kbps.

Analyzing digital data signals

There are two basic test instruments that can be used to detect loss of digital data: the **oscilloscope**, and the **data bus analyzer**. The oscilloscope is basically a sophisticated voltmeter with a two-dimensional graph display that can be used to measure voltage (amplitude) and frequency (time) of an electric signal. This allows the operator to view voltage and/or frequency changes over time; hence, this instrument is ideal for analysis of digital data. An oscilloscope is typically used to analyze signals, such as digital data that change too fast to be monitored by a common multimeter.

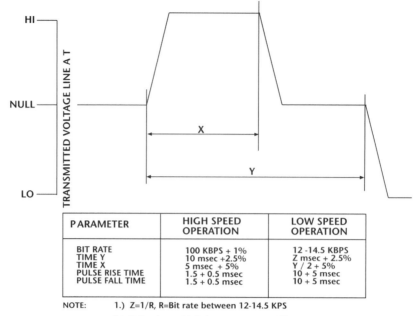

PARAMETER	HIGH SPEED OPERATION	LOW SPEED OPERATION
BIT RATE	100 KBPS + 1%	12 -14.5 KBPS
TIME Y	10 msec +2.5%	Z msec + 2.5%
TIME X	5 msec + 5%	Y / 2 + 5%
PULSE RISE TIME	1.5 + 0.5 msec	10 + 5 msec
PULSE FALL TIME	1.5 + 0.5 msec	10 + 5 msec

NOTE: 1.) Z=1/R, R=Bit rate between 12-14.5 KPS

2.) Pulse rise and fall times are measured between 10% and 90%. Voltage amplitude points are on the leading and trailing edge of the pulse.

Fig. 2-46   ARINC 429 digital signal timing tolerance. Notes: 1) Z = 1/R; where R = bit rate selected from 12 - 14.5 Kbits/s. 2) Pulse rise and fall times are measured between the 10% and 90% voltage amplitude points on the leading and training edges of the pulses.

The storage scope

A **storage scope** (or storage type oscilloscope) allows the technician to hold a given portion of digital signal on the oscilloscope's CRT display even though the signal may continue to change. This feature is extremely helpful for data bus troubleshooting since digital data is transmitted in a series of changing voltages. A typical storage type oscilloscope is shown in figure 2-47. Whenever connecting any oscilloscope to a digital data bus, the scope probe should be connected between the two twisted wires of the data bus as shown in figure 2-48. This connection will typically be made at the plug to an LRU. In most cases, the plug must be disconnected and two extra pins slipped into the correct sockets of the connector. The oscilloscope can then be connected to the "extra" pins.

The data bus analyzer

The **data bus analyzer** is a common carry-on piece of test equipment used to troubleshoot digital systems. There are many types of data bus analyzers, but the basic functions are quite similar. Bus analyzers are used to: 1) receive and review transmitted data, or 2) transmit data to a bus user. Before using any analyzer, one must first be sure the bus language is compatible with the bus analyzer. For example, the **DATATRAC 400** shown in figure 2-49 can monitor, simulate, and record data transmissions for avionics equipment using ARINC 429 or CSDB standard data buses. Data bus analyzers are also incorporated into the software of many modern digital aircraft systems. For example, the Boeing 747-400 central maintenance computer system input monitoring function can be used to detect ARINC 429 bus problems similar to a carry on data bus analyzer. One major advantage of the built-in test equipment (BITE) systems is that they often record the exact time when the fault occurred. This can be of great benefit when dealing with intermittent problems.

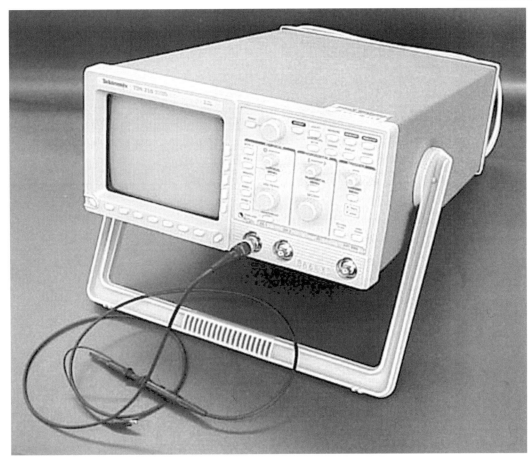

Fig. 2-47   A typical storage oscilloscope for display of digital data.

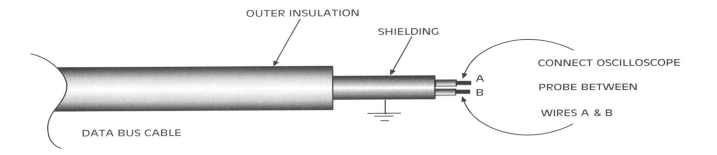

Fig. 2-48   Connecting an oscilloscope to a data bus cable.

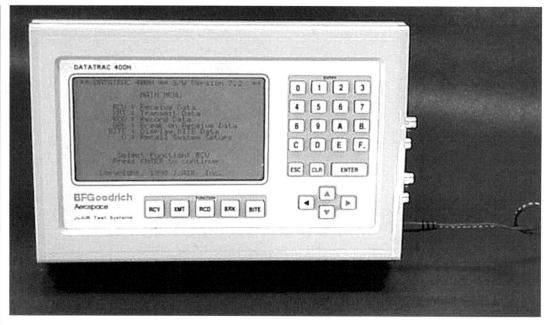

Fig. 2-49   A typical data bus analyzer.

Using a typical data bus analyzers

During system troubleshooting a data bus analyzer is often used to capture a stream of data being transmitted between digital devices. The recorded data can then be displayed by the analyzer for evaluation. If inconsistencies are detected, the transmitter or the data bus system is faulty. Remember, a data bus analyzer should be used to check the input and output signals of the transmitter, the receiver, and the data bus. Use the bus analyzer to determine if the correct signal is being transmitted and/or reaching it's destination. Some data bus analyzers are capable of reading several transmission lines or **channels** at one time. This allows for comparisons to be made which might expedite troubleshooting. Whenever using a data bus analyzer for ARINC 429 data, always be sure the word label and equipment identifier are both programmed into the test unit. The label/identifier combination is needed to ensure the bus analyzer monitors the data correctly.

For the DATATRAC 400 (figure 2-49) there are five basic modes of operation:

> **Receive:** This allows the technician to receive either high or low speed ARINC 429 data. The data received can be displayed in various formats, such as hexadecimal, decimal, or binary. The data may also be down loaded via an RS-232 port.

> **Transmit:** The data bus analyzer is capable of sending digital data in order to simulate communications between avionics equipment. This mode is often used to test the operation of a specific LRU when given the correct data bus signal.

> **Record:** A particular data label and record rate are selected and the avionics bus analyzer collects information sent to that address. All the data can then be displayed in a numerical format.

Break: This mode allows for "trapping" intermittent data in memory and reviewing the data as time permits.

BITE: This mode supports the Built In Test Equipment (BITE) for various transport category aircraft, such as, the B-747-400, A-320, and MD-11. Up to 240 BITE displays can be saved in the analyzer memory and viewed or downloaded at a later time.

In most cases, the data bus analyzer must be connected to the system at the connector plug of an LRU. For example, if you plan to receive data transmitted to the generator control unit (GCU), you would unplug the GCU from the bus and connect the analyzer to the data bus cable at the GCU connector plug. If the correct connections were made and the test equipment properly adjusted, the analyzer will display all messages sent to the GCU on that bus.

On some aircraft, a specific connector is available to allow for direct connection into the data bus system. Figure 2-50 shows a carry-on bus analyzer connected to the data bus. In this case, the technician is monitoring the fuel quantity. Note: the selector is set to the CMC (Central Maintenance Computer) and the analyzer indicates 3,246 lbs.

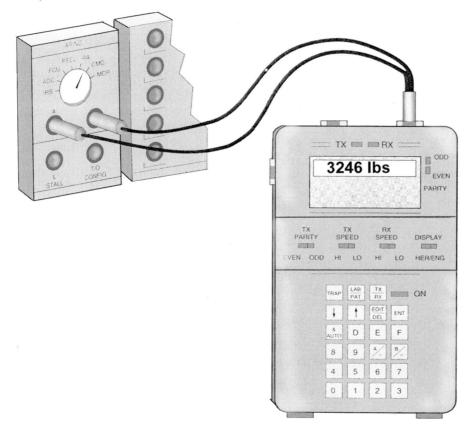

Fig. 2-50   Example of a typical bus analyzer installed on a transport category aircraft.

In the transmit mode, a digital data bus analyzer can be very helpful to simulate different avionics equipment. For example, the analyzer can be used to send a signal to an Electronic Flight Instrument System (EFIS). The operator can then verify that the correct display appears on the CRT. Some analyzers provide for dynamic transmission of data. For example, with dynamic data, an EFIS course display can be made to rotate 360^0, or the auto flight system can be tested through simulation of aircraft movement.

MULTIPLEXING AND DEMULTIPLEXING

Many avionics systems receive data inputs from a variety of sensors or subsystems located throughout the aircraft. In order for these remote devices to communicate with the avionics' computer circuitry, the input data must be received in the correct format. Most digital data is transmitted in a **serial** form, that is, only one binary digit at a time. However, many subsystems or individual sensors output data in parallel form. Therefore, some means are needed to convert parallel data into serial data and visa verse.

As discussed earlier in this chapter, transmission of data in serial form means each binary digit is transmitted for only a very short time period. After one bit of information is sent, the next bit follows; this process continues until all the desired information has been transmitted. This type of system can be thought of as **time sharing**, because each transmitted signal shares the wires for a short time interval. **Parallel data transmission** is a continuous-type transmission requiring two wires (or one wire and ground) for each signal to be sent. Parallel transmission receives its name because each circuit would be wired in parallel with respect to the next circuit. One pair of transmitting wires may be used to handle enormous amounts of serial data. If the information signals were transmitted in parallel form, hundreds of wires may be required to perform a similar task.

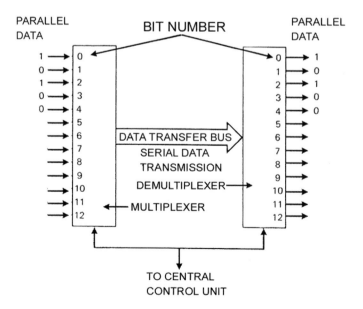

Fig. 2-51   Parallel data, converted to serial data, transmitted, and converted back to parallel data.

Serial data transmission requires less wire than a parallel system; however, an inter-pretation circuit is needed to convert all parallel data to serial type information prior to transmission. The device which converts parallel data into serial form is called a **multiplexer** or (MUX) circuit. The unit used to convert serial data into parallel form is called a **demultiplexer** or (DEMUX) circuit. As illustrated in figure 2-51, parallel data is sent to a multiplexer, where it is converted to serial data and sent to the data transfer bus. The **data transfer bus** is a two-wire connection between the multiplexer and the demultiplexer. The demultiplexer receives the serial data and reassembles it into parallel form. In this example, the byte 10100 is being received by the multiplexer in parallel form. Starting at the top and working down, the mul-tiplexer transmits each digit individually. Bit number 0 is the first to be transmitted. Bit number 1 is the next digit transmitted, bit number 2 next, and so on. This system repeats until all the parallel data are converted to serial form and individually con-nected to the data transfer bus. The demultiplexer receives a serial data input from the transmission bus and reassembles it into a parallel form. The output of the demultiplexer is identical to the input of the multiplexer (10100). It should be noted that the circuitry of figure 2-51 contains both a multiplexer and demultiplexer. This arrangement is used only when both the transmitter and receiver require parallel data formats. If the transmitter uses parallel data and the receiver can only accept serial data, a multiplexer would be used (see figure 2-52a). If the LRU transmits serial data and the receiver requires parallel data a demultiplexer would be used (see figure 2-52b).

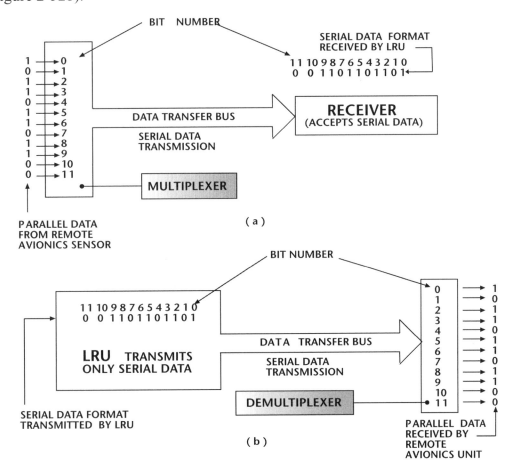

Fig. 2-52 Serial and parallel data transmissions. (a) multiplexer located in the transmitter; (b) demultiplexer located in the receiver.

On systems which share a serial data bus, some means of control must be used to coordinate the MUX/DEMUX arrangement. As shown in figure 2-51, a CCU (central control unit) can be used to coordinate the transmission and reception of data. This control is essential to ensure all serial data is transmitted and received at the proper time intervals. This system of serial data transmission may seem somewhat complex; however, the alternative, parallel data transmission would require one wire for each data bit to be transmitted. Since thousands of bits of information are transferred among various airborne systems, serial data transmission techniques are the obvious choice.

The use of multiplexers and demultiplexers is only necessary when a change from serial to parallel (or visa versa) is required. In many cases, serial data is transmitted to another component which can "read" serial data. In this case, no change in format is required.

When multiplexers or demultiplexers are required for operation of a system, they are typically contained within another unit. For example, figure 2-53 (top left and top right side of diagram) shows the data from the RSSs (radio sensor system) to the SDD (sensor display driver) is sent directly into a multiplexer (MUX) circuit. In troubleshooting this system, the line technician may never know the MUX circuit exists. If the internal multiplexer fails, the line technician may simply detect a defective SDD. The bench technician, on the other hand, would determine the MUX circuit inside the LRU was inoperative.

MUX/DEMUX Systems For In-Flight Entertainment

On many transport category aircraft, MUX/DEMUX circuits are used for the control of passenger entertainment. Each passenger has a control (installed in the armrest of the seat) which is used to select the listening channel and volume for the in-flight entertainment. In this case, multiplexers are often stand-alone units and can be replaced individually by the line technician. Each individual armrest control would transmit parallel data. A group of controllers would be sent to an area multiplexer where the data is assembled and transmitted to the passenger entertainment control center, or main multiplexer. Data from the entertainment control center is transmitted in serial form and demultiplexed for a group of seats.

On some aircraft, the multiplex system controls passenger entertainment, passenger reading lights, and flight attendant call lights. The MUX/DEMUX units are often located beneath a group of seats or in the head liners above the seats. Figure 2-54 shows a diagram of the passenger entertainment system on a Boeing 767. Notice, this system contains both zone and main multiplexers.

ANALOG TO DIGITAL CONVERTERS

As the popularity of digital avionics grew, many aircraft employed both analog and digital systems. This created communication problems because pure digital systems are incapable of reading analog data and visa versa. To allow communication between analog and digital systems **analog to digital** and **digital to analog** circuits were developed. Often referred to as **A/D converters** or **D/A converters,** these circuits can be found on virtually any system which employs both analog and digital devices.

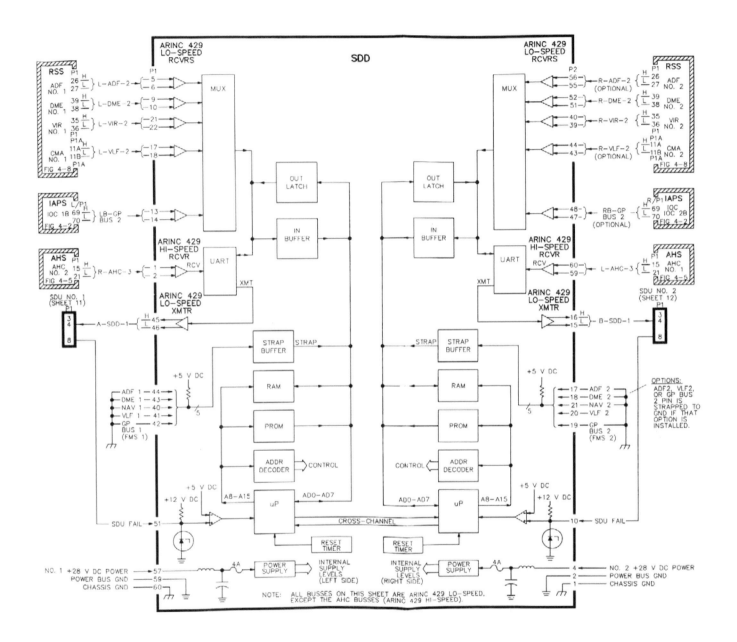

Fig. 2-53   Example of multiplexers used within an LRU. (Rockwell International, Collins Avionics Divisions) For Training Purposes Only.

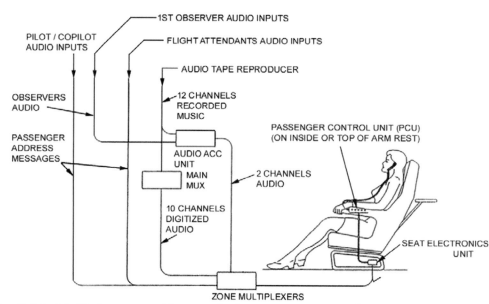

Fig. 2-54 **Multiplexers used in a typical passenger entertainment system.**

A/D converters utilize microprocessor circuitry to convert a given analog signal into a digital format. A/D converters are typically designed for a specific purpose. That is, the conversion circuitry must be programmed to analyze a given analog input and produce a specific digital output. Figure 2-55 shows the conversion of an analog signal from a temperature transducer. This A/D converter must be programmed to accept the analog signal with voltage limits of 0 to +20 V DC and output a digital signal of +5 V DC for logic 1, and 0 V DC for logic 0.

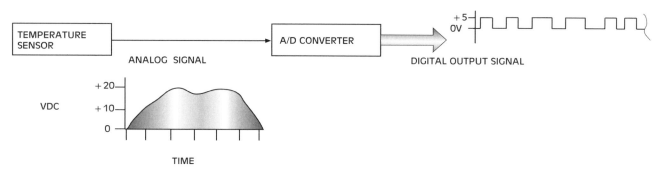

Fig. 2-55 **Conversion of an analog signal to a digital signal.**

Like multiplexers, A/D converters are most often found incorporated in the circuitry of various LRUs. For example, the inputs to a digital autopilot may include some analog signals; these signals would be converted to a digital format inside the auto-pilot computer (LRU). Since the outputs from the digital autopilot to the servo motors are analog signals, the autopilot circuitry must also incorporate a D/A converter.

For the most part, a line technician would not replace a specific A/D or D/A converter. These units are typically incorporated in a given LRU. If the output signal from an LRU is not correct, replace the unit, including the A/D and D/A converter.

FLAT PANEL DISPLAYS

Over two decades ago, electronic instruments were introduced for display of aircraft systems data. On all types of aircraft new electronic instruments, referred to as EFIS (electronic flight instrument systems), were slowly replacing traditional electromechanical instruments. The electronic instrument displays consisted of a CRT (cathode ray tube) similar to a typical television or computer monitor. It was often said that newer aircraft had a *glass cockpit*, describing the lack of traditional instruments and the glass face of the CRT. The CRT revolution helped to improve both crew awareness and system reliability. Unfortunately, CRT displays are relatively heavy, large, and consume a lot of power. The many benefits of the CRTs made it very popular; however, the industry has long since looked for a replacement. The answer came in the form of the flat panel display.

A **flat panel display (FPD)** is a solid-state device used to display various formats of video information. There are several versions of flat panel displays currently available and each type is lighter, smaller, consumes less power, and produces less heat than a typical CRT display. Early FPDs had several limitations. Color FPDs were hard to manufacture and had limited color range. Early FPDs had a slow response time and poor resolution. When viewed from an angle, the display seemed to go blank, and in general these displays were poorly lit. For years, flat panel displays have been widely used on lap top computers. Computer designers accepted the limitations of the flat panel display since they consume very little power. (A very important consideration for portable computers.) In fact, it would be safe to say that lap top computers would be virtually impossible without the development of flat panel displays.

Today the technology exists to produce a flat panel display, which overcomes previous limitation. Modern FPDs equal the brightness, resolution, and update speeds of a typical CRT. Improved production techniques have enhanced reliability and lowered the cost of FPDs. Flat panel displays are now the display of choice for new aircraft instrument systems. Virtually every aircraft instrument manufacturer has committed to flat panel displays. The Boeing 777 was the first transport category aircraft to employ all flat panels for primary flight and systems data. Now FPDs are being installed into new business jets, commuter aircraft and helicopters. Even older aircraft will soon be able to take advantage of flat panel displays. Units, which are virtually *drop-in* replacements are being developed for aircraft such as the DC-10, B-727, and B-737. Many of these aircraft will soon replace their older electromechanical instruments with flat panel displays.

There are several types of flat panel displays currently available or under development that show promise for use in future aircraft. Today the **liquid crystal display (LCD)** is used on many aircraft in both monochrome and full color formats. In the future, technological developments will bring improvements to LCDs and introduce other types of flat panels.

Liquid Crystal Displays

One of the most popular flat panel displays is called the **liquid crystal display
(LCD)**. LCDs have been used on aircraft for several years. The Boeing 756 and 767
were two of the first large aircraft to employ LCDs. On these aircraft, LCDs are
used as backup instruments for engine data. These early displays were one-color,
poorly lit and had a limited viewing angle. Newer displays are full color, backlit and
have a wide viewing angle. The Boeing 777 employs the new technology full color
LCD displays.

LCD theory of operation

Liquid crystals are a substance that exhibits the properties of both a liquid and a
solid. As light passes through a liquid crystal the light follows the alignment of the
molecules of the crystal. A 19th century botanist named Friedrich Reinitzer first
discovered liquid crystals. In the 1960s it was discovered that an electrical charge
could be used to align the crystal molecules and the crystals could be used to control
light. Since the early 1970s, liquid crystals have been used for various types of
displays.

Every liquid crystal display contains several key elements sandwiched together. As
seen in figure 2-56, the LCD uses two polarizing filters on the outside of the sandwich. Just inside the polarizers are two finely grooved glass plates. Inside the glass
plates are the LCD electrodes and finally the liquid crystal material is located between the electrodes.

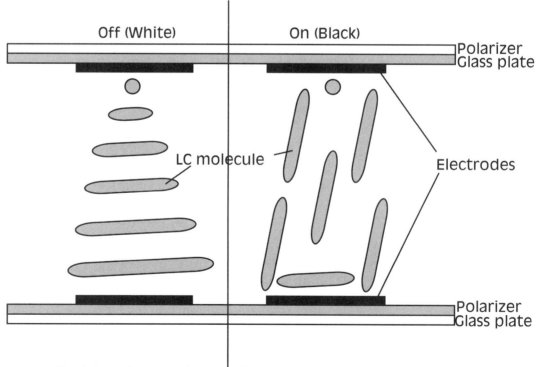

Fig. 2-56 Diagram of a typical liquid crystal display pixel.

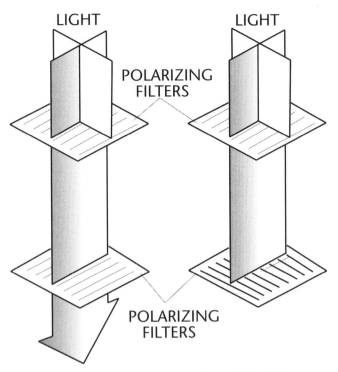

LIGHT   LIGHT

POLARIZING
FILTERS

POLARIZING
FILTERS

*Fig. 2-57   Two polarizing filters; when arranged parallel – light passes; when arranged
perpendicular – no light passes.*

A polarizer is a light filter made of extremely fine lines. The filter could be thought
of as an incredibly small blind used to control light through a window. As seen in
figure 2-57, if two polarizing filters are arranged perpendicular they block light
from traveling through the filters.

Within the LCD "sandwhich", just inside the polarizers are two grooved glass plates.
The grooved plates are used to align the liquid crystal molecules. As seen in figure
2-58, liquid crystal molecules naturally tend to align themselves with the grooves in
the glass plate. These glass plates are often referred to as alignment plates. The
crystal molecules twist light entering the top plate as it passes through the crystal.
By applying a voltage to the crystal, the alignment of the molecules can be controlled as seen in figure 2-59.

When the LCD sandwich is complete, light is controlled by the two polarizing filters and by the liquid crystal molecules. Figure 2-60 shows the operation of a basic
LCD. When there is no voltage applied to the LCD, light enters the top polarizer,
twists through the liquid crystal and passes through the bottom-polarizing filter.
When a small voltage is applied to the LCD electrodes, the crystal molecules align
and the polarizing filters block light. In a typical LCD, the glass alignment plates
and the electrodes are combined into one element. It should also be noted that only
a portion of the light passes through the LCD. LCDs always block some light and
hence create design challenges.

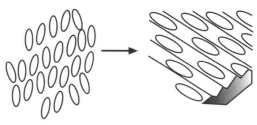

Fig. 2-58   Liquid crystal molecules aligned with the grooved glass.

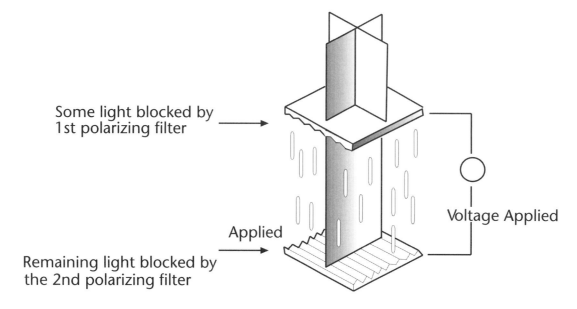

Some light blocked by 1st polarizing filter

Voltage Applied

Applied

Remaining light blocked by the 2nd polarizing filter

Fig. 2-59   When a voltage is applied to the LCD electrodes the LC molecules align.

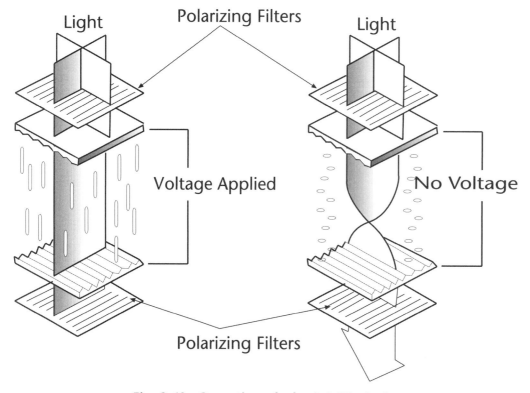

Light

Polarizing Filters

Light

Voltage Applied

No Voltage

Polarizing Filters

Fig. 2-60   Operation of a basic LCD pixel.

The previous discussion focused on one liquid crystal element commonly referred to as a pixel. In order to form a complete display with high resolution, thousands, or even millions of the individual pixels are ganged together. Each **pixel** is a distinct LCD acting independently of all other pixels in that display. Each pixel creates one dot on the flat panel display. The dots are used to form letters, numbers, and video images. A typical digital watch employs an LCD display with relatively large pixels. If one examines the watch display using a magnifier, the individual pixels can easily be identified (figure 2-61). The number of pixels per square inch determines the display **resolution**. A higher resolution creates a clearer image. Unfortunately the higher the resolution, the smaller the pixel and the more difficult the display is to produce. To obtain the resolution necessary for a typically full color primary flight display, millions of pixels are required.

Controlling the LCD

There are two basic methods used to control the operation of an LCD, static and dynamic drives. A **static drive** control system is used to control simple displays containing only a few pixels. The static drive system was first introduced on early versions of flat panel displays. Early displays had a very slow response time and were not well suited for displays requiring rapid change. Static drives basically consisted of a separate wire to each given pixel in the display. Static drives required so many wires that large displays containing thousands of pixels are impractical.

LOW
RESOLUTION

HIGH
RESOLUTION

Fig. 2-61   Pixel arrangement for a low and high-resolution display.

A **dynamic LCD drive** employs a system that shares wires to the various pixels on the display. The wiring is typically set up into a grid system that allows for large displays with fewer electrical connections. Today, the most common dynamic drives are the active matrix and the passive matrix. A **passive matrix** LCD employs a grid of conductors with pixels located at each intersection of the grid. Current is simultaneously sent through two conductors on the grid to control the illumination of each pixel. An **active matrix** has a transistor located at each pixel intersection, requiring less current to control the illumination of the pixel. Since less current is used, each pixel can be switched on and off more frequently and the screen response time is greatly improved. Active matrix displays are often referred to as **thin film transistor (TFT)** displays. Active matrix LCDs are used for aircraft instrument and other displays requiring a fast update rate. Passive matrix LCDs are suitable for calculators and other displays that do not require a rapidly changing image. Passive matrix LCDs are found on aircraft, and typically used for secondary instruments or simple data displays.

Transmissive vs. reflective

LCDs can be illuminated using reflected ambient light, or by passing light from a dedicated source through the liquid crystal. **Reflective LCDs** employ a mirror mounted to the rear of the liquid crystal sandwich (see figure 2-62). Ambient light enters the front of the display, passes through the crystal, is reflected off the mirror and passes once again through the crystal. This area of the LCD appears *illuminated*. When the crystal electrodes are energized, the light cannot pass through the crystal/polarizer layers and the mirror reflects no light. This portion of the display appears dark. The major advantage of a reflective LCD is its compact size and low power requirements. A reflective display, however, is difficult to see in some lighting conditions and is totally useless without ambient light or an additional light source.

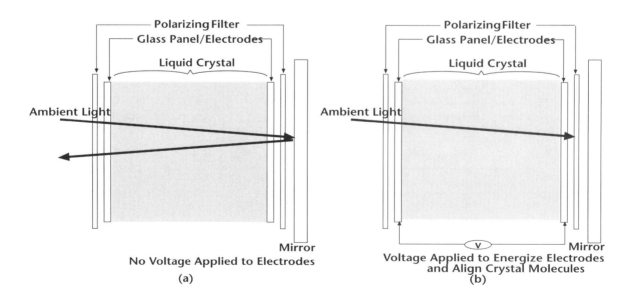

Fig. 2-62 A reflective LCD pixel; a) no voltage applied to the electrodes – light passes through the pixel; b) voltage applied to electrodes – light is blocked by the polarizing filters.

Transmissive displays are LCDs containing a dedicated light source mounted to the rear of the display. Transmissive FPDs are much brighter and easier to view than reflective displays. Unfortunately, transmissive displays are less efficient since they require energy to power the light source. The most common light source is known as a **cold cathode fluorescent tube** (CCFT). A CCFT is a highly efficient fluorescent tube that weaves back and forth across the back of the FPD to provide a bright even light. Most of the FPDs used on aircraft are transmissive displays. Figure 2-63 shows a typical transmissive color display.

Color LCDs

One of the more common LCDs employs a gray scale display. Blocking light through various pixels produces the LCD image. The active pixel looks gray (black) since no light travels through that pixel. In most cases this type of display relies on room light for illumination. Gray scale displays are commonly used for calculators and watches. Due to their poor quality, gray scale displays have limited use on aircraft. This type of LCD may take on virtually any color if a filter is added to the display. Amber is a common display color and is often referred to as a monochrome display.

Color LCDs are now the flat panel display of choice for most aircraft applications. Color FPDs are much more complex to produce since they require three pixels (red, blue, and green) to replace one pixel of a gray scale display. Most color FPDs contain an active matrix control system and require an additional lighting source for illumination. Figure 2-63 shows a cross sectional view of a typical color LCD. The display is illuminated using a light source located at the rear of the LCD. There are three color filters located on the viewing side of the FPD. As light passes through the LCD it also passes through the color filters. Each color pixel is controlled independently allowing the display to achieve a full color range.

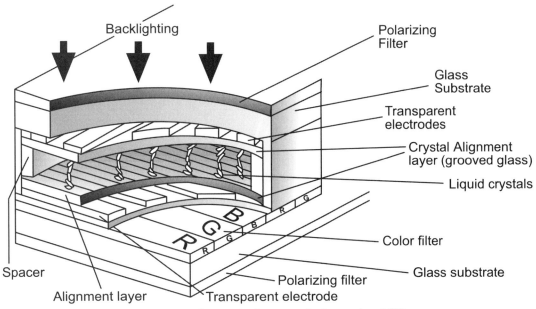

Fig. 2-63 Sectional view of a typical transmissive color LCD.

Viewing angle

Early FPDs had very limited viewing range. As the FPD was viewed from the side, the image would fade and lose contrast. For several years this problem made FPDs impractical for primary aircraft displays. Recently engineers have developed thinner LCDs and LCDs with a dual domain alignment layer. The thin glass alignment layer (figure 2-63) is modified to direct light into a wide viewing angle. This process creates a type of curved lens on each pixel. Modern aircraft displays have a wide viewing angle; however, they are costly to manufacture.

Repair and Maintenance of FPDs

For the most part, FPDs require very little maintenance. Cleaning the display should be done on a regular basis using a soft cloth. Be careful not to scratch the display. In the event an aircraft FPD fails it is typically replaced not repaired. However, replacing aircraft FPDs is extremely costly, so many manufactures and secondary repair facilities are developing methods to repair FPDs.

The most common defect seen with FPDs is one or more inactive pixels. An inactive pixel will always be set to the *on* condition or always set to *off*. This will create a small black spot on the display or create a pixel that is always illuminated. Most aircraft FPDs have a tolerance for defective pixels. A typical primary flight display can have up to 15 defective pixels before the display must be replaced. As of this time, repairing a defective pixel is not possible.

If the entire FPDs fails to operate, check all wiring and connectors to the display. This is an external repair and can often be accomplished in the field. If one or more horizontal or vertical lines appear on the display, the video driver, display-controller or related electrical connections have failed. The FPD can still be repaired; however, it is a difficult task best left to specialists.

The back lighting of the FPD may also become defective. With age, the CCFT will dim naturally creating a poorly lit display. The display illumination can be tested using a light meter. Although it is rare, the CCFT may fail completely. A defective CCFT can be replaced given the right equipment and a lot of practice.

ELECTROSTATIC DISCHARGE SENSITIVE COMPONENTS

Over the past decade, avionics systems have gone through a transformation of improved performance and reliability. This transformation was mainly the result of the extensive use of microelectronics in computer circuits. Microelectronic devices, such as integrated circuits and microprocessors, incorporate semiconductor material measuring only a few millionths of an inch thick. These subminiature structures are extremely vulnerable to damage from static electricity. Devices made of CMOS (Complementary Metal Oxide Semiconductor) components are especially vulnerable to static discharge.

Components which are ElectroStatic Discharge Sensitive are commonly referred to as ESDS parts. Since these components are becoming popular on all types of avionics equipment, the avionics technician must be aware of the potential dangers from static electricity. It is estimated that over five billion dollars in micro electronics and

P.C. boards are scrapped annually due to damage from static electricity. Thousands of warranty repairs and unnecessary failures are created in avionics shops annually. Awareness and proper precautions for ESDS components can increase productivity, customer satisfaction and of course improve air safety.

Electrostatic Discharge

Electrostatic discharge is the discharge (movement of electrons) created when any material containing a static charge comes in contact with a material containing a different or neutral charge. A static charge is present when a material possesses an excess or deficiency of electrons. A neutral charge is present when a material possesses an equal number of electrons and protons. Any time two components of different static charge get close enough together, a static discharge will occur. Typically, a distance of 10 mm or less is required to produce a discharge of static electricity that was generated due to normal maintenance activities.

Production Of A Static Charge

Friction between two different materials produces the majority of static electricity. As two dissimilar materials rub together, there is often a transfer of electrons from one surface to another. The amount of electron transfer is mainly a function of the type of materials making contact, the amount of friction between the two and the relative humidity of the surrounding air. Static electricity can also be generated on some materials, such as plastic, by subjecting them to heat or pressure.

Most everyone has experienced the production of a static charge while walking across a nylon carpet with plastic sole shoes. In this situation approximately 35,000 volts can be generated if the relative humidity is 20 % or less. Normally, you do not notice the static charge until you come in contact with another object, such as a door knob. At that time you are experiencing the discharge of the static electricity which was previously (before touching the door knob) stored on your body. This type of discharge will damage ESDS parts.

The minimum perception level to feel a static electrical discharge is typically 3,000 volts. For a discharge to be heard it requires approximately 5,000 volts; and the discharge cannot be seen if less than 10,000 volts. The current during discharge is extremely low and produces only minor discomfort to humans; however, a static discharge of as little as 100 volts can damage certain electronic components. Figure 2-64 shows an example of the voltage generated by common materials found in a maintenance facility. It is imperative that technicians realize that almost any situation can create enough static charge to damage sensitive components.

Identifying ESDS Components

In order to protect components from static discharge one must first identify ESDS parts and then take the necessary precautions. Technicians must be aware of the various type of labels which have been designed to identify sensitive components. The most common labels are shown in figure 2-65. One or more of these labels should be placed on all ESDS parts. The label should be easily visible to anyone who might come in contact with the component. The identification labels must be used on components installed in the aircraft, as well as components found in shipping, storage or in a repair facility.

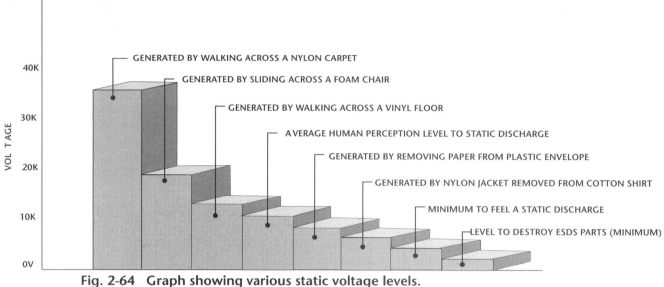

Fig. 2-64   Graph showing various static voltage levels.

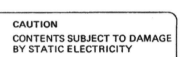

Fig. 2-65   Common labels used to identify ESDS parts.

ESDS parts come in all shapes and sizes. The LRUs found on virtually any modern aircraft are most likely electrostatic sensitive and should not be handled without proper precautions. Each equipment rack containing a sensitive component must be labeled with the appropriate decal. Other components which are often vulnerable to static discharge include computer cards (circuit boards) and certain integrated circuits. These components are most vulnerable during removal and installation, as well as shipping, handling and bench repair.

Protecting Components

The basic premise for protection of all ESDS parts is to prevent a static charge from entering that component. This is accomplished by 1) electrically neutralizing any-one or anything which comes in contact with or near the part, or 2) electrically insulating the sensitive part from potential static discharge. It should be noted that a charged body does not need to contact the ESDS component to damage it. If a sensitive component is close enough (less than approximately 1/2 inch) to the static charge, the part can be damaged.

Neutralization is typically accomplished by connecting all personnel and equipment to an electrical ground source. Prior to working with an ESDS part the technician must connect a grounding strap to bare skin and ensure its security. A wrist strap connected to ground through a flexible wire is typically used for this purpose (figure 2-66). All test equipment, including bench tops, must also be grounded. If equipment installed in the aircraft is labeled ESDS, always connect a wrist strap to the aircraft's ground prior to touching any component. A grounding socket is often provided in the aircraft's electrical equipment bay for this purpose.

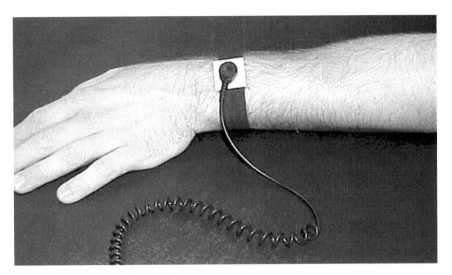

Fig. 2-66   A typical grounding wrist strap.

A SAFETY NOTE: Always use an approved wrist strap. Commercially available straps contain a resistor in series with the ground circuit to protect the user from electrocution if contact is accidentally made with a hot wire.

To electrically insulate a component, place all sensitive parts in the appropriate conductive container. These containers are often specially designed boxes for specific line replaceable units or conductive bags used for smaller items such as cards or integrated circuits. Once an item has been placed in its protective container, be sure it is labeled with an ESDS decal.

Technicians should be aware that there are two types of protective bags currently available: 1) the pink antistatic polyethylene bag is used to protect components from generating a static charge caused by movement while stored inside the bag, and 2) the dark gray antistatic bag will protect the component from static fields outside the bag, static discharge current, and static fields generated inside the bag (see figure 2-67). The gray colored bag offers the most protection and should be used under most circumstances. If in doubt, ask your manufacturer for more details concerning the correct protective container. Once the ESDS component is placed inside the correct protective container, the unit is safe from damage caused by static electricity and may be safely handled by ungrounded personnel.

Other means to protect components include the use of antistatic caps which should be placed over all connections of sensitive LRUs, and metal clips or conductive foam used to short together the leads of individual components or cards. Whenever using any protective device be sure to inspect it for defects or unacceptable wear.

At the work station, the ESDS components should not be set directly on to the table top. The work station should be equipped with either a **conductive** or **static dissipative** mat. The **conductive** mat may be used whenever a component is serviced in a power off condition. If the component is operated while placed on the conductive-type mat, the mat itself may bridge the connections between any components and cause component failure. In the case where components are operated while in contact with the mat surface, a **static dissipating** mat must be used. Be sure to check that your work stations are equipped with the correct type of protection.

Nonconductors at the work station are also a source of problems when working around ESDS parts. If plastic tools or pens, for example, come into contact with a sensitive component damage may occur. Nonconductive plastic does not dissipate its charge easily. A grounded technician holding that tool is not enough. Examine your work habits, and if you find nonconductors used at work stations, the best solution is to install an **Ionization Air Blower**. An Ionization Air Blower will safely delete any static charge formed on nonconductors. It is also helpful to maintain a relatively high humidity level in all work areas. High humidity helps to dissipate static charges before they become a problem.

Types Of Static Damage

Damage from static discharge can take two forms. They are commonly referred to as **hard** and **soft** failures. A **hard failure** caused by static discharge will create an immediate system defect. A **soft failure** caused by static discharge may only injure the component and cause erratic operations and/or eventual failure. A component which experienced a soft failure is often said to be **wounded.** A wounded part may pass all bench and/or preflight tests and still fail during flight. Wounded parts account for approximately 90% of all static damage. Often wounded parts cycle from aircraft to shop several times before the fault can be detected and repaired.

Modern avionics equipment contain a variety of ESDS parts. Every technician must be aware of the problems involved with static electricity and take the necessary precautions. The defect(s) caused by a static discharge to a sensitive component will not be visible to the naked eye; however, the results could be catastrophic. When in doubt, assume the component is ESDS and take the necessary precautions.

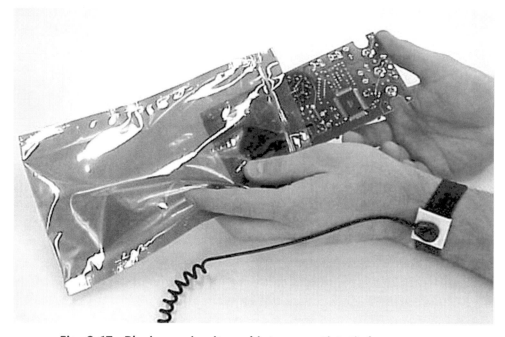

Fig. 2-67 Placing a circuit card into an antistatic bag.

CHAPTER 3
ELECTRONIC
INSTRUMENT SYSTEMS

INTRODUCTION

Over the past two decades, aircraft instruments have gone through a revolution. Older mechanical-type instruments have been replaced with electronic instruments offering improved accuracy and reliability. Electronic Flight Instruments Systems (EFIS) became possible with the development of a sunlight-readable CRT display and sophisticated aircraft computer systems. A digital data bus system is used to transfer the majority of information between the various components of EFIS. With large amounts of data to transfer, using digital (not analog) systems saves hundreds of pounds in additional wiring.

Electronic flight instruments began to enter the transport category and corporate type aircraft market in the early 80's. Due to their flexibility in data displays and proven reliability, EFISs have become commonplace in modern aircraft. However, modern integrated EFIS barely resemble the original systems. Original electronic flight instrument systems were designed to replace high maintenance electromechanical **HSIs (horizontal situation indicators)** and **ADIs (attitude director indicators)**. Newer electronic instrument systems also provide HSI and ADI data, and are thoroughly integrated with other aircraft systems. This allows for the presentation of a multitude of other display functions. In many cases, state-of-the-art systems are referred to as **Integrated Display Systems (IDS)** or **Electronic Instrument Systems (EIS)** to help distinguish them from the original EFIS. Keep in mind, the *F* in *EFIS* stands for *flight*. The early version electronic instruments were used to display only flight data, such as attitude, heading, and navigational references. Newer electronic display systems are used to show flight data and various other system data.

Electronic instrument systems have been in common use for nearly two decades. During that time, systems have grown from single display units to fully integrated systems. State-of-the-art EIS computers interact with autoflight, autothrottle, and integrated test equipment systems. This type of integration allows for more system flexibility, better redundancy, and a substantial weight savings over earlier stand-alone EIS.

It is the purpose of this chapter to present the basic principles of Electronic Instrument Systems (EIS). To do so, two different generations of EIS will be examined. The first system will be an early version stand-alone EFIS. The second is a fully integrated advanced system. Although only a few basic systems will be presented in detail, all EIS operate in a similar fashion. The information in this chapter will allow a technician to easily adapt to the specifics of any electronic instrument systems in use today.

Fig. 3-1 Instrument panel of a Dassault Falcon 900 showing a 5 tube EFIS.

ELECTRONIC FLIGHT INSTRUMENT SYSTEMS (1st generation EIS)

An **electronic flight instrument system (EFIS)** typically employs two or more CRT displays to present alphanumeric data and graphical representations of aircraft instruments. Each EFIS display replaces several conventional instruments, and various caution/warning annunciators. For the most part, EFIS was introduced on the Boeing 757 & 767 in the early 80's. About this time, similar systems began to find a home in corporate type aircraft, such as the Beechcraft King Air, Falcon jet, and Gulfstream G-3. Figure 3-1 shows the instrument panel of a Dassault Falcon 900 with a five tube EFIS. Note: the word "tube" is often used to refer to the CRT display units. This system, made by Honeywell, provides two EFIS displays for both the pilot and first officer. The fifth tube installed in the center of the panel is called a multifunction display. The **multifunction display (MFD)** is used to display weather radar data, course and flight plan information, and system checklist. Most importantly, the MFD provides backup functions in the event of a partial system failure.

Collins EFIS-85 & 86 System Components

The Collins EFIS-85 and 86 are two first generation electronic flight instrument systems, which are very popular in corporate and commuter-type aircraft. The EFIS-85/86 are part of the Collins Pro Line II series of avionics equipment and are designed to operate in conjunction with a Collins autoflight system. The EFIS-85/86 operate virtually the same; however, the CRT display of the 86 is larger. The Collins EFIS equipment will be examined in detail in the following pages; the autoflight system will be presented in the upcoming chapters.

The basic system components of the EFIS-85/86 systems are shown in figure 3-2. A five-tube system consists of four **electronic flight displays (EFDs)**, two **display processor units (DPUs)**, one **multifunction display (MFD)**, one **multifunction processor unit (MPU)**, a **course heading panel (CHP)**, and a **display control panel (DCP)**. On some aircraft a single unit **display select panel (DSP)** is used to

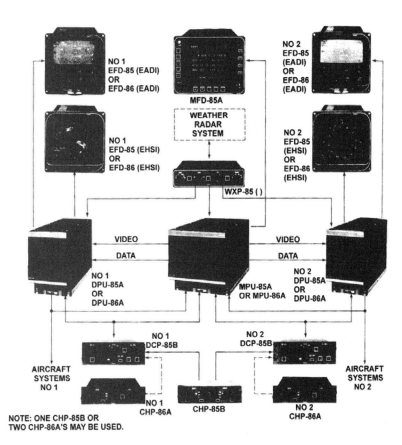

Fig. 3-2 Basic components of an EFIS. (Collins Divisions, Rockwell International)

replace the CHP and the DCP. Information from the radar feeds all three EFIS processor units, the weather radar system (center of figure 3-2) is not part of the EFIS system.

The displays

The **Electronic Flight Displays (EFDs)** are used to provide primary flight and navigation data on a high resolution color CRT. The EFIS-85 system employs a CRT approximately five inches measured diagonally; the EFIS-86 system employs a CRT approximately six inches measured diagonally. The EFDs are located directly in front of the pilot and copilot. The EFDs are typically arranged in a vertical fashion (EADI on top and EHSI on the bottom). The **EHSI** is an **Electronic Horizontal Situation Indicator** and the **EADI** is an **Electronic Attitude Director Indicator**. It should be noted some single tube systems are also in use. On a single tube EFIS the display is typically used as an **EHSI**. As shown in figure 3-3, the EHSI will provide three different formats: **Rose, ARC,** and **MAP**. In each of these formats the EHSI shows basic navigational data, such as compass and selected heading, navigational source and course deviation, selected course and track, and ground speed. Weather radar information can also be imposed on the EHSI in the approach or enroute modes. The EADI operates in one format as shown in figure 3-4. The EADI provides information such as heading, vertical speed, radio altitude, pitch and roll data, aircraft attitudes, and autopilot status.

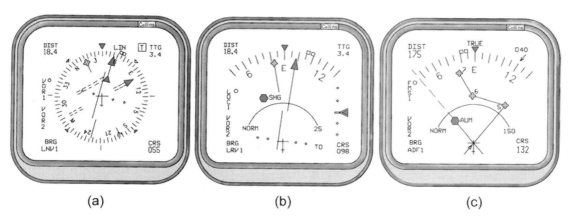

Fig. 3-3 Typical EHSI display. (a) Rose format; (b) ARC format; (c) MAP format. (Collins Divisions, Rockwell International)

Fig. 3-4 Typical EADI display. (Collins Divisions, Rockwell International)

In general, all EHSI and EADI formats are very similar. However, different EFIS manufacturers will present information in slightly different locations on the display or use different symbols or colors for some basic data. The EADI and EHSI shown in figure 3-5 represent the formats used on the Bendix/King EFIS-10. From this figure one can see differences in the display formats between the Bendix/King EFIS-10 and the Collins EFIS 85/86 shown in figures 3-3 and 3-4.

The Collins EFIS-85/86 EFDs consist of a CRT assembly, deflection and video amplifiers, a phosphor protection monitor, and a high-voltage power supply. As shown in figure 3-6, these basic functions are all contained on various cards within the EFD. The inputs to the EFD include: 1) a power monitor from the EFD circuit breaker, 2) a video intensity (VINT) signal from the DCP, 3) video information from the DPU, and 4) power inputs from the DPU.

The processor unit

The **DPU 85/86** is a line replaceable unit approximately 12.5in (316mm) x 7.6in (194mm) x 5in (128mm) and weighs 13.6lbs (6.2kg). The DPU receives inputs from various aircraft systems and other EFIS components. This information is processed within the unit to produce the deflection and video signals needed by the displays. During normal operation of the EFIS-85/86 each DPU is used to produce the signals necessary to drive one EADI and one EHSI. During isolated backup mode, one DPU can power four EFDs. In isolated backup mode both the pilot's and copilot's displays will show identical information since the same processor is used to drive all four displays.

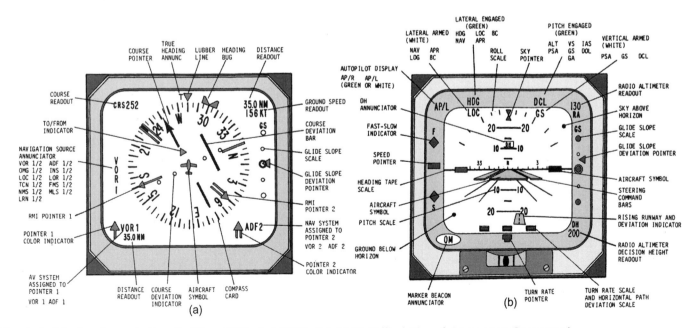

Fig. 3-5 Display formats Bendix/King EFIS. (a) EHSI; (b) EADI (Allied-Signal Aerospace Company)

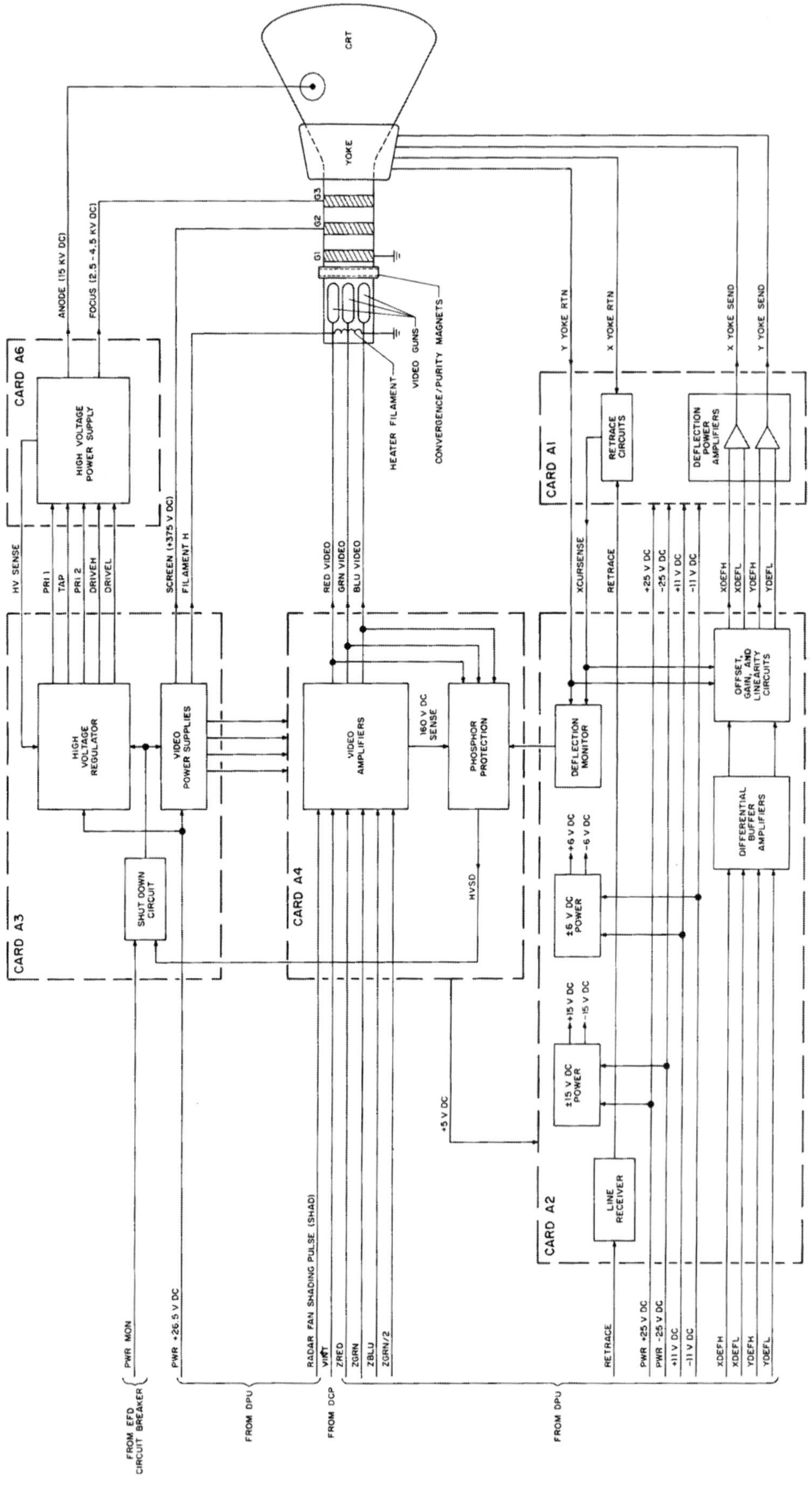

Fig. 3-6 Electronic flight display block diagram. (Collins Divisions, Rockwell International) For Training Purposes Only.

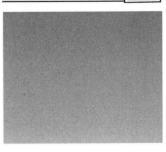

As seen in the block diagram of figure 3-7, there are eight separate cards within the DPU. The cards each have a specific function: 1) input/output interface (card A7 & A8), 2) display processor (card A4), 3) symbol generator (card A5), 4) ADI multiplexer (card A6), 5) power supply (cards A3 & A9), and 6) the relay card (A10).

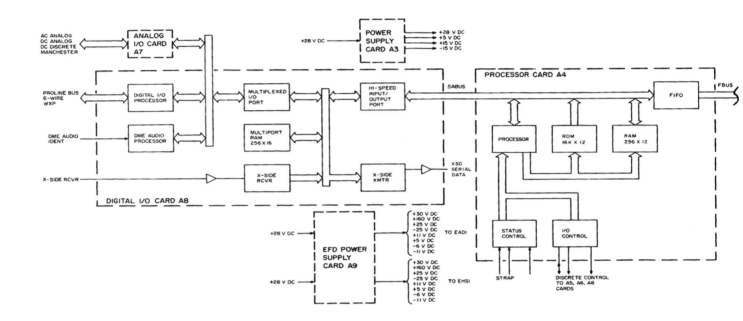

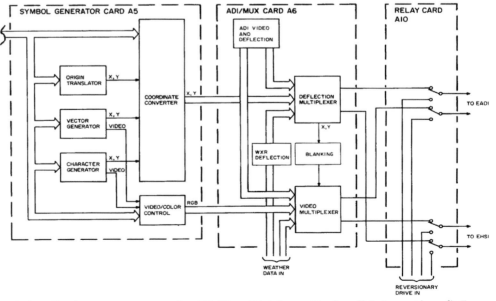

Fig. 3-7 Block diagram of the display processor unit. (Collins Divisions, Rockwell International) For Training Purposes Only.

	INPUTS	
AC Analog Inputs	*Desired track or track angle error & drift angle *Drift angle/bearing to waypoint *VOR/RNV bearing *Roll angle	*Pitch angle *Heading *ADF bearing
DC Analog Inputs	*Radio altitude *Localizer or MLS azimuth deviation *Glideslope, MLS glide path, or VNAV deviation *Crosstrack deviation *Roll steering command	*Pitch steering command *AOA/Fast/slow deviation *ADF bearing *Vertical deviation
DC Discrete Inputs	*To.From *Instrument flags *Instrument modes *EFIS and I/O straps *Marker beacon *FCS modes *CAT II request *Localizer mode	*VOR test *DME hold *DG mode *Master warn reset *Comparator reset *Comparator reset *Accelerometer monitor
Serial Digital Inputs	*DME distance *Long-range navigation information *DPU cross-side data *DCP/DSP data *Air data computer information	*Vertical navigation information *FCS modes *Altitude/heading *WXPdata *REF airspeed
Audio Identification Input	*DME audio identification	
AC Discretes	*Marker beacon (inner, middle, outer)	

Table 3-1   Display Processor Unit (DPU) Inputs. For Training Purposes Only.

	OUTPUTS	
AC Analog	*Heading error *Course datum	
DC Analog	*Crosstrack deviation	
DC Discrete	*Comparator warns *CAT II lights *Master warn *To/From *Back localizer	*Modes/status/flags *Over temperature *Decision height *Decision height *Radio Altimeter test
Serial Digital	*DCP/DSP control *FMS-90 WPT load	*Cross-side DPU data *FCS data

Table 3-2   Display Processor Unit (DPU) Outputs. For Training Purposes Only.

The input/output cards receive both AC and DC analog signals, DC and AC discrete signals, audio identification signals from the DME, and serial data from the cross-side DPU. Table 3-1 shows a list of DPU inputs. The I/O cards convert this data into a serial data signal that is sent to the processor card. The DPU also processes a variety of other data that is used by various system, such as the autopilot (see table 3-2).

The display processor card receives inputs from the I/O cards through a multiport RAM. The display processor uses discrete 12-bit TTL circuitry to control the operation of the symbol generator and ADI/MUX cards. The symbol generator produces the information needed to "draw" the picture on the EHSI and EADI CRTs. The symbol generator is divided into two separate functions, the character generator and the vector generator. A **character generator** is used to produce letters, numbers, and symbols. A **vector generator** is used to draw lines on the CRTs. The ADI/MUX card is used to produce the deflection and video signals needed for the EADI sky and ground background displays. There are three separate power supplies in the DPU contained on two power supply cards. One power supply (card A3) is used for DPU operations. The two power supplies on card A9 are used to power the EFDs. During troubleshooting of the EFIS 85/86 remember, the EFDs receive power from the DPU, the EFDs do not contain their own low-voltage power supplies. The A10 relay card provides switching to the EFDs during partial failure of the EFIS.

The multifunction processor

The **multifunction processor unit (MPU)** is used to process system data and provide video signals to the MFD (multifunction display). The MPU also stores checklist information which can be recalled and displayed on the MFD. The MPU has input/output capabilities to allow for redundant monitoring of various systems. The MPU can be thought of as a safety net, or backup processor for the DPUs. In the event of a single or dual DPU failure, the MPU can provide the deflection and video signals to the EFDs. The MPU cannot supply operational power to the EFDs.

The multifunction display

As the name implies, the **multifunction display unit (MFD)** performs multiple functions. The MFD provides weather radar data, navigational maps, and page data for checklists and maintenance functions. The MFD contains a CRT display surrounded by several control switches, making the MFD slightly larger than the EFDs (see the top portion of figure 3-2). The mode selector, line select keys, and display function controls for the MFD are installed on the face of the unit. Unlike an EFD, the MFD contains its own power supply circuitry. The MFD is typically located in the center section of the instrument panel so it is accessible by both pilots. It should be noted, there is no backup display for the MFD. If this unit should fail, most MFD functions would be lost; however, weather radar and some navigational data can be displayed on the EHSI.

The control panels

There are three types of control panels available for the EFIS-85/86; the CHP, DCP, and DSP (see figure 3-8). Any given EFIS will be equipped with one DSP or a combination of one CHP and one DCP for each DPU in the system. The specific configuration is a function of aircraft/EFIS design and will be shown under the system architecture later in this chapter. The control panels are used to select various EFIS display configurations, navigational data selection and other similar parameters. The DSP provides selection of various navigational parameters. As the pilot activates the various EFIS controls, the DSP produces a digital signal for transmission to the MPU and DPU(s). The MPU or DPU then changes this information on the appropriate display(s).

If a CHP and DCP are used instead of a single DSP, the control functions are distributed between two separate panels. As seen in figure 3-9, the CHP produces signals that are sent to the DCP. The DCP scans the front panel switches of the DCP and CHP for pilot inputs. The DCP then processes the data and sends a serial data transmission to the DPU and MPU.

Fig. 3-8 EFIS control panels. (a) CHP; (b) DCP; (c) DSP. (Collins Divisions, Rockwell International)

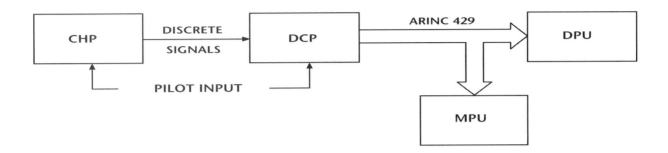

Fig. 3-9   EFIS control panel block diagrams. (Collins Divisions, Rockwell International)

Collins EFIS-85/86 System Architecture

The Collins EFIS-85/86 can be installed in several different configurations as specified by the type of aircraft, and owners request. The major configuration changes are found in the number of display tubes and type/number of control panels installed. The simplest EFIS-85/86 system is shown in figure 3-10. In this system, there are two EFDs (one EADI and one EHSI) controlled by a display select panel. The weather radar system (shown inside the dashed box) is an optional system separate from the EFIS. This two tube EFIS could also be controlled using one DCP and one CHP.

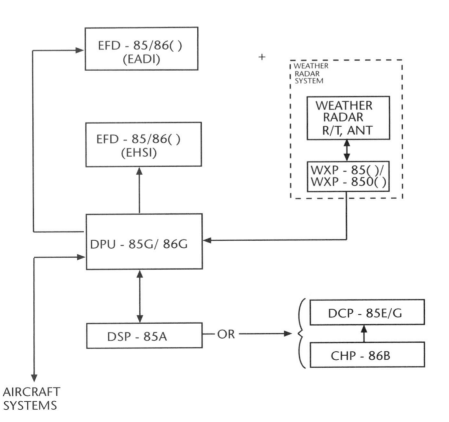

Fig. 3-10   Block diagram of a 2-tube EFIS. (Collins Division, Rockwell International)

Figure 3-11 shows the different types of data that are transmitted to and from the various EFIS components. The DPU receives six different types of data from the various aircraft systems and digital data from the DSP. The DPU receives control signals from the DSP through a CSDB digital data link. Weather data is sent to the DPU from the weather radar receiver/transmitter unit via a digital data bus. The DPU processes the data and sends video control signals to the EFDs. The DPU also sends all low voltage DC power signals to the EFDs.

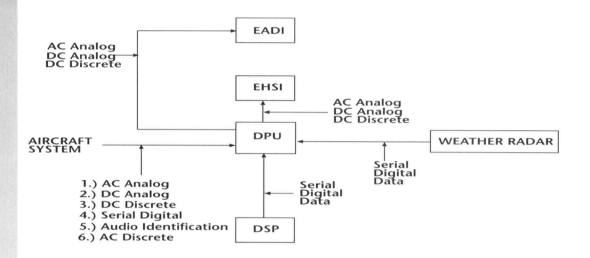

Fig. 3-11 Data interference to various EFIS components.

A **three** tube EFIS consists of two EFDs and one MFD (see figure 3-12). A multi-function processor unit (MPU) is needed to drive the MFD. This arrangement provides system redundancy since the MPU can be used to supply video data to the EHSI and EADI in the event of a DPU failure.

A **four** tube EFIS employs two EADIs, two EHSIs, and two DPUs. As seen in figure 3-13, the two DPUs "cross talk" video information and system data for comparison purposes. This system provides redundancy through the ability of one DPU to drive two EHSIs and two EADIs with identical displays in the event the opposite side DPU fails.

The **five** tube EFIS contains four EFDs, one MFD, two DPUs, and one MPU (see figure 3-14). This system provides the maximum redundancy of the Collins EFIS-85/86 series. With three processor units (two-DPUs, and one-MPU) the system is extremely unlikely to experience complete EFIS display failures. In a five-tube system each DPU will use the MPU as back up for video information in the event of a failure. As mentioned earlier, the relay cards in the DPUs control the output signals to the EFDs. Figure 3-15 shows the ability for the relay card to receive data from the MPU or on-side DPU.

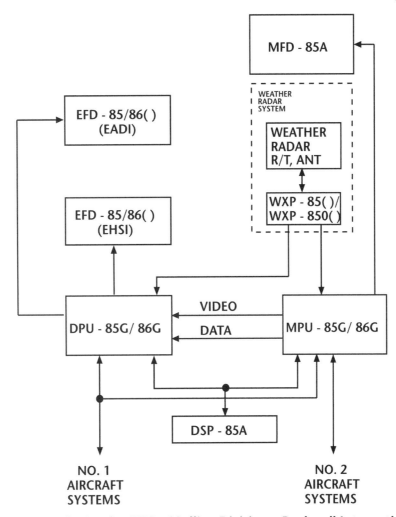

Fig. 3-12   Block diagram of a 3-tube EFIS.   (Collins Divisions, Rockwell International)

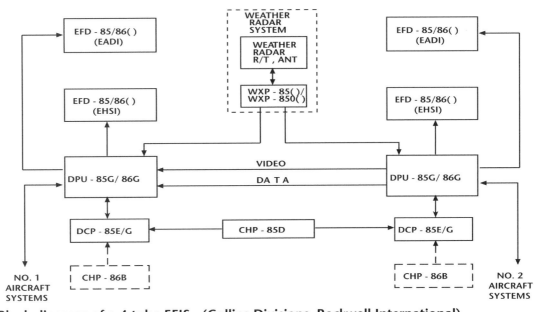

Fig. 3-13   Block diagram of a 4-tube EFIS.   (Collins Divisions, Rockwell International)

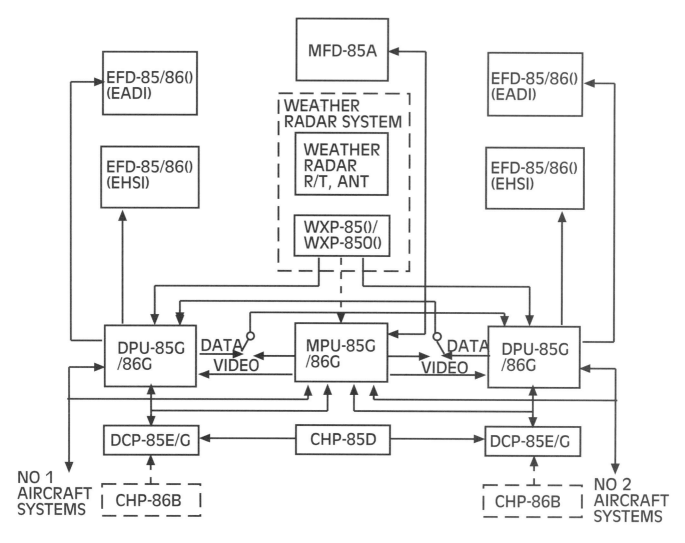

Fig. 3-14   Block diagram of a 5-tube EFIS.   (Collins Divisions, Rockwell International)

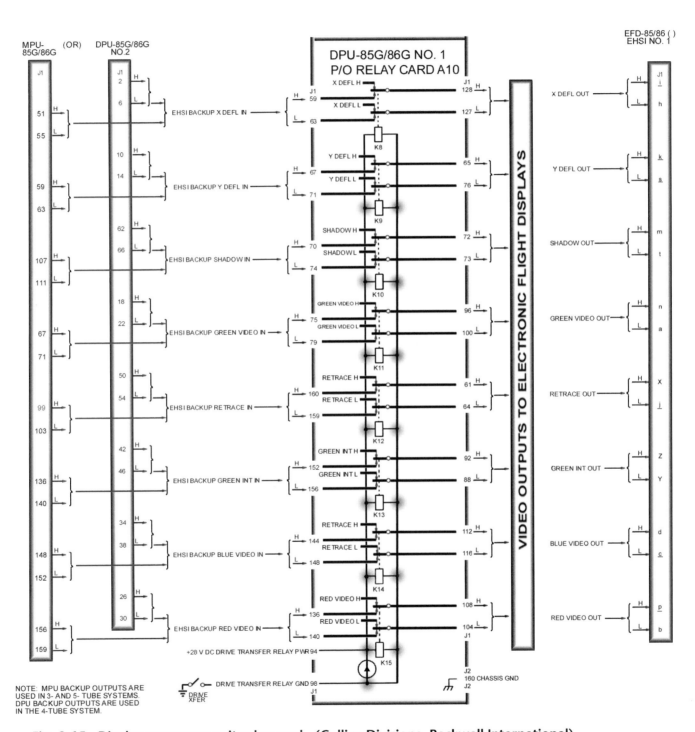

Fig. 3-15   Display processor unit relay card. (Collins Divisions, Rockwell International)

Collins EFIS-85/86 System Operation

The EFIS-85/86 is capable of a multitude of flight related functions that require a thorough understanding of aerial navigation. The materials presented here will provide a brief summary of the "flight" related controls and concentrate on operation of the system for maintenance purposes. Initial operation of the EFIS is quite simple; the system is active at any time the appropriate electrical busses are powered.

In most installations the EFIS equipment will be connected to the aircraft's avionics bus(ses). This will allow for isolation of EFIS during engine starting and other times when the ships electrical power is considered unstable. The system is capable of withstanding voltage spikes of +/- 600 volts at a rate of 50 spikes/minutes. However, it is considered that exposure to voltage spikes at any level and rate can compromise system reliability. The voltage required for normal operation of the EFIS-85/86 is between **29.5** Vdc and **22.0** Vdc. In emergency operations the EFIS will function with a voltage as low as **18.0** Vdc. The entire EFIS will require between 13.5 and 22.2 amps @ 27.5 Vdc for operation.

During operation the DPU, MPU, and EFDs require forced air cooling. If cooling is not adequate, EFDs will blank out and processor units may be damaged. The MPU has the capability to drive an *overheat* annunciator installed on the aircraft's panel. During any EFIS operation, be sure the proper cooling is provided, and discontinue operation if temperature limits are exceeded.

Configuration strapping

Many of the features available for display by EFIS are controlled through strapped inputs. *Strapping* is a term given to an electrical input which is permanently connected to a digital 1 or digital 0 signal. Strapping connections are typically installed any time a system is modified and during initial system installation. EFIS configuration strapping is used to determine various display, input/output signal formats and other system parameters. A specific system configuration is achieved by connecting a strap connection (or multiple connections) to ground or by leaving it (them) open. On the Collins EFIS 85/86, strapped inputs open (not grounded) are considered logic 1. Strapped inputs connected to ground are considered logic 0. For example, the DPU may be strapped to allow for operation of GPS equipment. If DPU plug 2 pin 136 is connected to ground (set to 0), and pin 116 is open (set to 1) it will allow for EFIS display of GPS operation. The specific configuration/operation of any EFIS is always a function of the strapped inputs.

It should be noted that logic 1 and logic 0 can be defined in various ways for different aircraft or different systems. For example, a logic 1 may be obtained as +28Vdc, open , or connection to ground. The voltage value for strapped inputs is determined during the design of any system and memorized by the operational software.

During maintenance and troubleshooting, strapped inputs should not be overlooked. If a system fails to operate properly, the fault may be caused by a poor strap connection. If a strapped input changes from binary 1 to binary 0, the EFIS processor may not recognize a given input signal, or may not process the signal correctly. As in the previous example, if plug 2 pin 136 (strapped connection for GPS) comes loose, the processor will no longer be configured for GPS display. Even if all systems are operating properly, the EFIS will not display GPS information until the strap is repaired.

Display control panel (DCP) operation

The DCP provides dimming for both the EHSI and EADI along with other controls used to select EHSI format and functions. Refer to the controls as numbered in figure 3-16 for the following discussion on the operation of the DCP. The *DH SET* (#1) is used to set the decision height displayed on the EADI. The *RDR* push-button (#2) is used to bring the radar display to the EHSI in the ARC or MAP formats. The *HSI/ARC/MAP* three-position rotary switch (#3) allows the pilot to select the EHSI display format. The *HSI* position displays the full compass rose similar to a conventional HSI. The *ARC* position displays approximately 80 degrees of an expanded compass rose on the HSI. The *MAP* position is similar to the *ARC* display except it also includes VOR and waypoint symbols. The three different modes are shown in figure 3-3.

The *SEL/RNG* knob (#5) is used to select the range of the ARC or MAP displays. Available ranges are 5, 10, 25, 50, 100, 200, 300, and 600 nautical miles. The *CRS ACT* button (#4) will allow for selection of the active course. When the *CRS ACT* button is pressed, the *SEL/RNG* knob is used to make the appropriate selections from a menu on the EHSI display. The *BRG* button (#6) allows the operator to set the bearing pointer using the *SEL/RNG* knob. The *CRS PRE* button (#7) is used in conjunction with the *SEL/RNG* to select the second (standby) navigational source parameters.

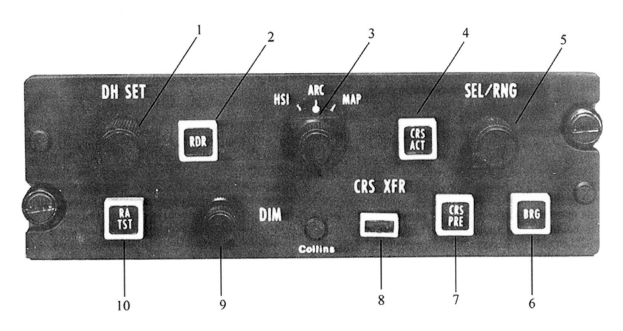

Fig. 3-16   DCP panel. (Collins Divisions, Rockwell International)

The *CRS XFR* button (#8) is used to select the preset second (standby) course. The *DIM* (#9) control is used to adjust the intensity of the EHSI and the EADI. There are two concentric controls used for dimming, each changes the brightness of one of the displays. During troubleshooting be sure these controls are in the correct position if the EFDs are not illuminating. The *RA TST* push-button (#10) is used to initiate the Radio Altimeter test sequence. If the system is operating normally the EADI should display a specified altitude (typically 600ft). The *RA TST* button is also used to control the BITS mode troubleshooting function, as will be discussed later.

Course heading panel (CHP) operation

The CHP is accessed by the flight crew to select course and heading functions. The CHP sends all data to the DPUs and MPU through the DCP. Refer to the controls as numbered in figure 3-17 during the following discussion of the CHP operation. It should be noted that this CHP is used to control a single EHSI; in a 4-tube EFIS two CHP-86Bs or one CHP-85D (a dual control version) would be required. The *HDG* knob (#1) is used to position the heading cursor on the HSI. The *PUSH /HDG / SYNC* button (#2) is used to synchronize the airplane heading and the heading cursor under the lubber line on the EHSI. The *NAV /DTA* button (#3) provides the displays of time-to-go, ground speed, elapsed time, and wind information in the upper corner of the EHSI. The *CRS /CNT* button (#4) determines whether the course knob (#5) is controlling the active or preselect (standby) course. The *CRS* knob (#5) is used to control the course arrow on the display. The *PUSH CRS DIRECT* is located inside the *CRS* knob and is used to zero the VOR course deviation by rotating the course arrow on the EHSI. The *ET* button (#7) controls the elapsed timer functions.

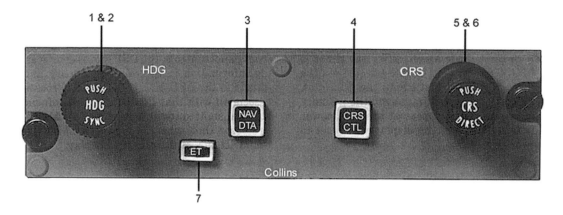

Fig. 3-17   A Course heading panel (CHP).   (Collins Divisions, Rockwell International)

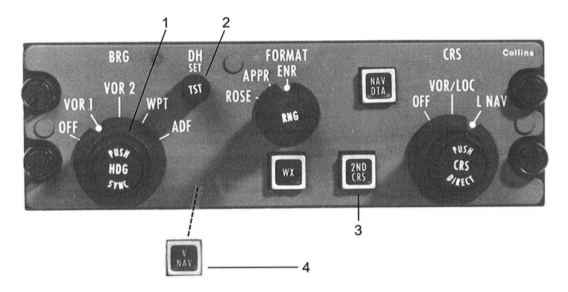

Fig. 3-18   A Display Select Panel. (Collins Divisions, Rockwell International)

Display select panel (DSP) operation

On many EFIS-85/86 systems, the DSP is used to replace a DCP/CHP combination. For the most part, the controls found on the DSP are similar to those of the DCP and CHP. Reference Figure 3-18 during the following discussion for some of the differences. Switch #1 is used to select a given navigational source. The five possible selections are OFF, VOR 1, VOR 2, waypoint, and ADF. The radio altimeter test switch (#2) is located in the center of the decision height knob. The *2ND CRS* push-button (#3) replaces the *CRS /CNT* button on the CHP. The *VNAV* push-button (#4) is used to allow the presentation of vertical navigation data on the EHSI.

Multifunction display (MFD) operation

The multifunction display (MFD) is used to access a variety of page data, as well as display weather and navigational maps. The controls located on the perimeter of the MFD are used to control the functions of the MFD (see figure 3-19). The *PWR* (power) switch in the upper right corner of the MFD is used to turn on/off the display. The *RDR* (radar) button will allow display of the weather radar data on the MFD (see figure 3-20). The *NAV* allows the navigational map data to be displayed (see figure 3-21). Both navigational and radar data can be selected simultaneously. The *RMT* (remote) button allows the MFD to display page data from four different remote sources. The *PGE* (page) and *EMG* (emergency) buttons allow for the selection, control, and data entry of up to 100 pages of "checklist" information (see figure 3-22).

Fig. 3-19 A multifunction display.
(Collins Divisions, Rockwell International)

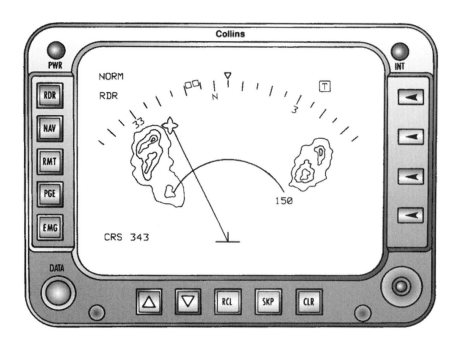

Fig. 3-20 Weather data shown on the MFD.
(Collins Divisions, Rockwell International)

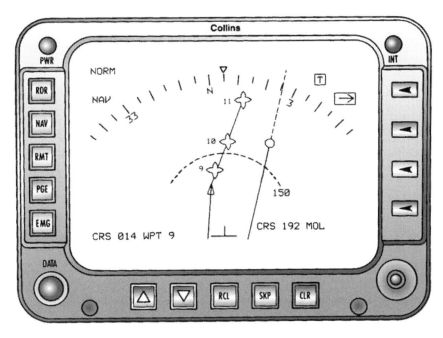

Fig. 3-21 Navigational data shown on a multifunction display.
(Collins Divisions, Rockwell International)

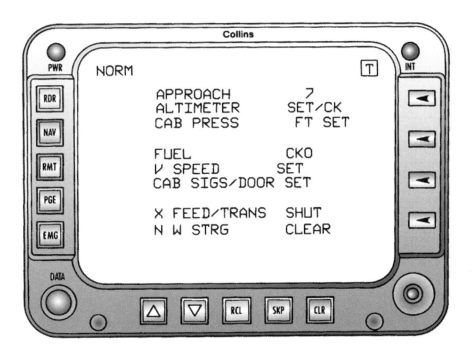

Fig. 3-22   MFD showing "PGE" (checklist) data.
(Collins Divisions, Rockwell International)

In the bottom left corner of the MFD is an electrical jack labeled *DATA*. This jack is used to connect the remote data programmer for page data entry. The data programmer will be discussed later in this chapter. The *INT* (intensity) control is used to adjust the brightness of the MFD display. On the right hand side of the MFD are four display select keys (DSKs). The labels that appear on the CRT during various MFD operations identify the push-button DSKs functions. Pressing the appropriate key will select a given menu function as displayed on the MFD (see figure 3-23).

The joystick found in the lower right corner of the MFD is used in the NAV, RMT, PGE and EMG modes. In the NAV mode the joystick is used to aid in the entry of waypoints. In the RMT, PGE, and EMG modes the joystick is used to scroll through pages or chapters of data. The *advance* and *reverse* arrows (located on the bottom left side of the MFD) are used to move a cursor though the list of page data. As the cursor is moved through the data field, each respective line changes from yellow to white to indicate the item on that line has been completed. The *RCL* (recall) switch brings back previously viewed data while operating in the PGE or EMG modes. The *SKP* (skip) mode moves the cursor past a given line on the data page. The *CLR* (clear) key resets all PGE and EMG data page lines to yellow.

Abnormal system operation

The EFIS equipment can experience two basic types of failures: 1) **failure causing a comparator warning**, or 2) **failure of a system element**. A **comparator warning** occurs whenever two signals compared by the processor units disagree. A comparator error will occur only in systems that employ two or more processors (i.e. 1-DPU and 1-MPU, 2-DPUs, or 2-DPUs and 1-MPU). Comparator monitoring is performed by each DPU during normal operation. Both DPUs convert on-side signals received from aircraft systems into digital data. This data is then stored in memory and also transmitted to the opposite-side DPU. Since both processors now have the data from redundant aircraft systems, a comparison can be made by the DPU circuitry. If the processor receives two input signals that disagree, a comparator flag is displayed on the EFDs. A comparator error will show up as a 2, 3, or 4 letter system identifier in a small box on the display (see figure 3-24). Comparator errors are displayed in yellow. In the event of a single DPU failure, the MPU is capable of comparator monitoring. The MPU would then compare its input signals with the operable DPU. Once again, if the comparison shows a disagreement a fault flag will be displayed.

In most aircraft, an installer supplied (i.e. not part of the EFIS equipment) comparator annunciator is used to alert the flight crew to a comparator disagree fault. A master warning light is also installed on some systems for both the pilot and copilot. The master warning light will most likely be a light/switch assembly that illuminates for all EFIS warnings. The switch is used to reset the warning signal activated by the DPU or MPU. Momentarily pressing the reset switch will extinguish the master warning. If the fault still exists, the flashing EFD warning message(s) will change to a steady display and the comparator annunciator will also remain lit. If the fault no longer exists after the reset is pressed, all warning messages will be extinguished.

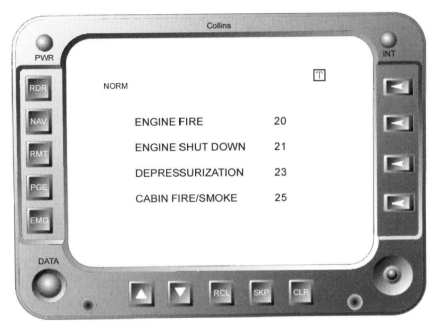

Fig. 3-23   MFD showing a typical menu function. (Collins Divisions, Rockwell International)

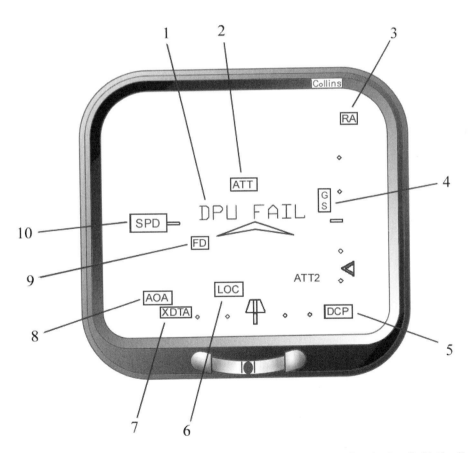

Fig. 3-24   Typical EADI fault flags; 1) Display processor unit (red) 2) Attitude (red) 3) Radio Altimeter (red); 4) Glide Slope (red); 5) Display control panel (red); 6) Localizer (red); 7) cross-side data bus (red); 8) Angle of Attack (red); 9) Flight Director (red); 10) Speed Command (red). (Collins Divisions, Rockwell International)

There are a variety of warnings that can be displayed by EFIS. The master warning is activated and the EFDs will display a specific message related to the failure. If the aircraft is approved for category II landing capabilities a separate CAT II annunciator will also be installed. Each system failure will have a specific message on one or both EFDs.

Failure of an EFIS **system element** can be caused by a loss of data from an aircraft system which reports to EFIS or it can be caused by a loss of one or more of the EFIS components. In either case, the warning associated with the fault will typically create a red warning message on the appropriate EFD. An example of the system element fault flags on the EADI is shown in figure 3-24. Fault flags are displayed in red. (Comparator errors are displayed in yellow.) EFIS **fault flags** are any indication displayed on one or more of the EFDs related to a complete or partial system failure. If a failure occurs in a system that feeds EFIS, the flag will be displayed until the fault is repaired. If one of the processor units fail, the flag *DPU FAIL* or *MPU FAIL* will be displayed on the CRT for five seconds. After five seconds the display will go blank except for the failure message. In this case, the pilot can activate a transfer switch to regain the display driven by another processor.

A **reversionary mode** is available on EFIS to allow for system operation in the event of a component failure. The reversionary switches are supplier installed (i.e. not part of the EFIS equipment) and should be located in some area accessible to both pilots. The reversionary switches can be installed to bypass failures of the EFDs, DPUs, MPU, DCP, and/or the DSP. Switching of input data, such as attitude and compass information, is also an available option.

In the event of a single EFD failure, a reversionary switch can be used to transfer the EADI to the EHSI or visa versa. In either case, the information typically shown on both displays will be compacted and displayed on the operable EFD. The EFIS compact mode is called the **composite format** (see figure 3-25). Another option allows external switching to move the full format EADI to the EHSI; the EHSI is then transferred to the MFD.

In the event of a DPU failure, the EADI and EHSI driven by that DPU will display the flag *DPU FAIL* in red. At this time the pilot would activate the reversionary switch to allow the MPU to drive the EADI and EHSI. In this configuration, no degradation of the EADI and EHSI display occurs. The MFD will only be able to display the EHSI or a composite format. The MFD controls (except power and brightness) are inactive. No backup is provided for an MFD failure. As usual, navigation and weather data are available on the EHSI; however, checklist data is lost.

On a 4- or 5-tube EFIS an external switch can be used to select a cross-side control panel if a DSP or DCP fails. If the cross-side DSP is selected, both DPUs are controlled by the selected DSP. Whenever a cross-side control panel is selected, the appropriate message *XDSP* or *XDCP* will be displayed on the EADI in yellow.

The EFIS-85/86 systems have the capability to allow for reversionary switching of attitude and heading information sent to the MPU or DPUs. The reversionary switches

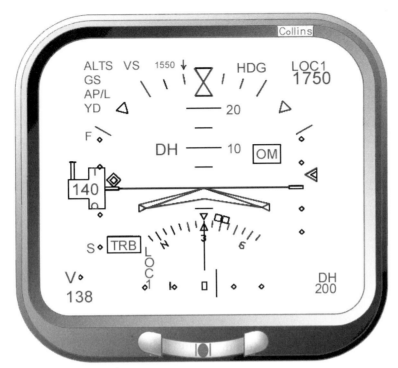

Fig. 3-25   EFIS composite format combining information from ADI and HIS on one display. (Collins Divisions, Rockwell International)

are external to EFIS, installed in an accessible area of the flight deck. Switching is typically done only in the event of an on-side failure of attitude or heading data. If the pilot's attitude data source fails, the pilot can retrieve attitude data from the copilot's source, and vise versa. If the pilot's on-side heading data source should fail, the information can be received from the copilot's source; and visa versa. The pilot's on-side sources are called *ATT1* and *HDG1*. The copilot's on-side sources are called *ATT2* and *HDG2*. Whenever receiving cross-side attitude data, the appropriate message will be displayed in yellow. For example, if the pilot has selected cross-side attitude data, the message *ATT2* will be displayed on the pilot's EADI.

Programming the MFD checklist data

The MPU can store up to 100 pages of data each containing 12 lines of up to 20 characters each. Remember the MPU stores the data that is displayed on the MFD. This data is used for checklists accessed by the *PGE* (page) or *EMG* (emergency) buttons on the MFD (see figure 3-19). Checklist data programmed into the MFD is stored in a nonvolatile memory. The Collins **remote data programmer** is used to install new checklist data or modify an existing checklist.

As seen in figure 3-26, the remote data programmer (RDP-300) is a calculator-type unit with an alphanumeric keyboard. The data programmer plugs into the jack found in the lower left corner of the MFD. To program the MPU, apply system power and turn on the MFD. Connect the data programmer, and allow 15 seconds warm up time. Pressing the *PGE* button will access the page checklist; pressing *EMG* will access the emergency checklist. The data loader is then used to enter the appropriate data to each page of the checklist. The EFIS installation manual offers detailed instructions for data entry.

TP6-0889-017

Fig. 3-26   Remote Data Programmer (RDP-300), used to enter checklist data into the MPU memory (Collons Divisions, Rockwell International)

Another method is available for entering data into the MFD checklist memory. EFIS checklist data can be entered into a checklist **entry unit** using a personal computer; then transferred to the MFD. Figure 3-27 shows a CEU-85/85A checklist entry unit. The CEU-85 is interfaced with Apple computers through a Collins Pro Line II interface card. The CEU-85A interfaces with a personal computer using the computer's RS-232 port. Using the *Checklist Editor* software the appropriate checklist can be made or an existing list modified. The checklist is then down-loaded into the checklist entry unit. The checklist entry unit is then plugged into the data jack of a powered MFD and the information is transferred from the entry unit to the MFD. It requires approximately 11 minutes to complete the data transfer. To avoid loosing data, be sure there are no power interruptions to the system during data loading.

TP6-4650-017

Fig. 3-27   The CEU-85/58A checklist entry unit. (Collins Divisions, Rockwell International)

EFIS-85/86 Maintenance

The Collins EFIS-85/86, like most other EFIS equipment, requires very little routine maintenance. Most of the maintenance is performed on an as needed basis (i.e. after a fault occurs). However, to ensure long life of any EFIS, it is essential that the components receive proper cooling. To ensure proper cooling, all air passages and equipment housings must be clean and free of dust and lint. It is good practice to include the cleaning of all cooling fans, and equipment housings in the aircraft's scheduled maintenance program. This will improve the reliability and longevity of all EFIS components.

Technicians should always take care when cleaning the EFIS display units. The face of the CRT is coated with a special non-glare film. This film is extremely thin and can easily be damaged by abrasive cleaners or dirty cleaning rags. Proper cleaning should be accomplished with a denatured alcohol cleaner and a soft dirt/lint free rag. Lens cleaning tissues may also be used for cleaning the CRT.

Proper brightness of the EFIS CRTs should be confirmed at regular intervals. The Collins Maintenance Manual suggests the formal test be done at approximately 5000 hours of EFIS operation; however, a quick visual check can easily be performed during routine inspections. A quick check is done by simply noting the brightness of the display and adjusting the appropriate brightness knob. The display should be easily read in full sunlight with the brightness control set to approximately 3/4 maximum. If the CRT is difficult to read with the control near the maximum position, perform the formal brightness test described below.

The formal brightness test is done using a Minolta LS-100 light meter. The aircraft must be located in an area of low ambient light for this test. The light meter is held in front of the CRT while operating in the appropriate test mode. The proper level of light must be indicated on the meter when the brightness control is turned to maximum. If the proper level is not achieved, the display should be replaced.

Fault monitoring

During normal operation the EFIS-85/86 performs **on-line monitoring** which is a continuous fault monitoring process designed to detect internal and external faults. Each processor in the system (MPUs and DPUs) contains an on-line monitoring software program. The program tests for a periodic *keep-alive* signal generated by the processor. If this signal is not detected at the proper time intervals, it is assumed that the DPU/MPU is not processing data correctly and a fault flag is displayed.

The data busses, which connect the aircraft subsystems to EFIS, are tested using a series of **test words**. The processor initiates a test word that is transmitted through a given data bus then returned to the processor. If the test is not successful, the appropriate fault flag is displayed. **Activity monitors** are used to monitor each EFIS data bus to ensure data transmission at regular intervals. If the activity monitor determines a given subsystem is not transmitting data, the appropriate flag is displayed.

On some systems, the voltage level transmitted to the processor is monitored for proper values. If the voltage is found to be out of limits, a flag is displayed. The I/O section of the processor and the display processors routinely compare their program checksums. If the **checksum monitor** determines an invalid calculation, the DPU/MPU fault flag is displayed. A **RAM monitor** circuit is activated during each power up of the EFIS processors to test the I/O and display processors RAM storage. If the test fails, the flag *DPU/MPU FAIL* will be displayed.

Self-test functions

The EFIS DPUs and MFDs contain a **self-test** software program, which is used to test the health of the system upon request by the pilot/technician. The self-test function is comprised of two modes: **confidence test** and **diagnostics/maintenance routines**. The self-test function is activated through a momentary contact switch supplied during EFIS installation. The confidence test is typically used during preflight to ensure reliance in the EFIS equipment, or during maintenance to verify proper system operation after a repair. Maintenance personnel typically use the diagnostics/maintenance routines during in-depth troubleshooting. The diagnostic/maintenance routines will be discussed under the *BITS mode* section of this chapter.

Pressing the test switch will connect a ground signal to the processor unit (DPU and/or MPU) and initiate the confidence test. During the confidence test, the pilot's EADI pitch and roll values will change by an increment of +10 degrees, +20 degrees respectively. The copilot's EADI will change -10 degrees and -20 degrees respectively. The word *test* and *pitch and roll miscompare* messages are displayed on the EADI while the confidence test is in progress. If the test switch is pressed again for an additional four seconds the pitch, roll and heading values return to normal and all EADI and EHSI test flags are displayed. This *flag* display will continue until the test switch is released.

Troubleshooting Techniques

Whenever troubleshooting any system that is comprised of several LRUs, it is very important to become familiar with the power distribution of the components. The input power to a 5-tube EFIS-85/86 system comes from over a dozen circuit breakers; all must be in operable condition before any troubleshooting can begin. As seen in figure 3-28, the power required to operate the EHSI and EADI comes through the on-side DPU. The DPU receives three separate power inputs, and the EFDs receive a +28 Vdc power monitor signal from the bus. Also make sure adequate power is available for proper operation. A weak battery or poor quality ground power unit may cause EFIS problems which are not system related. The EFIS-85/86 can operate successfully down to approximately 18 Vdc; however, CRT displays may dim and images may shrink at low voltage. Remember, 27.5 Vdc is the recommended voltage.

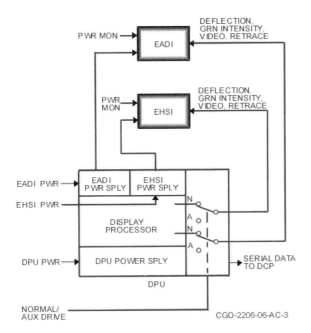

Fig. 3-28 Power inputs to the DPU. (Collins Divisions, Rockwell International) For Training Purposes Only.

Proper equipment cooling is imperative during EFIS operation and troubleshooting. The EFDs, DPUs, MFD, and MPU must be cooled by forced air produced by blower fans located in the appropriate areas. The EFIS-85/86 employs an over temperature monitor circuit in the MPU. If the temperature rises above a 135° C (275° F) the MPU will illuminate an installer-supplied annunciator. If the EFDs overheat, the displays will automatically blank to help cool the unit. A **fan motor module** is available to monitor the speed of the system cooling fans. If the fan motor module detects a low speed cooling fan, a supplier installed annunciator light is illuminated. Keep in mind many aircraft do not have an overheat indicator. If a display goes blank after prolonged operation and functions again when cool, suspect poor cooling. Look for inoperable fans, clogged filters, or disconnected air ducts. And don't forget that birds and rodents tend to build nests inside air inlet ducts.

The strap configuration of the EFIS equipment is also an important consideration during troubleshooting. A specific strap connection (plug and pin number) is defined by the system interconnect diagrams. Figure 3-29 shows a sample of the strapping definitions table for the EFIS-85/86. In this example it can be seen that the specific autopilot model can be selected using three strap connections, the radio altimeter setting requires two strap connections, and the DPU side is determined using one strap connection.

An intermittent or disconnected strap connection can create difficult troubleshooting problems. For example, if the straps which define *autopilot type* should become disconnected the DPU will **not** recognize the data transmitted from the autopilot and a flag will be displayed. The technician may fault the autopilot, the DPU or the associated bus without suspecting the strap configuration. Since the strap is simply a connection to ground (or not) they are typically not a problem; however, the technician should always

EFIS-85B (2/12) and EFIS-86B (2/12) Input Specifications

STRAP INPUTS	UNIT PIN NUMBER	STRAP NUMBER	SIGNAL CHARACTERISTICS
		12.0 DPU/MPU-85G/86G STRAPPING DEFINITIONS	
Autopilot Type	DPU (P2-61,115,121) MPU (P2-1,5,9)1 MPU (P2-4,8,12)2	1 2 3 0 0 0 1 0 0 0 1 0 1 1 0 0 0 1 1 0 1 0 1 1 1 1 1	FCS-80 dual FD Sperry A, FCS-A single FD Sperry B, (Citation) FCS-B APS-86 (dual FD) APS-65 FCS-80 single FD Sperry A, (GIII)FCS-A single FD) APS-85 (single FD)
Radio Alt Type	DPU (P2-133,134) MPU (P2-61,65)1 MPU (P2-64,68)2	4 5 0 0 1 0 0 1 1 1	AL T-50 (0 to 2000 feet) AL T-55 (0 to 2500 feet) AL-101 (0 to 2500 feet) Undefined
DPU Side	DPU (P2-140)	6 0 1 Note 0 = Ground 1 = Open	DPU use only Left DPU (left side I/O annunciated) Right DPU (right side I/O annunciated)

Fig. 3-29 Sample of strapping definitions. (Collins Divisions, Rockwell International) For Training Purposes Only.

be aware that strapping controls a variety of EFIS functions and if a strap is incorrect (or has a loose connection) that portion of the system is inoperative.

Isolating faulty LRUs,

In most cases, troubleshooting EFIS is simply a matter of finding the defective LRU. On 2-tube systems, EFDs can be switched or suspect LRUs can be swapped with the same component "off the shelf" or from another aircraft. Isolating defective LRUs on aircraft with a 4- or 5-tube system can often be done from the flight deck as follows:

Example 1

    IF - the pilot's EHSI has a blank display,

    THEN - the pilot's processors can be switched to drive the copilot's displays.

    IF - the fault shows up on the copilot's EHSI also, the processor is most likely defective.

IF - the fault does not show on the copilot's EHSI, the fault most likely lies in the pilot's EHSI. (Swapping the pilots EFDs would then verify if the fault was in the pilot's EADI.)

Example 2

IF - the copilot's side EADI is blank,

THEN - switch to MPU drive for the copilot's displays.

IF - The EADI operates normally, the copilot's DPU is most likely defective.

IF - The EADI is still blank, the display unit is most likely defective.

Be careful, in this example it is easy to be fooled. Remember the DPU supplies power to both on-side displays. A power supply problem in the DPU could cause a blank EADI. In this case, the EADI would stay blank when driven by either the MPU or the DPU. (The MPU only supplies video signals to the EADI and EHSI; the MPU does not supply power to these displays.)

Example 3   (assumes the aircraft has reversionary switching of attitude and heading data)

IF - the pilot reports inaccurate attitude data on left side EFDs,

THEN - use the reversionary switching to drive the left side displays with the right side attitude sensors.

IF - the problem is corrected, the left side sensors (which normally drive the left side displays) are defective.

IF - the problem still exists, the left side DPU or associated wiring is most likely at fault.

THEN- Swap the left and right DPUs.

IF - the fault moves to the right side displays, the DPU now in the left side (previously in the right side) is defective.

The procedure for reversionary switching of various EFIS LRUs can vary from aircraft to aircraft. It is always best to refer to the aircraft's manuals to verify proper LRU reversionary switching procedures. To aid in troubleshooting, always become familiar with the maintenance manuals for the system in question. The EFIS-85/86 maintenance manual contains a flow chart for isolating defective LRUs (see figure 3-30).

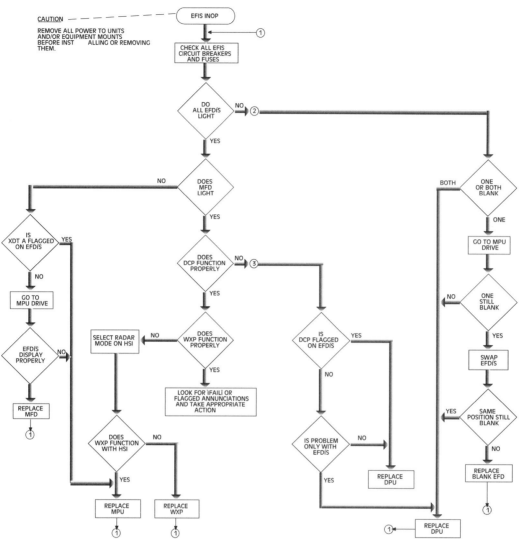

Fig. 3-30 EFIS troubleshooting flow chart. (Collins Divisions, Rockwell International)

If replacement of an LRU is ineffective at repairing a system fault, the system's wiring should be considered a primary suspect. Connector plugs become worn or dirty as the LRUs are removed or replaced. Be sure to check the plugs and sockets carefully. Employ standard techniques for testing continuity of system wiring. Also be sure to test for grounded wiring due to failed insulation or a shorted connector. Oscilloscopes or data analyzers can be used to test digital data signals.

Remember one of the best tools for troubleshooting defective pin connectors is a flashlight, a magnifying glass, and a good pair of eyes. Defective connections are typically discolored, burnt, pitted, bent, or corroded. These traits can often be detected visually. The hardest part for a visual inspection is getting into a position where the connections can be viewed properly. During troubleshooting, if a wiring problem is suspect, remove the LRU and inspect the connections related to the faulty system. First look at the LRU connections; if a defect is found, remember the mating contact in the LRU rack is most likely also defective. For proper repair, the faulty pin **and** mating socket must be repaired. In many cases, a through cleaning will correct the problem. However, when cleaning contacts be extremely careful; it is easy to bend connections and cause greater damage.

BITS mode

The diagnostic/maintenance routines available under the *self-test* function of the Collins EFIS-85/86 are also referred to as the **BITS mode**. The EFIS **BITS mode** is used to access **real time** data from the RAM memory of the DPUs and/or MPUs. As a processor receives data from the various systems and sensors throughout the aircraft, the information is stored in the processor's **multiport RAM (MPR)**. The MPR stores data from both on-side and cross-side systems, some of which is transmitted due to internal EFIS operations. The MPR stores the data in a digital format and updates the information as new data arrives. This data can be displayed on the EFDs or MFD using the EFIS BITS mode. In other words, BITS mode allows you to *look inside* the processor circuitry at the digital signals being processed. This function is a type of built-in data bus analyzer for the EFIS processors. Keep in mind that BITS mode is an in-depth troubleshooting tool and is typically used by only the most experienced technicians.

BITS mode functions

There are five functions available in the EFIS BITS mode; 1) **multiport RAM (MPR)**, 2) **RAM**, 3) **MFD**, 4) **alignment tests**, and 5) **FCS (flight control system) diagnostics**. The MPR is most useful for operational verification after system installation and in-depth troubleshooting. MPR is the BITS mode most often used by line technicians for EFIS troubleshooting. The RAM function allows access to the display processor RAM. The MFD function displays data stored in the MFD page data (checklist) RAM. The alignment tests function displays a test pattern generated by the processor to verify symbol generator and EFD operation. The RAM, MFD, and alignment tests are most often used during *in-shop* repair of an LRU. The FCS (Flight Control System) diagnostics are available only if the aircraft is equipped with the Collins APS-85/86 flight control system. The MPR function of the BITS mode will be discussed here since it relates specifically to flight-line EFIS maintenance.

The MPR function of the BITS mode can be displayed for either the left or right DPU, or the MPU. To activate the pilot's side DPU BITS mode, press and hold the EFIS test switch **and** the radio altimeter test (*RA TEST*) switch on the pilot's DSP or DCP. This will cause the BITS mode to be displayed on the pilot's EFD. The test and RA switches can now be released. To page through the BITS data, press the RA test switch sequentially. The copilot's DPU multiport RAM is accessed similarly using the on-side EFIS test and RA switches.

On a five-tube system, the BITS mode for the MPU memory can be accessed by pressing the pilot's EFIS test switches while simultaneously pressing the copilot's RA test switch. The BITS mode will be displayed on the MFD. To exit from the BITS mode, system power can be cycled, or repeatedly press the RA test switch for the appropriate processor (DPU or MPU).

Interpreting BITS mode

Figure 3-31 shows a typical BITS mode display. BITS mode can monitor 256 EFIS processors inputs. Each input is given a specific address. Address numbers 000 through 127 are parameters from left-side (pilot's-side) systems. Address numbers 128 through 255 are parameters from right-side (copilot's-side) systems. Figure 3-32 shows the relationship of the on-side/off-side locations for the pilot's DPU and the MPU. The BITS mode address numbers are simply identifiers for the given systems or sensors which feed the EFIS processor. The address numbers may also be referred to as *locations*. In other words, each system is assigned a given BITS mode number. Figure 3-33 shows a partial alphabetized list of systems and their associated addresses. This information would be found in the maintenance manual. In this figure, it is shown that address 097 is IAS (indicated airspeed) from the ADC (air data computer). We will use address 097 for the following example on decoding the data field.

To interpret BITS mode data, reference an EFIS maintenance manual while access-

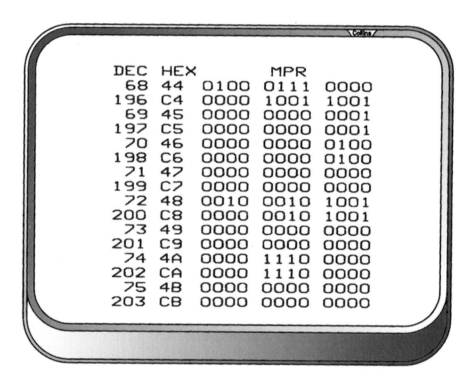

DEC	HEX		MPR	
68	44	0100	0111	0000
196	C4	0000	1001	1001
69	45	0000	0000	0001
197	C5	0000	0000	0001
70	46	0000	0000	0100
198	C6	0000	0000	0100
71	47	0000	0000	0000
199	C7	0000	0000	0000
72	48	0010	0010	1001
200	C8	0000	0010	1001
73	49	0000	0000	0000
201	C9	0000	0000	0000
74	4A	0000	1110	0000
202	CA	0000	1110	0000
75	4B	0000	0000	0000
203	CB	0000	0000	0000

Fig. 3-31   Electronic flight display BITS mode.   (Collins Divisions, Rockwell International)

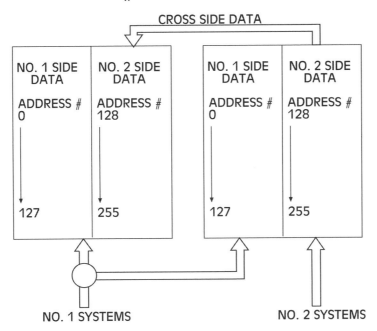

DPU #1 MPU

CROSS SIDE DATA

NO. 1 SIDE DATA	NO. 2 SIDE DATA	NO. 1 SIDE DATA	NO. 2 SIDE DATA
ADDRESS # 0	ADDRESS # 128	ADDRESS # 0	ADDRESS # 128
↓	↓	↓	↓
127	255	127	255

NO. 1 SYSTEMS   NO. 2 SYSTEMS

Fig. 3-32   Diagram showing on-side and off-side date to the DPU and MPU.

EFIS-85B Multiport RAM (MPR) Alphabetical Index

ON-SIDE ADDRESS (DECIMAL)	MPR DESCRIPTION
124	FD ANNS
075	FD MODES
017	FLAG INPUTS, DISCRETES
121	FULL-TIME GS DEVN (DOTS)
120	FULL-TIME LOC DEVN (DOTS)
010	GLIDESLOPE DEVN (DOTS)
1 10	GND SPD (BCD-1000, 100,10), LRN
028	GND SPD (BINARY), ACT LATL SOURCE
01 1	GND SPD (BINARY), DME
121	GS DEVN (DOTS), FULL-TIME
102	HDG ANGLE (ACT)
071	HDG ANGLE (RAW)
031	HDG ERROR (ACT FILTERED)
109	HDG ERROR (ACT), SELECTED
067	HDG ERROR (RAW)
089	HDG, DCP SELECTED
039	HSI MODE DISCRETES
123	HSI VNAV DEVN, SCALED
097	IAS (FIL TERED), ADC
107	IAS (RAW), ADC
1 19	IAS DISPLAY BASE OFFSET IN VGUíS
1 13	IDENT, DME
055	JOYSTICK CURSOR BRG
056	JOYSTICK CURSOR DIST
019	LATL ACT ANNS
1 16	LATL DEVN (DOTS), ACT
020	LATL PRESET ANNS
022	LATL SOURCE TTG (BINARY), ACT
043	LATL SOURCE BRG-TO-WPT (ACT)
009	LATL SOURCE DEVN (VGU), ACT
046	LATL SOURCE DEVN, PRESET
040	LATL SOURCE DIST (BINARY), ACT
028	LATL SOURCE GND SPD (BINARY), ACT

INDICATED
AIRSPEED
(IAS)

Fig. 3-33   Alphabetical list of BITS mode address and system titles.

ing the BITS mode display. It should be noted that different systems may have slightly different display formats; however, the general concepts are the same. Refer to figure 3-34 during the following discussion. The address of each system is shown in the two left columns of the BITS mode display. The left-most column is a decimal (DEC) address; the next column is the hexadecimal (HEX) address. To the right of each address is the data field for that system. The data field shows the digital data of the given system (1s and 0s). Each data field is made up of 12 bits shown in three groups of four. The least significant data bits are shown in the right-most column (bits 3, 2, 1, & 0). The most significant digits are shown on the left.

Address 097 and 225 are displayed on the top two lines (figure 3-34). Remember the DH knob can be used to scroll through different address numbers. Address 097 defines IAS (indicated airspeed) for the right side and 225 is IAS for the left side. Table 3-3 is an example of a typical MPR RAM description table for a Collins EFIS-85/86. This table shows the on-side and off-side decimal addresses and the values of the related binary data field bits. This table must be used to decode the data. In our example, IAS (address 097/225) is shown at the top of the table.

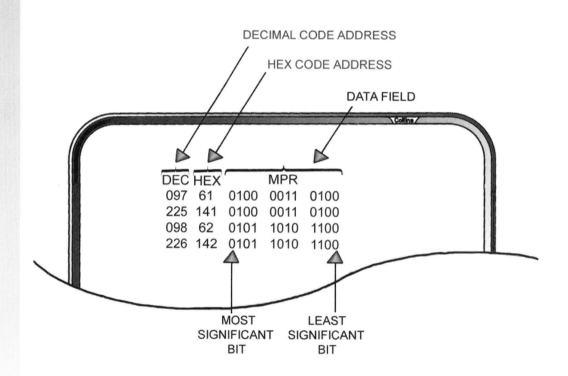

Fig. 3-34 Example of BITS mode display showing indicated air speed (IAS) data.

LOCATION (ADDRESS) ON-SIDE/OFF-SIDE	DESCRIPTION
097/225	IAS 11 - (-512) KNOTS 10 - 256 KNOTS 9 - 128 KNOTS 8 - 64 KNOTS 7 - 32 KNOTS 6 - 16 KNOTS 5 - 8 KNOTS 4 - 4 KNOTS 3 - 2 KNOTS 2 - 1 KNOTS 1 - 0.5 KNOTS 0 - 0.25 KNOTS
098/226	TAS 11 - (-1024) KNOTS 10 - 512 KNOTS 9 - 256 KNOTS 8 - 128 KNOTS 7 - 64 KNOTS 6 - 32 KNOTS 5 - 16 KNOTS 4 - 8 KNOTS 3 - 4 KNOTS 2 - 2 KNOTS 1 - 1 KNOTS 0 - 0.5 KNOTS

Table 3-3 Multiport RAM numerical index. For Training Purposes Only.

To interpret the meaning of a specific data field, simply determine the value of each bit as given in the table. The data for address 097 is 0100 0011 0100 (figure 3-34). Only bits set to binary 1 will have value. Bits set to binary 0 will have no value. Figure 3-35 shows a sample of the decoded data for IAS (address 097). To determine the numerical value of a given data field add all the values of each bit set to binary 1. In this example bits 10, 5, 4, and 2 are set to binary 1. Therefore, the values of 256, 8, 4, and 1 are added to find the total value of the data field (269 knots). The indicated airspeed value calculated by the on-side processor is therefor 269 knots. Decoding information in this manner may be extremely helpful during system troubleshooting. If the pilot reports an indicated airspeed error, the technician can use BITS mode to look *inside* the processor and determine if the correct signal has reached the DPU or MPU. BITS mode is another tool for technicians to use during flight line troubleshooting.

Other MPR data can be analyzed in a manner similar to the one just described. Table 3-4 shows the MPR locations and descriptions for the flight director modes (location 074/202). A binary 1 displayed for a given bit indicates the description is there as labeled. For example, if location 074 bit number 13 is "1" it would indicate an *ELE OUT OF TRIM* (elevator out of trim) condition is detected by the left side flight director. If bit 13 was set to "0" it would indicate an elevator out of trim condition did not exist (i.e. the elevator trim was within tolerance).

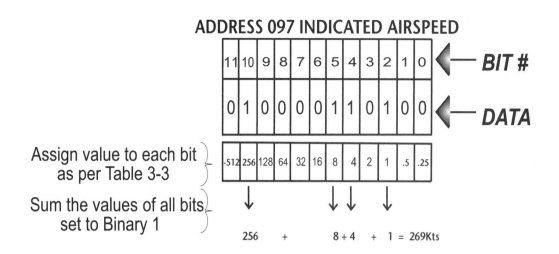

Fig. 3-35 Method used to decode BITS mode data, in this example IAS equals 269 Knots.

DPU/MPU MULTI PORT RAM (MPR) LOCATIONS WITH DESCRIPTIONS

DECIMAL LOCATION NO 1 SIDE/NO. 2 SIDE	HEX LOCATION NO. 1 SIDE/NO. 2 SIDE	SOURCE	DESTINATION	DESCRIPTION
074/202	4A/12A	DIGITAL I/O PROCESSOR	ANALOG I/O PROCESSOR	FLIGHT DIRECTOR MODES APS-65/85/86 MODES 15- RUD OUT OF TRIM 14- AIL OUT OF TRIM 13- ELE OUT OF TRIM 12- YD DISCONNECT 11- AP DISCONNECT 10- V-BAR IN VIEWS 9-STEERING VALID 8-NOT DEFINED 7-HEADING SYNC 6-HEAVY 5-MEDIUM 4-LIGHT 3-YD ENGAGE 2-AP ENGAGE 1-DESCENT 0-NOT DEFINED

Table 3-4 Multiport RAM locations and descriptions. For Training Purposes Only.

Fault isolation using BITS mode

During troubleshooting if the technician suspects a faulty component transmitting to EFIS it can be verified using the BITS mode. Let us assume the pilot reports an indicated airspeed (IAS) disagree message on the EFIS display. (Remember a disagree flag will be displayed in yellow.) The technician could take the following steps:

1) Find the location code for IAS in the MPR alphabetical index (see figure 3-33). The IAS data comes from the Air Data Computer (ADC).

2) Enter the MPR test function for the pilot's side DPU. Page through the data until location 097/225 is displayed.

3) Determine if the binary bits for the right and left side IAS agree or disagree. (Keep in mind the data transmitted from various sensors will most likely have some tolerance. To be considered in disagreement, the data must differ by more than the tolerance allows.)

4) If the values agree, configure the aircraft to various airspeeds (using ground test equipment) and determine if the values agree at various airspeeds. If the bit values for IASs consistently agree, the fault must be in one of the EFIS processors.

5) To determine which processor may be faulty, use the switching procedures previously discussed under troubleshooting techniques.

6) If the right- and left-side IASs disagree as measured in step 4, the fault must be in one of the ADCs (or the related wiring).

7) To determine which ADC is faulty, the IAS values on the ground test equipment can be compared to the values presented on the BITS mode display. The faculty LRU is whichever system (left or right ADC) disagrees with the ground equipment.

The BITS mode is an extremely valuable troubleshooting tool, which can be used for in-depth troubleshooting of EFIS faults. In many respects, the BITS mode is similar to a "built-in" data bus analyzer. Whenever using BITS mode, remember that the data is real time **and** there is no data memory if power to the system is interrupted. If an in-flight fault occurs, the power to EFIS should be left on so a technician can examine the BITS mode or fault flag caused by the fault. If the fault is intermittent the bits data and fault flags will only appear when the fault is active.

Also, keep in mind the BITS mode looks at data stored in the MPR. This data has been manipulated by the I/O section of the processor and is not necessarily identical to what is being transmitted to the processor. In many cases, what is transmitted to the DPU/MPU is an analog signal converted to digital for use in the processor. If you wish to look specifically at what is being transmitted to the I/O section of the DPU/MPU, measure the transmitted signal directly at the DPU/MPU connector plug. Collins breakout box (CTS-9) has all the necessary connections to allow access to the 160 pin connectors found on the back of the EFIS processors. Using a breakout box, the technician can keep EFIS operable and still access the pin connectors on the back of the processors.

BENDIX/KING EFS-10

The Bendix/King EFS-10 was a first generation system designed for corporate and commuter type aircraft. The EFS-10 was the first in a series of stand-alone electronic flight instrumentation systems designed by Bendix/King. The EFS-40 and EFS-50 were the next systems to be developed and incorporated several changes to improve performance, cooling and reliability. The general architecture of the EFS-40/50 remain similar to the EFS-10. Bendix/King chose the acronym EFS for the designator of their systems. (**EFS** stands for **Electronic Flight System**). The EFS-10, -40, and -50 are all similar to the Collins EFIS-85/86 discussed in the previous sections of this chapter. Due to the overwhelming similarities between the two systems, the Bendix/King EFS discussion will be brief and concentrate on the unique features of the system. It should be noted that Bendix/King has gone through several corporate mergers since the development of the EFS-10. Originally the company was a sole entity of Bendix/King Avionics; the company was sold to AlliedSignal Aerospace division. In late 1999 AlliedSignal merged with Honeywell, and at the time this text was written Honeywell was in the process of merging with General Electric. Most people refer to the EFS-10 as a Bendix/King system. Under any name, the EFS-10 and its predecessors the EFS-40/50 are reliable systems, which are extremely popular in a variety of corporate aircraft.

System Description

The EFS-10 can be configured to match the needs of the aircraft, and the owner's bank account. The system is available in a simplified version containing only one processor and two displays (one-EHSI, one-EADI). This system provides very limited backup; however, it is the least expensive. The processor unit for the EFS-10 is called a **Symbol Generator (SG)**. The SG receives inputs from various aircraft systems. The SG processes the data and creates the video and deflection signals, which are sent to the display units (see figure 3-36). The SG can also process data, and supplies information to the autopilot and other navigational systems. The majority of the data transmitted to and from the SG is digital ARINC 429 data. Radar data is transmitted in the ARINC 453 format. This system can be configured with either one or two control panels.

Another popular version of the EFS-10 is configured with two symbol generators. This is often referred to as a four-tube system since there are four display units. Bendix/King refers to their display units as **Electronic Displays (ED)**. The EFS-10 is typically equipped with the ED-102 display. This configuration allows both the pilot and copilot to have an electronic HSI and ADI. The four-tube system employs two symbol generators (see figure 3-37). The installation of a second SGs provides for enhanced reliability. The SGs cross talk with information on pertinent data. Comparison checks are made and the system can flag any discrepancies.

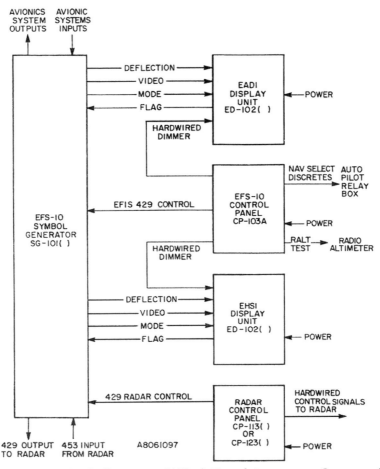

Fig. 3-36 EFS-10 three-tube block diagram. (Allied-Signal Aerospace Company)

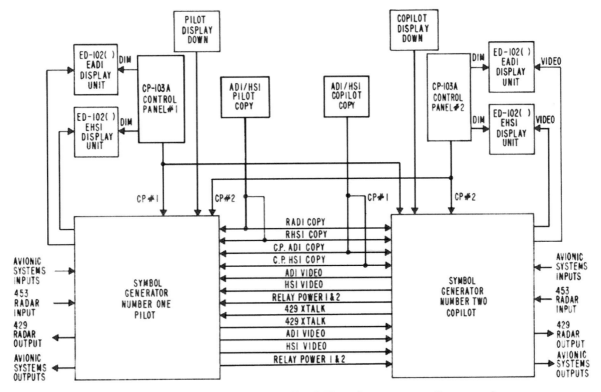

Fig. 3-37 EFS-10 four-tube block diagram. (Allied-Signal Aerospace Company)

The most popular configuration for the EFS-10 is the five-tube system. In this arrangement there are three SGs, two EADIs, two EHSIs, and one MFD. The MFD is used to display weather radar, checklist data, and provide backup for the EHSI. The MFD receives video information from the third (center) SG. As seen in figure 3-38, the five-tube EFS-10 enhances reliability through the use of cross talk busses for all three SGs, and a variety of switches used to control backup operations. These switches are called *Standby Switches*. Near the middle of the diagram notice the SGs are represented by three rectangles. The left SG is #1, the center is #3, and #2 is on the right. Look carefully to find the SG number at the top of each rectangle. Notice the cross talk (*XTALK*) bus is located just below the SGs and is a double wide shaded arrow. The cross talk bus is currently switched to connect SG #1 and SG #2. SG #3 is currently being used for MFD operations and provides no comparator monitoring. Below the SGs are the navigational radios and the flight directors (FD #1 and #2 at the bottom of the diagram). Notice the flight directors can receive input data from either SG #1 or SG #2. The five displays are located at the top of figure 3-38. On the diagram, the four main reconfiguration switches are located between the displays and the SGs. The two *DISPLAY DOWN* reconfiguration switches are used to move the pilots or copilots EADI into the EHSI position in the event of a system failure. If you look at the switch carefully, you will notice that 28 VDC power is removed from the EADI when the *display down* switch is activated. This turns off power to the EADI. In that same general area of the diagram are the two symbol generator reconfiguration switches. The switch labeled SG 1/3 enables the pilot to transfer left side EADI and EHSI operations from SG #1 to SG #3. The switch labeled SG 2/3 enables the copilot to transfer right side EADI and EHSI operations from SG #2 to SG #3. Reconfiguration switches are typically lighted push-button switches located on the instrument panel near the EFS displays. The reconfiguration switches are **not** part of the EFS control panels.

Comparing EFS-10 and EFIS-85/86
Major differences between the Bendix/King EFS-10 and the Collins EFIS 85/86 are apparent when comparing both companies five-tube systems.

· The EFS-10 contains no push-button controls on the MFD. The EFIS 85/86 MFD has line select keys and a joystick control mounted on the perimeter of the MFD. On the EFS-10 system all MFD controls are installed on a separate EFS control panel; therefore, all five displays are typically interchangeable.

· The Bendix/King (EFS-10) system uses three identical symbol generators that are completely interchangeable. The Collins (EFIS 85/86) employs only two interchangeable processors, the DPUs (display processor unit); the third processor is an MPU (multifunction processor unit). The Collins MPU is not interchangeable with the DPUs.

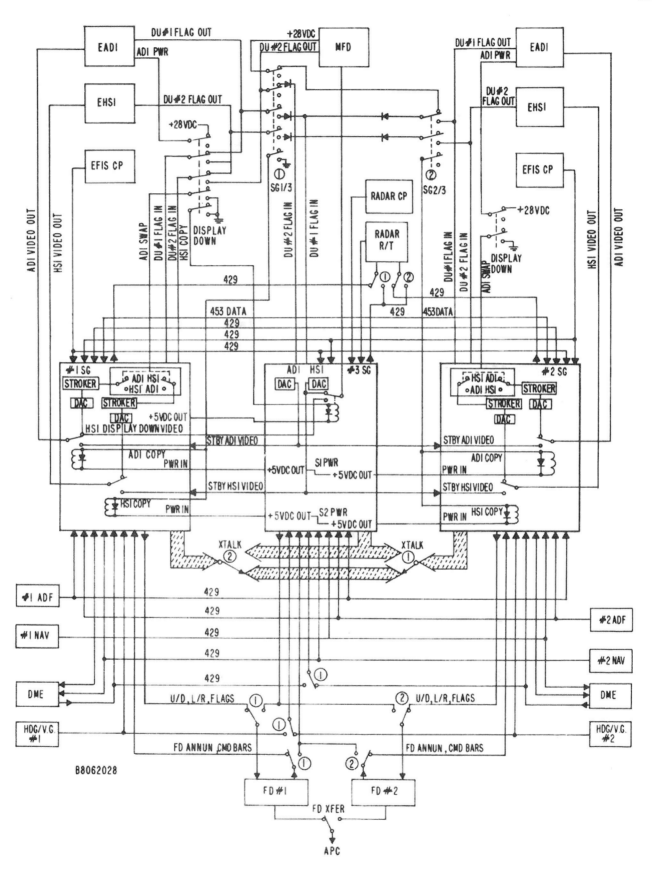

Fig. 3-38 EFS-10 five-tube block diagram. (Allied-Signal Aerospace Company)

· The two systems differ in the event of a display (EHSI or EADI) failure. If a display unit fails on the EFS-10, the crew would press the *display down* switch and the EADI would move to the lower display. On the EFIS 85/86 when a display unit fails the crewmember would select the *compact mode*, which shows both the EADI and EHSI information on the operable CRT in a compact format.

EFS-10 Troubleshooting

The EFS-10 has three basic levels of fault isolation. 1) The system performs continuous fault monitoring and shows any appropriate fault flags on the display units. 2) The pilot or technician can activate a preflight self-test that runs a sequence of system tests to verify correct operation. 3) EFS-10 has a *Maintenance Page* format that allows the technician to check system configuration, system strapping, radar straps, software codes, LRU status tests, and activate a test pattern. The various troubleshooting modes are activated through the EFS control panel by rotating the function select switch to the test (*TST*) position. The various test options are then displayed on the CRT. Simply follow the prompts on the display to retrieve the appropriate data.

The EFS-10 employs strapped connections to ensure proper system configuration. Like the Collins EFIS 85/86, the EFS-10 can interface with a variety of different systems. During installation or system modification, the correct strap connections must be made to ensure the EFS-10 operates properly. Be sure to consider improper strapping when troubleshooting defects. If a strap wire comes loose, the system is no longer configured properly and may not function. Configuration strapping can be checked using the *Maintenance Pages* option. Like all electronic systems, cooling is critical for the EFS-10. All displays and all symbol generators contain a cooling fan. During inspection and/or maintenance, be sure these fans are clear from obstruction and operating properly.

TRANSPORT CATEGORY FIRST GENERATION EFIS

The first generation EFIS found on transport category aircraft are very similar to the EFIS 85 and 86 systems described earlier in this chapter. The transport category system studied here will be the EFIS found on the Boeing 737-300. Other transport category aircraft that employ first generation EFIS include the Boeing B-757, B-767, some B-737s, the Airbus A-300, and the McDonnell Douglas MD-80. Keep in mind; general concepts of the B-737 EFIS can be applied to virtually all first generation EFIS found on transport category aircraft. Only a brief description of the systems will be presented to avoid repetition from previous sections of this chapter.

B-737-300 EFIS General Description

The B-737 EFIS employs four display units used for the pilot's and copilot's EADI and EHSI. The displays employ a typical color CRT design each weighing approximately 23 pounds and powered by 115VAC @ 400HZ. Located on the rear of each display is an inlet and outlet port for cooling air. The ship's forced-air cooling sends air through each display. The inlet port contains a fine mesh filter assembly which must remain clean for proper unit cooling. As with the Collins 85/86 systems, only flight data is displayed; the B-737 EFIS cannot display engine parameters or airframe systems information. The 737-300 contains a separate electronic display system for engine and airframe system data. As seen in figure 3-39, the main components of the system include two EADIs, two EHSIs, and two symbol generators. The **symbol generators** are LRUs, which process data from the various aircraft systems and provide data to the four EFIS displays. The symbol generators perform basically the same functions as the display processors units (DPU) found on the Collins EFIS 85/86 or the symbol generators of the Bendix /King EFS-10.

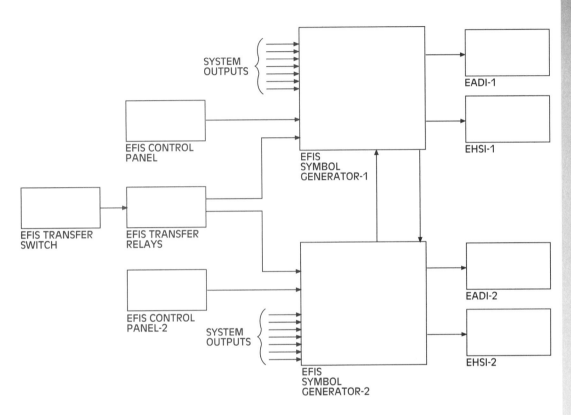

Fig. 3-39 Boeing 737-300 EFIS block diagram. (Boeing Commercial Airplane Company)

The symbol generators receive data from a variety of aircraft systems, such as attitude sensors, air data computers, navigational radios, and the EFIS control panels (figure 3-40). Most of the information sent to the symbol generators is transmitted as an ARINC 429 digital data signal. The symbol generators also receive weather radar (ARINC 453) and discrete signals from various sources. The symbol generators monitor the **Built In Test Equipment** (BITE) of each display to ensure the health of the four CRTs. The symbol generator monitors the cooling function of each display unit. If a display reaches the first over-temperature threshold the symbol generator turns off the raster signal to that display and part of the video is lost. This helps to cool the display unit since the CRT is using less power. If the second over-temperature threshold is reached, the symbol generator will turn off both the raster and the stroke signals and the display goes blank. (The raster generator provides the background video, such as the attitude ball and weather images. The stroke generator provides the signals to generate lines and characters, such as arrows, letters, and numbers.)

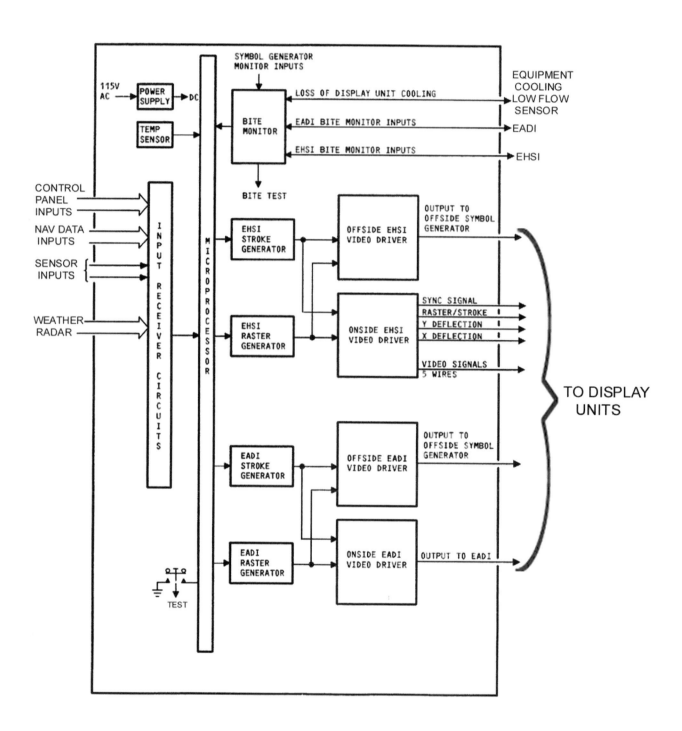

Fig. 3-40 Internal block diagram of a 737-300 EFIS symbol generator.
(Boeing Commercial Airplane Company)

Pin programming is used to set various display parameters and to identify each unit as symbol generator number one or number two. **Pin programming** is identical to configuration strapping mentioned during the discussion of the Collins 85/86 EFIS. For more information on pin programming (configuration strapping), refer to the previous sections of this chapter. The symbol generators also contain BITE circuitry to record EFIS faults. BITE access will be discussed further during system troubleshooting.

Component Location and Operation

The symbol generators are located in aircraft's electrical equipment compartment, weigh approximately 21 pounds and receive power from a 115V, 400 Hz bus. Symbol generator number one normally drives the left side EADI and EHSI. The number two symbol generator normally drives the right EADI and EHSI. In the event of a unit failure, either the right or left symbol generator can drive all four displays. The switching for the symbol generators is controlled by two EFIS transfer relays located in the electronic equipment compartment installed behind the flight management computer. The EFI switch located on the flight deck controls the relays.

During the following discussions refer to figure 3-41 to determine EFIS component locations. The EFI switch (used to select a symbol generator) is located on the pilot's overhead panel. The main EFIS control panels are located on the pedestal between the pilot and copilot seats. There are two EFIS control panels; the pilot and copilot can each select a different format for their EFIS displays. The EFIS control panel is divided into two sections. On the left are the controls for the ADI. The decision height (DH) selector and display are located on the top left side of the panel. The ADI brightness control is located in the lower left of the EFIS control panel. The EHSI controls are located on the right side of the EFIS control panel. The mode selector allows the flight crew to select the EHSI display mode. The range switch determines the weather radar and TCAS operational range. The WXR switch turns on/off the weather radar. The five map buttons on the lower left of the display determine the map mode of the EHSI.

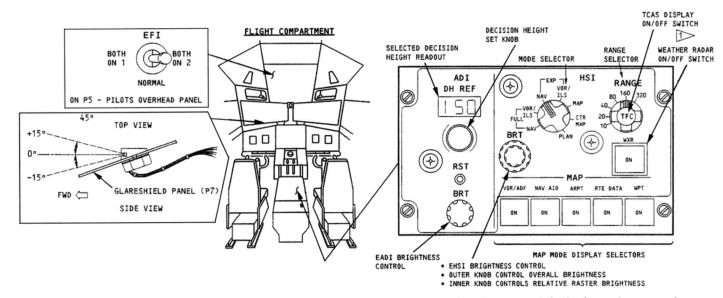

Fig. 3-41 Locations of various EFIS components on the 737-300. (Boeing Commercial Airplane Company)

During normal operation each display operates using a three-color CRT. In the event one of the color guns fails, the system reverts to a monochromatic mode. If a system which feeds the EFIS symbol generators should fail, that data will be removed from the display and the appropriate fault flag will be presented. Most fault flags are short abbreviations of the failed system, such as ATT (displayed in yellow) for loss of attitude data. Both the pilot and copilot's displays are equipped with a remote light sensor (photo diode), which is used to ensure proper brightness of the CRT. Figure 3-41 shows the light sensors located on the glare shield of the aircraft facing forward in order to monitor ambient light. Each display unit also contains a local light sensor to monitor cabin light. The symbol generator uses inputs from both the remote and local light sensors and adjusts the CRT brightens accordingly.

System Maintenance and Troubleshooting

On most aircraft, the display units are identical and can be interchanged. An inclinometer must be installed on any display in the EADI location and removed if the display is used as an EHSI. Pin programming is used to "tell" the display where it is installed. To remove a display, depress the handle firmly against the front of the display and remove the two screws located in the handle (see figure 3-42). Lower the handle to unlock the display and pull firmly. To reinstall the display, reverse the procedure. Be sure to depress the handle whenever installing or removing the screws. This will help protect the screws from being stripped. Both symbol generators are also identical and can be swapped for maintenance or troubleshooting purposes.

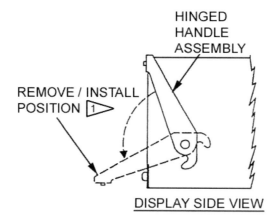

Fig. 3-42 Removal and installation techniques for a typical Boeing display unit. (Boeing Commercial Airplane Company)

The 737-300 EFIS is equipped with BITE circuitry, which allows for access to fault data for maintenance purposes. The BITE information is accessed through the right or left Control Display Unit (CDU) located in the center console between the pilot and copilot's seats. The CDU is an alphanumeric keypad and display used primarily for operation of the Flight Management System (FMS). The CDU will be discussed in greater detail later in this text. The EFIS BITE is part of the symbol generator circuitry; however, it is accessed through the FMS computer and only when the aircraft is on the ground. The BITE circuitry contains four categories: 1) in-flight faults, 2) ground faults, 3) current status, and 4) discrete status.

In-flight faults are faults, which occur whenever the aircraft is airborne. The EFIS symbol generators continuously monitor the system and it's various input/output signals. If a fault occurs, it is stored in a nonvolatile memory for up to nine consecutive flights. The in-flight faults display will indicate whether the fault was intermittent or continuous. **Ground faults** can store up to seven pages of faults, which occurred while the aircraft was on the ground. The faults are listed in alphabetical order. **Current Status** is used to test the various components of the EFIS and related interfaces. It should be noted that if an in-flight fault shows as intermittent, it might appear to be functioning properly during the current status check. **Discrete Status** is used to test any discrete signals that connect to EFIS. Assess of the B-737 EFIS BITE information is a menu driven process similar to BITE data from other systems. BITE access will be covered in greater detail in chapter five of this text.

If the EFIS symbol generator should fail, the BITE information for that system is inaccessible. During a failure of the symbol generator processor circuitry, the message *SG FAIL* will be displayed on the EADI and EHSI. If symbol generator power should fail, any related EADI and EHSI will go blank.

During initial power-up of the EFIS, a self-test is initiated. If a fault is detected, the appropriate fault flag will be displayed. During this self-test, the EFIS pin programming is also checked for accuracy. A parity pin is used to verify the correct pin programming of the system. If a pin programming error exists, both displays go blank and a message *PARITY ERROR* is displayed. If a pin programming error exists, the BITE for that symbol generator remains operational.

INTEGRATED ELECTRONIC INSTRUMENT SYSTEMS (advanced EIS)

In the late 1980's a new generation of electronic flight instrument systems was beginning to emerge. These *2nd generation EFIS* were designed to allow for better flight deck management, and improved reliability and maintainability. These improvements were made possible through the total integration of the aircraft's flight management, engine indicating, central maintenance computer, and the electronic flight instrument systems. Improved computer software allowed this integration to become successful. Although many of these systems take on different names or features for different aircraft, the end result is still the same: a fully integrated electronic instrument system.

Today, advanced electronic instrument systems (EISs) can be found on a variety of transport category aircraft; the Boeing 747-400, the McDonnell Douglas MD-11, and the Airbus A-320. Many corporate- and commuter-type aircraft also employ the advanced EIS, such as, the Gulfstream G-5 and the Cessna Citation X. From the flight deck, these systems have an obvious change over early EFIS systems; the displays are larger and the instrument panel less cluttered. Figure 3-43 shows the instrument panels of the A-320; note the six IDS displays.

Some of the latest aircraft have gone even one step further toward integrating avionics systems. The Boeing 777, for example, employs an **Airplane Information Management System (AIMS)** which allows for subsystems of various LRU computers to share both hardware and software components. The Collins Pro Line 4 system, developed for general aviation aircraft, uses an **Integrated Avionics Processor System (IAPS)** which provides integrated functions for various avionics systems.

In this portion of the chapter, the EIS of two aircraft will be examined. The Beechjet 400A which uses the Collins Pro Line 4 system, including IAPS, will be presented. Also, the McDonnell Douglas MD-11 EIS will be introduced. This will provide a brief look at both a transport category and corporate-type second generation electronic flight instrument systems.

The Beechjet 400A Electronic Instrument System

The electronic instrument system used in the Beechjet 400A is a fully integrated system which incorporates state-of-the-art CRT displays. The Beechjet can be configured for a 2-, 3- or 4-tube system. In the 2-tube system, one electronic flight display (EFD) and one multifunction display (MFD) are installed on the pilot's side of the instrument panel. In a 3-tube system, one EFD is installed for the copilot; one EFD and one MFD are installed on the pilot's side. The four-tube system can be configured with three EFDs and one MFD or two EFDs and two MFDs. It should be noted this system also employs two monochrome backup displays located outboard of the main displays. The material presented in this portion of the text will be concerned with the 4-tube (3-EFD, 1-MFD) configuration as seen in figure 3-44.

The Pro Line 4 systems installed in the Beechjet 400A have many similarities to the earlier version electronic flight instrument systems (Pro Line II). For example, both systems: 1) utilize CRT displays for navigational and attitude data, 2) employ MFD for display redundancy, 3) interface with various aircraft systems using digital data busses, 4) interface with the aircraft's autoflight system, and 5) have a built-in diagnostics. The advanced EIS has improved on many of the early system's shortcomings. Some of the changes made to the advanced EIS include: 1) the new CRT displays that are larger and more easily read, 2) Control Display Units (CDUs) are used as a central control center for EIS and flight guidance data selection, 3) an Integrated Avionics Processor provides system integration to improve interfacing capabilities, and 4) the built-in diagnostics are more easily interpreted and provide improved reliability. The major differences between the first and second generation systems are presented during the following discussions.

Fig. 3-43 Instrument panels of the A-320.   (Airbus Industrie)

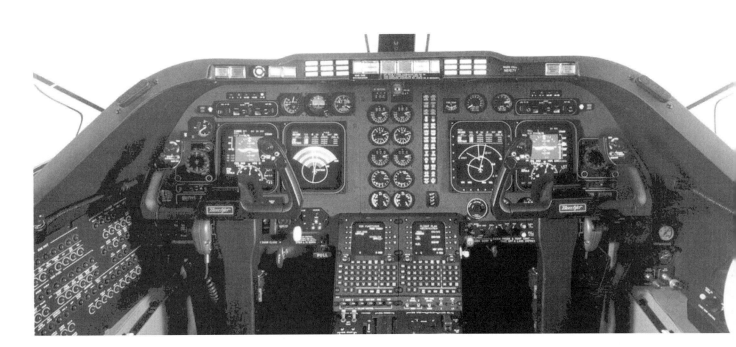

Fig. 3-44 Typical 4-tube EFIS system in a corporate jet aircraft. (Collins Divisions, Rockwell International)

System Components

In general, the individual components of the Pro Line 4 system are lighter, more powerful and use less electrical power than their predecessors. Some of the components' internal functions have been combined into IAPS. Many of the redundant LRUs, such as the display units are still interchangeable which is very helpful when troubleshooting the system.

The displays

The Beechjet 400A system is referred to as the **IDS (Instrument Display System)**. The IDS contains four main displays as shown in figure 3-45. The PFD is the **primary flight display**. There is one PFD for both the pilot and copilot. The MFD is the **multifunction display**. This unit is typically installed on the copilot's side only; however, the pilot's side may also employ an MFD instead of an ND. The MFD is used as a back up for the PFD in the event of a failure. The ND is the **navigational display**. There is one ND installed on the pilot's side of the instrument panel.

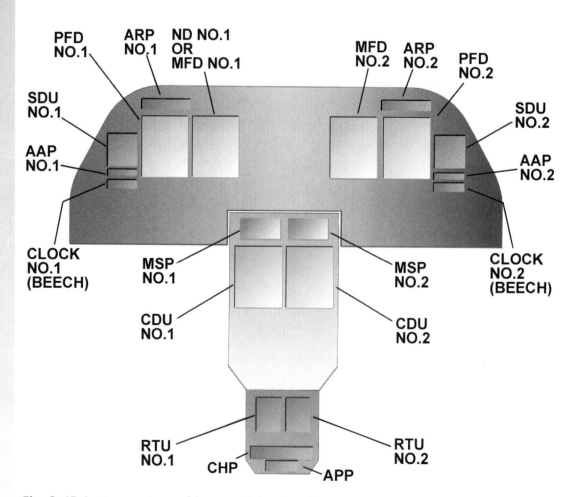

Fig. 3-45 Instrument panel layout of the Beechjet 400A. (Collins Divisions, Rockwell International)

Any of the PFDs and NDs are completely interchangeable for maintenance purposes. If two MFDs are installed, they are also interchangeable. The specific installation locations are determined by configuration strapping. The displays on this advanced system differ from the EFIS-85/86 in that they contain their own power supplies as well as video circuitry. (The EFIS-85/86 displays operated in conjunction with a DPU power supply.) Figure 3-46 shows an internal block diagram of a display unit. The top left corner of the diagram shows the dedicated fan circuitry. Each display unit has a cooling fan that operates any time the display receives power.

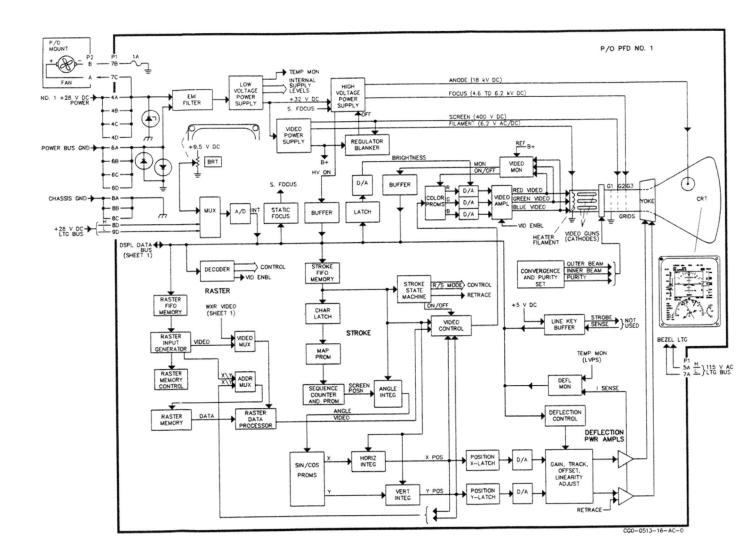

Fig. 3-46 Internal block diagram of a display unit. (Collins Divisions, Rockwell International) For Training Purposes Only.

As can be seen in figure 3-47, the PFD combines the HSI and ADI displays on one CRT. This provides the pilot with complete navigational data in one location. The ND is used to display lateral navigation data, course map displays and radar information. For both the PFD and ND, several variations of the display formats are pilot selectable.

On the Pro Line 4 system there are two backup displays called **Sensor Display Units (SDUs)**. These displays are located outboard of the pilot's and copilot's PFDs (see figure 3-45). The SDUs are monochrome displays used to show basic compass and navigational data. The SDUs receive their input data from an independent processor; allowing for operation free of any possible main display system failures. A sensory display unit showing the RMI format is shown in figure 3-48.

The multifunction display units contain 18 line select keys (LSKs) located around the perimeter of the CRT. The LSKs are used to select a given function from the MFD display. For example, to select the *AVIONICS STATUS* function (figure 3-49) press the third LKS from the top on the right side of the display. The LSKs will perform different functions according to the information currently displayed by the MFD. Similar to the PFD or ND, the MFD contains its own processor and power supply circuitry. A dedicated cooling fan is also used for the MFD.

It should be noted that many of the newest electronic instrument systems are being equipped with flat panel displays. These displays offer several advantages over a conventional CRT display. Flat panel displays will soon be used for all aircraft instruments; however, today the industry standard is still the CRT. Flat panel displays are discussed in detail in chapter two of this text.

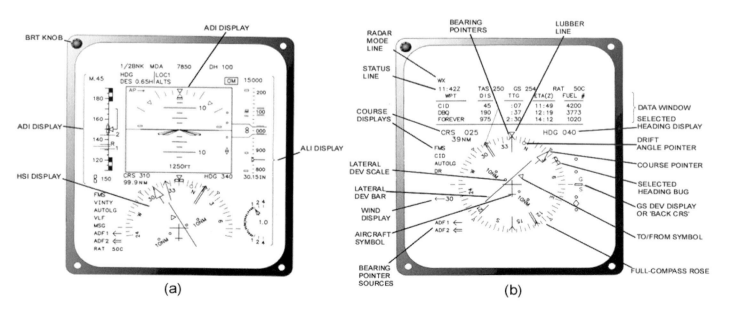

(a)   (b)

Fig. 3-47 Typical PFD and ND. (Collins Divisions, Rockwell International)

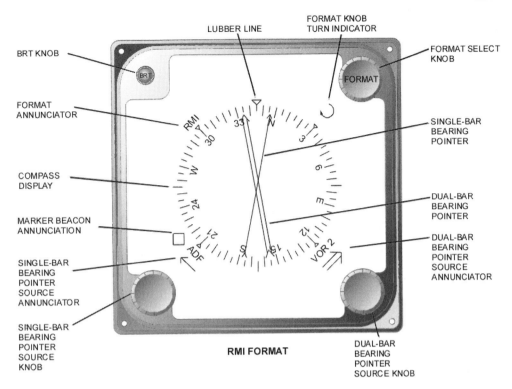

Fig. 3-48 Sensory display unit. (Collins Divisions, Rockwell International)

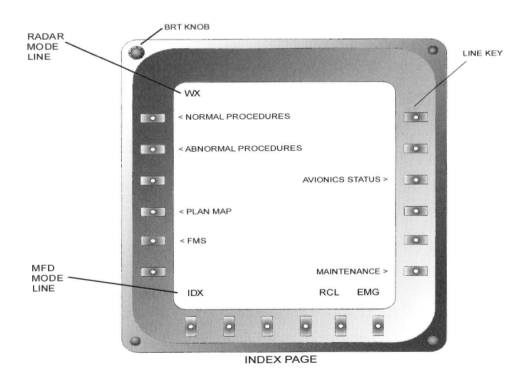

Fig. 3-49 A multifunction display unit. (Collins Divisions, Rockwell International)

Integrated avionics processor assembly

As mentioned earlier, the Beechjet 400A employs an **Integrated Avionics Processor System (IAPS)** which provides a central location for integration and interfacing of the avionics equipment. The IAPS functions similar to a distribution network. Data from the various avionics equipment is sent to IAPS. The input data is monitored, checked for integrity, sequenced in the correct order, and transmitted to the display units and flight management/control system. The integrated avionics processor assembly consists of an **IAPS Card Cage (ICC)** and several line replaceable modules (see figure 3-50). This configuration allows for replacement of individual modules without effecting the entire processor unit. The modules perform various power supply, computing and data distribution functions. The six types of modules found in the IAPS include: 1) **FMCs** (flight management computers), 2) **CDC** (control display coupler), 3) **FCC** (flight control computer), 4) **ATC** (automatic trim coupler), 5) **IOC** (input/output concentrator), 6) **PWR** (power module).

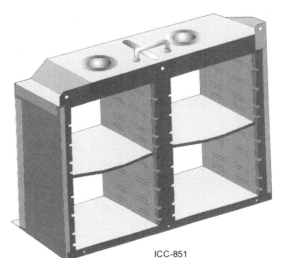

IAPS CARD CAGE

ICC-851

VARIOUS LINE REPLACABLE MODULES

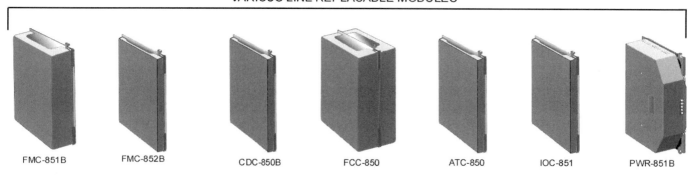

| FMC-851B | FMC-852B | CDC-850B | FCC-850 | ATC-850 | IOC-851 | PWR-851B |

Fig. 3-50 Components of the integrated avionics processor assembly. (Collins Divisions, Rockwell International)

The IAPS is divided into four quadrants: 1A, 1B, 2A, and 2B. Each quadrant contains its own power supply and I/O data concentrator (see figure 3-51). The IAPS card cage is comprised of several slots which accept the modules. Each module slides into its respective slot and is held in place by a cam-latch assembly. The connectors for each module are located at the rear of the ICC and engage as the module is installed. The ICC is located in the aircraft's equipment bay. It should be noted that each power supply card contains an *ON* lamp which can be seen through the ICC cover. Each lamp should be illuminated any time it's respective power supply is operating. This is an extremely handy feature for basic troubleshooting.

The ICC contains two **environmental control cards** that ensure a stable operating temperature for the line replaceable modules. The environmental control card heaters and fan assemblies are integrated into the ICC. If the IAPS temperature drops below limits the heater is turned on and the fan operates on low speed. If the temperature rises above normal operating limits, the fan operates on high speed.

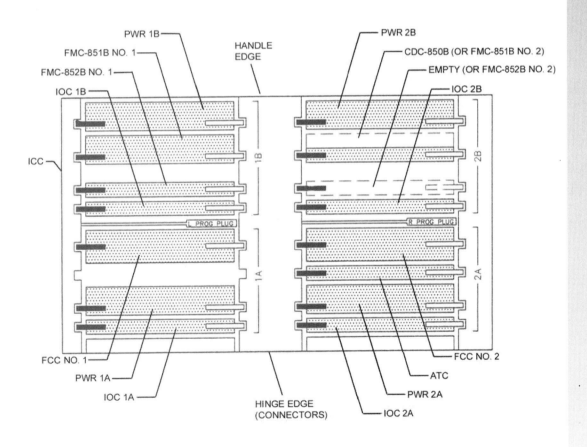

Fig. 3-51 IAPS module locations. (Collins Divisions, Rockwell International) For Training Purposes Only.

IDS controls

There are seven different panels used to control the complete integrated display/ autoflight system. The **altitude awareness panel (AAP)**, the **autopilot panel (APP)**, the **air data reference panel (ARP)**, **course heading panel (CHP)**, and the **mode select panel (MSP)** all deal with various navigation and autoflight functions. These panels are used to make changes to the EFD formats according to the appropriate pilot selection. The PFDs and ND are on any time the avionics bus is active. Their brightness is controlled through a knob located in the top left corner of each display unit.

The Beechjet 400A employs a **Control Display Unit (CDU)** that is the primary controller for the PFDs and ND. There are two CDUs (1-pilot's and 1-copilot's) located in the center console of the aircraft; each controls their respective displays. The CDU contains eight line select keys and 26 dedicated push-buttons (see figure 3-52). The *DATA* key is used to select FMS data for display on the MFD and ND. The *MAP* key is used to display a dynamic present position map on the ND and MFD. When the *HSI* button is pressed the conventional "full-compass rose" format is displayed on the MFD and ND. The CDU is also used to control a variety of flight navigation management functions.

System Architecture

The Pro Line 4 electronic instrument system is thoroughly integrated with the flight management system as can be seen in figure 3-53. The major integration units are the IAPS side 1 and 2. The IAPS receives data from the various aircraft systems through digital data busses, manipulates the data, and transmits the necessary data to the IDS. The IAPS contains four I/O concentrators (IOCs), two right side and two left side. Each display unit receives digital data from its respective on-side IOCs. All digital data busses to the IDS use ARINC 429 format; except the weather radar,

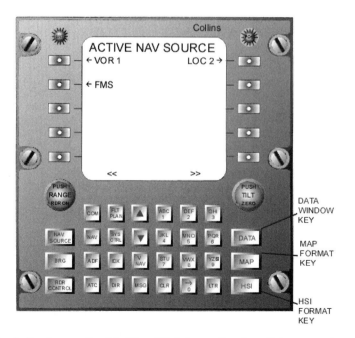

Fig. 3-52 A control display unit. (Collins Divisions, Rockwell International)

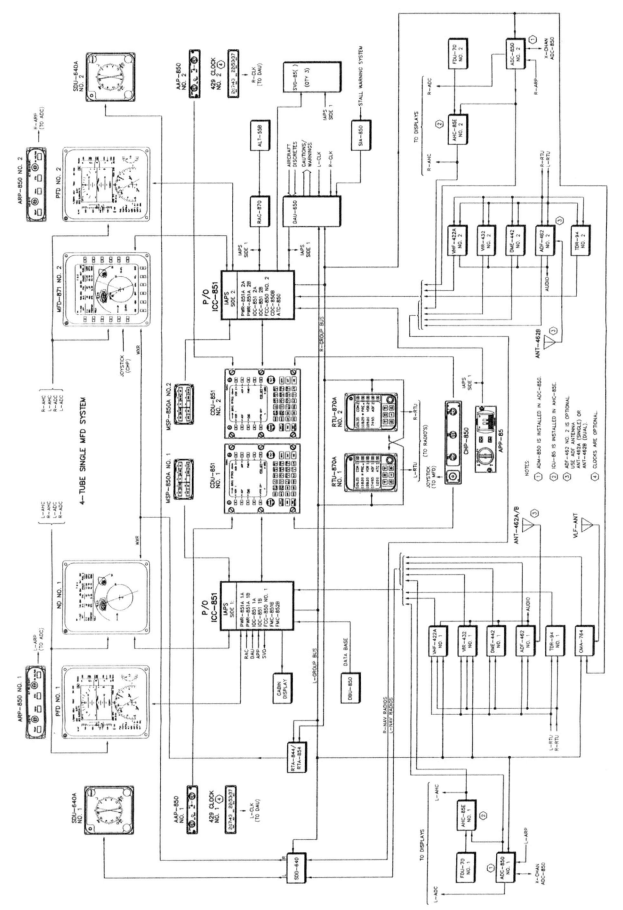

Fig. 3-53 Beechjet 400A integrated avionics system block diagram. (Collins Divisions, Rockwell International) For Training Purposes Only.

which transmits an ARINC 453 digital signal. As seen in figure 3-54, the IOCs receive inputs from the various aircraft LRUs. The output data is sent directly to the flight deck displays, the flight control system (FCS) and the flight management system (FMS). IOC output data is also transmitted to other avionics systems on three general purpose (GP) busses.

Displays

The PFDs (shown in the upper portion of figure 3-53) each receive critical data directly from the R/L (right and left) **Attitude Heading Computers (AHCs)** and the R/L **Air Data Computers (ADCs)**. Since the PFDs also receive data from the IOC (in IAPS), this provides a redundancy for critical data. The ND also receives data from the AHCs, ADCs, and IAPS. The PFDs, ND and MFD each have an input data bus returned to IAPS to monitor system health. The MFD receives inputs from the IOCs and FMCs within IAPS. The MFD also receives direct data from the AHCs, ADCs, **CHP (course heading panel)** and the **weather radar system (WXR)**.

Two **Sensor Display Units (SDUs)** are installed in the Beechjet 400A IDS to provide backup to navigation and sensor data. Each SDU receives information through a digital data bus from the **Sensor Display Driver (SDD)**. The SDD receives inputs from R/L navigation radios, both AHCs, and an IAPS IOC. The SDD is divided into two channels, each channel drives one SDU.

Discrete inputs from various aircraft systems to the IDS are routed through the **Data Acquisition Unit (DAU)**. The DAU is shown in figure 3-53 just to the right and below the right side IAPS. The DAU converts all discrete signals to an ARINC 429 format and transmits that data to IAPS. The DAU also converts digital data received from IAPS for operation of the discrete caution and warning annunciators.

Reversionary switching

The IDS employs **reversionary switching** to provide for backup of PFD displays, AHC and ADC inputs, and the CDU. The reversionary switches send discrete signals directly to the IDS as shown in figure 3-55. The reversionary switches allow the flight crew to select which AHRS (AHRS to be discussed in chapter 6) and AIR DATA source (1 or 2) will be used for inputs to the PFD, ND and MFD. The PFD backup can be selected, which moves the PFD display to the ND or MFD. The CDU switch allows for selection of the on-side or off-side control display unit.

Separate display control switches are supplied for both the pilot and copilot. Therefore, the pilot and copilot's IDS can be operated from independent attitude/heading and air data sources. To select the off-side CDU, press the *REV* button, press *NORM* to select the on-side CDU. To move the PFD to the MFD or ND, press the *REV* button on the PFD reversionary switches.

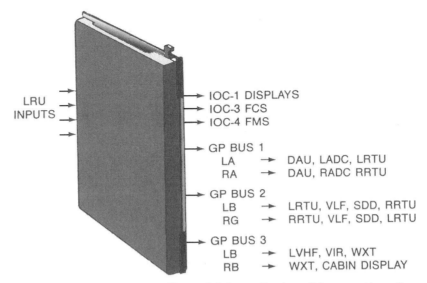

Fig. 3-54 IOC interface diagram. (Collins Divisions, Rockwell International)

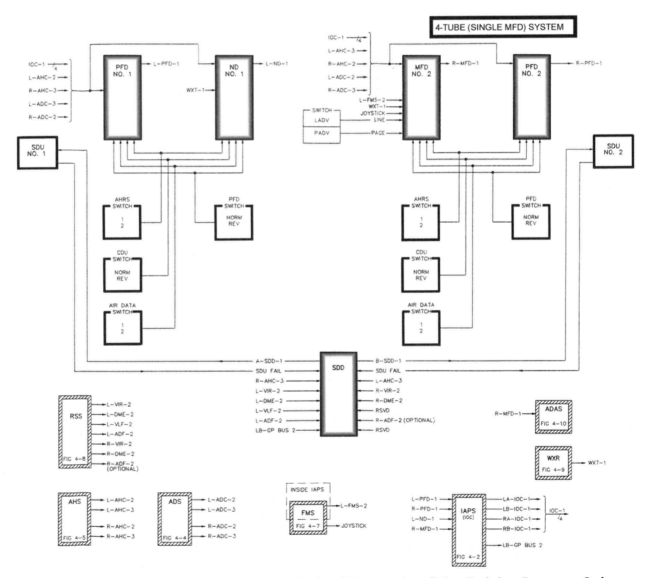

Fig. 3-55 IDS interface diagram. (Collins Divisions, Rockwell International) For Training Purposes Only.

System Troubleshooting

The Collins Pro Line 4 equipment has significant improvements over the Pro Line II series. System diagnostics are still accessed through the MFD as they were on earlier systems; however, a more detailed troubleshooting menu is available. The Pro Line 4 IDS diagnostics resemble the diagnostics systems, such as, EICAS or ECAM, found on transport category aircraft. That is, Pro Line 4 avionics troubleshooting is selected from the diagnostics menu, coordinated through a central location and displayed on a CRT.

Diagnostics architecture

With any built in diagnostics system, the technician should become familiar with capabilities of the system prior to fault isolation. The diagram in figure 3-56 shows the diagnostics architecture for the 4-tube IDS. The arrows on the diagram indicate the flow of information within the system as a request for diagnostics data is made. Refer to figure 3-56 during the following discussion on the diagnostics architecture.

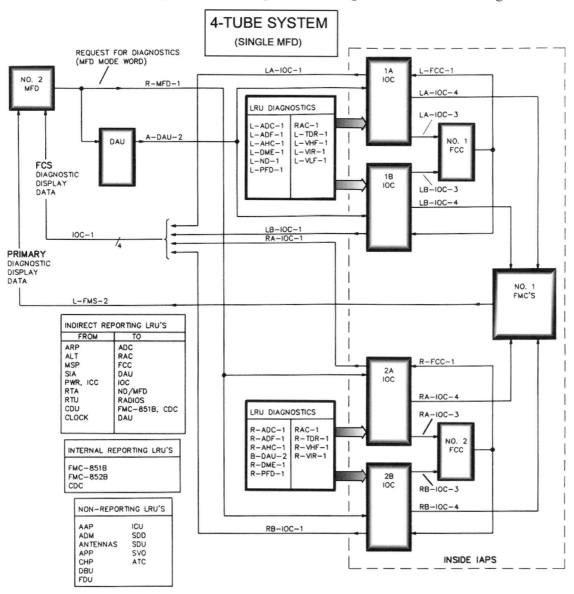

Fig. 3-56 Diagnostics configuration four-tube IDS. (Collins Divisions, Rockwell International) For Training Purposes Only.

The flight management computer (FMC), shown in the center right side of the diagram, is the core of the diagnostics system. Each avionics system and subsystem that has internal monitoring reports to the FMC. The FMC contains the central monitoring circuitry. The FMC monitors the IOC data busses for fault messages caused by system failures. The FMC will also detect any system/subsystem that fails to transmit data. If a failure is detected the FMC stores the information in a nonvolatile memory for retrieval at a later date. If the FMC detects a failure, the yellow message (*MSG*) light on the CDU will flash. The PFD also displays the term *MSG* displayed in yellow.

When a selection is made for diagnostics using the correct multifunction display LSK, the data word is transmitted on bus R-MFD-1 to the 2A-IOC, 2B-IOC, and the DAU (data acquisition unit). The DAU transmits the request on data bus A-DAU-2 to the opposite side IOCs. The IOCs then send the request for diagnostics data to the number 1 FMC. The FCC (flight control computer) also receives the request since the FCC manages diagnostics for the flight control system.

After receiving a request for diagnostic data, the FMC processes the request and transmits the requested data back to the MFD. The diagnostics data is transmitted on the high-speed bus L-FMS-2 from the FMC to the MFD. If a request is made for additional diagnostic information, the cycle repeats. Recognizing this flow of data becomes important when the diagnostics request fails. For example, if the data bus L-FMS-2 should fail, the diagnostics data would never reach the MFD. In any situation where a request for diagnostics data is not displayed, check the bus structure and LRUs that interact during diagnostic requests.

Troubleshooting levels

The Beechjet 400A avionics troubleshooting is divided into two categories, **Level 1** and **Level 2. Level 1** troubleshooting is considered the initial fault isolation procedure used when a problem first arises. Level 1 troubleshooting consists of a preliminary review of diagnostics data to see if the system has detected a faulty LRU. **Level 2** troubleshooting procedures encompass a detailed study of the diagnostics data to find the failed system or component. Level 2 troubleshooting requires the technician to reference the avionics maintenance manual for analysis of fault code data. This analysis will reveal problems created by faults both internal and external of the LRUs.

Whenever a fault occurs in the avionics system, the technician should begin with Level 1 troubleshooting. The first step in the troubleshooting procedure is to access the MFD Index Page. To access the index page press the line select key adjacent to *IDX* (Index) on the MFD (see figure 3-57). The index page can be accessed from any MFD display format. There are three different formats used to display diagnostics data on the Pro Line 4 system: **Avionics Status, LRU Diagnostic Data**, and **LRU Fault History**. The avionics status data is accessed directly from the MDF index page. To access the LRU diagnostics data or fault history, press the *Maintenance* LSK on the MFD index page. The LRU diagnostics is then selected from the MFD maintenance menu page as shown in figure 3-58.

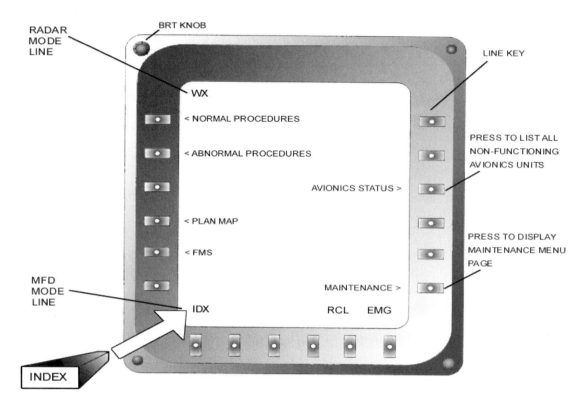

Fig. 3-57 MFD display used to access troubleshooting of avionics systems. (Collins Divisions, Rockwell International)

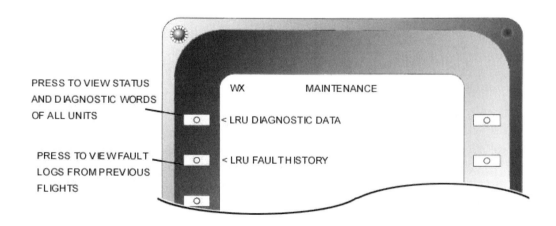

Fig. 3-58 MFD, Maintenance menu page. (Collins Divisions, Rockwell International)

Avionics status page

The **Avionics Status** page is a real time display of avionic systems/subsystems that experience a complete or partial failure. The Avionics Status page is used for Level 1 troubleshooting, and should locate the defective LRU for approximately 90% of the possible faults. To access the Avionics Status page press the LSK adjacent to *AVIONICS STATUS* > on the MFD (see figure 3-57). If more than one page exists in the report, access additional pages by using the joystick on the CHP (course/ heading panel). To exit from Avionics Status, press the *IDX* LSK on the MFD.

A typical Avionics Status page is shown in figure 3-59, refer to this figure for the following discussion. In this example, three faults are currently active. If there are no active faults the page will display *"NO FAULTS"*. To keep nuisance messages to a minimum, a fault must be active for 60 seconds prior to posting on the Avionics Status page. The top line of the page contains the radar mode. In this example, the weather radar is on. The LRU column is used to list the name of the LRU suspected of failure. The MFD mode line is used to identify the function of the adjacent LSK. *IDX* is used to return to the MFD index page, *RCL* will recall the previous page, and *EMG* is used to display the emergency checklist. The *Status/Code* column is used to describe the failed condition and list the fault code.

The **status field** on the Avionics Status diagnostics page is a plain English message describing the suspected LRU fault. In figure 3-59, the second failure listed is "PFD DISPLAY 2" and has a status field of "OFF-CHK BREAKR". This message tells the technician that the diagnostics system suspects there is no power supplied to PFD 2. The most likely place to start troubleshooting is at the circuit breaker. If the circuit breaker is found to be OK, the next logical step would be to check the power feeding PFD2. If no power is detected at PFD2 the logical conclusion is defective wiring. Carefully inspect all related wires and connectors between the circuit breaker and PFD2. Also, be sure to verify proper ground connections.

Fault codes

The code field of the Avionics Status page contains a five-digit **fault code**, which is used to help isolate the cause of a given fault. The fault code is typically used to solve a problem only if the suggestions given by the status field were ineffective. To determine the specific fault code, the diagnostic circuitry monitors the transmitting busses of various LRUs. From this information, the diagnostics will make a determination as to the problem with the system.

An explanation of the fault codes is given in the avionics maintenance manual. As seen in figure 3-60, the explanation table contains the fault code number; a description of the fault and the reason the diagnostics reported the failure. In the example "PFD DISPLAY 2" the fault code is 2PFD1. According to the explanation table, the reason this fault was reported was "R-PFD-1 bus to the 2A and 2B IOC's inactive at both units." The most likely reason both busses would have no data is the PFD is not transmitting. The most likely cause the PFD is not transmitting is a failed PFD or no power to the PFD. Hence the associated status field for this failure is "OFF-CHK BREAKER." Note, fault codes offer more information than the status message, i.e. the fault code identifies which specific bus has failed. This information can be very helpful if the problem is wiring and not an LRU problem.

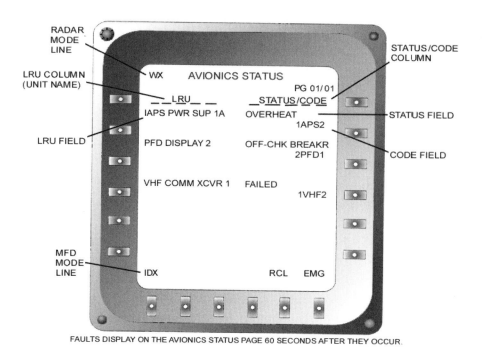

FAULTS DISPLAY ON THE AVIONICS STATUS PAGE 60 SECONDS AFTER THEY OCCUR.

Fig. 3-59 MFD, Avionics Status Page. (Collins Divisions, Rockwell International)

FAULT CODE	DESCRIPTION	REASON
PFD DISPLAY 1 (Cont)		
1PFD5	IOC 2A input failed.	RA-IOC-1 bus from the 2A IOC to the large displays is inactive at the PFD input but active elsewhere.
1PFD6	IOC 2B input failed.	RB-IOC-1 bus from the 2B IOC to the large displays is inactive at the PFD input but active elsewhere.
1PFD7	ADC 1 input failed.	L-ADC-3 bus from the #1 ADC to the left large displays is inactive at the PFD input but active at the ND/MFD input.
PFD DISPLAY 2		
2PFD1	No output data; check circuit breaker	R-PFD-1 bus to the 2A and 2B IOC's inactive at both units.
2PFD2	Various circuits failed.	An internal PFD fault is detected.
2PFD3	IOC 1A input failed.	LA-IOC-1 bus from the 1A IOC to the large displays is inactive at the PFD input but active elsewhere.
2PFD4	IOC 1B input failed.	LB-IOC-1 bus from the 1B IOC to the large displays is inactive at the PFD input but active elsewhere.
2PFD5	IOC 2A input failed.	RA-IOC-1 bus from the 2A IOC to the large displays is inactive at the PFD input but active elsewhere.
2PFD6	IOC 2B input failed.	RB-IOC-1 bus from the 2B IOC to the large displays is inactive at the PFD input but active elsewhere.
2PFD7	ADC 2 input failed (4-tube system only).	R-ADC-3 bus from the #2 ADC to the right large displays is inactive at the PFD input but active at the MFD.

FAULT CODE 2PFD1 ⟹ (points to row 2PFD1)

Fig. 3-60 Fault code explanation. (Collins Divisions, Rockwell International)

For the Beechjet 400A, there are three messages which can appear in the status field of the Avionics Status page. 1) *OVERHEAT* indicates an IAPS power supply overheat. 2) *OFF-CHK BREAKR* indicates all transmitting busses from the LRU are inoperative; the diagnostics circuit assumes no power to the unit and the circuit breaker should be checked first. 3) *FAILED* is displayed when the LRU is still transmitting but not on all busses, the diagnostics circuitry assumes a failed LRU.

LRU fault history

The **Fault History Page** of the IDS diagnostics is used to access a record of faults that occurred during previous flights. The Fault History Page is helpful when troubleshooting a reoccurring fault or intermittent problem. To view the LRU fault history page, press the following line select keys on the MFD: 1st press *IDX*, 2nd press the *MAINTENANCE* key, 3rd select *LRU FAULT HISTORY*. The LRU Fault History can display up to 40 faults which are stored in the FMC nonvolatile memory. The faults are organized into individual logs. Each log represents a previous flight. The most recent flight is log number 1.

As seen in figure 3-61, the LRU fault history page includes: 1) the flight log number, 2) time that flight began, 3) the name of the suspect LRU, 4) time at which the fault was first diagnosed, 5) the number of times (count) the fault occurred during that flight, 6) the diagnostic word, and 7) the fault code. The fault code is identical to the code given for the same fault on the Avionics Status page. The **diagnostic word** is a six-digit code used for Level 2 troubleshooting of the Pro Line 4 avionics systems. A thorough examination of the diagnostic word will be presented later in this chapter.

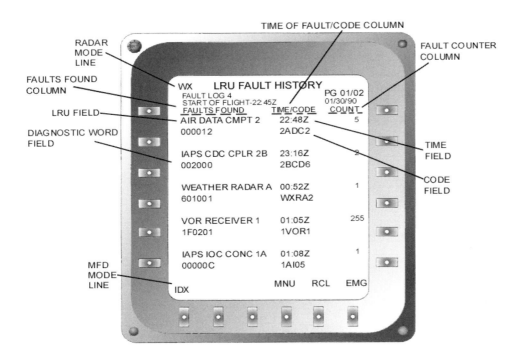

Fig. 3-61 MFD, Fault History Page. (Collins Divisions, Rockwell International)

LRU diagnostic data

The Pro Line 4 **LRU diagnostic data** page provides a real time in-depth analysis of LRU status. Each LRU that reports to the FMC is listed in alphabetical order on the LRU diagnostic data pages. The LRU diagnostics page provides the technician with a list of LRUs along with their status message and diagnostic code (see figure 3-62). This diagnostics page is typically accessed only if the Avionics Status page data does not provide the needed information to repair a fault.

The LRU diagnostics page is accessed from the MFD maintenance page. The CHP joystick can be used to scroll through different pages of the LRU diagnostics data. To exit the LRU diagnostics pages, press the index LSK on the MFD.

Diagnostic words

Diagnostic words are 6-digit hexadecimal codes, which are displayed on the LRU diagnostics page and LRU fault history page. Each digit of the diagnostic code has specific meaning for the particular LRU in question. The FMC diagnostic circuitry monitors the health of an LRU's internal circuitry and its associated busses to determine a diagnostic code.

In general, the avionics maintenance manual must be used to analyze a given diagnostic code. For normally operating systems the code always contains a 6 in the left most digit. For example the code *600001* indicates normal operation of the # 1 unit. The right-most digit of the code displays the LRU number (1 or 2). For example in figure 3-62 the AHRS computer 1 is reporting a code of 600001. This indicated the number 1 AHRS computer is operating normally. If the most significant digit of the code is *0* the unit has failed. The fault code in figure 3-62 shows that the AHRS computer 2 has failed.

Refer to figure 3-63 and 3-64 during the following discussion on diagnostic code

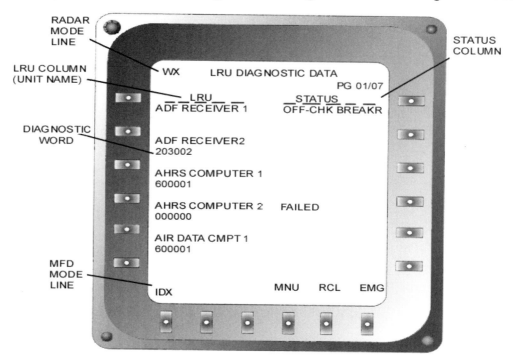

Fig. 3-62 MFD, LRU diagnostics page. (Collins Divisions, Rockwell International)

analysis. Assume we have a diagnostic code of *804020* for PFD #1. To analyze a diagnostics code simply determine which bits are set to 1 then identify the description for that bit. To determine which bits are set to 1 or 0, the hexadecimal diagnostics code (*804020*) must be converted to binary and applied to the diagnostic word explanation table in the maintenance manual. To convert the code:

First, convert the hexadecimal fault code to binary (figure 3-63).

Second, assign each binary digit to a specific bit number. The least significant digit (*9*) represents bits 9 to 12; the most significant digit (*8*) represents digits 29 to 32. In this case, the hex code (*804020*) is equivalent to the binary number *1000000001000000000100000*. Only digits 14, 23, and 32 are binary 1; all other digits are binary 0.

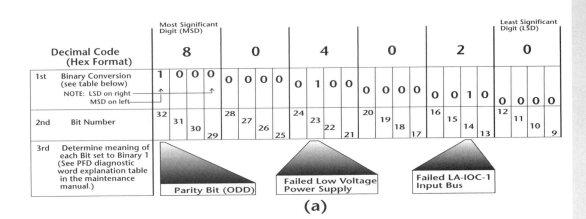

(a)

HEXADECIMAL TO BINARY CONVERSION TABLE

	HEXADECIMAL DIGIT															
BINARY VALUE:	0	1	2	3	4	5	6	7	8	9	A (10)	B (11)	C (12)	D (13)	E (14)	F (15)
1	0	1	0	1	0	1	0	1	0	1	0	1	0	1	0	1
2	0	0	1	1	0	0	1	1	0	0	1	1	0	0	1	1
4	0	0	0	0	1	1	1	1	0	0	0	0	1	1	1	1
8	0	0	0	0	0	0	0	0	1	1	1	1	1	1	1	1

(b)

Fig. 3-63 Diagnostic word interpretation; (a) analysis of code 804020, (b) Hex to binary conversion table.

Third, determine the English meaning of each bit set to binary 1. This is done using the diagnostics word explanation table from the avionics manual (see figure 3-64). In this example, the diagnostics have detected a failed LA-IOC-1 input bus (bit #14), and a failed low-voltage power supply (bit # 23). The parity (bit # 32) is set to binary 1 to achieve odd parity.

From this example one can see how to decode the PFD diagnostics code *804020*. The hex number shows a failure of the low-voltage power supply and an IOC input bus. The technician would then refer to the system manuals to determine the appropriate action. The analysis of a diagnostics code is somewhat time consuming; however, the process provides detailed fault isolation data. This information can be used during troubleshooting, or to verify correct system operation.

Level 2 troubleshooting procedures

As mentioned earlier, level 2 troubleshooting is an in-depth look at a given portion of the avionics system. In the avionics maintenance manual, there are specific test procedures for 65 LRUs. Each of these one-page tests are considered level 2 troubleshooting. Figure 3-65 shows the troubleshooting procedures for the number 1 MFD. In this example, the procedures direct the technician through a series of tests including reference to the avionics status page, checking data bus outputs, and swapping displays.

General Troubleshooting Techniques

Most integrated display systems are very symmetrical. The LRUs from the right side can be interchanged with the LRUs from the left side. If a fault occurs in a redundant system, simply swap the LRU with the identical unit on the opposite side of the aircraft. This procedure is simple; but be careful not to damage connector pins; take all necessary ESDS precautions, and be sure power is **off** to both LRUs before swapping.

In some cases, "swapping" of units for troubleshooting purposes can be done using the reversionary switches. Selecting the PFD reversionary mode blanks the normal PFD, and the ND now displays the PFD format. Reversionary switching can also be used to select different attitude and heading data sources.

If a failed bus is detected through the diagnostics program, the bus structure must be investigated. The transmitting LRU, the receiving LRU, or the bus wiring may cause a failed bus. Remember bus wiring includes all connectors, wire and related shielding. Systematically swapping LRUs that transmit and receive on the failed bus may be helpful to isolate a component problem. The bus (both wires) can also be tested for continuity, and isolation from ground. A data bus analyzer or oscilloscope can be used to test voltage levels. An ARINC 429 bus should have +/-10 volts from A to B for binary 1 and 0 respectively. Zero volts should be present for null conditions. A RS-422 bus should have +/-5 volts from A to B for binary 1 and 0 respectively. Also, be sure the data bus shielding is properly grounded at all termination points. See chapter two for more details on data bus troubleshooting.

OCTAL 350 BIT NUMBER	DESCRIPTION
	PFD DISPLAY Diagnostic Word
9	0
10	0
11	Failed R-AHC input bus
12	Failed L-AHC input bus
13	Failed LB-IOC-1 input bus
14	Failed LA-IOC-1 input bus
15	Failed RB-IOC-1 input bus
16	Failed RA-IOC-1 input bus
17	Failed L-ADC input bus
18	Failed R-ADC input bus
19	0
20	0
21	ARINC wraparound failed
22	High-voltage power supply failed
23	Low-voltage power supply failed
24	MEM memory module failed
25	I/O processor failed
26	Display processor failed
27	Video amplifier failed
28	Deflection amplifier failed
29	Graphics generator failed
30	* SSM code
31	* SSM code
32	Parity (odd)

If bit 14 is set to ë1í the diagnostic software has detected a failed LA-IOC-1 input bus.

If bit 23 is set to ë1í the low-voltage power supply has failed.

**SSM code:	Bit 31	Bit 30	LRU Status
	0	0	Failed
	0	1	No computed data
	1	0	Functional test
	1	1	Normal

EXAMPLE: Diagnostic word = 600000		
6	Bits 29-32	Bits 30 and 31 are set; SSM = Normal
0	Bits 25-28	No bits are set
0	Bits 21-24	No bits are set
0	Bits 17-20	No bits are set
0	Bits 13-16	No bits are set
0	Bits 9-12	No bits are set

Fig. 3-64 Diagnostic word explanation. (Collins Divisions, Rockwell International)

STEP	PROCEDURE
	Note
	The MFD 1 may be swapped with either of the PFDs to verify aircraft wiring and to isolate a failed unit. Each unit will display (MFD/PFD) information according to its mount; line key operation is enabled only at the MFD mount.
1.0	Check the AVIONICS STATUS page and troubleshoot according to table 5-3.
	If the AVIONICS STATUS page will not display, check the L-MFD-1 bus.
2.0	If the display is blank, check the circuit breaker. Then swap the MFD with an (operational) PFD to isolate a failed MFD or aircraft wiring problem.
3.0	If DISPLAY TEMP annunciates on the MFD, check the cooling fan installed in the MFD mount. The MFD provides fused power and ground for this fan.
4.0	Select the following formats and verify correct MFD display response. On the left CDU select the radar format, the plan map (selected from MFD IDX page), the present position map, and then the HSI format.
5.0	Check that the heading indication is valid and correct (agrees with PFD and SDU headings). If not, check for L-AHC-2 bus activity at P1 pins 3K/3J.
6.0	Check for a TAS readout on the MFD status line. If TAS is dashed or is not displayed, check for L-ADC-3 bus activity at P1 pins 2F/2E.
7.0	Set left AHRS reversion switch to 2, and verify that MAG 2 annunciates on the MFD. Repeat step 5.0. If heading is not correct, check the AHRS switch and the R-AHC-3 bus at P1 pins 11B/11A. Set AHRS switch to 1.
8.0	Set left AIR DATA reversion switch to 2, and verify that ADC 2 annunciates on the PFD 1. Repeat step 6.0. If TAS is not correct, check the AIR DATA switch and the R-ADC-2 bus at P1 pins 12G/13G. Set AIR DATA switch to 1.
9.0	Set left CDU reversion switch to REV, and verify that CDU 2 annunciates on the MFD. Check that the left CDU blanks and that the right CDU now controls the left PFD and MFD displays. Set CDU switch to NORM.
10.0	Set the left PFD reversion switch to REV. Verify that the MFD now displays PFD data and that the left PFD blanks. Set PFD switch to NORM.
11.0	Display the ACTIVE NAV SOURCE page (on CDU 1). Select each NAV source and verify that the MFD course display is white for FMS, green for VOR 1 (or LOC 1), and yellow for VOR 2 (or LOC 2).
12.0	Press the IDX line key to display the MFD index page. Then press the NORMAL PROCEDURES line key and display a checklist. Press the control wheel LINE ADV button to check the cursored line; press the optional PAGE ADV button to exit the checklist and select the next checklist on the NORMAL PROCEDURE menu.
13.0	A checklist should still be displayed on the MFD. Move the CHP joystick up to select the previous checklist page, down to select the next checklist page, right to exit the checklist and select the next one on the menu, and left to exit the checklist and select the previous one on the menu. Then press the IDX line key.
14.0	Select various MFD displays and check all line keys for sticky or improper operation.
15.0	If a problem is suspected with the L-MFD-1 bus to the DAU, display the LRU DIAGNOSTIC DATA page showing the DATA ACQ UNIT B word. Refer to table 5-5. If bit 18 is set, check for bus activity at DAU pins P2-48/47 to isolate a failed DAU input or wiring problem.

Fig. 3-65 MFD 1 test procedures. (Collins Divisions, Rockwell International)

TRANSPORT CATEGORY AIRCRAFT INTEGRATED ELECTRONIC INSTRUMENTS

This section of the text will take a brief look at three popular transport category aircraft. Their electronic instruments are advanced systems, which integrate with airframe and engine displays as well as warning systems, flight management systems, the flight data recorder, and the central maintenance computer. The integration of these systems allows for greater redundancy, improved reliability, and greater maintainability. Only a brief discussion of the electronic instruments will be given here since the integrated systems are discussed in chapters four and five of this text.

THE MCDONNELL DOUGLAS MD-11 ELECTRONIC INSTRUMENT SYSTEM

The McDonnell Douglas MD-11 aircraft employs a state-of-the-art **electronic instrument system (EIS)**, which displays flight and navigational data, as well as engine parameters and warning information. (Keep in mind McDonnell Douglas is now part of the Boeing Corporation, so this aircraft is actually a Boeing product.) As with the B-747-400, EIS is an integrated system that operates in conjunction with a variety of aircraft systems. The following discussion will briefly examine the MD-11 EIS.

System Components

As can be seen in figure 3-66 there are six CRT display units arranged in a horizontal fashion on the MD-11 instrument panel. The EIS contains three computers, called the **display electronic units (DEUs)**, that process system data and send output signals to the displays. There are three control panels used to select various functions of the EIS: the **source input select panel**, the **electronic flight instrument control panel**, and the **systems control panel**.

Display units

The CRTs display four basic formats: 1) **PFD (primary flight display)**, 2) **ND (navigational display)**, 3) **EAD (engine alert display)**, 4) **SD (system display)**. There are two each of the PFDs and NDs, one each for the pilot and copilot. The EAD and SD are normally used to display engine and systems parameters and to display alert and warning messages similar to the Boeing EICAS (discussed in chapter four). The main advantage of integrating all six displays is to provide greater flexibility for backup operation in the event of a display failure.

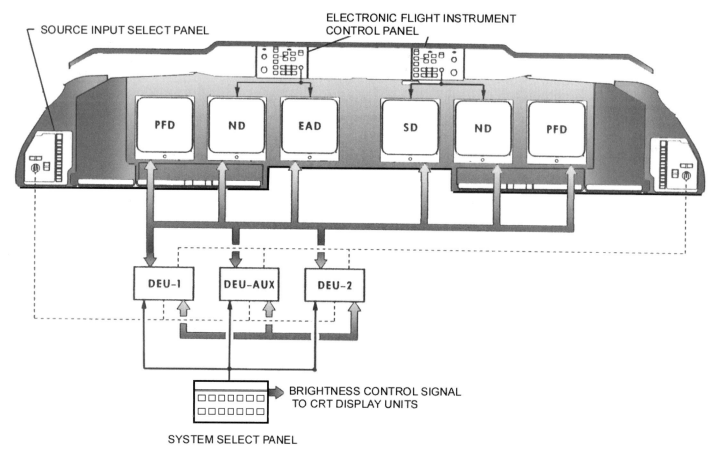

Fig. 3-66 MD-11 EIS display configuration and DEU interface. (Boeing Commercial Airplane Company)

The displays are fully interchangeable and contain 8 x 8 inch color CRTs. The CRT brightness is controlled manually by six individual controls on the system select panel (see figure 3-67). These controls can also be used to turn off each display if desired. It is standard operating procedure to turn off a display if it fails during flight. The displays can also be turned off if the aircraft is to be powered up for long periods during maintenance. A remote light sensor is used to adjust the display brightness automatically. The remote light sensor is located on the flight deck glare shield and monitors ambient lighting conditions. The PFD and ND formats are nearly identical to the formats of the displays on the Collins Pro Line 4 system discussed previously. The display units on most transport category aircraft receive power from several different busses to provide redundancy in the event of a bus failure.

Display electronics units

Typical of other transport category EISs, this system employs three DEUs (see figure 3-66). The **display electronic units (DEUs)** provide the processing power for the six displays and also feed various other aircraft systems. The DEUs receive digital, analog, and discrete inputs from a variety of aircraft systems and sensors. The pilot and copilot can use reversionary switching to select which DEU drives which display. During normal operation, DEU-1 drives the three left-side displays;

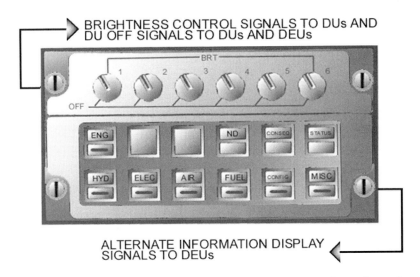

BRIGHTNESS CONTROL SIGNALS TO DUs AND
DU OFF SIGNALS TO DUs AND DEUs

ALTERNATE INFORMATION DISPLAY
SIGNALS TO DEUs

Fig. 3-67 MD-11 system control panel. (Boeing Commercial Airplane Company)

DEU-2 drives the three right-side displays. The auxiliary DEU is ready in a hot-standby mode. To provide redundancy, three different AC busses supply power to each DEU.

System testing and troubleshooting

The EIS found on the MD-11 contains a self-monitoring program to detect system faults. Duplicated input signals are compared for accuracy. DEUs cross-talk to compare data sent to the various displays. If an inconsistency is detected, a flag is displayed on the appropriate display to alert the flight crew. If a fault in the EIS is detected by onboard diagnostics, the technician can access fault information from a nonvolatile memory within the **centralized fault display system (CFDS)**. The CFDS is a fault monitoring system used on transport category aircraft to aid in troubleshooting and repair of various systems. The CFDS is covered in detail in chapter five of this text.

THE BOEING 747-400 INTEGRATED DISPLAY SYSTEM

The B-747-400 flight instrument displays are part of a complete aircraft monitoring system called the **Integrated Display System (IDS)**. The integrated display system monitors a variety of aircraft and engine system parameters as well as various flight parameters through three **Electronic Interface Units (EIU)**. As shown in figure 3-68, three EIUs manipulate incoming data and send outputs to six display units. The EIUs are the main processing units for the IDS. The EIUs receive information from 16 different systems, process the data, and then output signals to the flight deck displays and the other users such as the flight management system and flight data recorder. One major difference between transport category aircraft and light aircraft (such as the Beechjet 400A previously discussed) is the number of processors used to drive the displays. On transport category aircraft, three processors are required to provide added redundancy; most light aircraft only have two processor units.

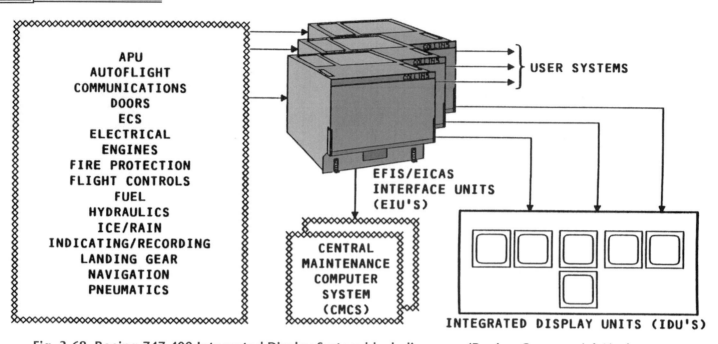

Fig. 3-68 Boeing 747-400 Integrated Display System block diagram. (Boeing Commercial Airplane Company)

The display units on the B-747-400 are called **Integrated Display Units (IDUs).** The six IDUs are located in a T configuration as shown in figure 3-69. During normal operation, the **Primary Flight Displays (PFDs)** are used to show flight data such as airspeed, attitude, and altitude. The **Navigational Displays (NDs)** are used for the display of navigational data, such as the compass rose, course map, and weather radar information. The two EICAS displays, located in the center of the instrument panel, are used to display engine and airframe systems data. EICAS will be discussed in chapter four of this text.

Since IDS is an integrated system, it provides greater flexibility in the event of a single or multiple display failure. In other words, there are six displays (not four) driven by the same processors; hence, an integrated system provides more options in the event of a display failure. On integrated systems, the EIUs also talk directly to the central maintenance computers. The **Central Maintenance Computer System (CMCS)** is used to monitor the health of various aircraft systems. All IDS BITE tests are accessed through the CMCS.

The three EIUs are located in the main equipment center of the aircraft. These computers are static discharge sensitive and should be handled only when using proper precautions. The electrical leads of the EIUs are made through an 800-pin connector mounted on the rear of the unit. Take special care not to damage the pins or sockets during removal and replacement of the computers. Maintenance of the B-747-400 IDS will be discussed in further detail in chapters four and five.

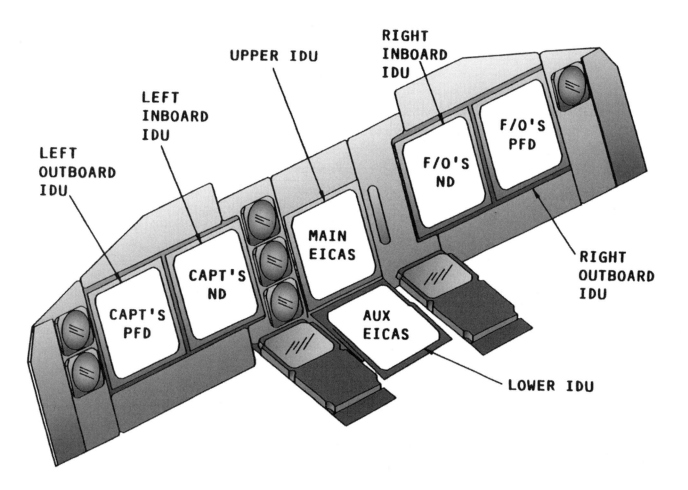

Fig. 3-69 Boeing 747-400 IDS display configuration. (Boeing Commercial Airplane Company)

THE AIRBUS A-320 EFIS

The flight displays found on the Airbus A-320 are part of an integrated display system called the **Electronic Instrument System (EIS).** The EIS contains six CRT display units arranged in a T configuration similar to the Boeing 747-400. The four outboard displays are used primarily for flight data (EFIS). The remaining two displays are used primarily for the **Electronic Centralized Aircraft Monitoring System (ECAM)**. ECAM is similar to the Boeing EICAS and will be discussed in chapter four. On the A-320 the two captain's side EFIS displays are called the **Primary Flight Display 1 (PFD 1)** and the **Navigational Display 1 (ND 1)**. The first officer's (copilot's) side consists of PFD 2 and ND 2. The two ECAM displays are referred to as the upper and lower ECAM displays (see figure 3-70).

As with other integrated systems, the major advantage of the EFIS/ECAM system is the flexibility in the display of data. In the event a display fails, one of the operational displays can be used to show the missing data. The flight crew can switch the PFD and ND data manually. This is accomplished by the PFD/ND XFR switch located just outboard of the captain's and first officer's EFIS displays. The data from the upper ECAM display is automatically transferred to the lower ECAM display in the event of a single ECAM display failure. The upper ECAM data moves to an ND in the event both ECAM displays fail.

197

Figure 3-70 shows the relationship of the EIS processors and the six displays. The main processors for the A-320 EIS are known as the **Display Management Computers (DMCs).** DMC #1 and #2 drive the displays during normal operation. DMC #3 operates in hot standby and provides back up for DMC #1 & 2. The DMCs receive inputs from four computers, the **Flight Warning Computer (FWC)** #1 and 2, and the **System Data Acquisition Concentrators (SDAC)** #1 and 2. The FWCs and the SDACs receive information from the various aircraft systems. Figure 3-70 shows the various input/output signals to the DMCs. Here it can be seen the DMC receives ARINC 429, ARINC 453, RS-422 and DSDL inputs. The **dedicated serial data line (DSDL)** is a digital bus structure unique to Airbus aircraft. Each DMC normally feeds three displays through a DSDL interface. Each display sends a feedback signal (used for health monitoring) to the DMC. The EIS BITE is monitored by the aircraft's centralized fault display system. More details on the various system interfaces and system troubleshooting will be given in chapters four and five.

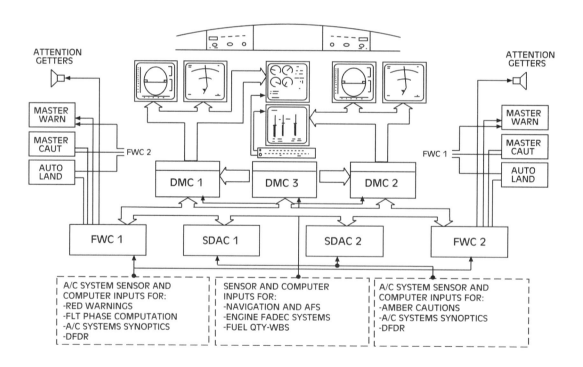

Fig. 3-70 A-320 Electronic Instrument System (EIS) Block diagram. (Airbus Industrie)

CHAPTER 4
INTEGRATED MONITORING
AND WARNING SYSTEMS

INTRODUCTION

Early in the history of aviation it became quite clear that certain systems should be monitored during flight. Even the most basic airplane had a means to determine engine RPM, engine oil pressure, and fuel quantity. As aircraft grew in complexity, many systems were added which required constant or periodic observations. Air-speed indicators, altimeters, and manifold pressure gauges soon cluttered the instrument panel. Twin engine and turbine engine aircraft brought more engine instruments. Faster, higher and longer distance flights required more systems and more instruments.

The aircraft grew from the simple machine first launched at Kitty Hawk to a complexity that required three or more flight crewmembers. The pilot, the copilot, a navigator, and the radio operator were all standard crew on some aircraft. Then, some time in the late 50's or early 60's, aircraft systems began to simplify. With the help of modern electronics (modern for the time period that is), the radio operator's job was combined with the navigator's responsibilities. The Boeing 727 required a three-person crew, and the newer B-737 required only a two-person crew.

At the same time when crew size seemed to decrease, more systems were added to the aircraft to improve flight safety. Landings with reduced visibility were now common place. Radar was onboard almost every commercial airliner. Hydraulics, pneumatics, and electrical systems grew in size and complexity. During this time period, aircraft designers and engineers realized that pilot workload was becoming overwhelming and systems monitoring had to be reduced. The first step was to reduce the number of instruments, gauges, and lights that required constant attention.

The invention of the horizontal situation indicator (HSI) and attitude director indicator (ADI) was a major step toward reducing pilot workload. The information on these indicators combined a multitude of instruments to reduce instrument panel clutter. Much of the information contained on these indicators was out of view until needed. This helped to reduce confusion and pilot workload. Today, information is presented to the flight crew on a "need to know" basis. Only necessary information, critical at that time, is displayed, once again reducing confusion and workload.

It was soon possible to combine many navigation and communication radio systems into a common control panel, and many of the systems that required manual operation could be operated automatically. But still, automation required more equipment. Size and weight were limiting factors as to how far automation could go. More electronics equipment also generated more heat, which presented cooling problems.

Then it happened, with the advent of the Boeing 757 & 767 aircraft, many of the older analog systems were replaced with lighter, faster and more reliable digital technologies. Conventional gauges and instruments, with a multitude of moving parts, were replaced with CRT displays. Now for the first time, a real reduction could be made in the flight crew workload. Automatic engine and system monitoring computers coupled to CRT display systems were introduced on the B-757/767. These aircraft contained the first generation of **Engine Indicating and Crew Altering Systems (EICAS)** available on transport category aircraft.

From the late 1970's to the present, improvements in computer technologies have allowed aircraft manufacturers to improve safety and reduce pilot workload. Modern aircraft such as the B-747-400, the A-320, or the B-777 monitor virtually every system that uses electricity. On modern aircraft, engine and systems monitoring is accomplished through two basic systems: EICAS and **ECAM (Electronic Centralized Aircraft Monitoring)**. EICAS is used on Boeing and McDonnell Douglas aircraft, while ECAM is used on the Airbus Industrie aircraft.

The EICAS and ECAM systems have also evolved over their life span. For example, early EICAS were more or less dedicated systems. They used computers dedicated to EICAS and did little else. Today, both EICAS and ECAM are integrated with other aircraft systems, such as the central maintenance system and electronic flight instrument system. The new Boeing 777 incorporates an **Airplane Information Management System (AIMS)** which further integrates electronic systems. AIMS actually shares some of the computer software between various systems of the aircraft. On many modern aircraft, system monitoring can even be downlinked to ground facilities. This allows maintenance crews to determine fault activity while the aircraft is still in flight, hence, reducing repair time and unscheduled delays.

Aircraft monitoring systems have come a long way. Today's computers allow for more control with less pilot interaction. Many corporate type aircraft now employ automated monitoring systems. Soon this type of technology will be used on all aircraft.

This chapter will examine the automated monitoring systems currently used on transport category aircraft. We will look at EICAS on two aircraft, the Boeing 757 and the 747-400. A state-of-the-art ECAM system will be examined for the Airbus A-320. Also presented in this chapter is the new integrated AIMS used on the B-777.

ENGINE INDICATING AND CREW ALERTING SYSTEMS (EICAS)

The automated engine and systems monitoring process used on modern Boeing aircraft is called **EICAS (Engine Indicating and Crew Altering System)**. This system incorporates two CRTs used to display engine data, airframe systems data, and warning messages to the flight crew. Early systems such as those found on the Boeing 757 & 767 operate more or less independent of other major aircraft systems. The EICAS found on the Boeing 747-400 and 777 are fully integrated systems that share components and information with the **electronic flight instrument system (EFIS), the central maintenance computer system (CMCS), and the flight management system (FMS)**.

The following discussion will focus on the features common to all first generation EICAS. This chapter will present the specific design features for the B-757 (an early version EICAS). The later portion of this chapter will cover the design and operation of an advanced EICAS, such as the system found on a B-747-400.

General Description

EICAS has many similarities to the electronic flight instrument system discussed in chapter three. Like EFIS, EICAS employs digitally controlled CRT displays; however, EICAS is used to display various system parameters, such as engine pressure ratio, RPM, and exhaust gas temperature. Other systems, such as hydraulic or pneumatic pressures and electrical system parameters can also be displayed, or in some cases, removed from the screen at the discretion of the pilot. Another vital function of EICAS is to monitor the various aircraft systems and to display caution and warning information in the event of a system failure. Although some of the latest corporate and commuter type aircraft employ EICAS, most of these systems are found on transport category aircraft made by the Boeing Commercial Airplane Company.

The EICAS displays certain aircraft system and engine parameters on a *need to know* basis. That is, not all the systems data are displayed continuously. In the event of a system malfunction, any vital information automatically appears on a CRT and the appropriate caution or warning signals are activated. During normal operation, only a minimum of engine data is displayed; additional system data may be displayed if the pilot activates the appropriate EICAS control.

EICAS displays

EICAS contains two CRT displays typically placed vertically as shown in Figure 4-1. During normal flight configurations, the lower CRT is typically blank. The lower CRT is used to display status of any malfunctioned system. The lower CRT is also a "backup" display in the event the top CRT fails. The upper EICAS CRT is called **the main EICAS display** and the lower CRT is called the **auxiliary EICAS display**. During normal operation, the upper CRT is used to display engine data, such as exhaust gas temperature (EGT), rotor RPM (N_1), and (N_2).

On any given aircraft, the EICAS displays are interchangeable and commonly referred to as line replaceable units (LRUs). On most aircraft, the EICAS CRTs and the EFIS CRTs are also completely interchangeable. Figure 4-2 shows a typical CRT, the removal handle and the location of the installation screws. To remove the CRT: 1) turn system power off, 2) hold the handle flat against the CRT and remove the retaining screws located at the top of the display, 3) pull the handle outward and the CRT will slide from its rack.

> **Caution**: 1) Most display units are electrostatic discharge sensitive. Be sure to take the necessary precautions to prevent damage to the unit. 2) The handle is spring loaded away from the face of the display when the unit is fully engaged. While removing or installing the retaining screws always release the spring pressure by holding the handle flat against the face of the display. Failure to do so will most likely damage the fasteners. 3) Some displays have an antiglare film on the face of the CRT. These displays are easily

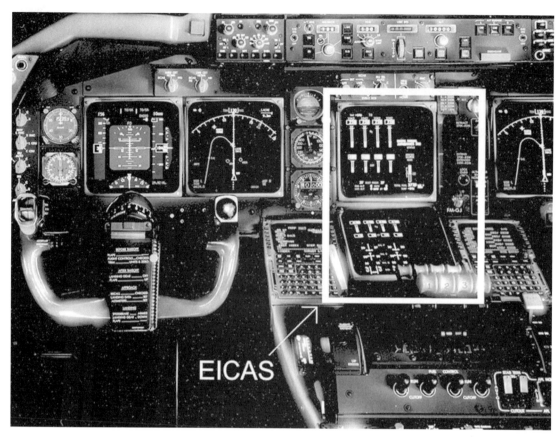

Fig. 4-1 Two EICAS displays from a Boeing 757.

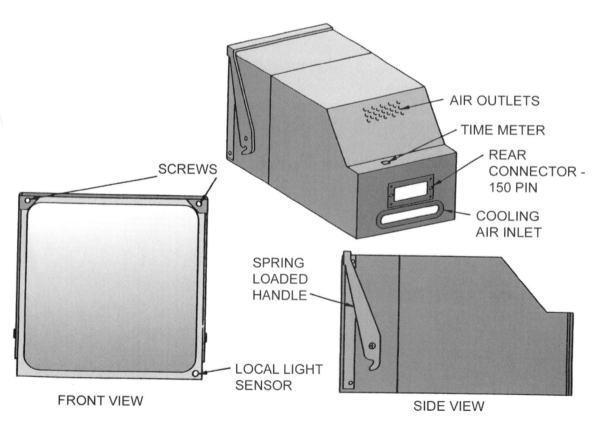

AIR OUTLETS

TIME METER

REAR
CONNECTOR -
150 PIN

COOLING
AIR INLET

SCREWS

SPRING
LOADED
HANDLE

LOCAL LIGHT
SENSOR

FRONT VIEW

SIDE VIEW

Fig. 4-2 Typical EICAS CRT.

scratched and require special cleaning procedures. Always be careful when using tools around the face of the display. Protect the display from rough handling and never set the unit face down.

In general, installation is preformed in reverse order of removal. During installation be sure the handle is in the full-out position and gently push the display into position. Then press the handle slowly toward the CRT. This will seat the display and engage the electrical connector pins at the rear of the unit. After a proper installation, the handle should lay flat against the display held by the retaining screws. It should be noted that in some units the installation screws and handle are located on the side of the CRT not on top.

EICAS formats

There are several display formats used by the different versions of EICAS. These formats vary with aircraft model. Some common formats include primary, secondary, and compact modes. The primary format shows on the upper CRT during normal operation. Four different colors are used to display information. A change in color indicates a change in system status. Six colors are used by most EICAS displays. White is used for display of various scales, pointers and digital readouts when the system is in the normal operating range. Red is used to indicate that a system has exceeded a predetermined value (Redline). If a system exceeds its limits, the entire scale, pointer, and/or digital readout will turn red. Red is also used to show certain warning messages. Green is used to indicate normal operating of a system. Amber is used for certain warning messages and some scale markings. In some cases, a system display will change to amber if that system enters the caution range. Magenta (pink) is used for certain messages and display parameters. Cyan (light blue) is used for various labels and messages.

Alert messages

During normal operation, EICAS is used to alert the flight crew as to any abnormal powerplant or airframe system operation. There are four types of alert messages used with EICAS. Level A are **warning** messages shown in red. These are the most important messages. Level B are **caution** messages displayed in amber. Level C are **advisories** displayed in amber or cyan depending on the specific model EICAS. Level D messages, called **memos**, are displayed in white. Status and maintenance messages are also shown on some systems, and are typically displayed in white.

Warnings are known as level A messages. Warnings require immediate attention and immediate action by the flight crew. Level A faults include very serious failures, such as engine fire or cabin depressurization. Warnings will also activate additional aural and visual annunciators on the flight deck. For example, a fire bell sounds in the event of an engine fire. There are very few faults that require level A warnings. Caution messages are known as level B alerts. Cautions appear on the EICAS CRT in amber just below any level A message. Level B alerts also create a distinct audio tone and illuminate a discrete annunciator. Cautions require immediate crew awareness and future crew action. Advisory messages (level C) require immediate crew awareness and possible future action. Memos are used for crew reminders. Typically there is no aural tone or master caution/warning associated with advisories or memos. For all levels of alerts, the most recent message appears at the top of its category. Level A messages appear at the top of the display, level B is below level A, level C is next and level D is shown on the bottom of the list. Status and maintenance messages are used to aid the flight crew and maintenance technicians in determining the aircraft's status prior to dispatch. Maintenance messages used in conjunction with the minimum equipment list for the aircraft will determine what repairs (if any) are required prior to the next flight.

All EICAS equipped aircraft contain two **master warning/caution annunciators.** These annunciators consist of an illuminated switch assembly located on the instru-

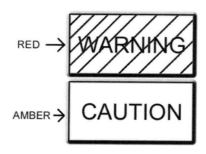

Fig. 4-3 EICAS master warning/caution light assembly.

ment glare shield directly in front of the pilot and copilot. The red warning is located in the top half of the assembly and the amber caution is located on the lower portion of the assembly (see figure 4-3). The warning and caution lights illuminate whenever EICAS presents a level A and B message. If the assembly is pressed, the lamps are extinguished and any aural warning is canceled; however, the related EICAS message remains on the CRT display.

At various times during the flight it is best not to distract the pilots with alert messages. Therefore, EICAS incorporates inhibit software to keep alert messages from

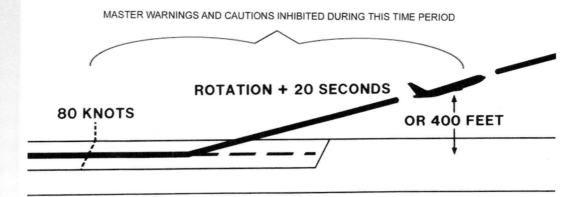

Fig. 4-4 EICAS take off caution inhibit criteria.

being displayed during crucial flight periods. Most of the inhibits are initiated during takeoff and landing. Figure 4-4 shows an example of a typical takeoff inhibit. In this case, the master warning and caution lights and related audio tones are inhibited from operation between the time the aircraft reaches 80 knots until the aircraft climbs to 400 feet radio altitude (or 20 seconds after liftoff, which ever comes first). If the problem still exists after the inhibit period, the appropriate EICAS message, discrete annunciator and aural tone will be activated.

Compact mode

The EICAS incorporates a compact mode, which is used when one display is inactive, or being used to display maintenance pages. In the compact mode the data is typically displayed in digital format only. In other words, vertical or round dial instrument representations are eliminated. The compact mode is used during flight if one display becomes inactive. Compact information can be shown on either the main or auxiliary displays. While in the compact mode, EICAS is said to have "degraded operation".

Event recording

All engine indicating and crew alerting systems have the capability to record system data in a nonvolatile memory. The memory can then be accessed by maintenance personnel for verification of system malfunctions and excedances. There are two types of data stored in the EICAS nonvolatile memory: **manual events**, and **automatic events**.

If a component fails causing a system to exceed limits or parameters, the EICAS will automatically record the event in the **auto event** mode. The EICAS **manual event** mode is activated whenever the *event record* button on the EICAS control panel is pressed. On early systems this nonvolatile memory was quite limited; later versions increased the storage capacity for recording fault information.

Maintenance pages

For maintenance purposes, EICAS employs a series of displays called **maintenance pages** to show data related to the various systems monitored by EICAS. For most aircraft, these pages can be accessed on the ground and/or during flight. Maintenance pages are displayed on one of the system's CRTs. The access of a specific page is usually done to verify the system performance during a previous flight. The maintenance technician will typically receive a "write up" from the flight crew, which indicates a system malfunction. The faulty system is then accessed through the EICAS maintenance pages, which provides more data on the system parameters.

FIRST GENERATION ENGINE INDICATION AND CREW ALERTING SYSTEMS

The first practical use of EICAS on transport category aircraft came with the introduction of the Boeing 757 and 767 aircraft. First generation systems are also found on some B-737s. The Boeing 757 engine indicating and crew alerting system will be discussed here as an example of a *first generation* EICAS.

System Architecture

A block diagram of the B-757 EICAS is shown in figure 4-5. In this type system, two EICAS computers monitor inputs from engines and airframe systems to display system data and alert the crew in the event of a malfunction. The EICAS computer uses both **discrete analog** and **digital** data to communicate with the various components of the system. The ARINC 429 data bus system is used for most EICAS digital data transmission and reception. But, be aware, other data bus formats may be used on certain systems.

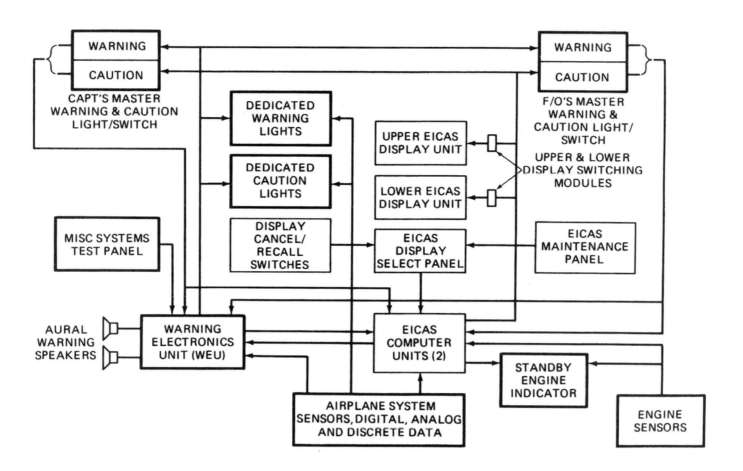

Fig. 4-5   Block diagram of a B-757 EICAS. (Boeing Commerical Airplane Company)

The major EICAS components are two computers, two CRT displays, one display
select panel, one maintenance panel, and one display switching modules. EICAS is
also made up of the several discrete components shown surrounded by a bold rectangle in figure 4-5. As shown in figure 4-6, the displays and switching panels are
located on the instrument panel. The discrete annunciators are located on the pilot's
and copilot's glare shield. The EICAS computers and switching modules are located in the main equipment rack.

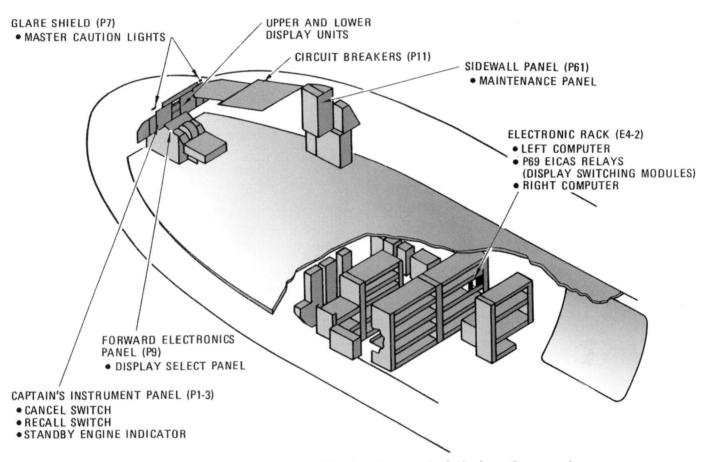

Fig. 4-6 Location of EICAS system components. (Boeing Commerical Airplane Company)

First generation systems contain several discrete annunciators and a standby engine
indicator as shown in figure 4-7. The discrete annunciators are used as "attention
getters" for the flight crew during certain system failures, such as ground proximity
warning or autopilot disconnect. The standby engine indicator is comprised of two
independent liquid crystal displays (LCDs). This system will be discussed in more
detail later in this chapter.

A more detailed block diagram for the B-757 EICAS is shown in figure 4-8. All
digital data connections are shown by a wide, double line, arrow; analog data is
shown by the narrow, single line, arrows. Reference this diagram during the remainder of the discussion on system architecture.

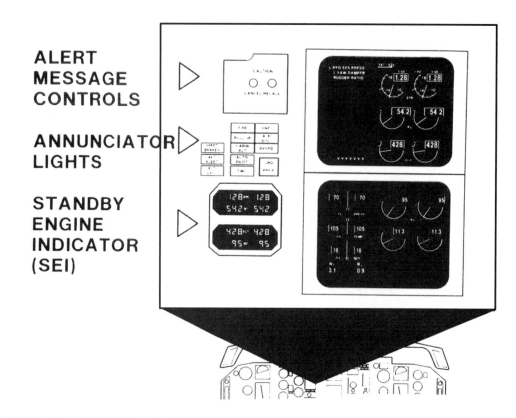

Fig. 4-7   EICAS standby indicators and discrete annunciators. (Boeing Commerical Airplane Company)

EICAS receives power from both 115 VAC and 28 VDC sources run through various independent circuit breakers. This system allows for redundant power inputs to EICAS. The right side computer, lower CRT display, and display select panel are powered by the right main AC bus. The left computer and upper CRT display receive power from the left main AC bus. The right 28 VDC bus powers the upper and lower display switching modules. To offer the least likely possibility of failure, the standby engine indicators receive 28 VDC from the hot battery bus. The master warning/caution lights are powered by the master dim and test circuits, receiving 28 VDC.

The two EICAS computers communicate with each other (cross talk) via a high speed ARINC 429 digital data bus for the purpose of comparing input data. This cross talk feature is essential to provide redundancy for the system. The EICAS computers receive over 400 analog inputs from various systems and specific switches throughout the aircraft. Discrete signals include information, such as ground/air and resets for the master warning/caution annunciators. Pin programming for each EICAS computer is also done through discrete input signals. Pin programming will be presented in more detail later in this chapter.

As shown in the upper left corner of figure 4-8, the discrete signals from the air/ground relay, maintenance control panel, and the cancel/recall switches are sent to the computers through the display select panel. In the lower portion of the diagram it can be seen that the discrete right and left engine data is paralleled to both the computers and the standby engine indicator (bottom right).

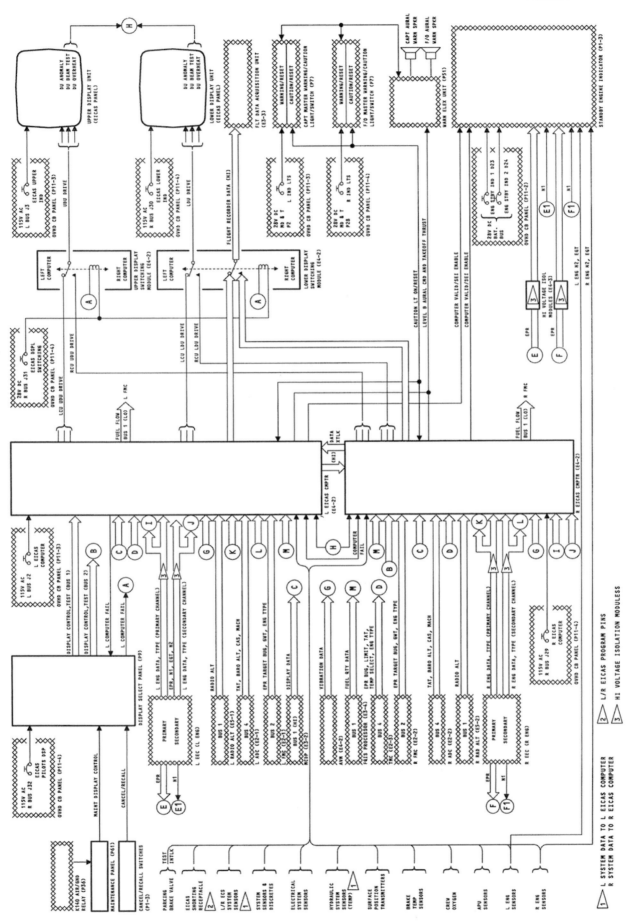

Fig. 4-8 Detailed block diagram of the B-757 EICAS. (Boeing Commerical Airplane Company)

209

Digital inputs for the EICAS computers include engine data from the left and right electronic engine controls (EECs). Due to the high priority of engine data, the EECs transmit on a dual channel directly to both EICAS computers. Some items, such as, the display select panel (DSP) and the fuel quantity indicating system (FQIS) transmit on a single digital bus to both EICAS computers. A third type of digital input is transmitted to only one computer and that data is shared through the computer cross talk bus. For example, the left flight management computer (FMC) transmits only to the left computer, the right FMC transmits to the right computer.

Switching modules

On the B-757, two EICAS **switching modules** are used to control the output data signals from the EICAS computers to the upper and lower CRT displays and to the flight data acquisition unit. These modules are made up of circuit cards located in the equipment rack between the left and right EICAS computers. The switching relays are located on the circuit cards. The switching relays (see the upper right portion of fig. 4-8) receive input signals from the display select panel.

The EICAS computer outputs are comprised of both digital and analog signals. The digital data is sent to the flight data recorder or flight data aquisition unit, through the lower display switching module. The left and right FMC receive digital data from the left and right computers respectively.

Analog outputs include signals which are supplied directly to specific systems and those transmitted through the upper and lower display switching modules. Typically the left computer output is transmitted to both EICAS displays through the normally closed contacts in the two switching modules. The normally open contacts connect the right computer to the upper and lower displays. The switching modules connect the right computer to the displays only when the display select panel computer switch is in the *AUTO* position and the left computer is inoperative, or when the *R* (right) computer is selected.

System monitoring

The diagram (figure 4-8) shows some of the monitoring circuitry of the EICAS. For example, the output from the upper and lower display units labeled *H* is used to input a validity signal to the EICAS computers. The circuitry within the displays will monitor the health of the incoming data and the display itself. Then, the display health is reported back to the computers. This type of self monitoring system allows EICAS to react correctly when a problem occurs.

EICAS Controls

There are two major control panels used for the engine indicating and crew alerting system on the B-757, the **display select panel (DSP)** and the **maintenance control panel (MCP).**

Display select panel

The DSP, shown in figure 4-9, is used during normal flight to set various system parameters. The two display push buttons *ENGINE* and *STATUS* are used to select

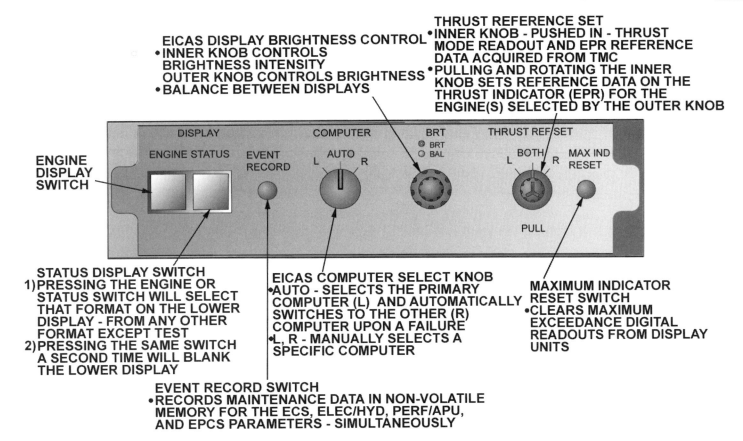

EICAS DISPLAY BRIGHTNESS CONTROL
- INNER KNOB CONTROLS BRIGHTNESS INTENSITY
  OUTER KNOB CONTROLS BRIGHTNESS
- BALANCE BETWEEN DISPLAYS

THRUST REFERENCE SET
- INNER KNOB - PUSHED IN - THRUST MODE READOUT AND EPR REFERENCE DATA ACQUIRED FROM TMC
- PULLING AND ROTATING THE INNER KNOB SETS REFERENCE DATA ON THE THRUST INDICATOR (EPR) FOR THE ENGINE(S) SELECTED BY THE OUTER KNOB

DISPLAY   COMPUTER   BRT   THRUST REF SET

ENGINE STATUS   EVENT RECORD   AUTO L R   ○ BRT ○ BAL   L BOTH R   MAX IND RESET

ENGINE DISPLAY SWITCH

PULL

STATUS DISPLAY SWITCH
1) PRESSING THE ENGINE OR STATUS SWITCH WILL SELECT THAT FORMAT ON THE LOWER DISPLAY - FROM ANY OTHER FORMAT EXCEPT TEST
2) PRESSING THE SAME SWITCH A SECOND TIME WILL BLANK THE LOWER DISPLAY

EICAS COMPUTER SELECT KNOB
- AUTO - SELECTS THE PRIMARY COMPUTER (L) AND AUTOMATICALLY SWITCHES TO THE OTHER (R) COMPUTER UPON A FAILURE
- L, R - MANUALLY SELECTS A SPECIFIC COMPUTER

MAXIMUM INDICATOR RESET SWITCH
- CLEARS MAXIMUM EXCEEDANCE DIGITAL READOUTS FROM DISPLAY UNITS

EVENT RECORD SWITCH
- RECORDS MAINTENANCE DATA IN NON-VOLATILE MEMORY FOR THE ECS, ELEC/HYD, PERF/APU, AND EPCS PARAMETERS - SIMULTANEOUSLY

Fig. 4-9 EICAS display select panel. (Boeing Commerical Airplane Company)

what information will be displayed on the lower EICAS CRT. The *EVENT RECORD* button is used to store system data in the EICAS nonvolatile memory. The *COMPUTER* knob is used to select which EICAS computer will be used to power the displays. In the AUTO mode, the system will automatically switch computers in the event the primary computer fails. The *BRT* control is used to adjust the EICAS display brightness. It should be noted that brightness is also automatically adjusted to compensate for changes in flight deck light levels. The *THRUST REF SET* is used to select which thrust management computer communicates to EICAS. The engine reference data can also be set by this control. The push-button labeled *MAX IND RESET* is pressed to remove any exceedance values on the EICAS displays.

Maintenance control panel

The controls of the MCP are located on the P61 panel installed aft of the first officer's seat. This panel is used during ground operations and is inhibited from certain functions unless the parking brake is set. The six *DISPLAY SELECT* push buttons located on the left of the panel are used to select which maintenance data will be displayed by EICAS. The label on each button represents the following systems: *ECS/MSG*, environmental control system/messages (miscellaneous); *ELEC/HYD*, electrical/hydraulic; *PERF/APU*, performance/auxiliary power unit; *CONF/MCDP*, configuration/maintenance control display panel; *ENG EXCD*, engine exceedances; *EPCS*, electronic propulsion control system (engine parameters).

A third EICAS control panel containing the cancel and recall switches is located on the pilot's instrument panel just to the right of the upper EICAS display. The *cancel* push-button is used to cancel any caution or advisory message currently displayed by EICAS. The *recall* switch is used to retrieve previously canceled messages.

Normal Operation Of The B-757 EICAS

During regular operation the early versions of the engine indicating and crew alerting systems can display three basic formats: **normal, secondary and status**. Figure 4-10 shows the typical EICAS displays for each of the three basic formats. Upon starting the first engine, the secondary format automatically appears on the auxiliary EICAS display. The secondary format displays engine parameters, such as engine pressure ratio (EPR) N_1 and N_2 rotor speeds, fuel flow, oil pressure, oil temperature, oil quantity, and engine vibration. This information is displayed in both a graphic and digital format.

Status page format

The status page format is displayed on the lower EICAS CRT when the status switch on the display select panel is momentarily pressed. The crew is alerted to the presence of a new status message by the word *STATUS* presented in cyan on the lower EICAS display. As shown in figure 4-11 the status format can display up to 11 status messages in the upper right of the display. If more than 11 messages are present, ten status messages appear followed by the term *page 1*. This indicates that a second status page is available by pressing the status button again. The system and control position information (items numbered 1-23, and 26-31 in figure 4-11) will appear on the status page whether status messages are present or not. System information includes hydraulic fluid quantity and pressure, APU data, oxygen system pressure, fuel flow and break temperature. The control surface positions are shown in the lower left corner of the status page.

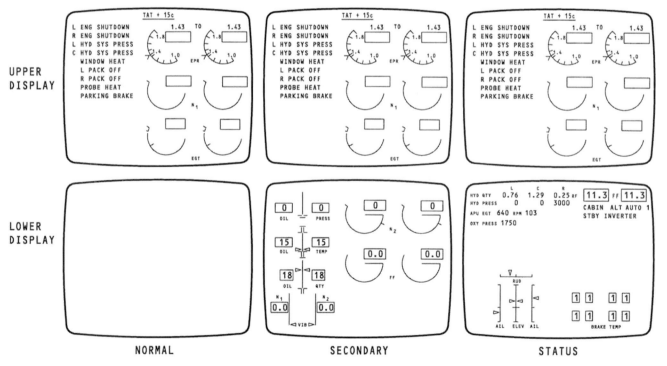

Fig. 4-10   Typical EICAS displays for the Normal, Secondary Engine, and Status formats. (Boeing Commerical Airplane Company)

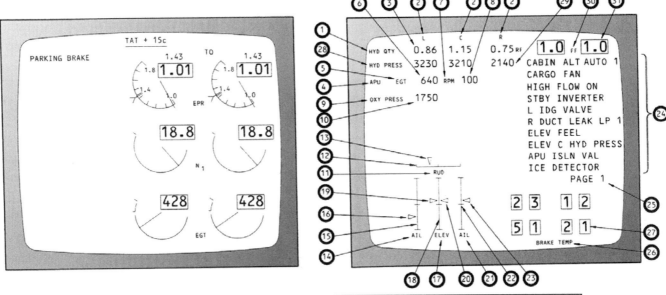

STATUS PAGE DISPLAYS				
NO.	DISPLAY	RANGE	SOURCE	COLOR
	SUBSYSTEM DISPLAYS			
1	SUBSYSTEM NAME - HYD QTY	N/A	EICAS COMPUTER	CYAN
2	SUBSYSTEM IDENTIFICATION - HYD QTY	N/A	EICAS COMPUTER	CYAN
3	ACTUAL HYD QTY READOUT	0.00 TO 1.50 (2)	HYD QTY SENSORS	(1)
4	SUBSYSTEM NAME - APU	N/A	EICAS COMPUTER	CYAN
5	SUBSYSTEM PARAMETER - EGT	N/A	EICAS COMPUTER	CYAN
6	ACTUAL EGT READOUT	0 TO 900°∞ C	APU SENSOR	WHITE
7	SUBSYSTEM PARAMETER - RPM	N/A	EICAS COMPUTER	CYAN
8	ACTUAL RPM READOUT	0; 3-120% RPM	APU SENSOR	WHITE
9	SUBSYSTEM NAME - OXYGEN PRESSURE	N/A	EICAS COMPUTER	CYAN
10	ACTUAL OXYGEN PRESSURE READOUT	0 TO 2500 PSI	CREW OXY PRESS SENSOR	WHITE
	CONTROL SURFACE POSITION DISPLAY			
11	CONTROL SURFACE NAME - RUDDER	N/A	EICAS COMPUTER	CYAN
12	HORIZONTAL SCALE	-36 T O +36°∞	EICAS COMPUTER	WHITE
13	ACTUAL RUDDER POSITION POINTER	-36 T O +36°∞	POSITION TRANSMITTER	WHITE
14	CONTROL SURFACE NAME - AILERON	N/A	EICAS COMPUTER	CYAN
15	VERTICAL SCALE	-22 T O +22°∞	EICAS COMPUTER	WHITE
16	LEFT OUTBOARD ALL POSITION POINTER	-22 T O +22°∞	POSITION TRANSMITTER	WHITE
17	CONTROL SURFACE NAME - ELEVATOR	N/A	EICAS COMPUTER	CYAN
18	VERTICAL SCALE	-24 T O +34°∞	EICAS COMPUTER	WHITE
19	ACTUAL LEFT ELEV POSITION POINTER	-24 T O +34°∞	POSITION TRANSMITTER	WHITE
20	ACTUAL RIGHT ELEV POSITION POINTER	-24 T O +34°∞	POSITION TRANSMITTER	WHITE
21	CONTROL SURFACE NAME - AILERON	N/A	EICAS COMPUTER	CYAN
22	VERTICAL SCALE	-22 T O +22°∞	EICAS COMPUTER	WHITE
23	RIGHT OUTBOARD AIL POSITION POINTER	-22 T O +22°∞	POSITION TRANSMITTER	WHITE
	STATUS MESSAGE DISPLAY			
24	STATUS MESSAGES	N/A	REFER TO TEXT	WHITE
25	OVERFLOW INDICATOR	N/A	EICAS COMPUTER	WHITE
	BRAKE TEMPERATURE DISPLAY			
26	SUBSYSTEM NAME - BRAKE TEMP	N/A	EICAS COMPUTER	CYAN
27	ACTUAL BRAKE TEMP READOUT	0 TO 9 UNITS	BRAKE TEMP SENSORS (8)	(3)
	HYDRAULIC PRESSURE		EICAS COMPUTER	CYAN
28	SUBSYSTEM NAME - HYD PRESS	N/A	FUEL LOW SENSORS	WHITE
29	ACTUAL BRAKE TEMP READOUT	0 TO 4000 PSI		
	FUEL LOW DISPLAY			
30	SUBSYSTEM NAME - FUEL FLOW	N/A		CYAN
31	ACTUAL FUEL FLOW READOUT	180 TO 12.2K KG/HR		WHITE

(1)	READOUT	WHITE
	REFILL	MAGENTA

| (2) | HYD QTY < 0.75,
RF MESSAGE APPEARS | |

(3)	NORMAL	0 TO 2 UNITS	BOX/NUMBERS CYAN
	THRESHOLD	3 TO 4 UNITS	BOX WHITE NUMBERS CYAN
	ABNORMAL	5 TO 9 UNITS	BOX/NUMBERS WHITE

Fig. 4-11   EICAS Status Page format. (Boeing Commerical Airplane Company)

213

Normal format

The normal format is displayed any time the status page and secondary engine displays are not active. In the normal format, the lower EICAS display is blank and the upper display shows primary engine data (EPR, N_1, and EGT). This information also remains on the upper CRT during display of secondary engine and status information. To access secondary engine or status page data while in the normal format, simply depress the corresponding button on the EICAS display select panel.

It should be noted, the EICAS computers use pin programming to determine status page format. For example, if pin number 27 is grounded, the hydraulic pressure for all three systems is displayed. If pin number 27 is open, hydraulic pressures will not be displayed. (Note: The pins being referred to are found in the electrical connector on the rear of the EICAS computer.) Fuel flow is displayed only if pin 10 is grounded; and the readout is in kilograms/hour if pin 13 is grounded, pounds/hour if pin 13 is open. There are literally hundreds of items set through pin programming. Program pins can be used to set display formats or used to configure input/output signals. The use of pin programming allows the EICAS computer to be configured to a variety of different aircraft. Pin programming is typically done during initial system installation; however, technicians should always consider program pins during troubleshooting. If a system fails, it may be due to a loose program pin connection.

Abnormal Operation Of The B-757 EICAS

In general, the EICAS can fail in four basic ways and still remain operational: 1) loss of one or more parameters, 2) loss of one CRT display, 3) loss of one or both EICAS control panels, and 4) loss of one system computer. In each of these four situations, the EICAS will be operational in some form. The first priority for EICAS is to display primary engine data; the second priority is for the system to display secondary engine data. In any failure mode, the system will automatically try to adhere to these priorities.

The loss of one parameter will result in an EICAS display with a blank in place of the missing parameter(s). Figure 4-12 shows an example where the right engine (EGT) data is invalid; that information is therefore not displayed.

A problem with the display unit, a wiring defect, or a problem with the EICAS computer can cause the loss of one CRT display. In any case, all EICAS information except the status page is available on the operable CRT. As discussed earlier during system architecture, a feedback circuit is used to monitor the CRT status. If the EICAS computer senses a CRT failure (or failure of data to the CRT), the system software automatically diverts all display data to the operable CRT. With only one EICAS CRT operable the **compact** mode will be used to display systems data. Figure 4-13 shows an example of the EICAS compact mode. In this figure you will notice there is an oil pressure exceedance on the right engine. This exceedance valve (65) is displayed in red. Note: the oil system parameters for both engines are displayed. In the compact mode it should also be noticed that the EGT data is displayed in digital format only, and fuel flow (FF) is displayed below EGT.

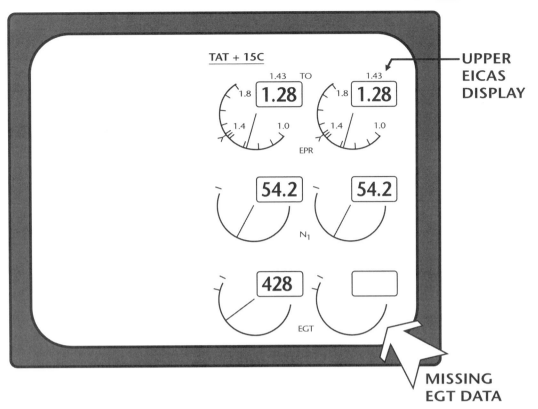

Fig. 4-12 Example of invalid data on the upper EICAS display. (Boeing Commerical Airplane Company)

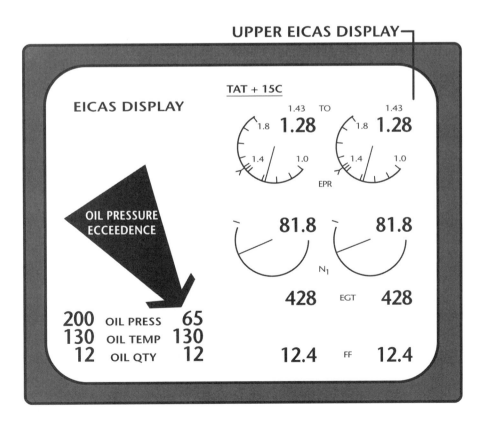

Fig. 4-13 Example of EICAS compact mode. (Boeing Commerical Airplane Company)

In the event that the EICAS maintenance control panel (or related wiring) fails, the systems accessed through that panel are inoperable and all other functions will operate normally. If the display select panel (DSP) fails, the following will occur:

1) The maintenance control panel and cancel/recall switches will become inoperative since their signal feeds through the DSP to the EICAS computers (see figure 4-8, upper left corner).

2) The display select panel functions become inoperative except for the brightness/balance controls.

3) The computers automatically display primary engine data on the upper CRT and secondary engine data on the lower CRT.

4) The message *EICAS CONT PNL* will appear on the upper CRT to inform the flight crew of the DSP failure.

Standby engine indicator

On the B-757 if both EICAS displays are inoperable, the **standby engine indicator** is used to display primary engine parameters, but all other EICAS information is not available. This type of failure occurs when both EICAS computers fail, both EICAS displays fail, or the associated wiring incurs a catastrophic failure. The standby engine indicator displays can be activated manually by placing the selector to the *ON* position (see figure 4-14), or the LCDs will turn on automatically in the event both EICAS computers or displays fail. The standby engine displays are also automatically activated whenever the EICAS test function is in operation.

The standby engine indicator is a LRU, which operates with two completely independent power supplies. Each power supply is fed through its own circuit breaker and related wiring. The indicator is fully functional with only one power supply operable. The indicator contains microprocessor circuitry to analyze both analog and ARINC 429 inputs. As seen in the lower right corner of figure 4-8, the standby indicators receive all vital engine data directly from the appropriate sensors. Even if both EICAS computers fail, the standby indicators will still function properly.

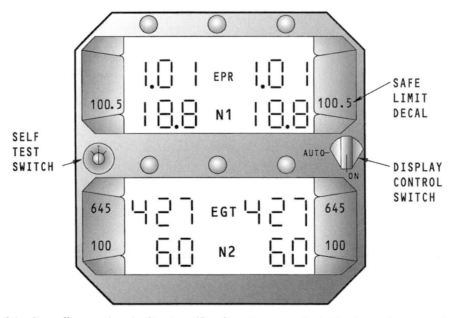

Fig. 4-14   Standby engine indicator. (Boeing Commerical Airplane Company)

B-757 maintenance pages

The **maintenance pages** of the B-757 are used to verify malfunctions or exceedances, and analyze the systems monitored by EICAS. The maintenance pages are displayed on the lower EICAS display while the aircraft is on the ground. The maintenance control panel (MCP) is used to control the display of maintenance pages. The MCP is used extensively by technicians during system troubleshooting and analysis.

MCP controls

The six display select switches on the MCP are used to display real time data for their associated systems (see figure 4-15). The *AUTO* and *MAN* push-buttons on the MCP are used to access data that was stored in nonvolatile memory during recording of an auto or manual event. As discussed earlier, auto events are recorded by EICAS during system malfunctions or exceedances; manual events are recorded whenever the record (*REC*) button on the DSP or MCP is pressed. When recording from the DSP, all systems are recorded simultaneously. When recording from the MCP, only the system currently shown on the lower EICAS display is recorded. The *ERASE* switch is used to clear the memory for the auto and manual events. The *TEST* switch is used to initiate the EICAS built in test circuitry.

It should be noted that the manual record button is a handy tool, which can be used to document system data during aircraft maintenance. The manual record function

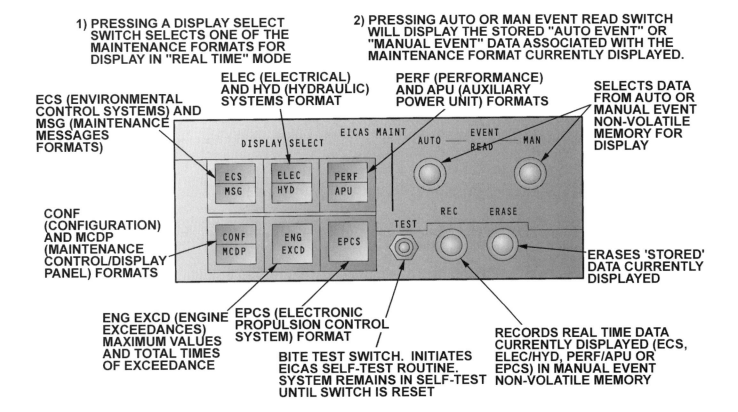

Fig. 4-15 Maintenance control panel switch layout. (Boeing Commerical Airplane Company)

can be thought of as a "snap shot" of the system being recorded. That is, the EICAS manual record function will store the system parameters at the instant you press the *REC* button. For example, if the technician wishes to record the operating parameters of the left AC generator, he/she would select the electric/hydraulic display, press *REC* in the MCP and EICAS will store the data. The information can be retrieved at a more convenient time and recorded in the maintenance records.

Access to maintenance pages

Whenever a specific maintenance page is selected, the engine data page goes into the compact format and the maintenance page is shown on the lower EICAS display. The various maintenance pages and their associated control buttons are shown in figure 4-16. To access the current status of a system using the EICAS maintenance pages, the following sequence should be followed:

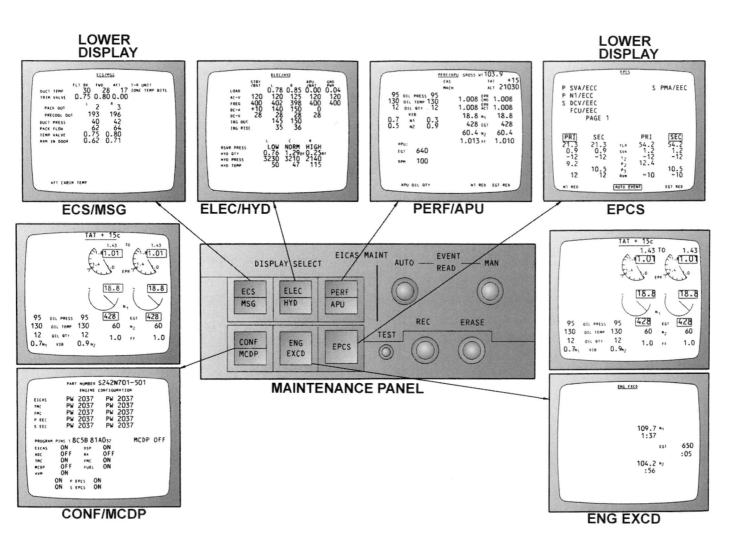

Fig. 4-16 Maintenance pages associated with the MCP switches. (Boeing Commercial Airplane Company)

1) Aircraft power must be available through ground power or onboard generators.

2) Any system that is to be monitored must be set into an operating condition.

3) Select the proper system using the display select switches on the EICAS maintenance panel.

The lower EICAS display will now show real time data for the system selected. If an auto event is stored in memory the white display *AUTO EVENT* will appear at the bottom of the page. To access the auto event information, simply press the *AUTO* button on the MCP. The current system display will change to the data stored during the auto event. During all auto event presentations, the term *AUTO EVENT* will appear in cyan at the lower portion of the page to indicate that the data currently presented is from memory (not real time) data. To access any manually recorded event, the *MAN* push-button is pressed. To return back to the primary engine display, exit the event mode by pressing the *MAN* or *AUTO* buttons, and press the system page button.

To erase an auto or manual event, simply press and hold the *ERASE* button on the MCP. This must be done while the event to be erased is presented on the EICAS display. Once the data presented on the display is removed, the memory has been erased. Only the data (manual or auto event) for the system currently on display has been erased. The recorded events for each system must be erased individually. After viewing (and making records as needed) of any auto event, it is important to erase the data. A new auto event can only be stored in memory after the previous event has been erased. The recording of any new manual events will automatically erase any manual event in memory.

System displays

Taking a closer look at a given system display shows the type of data presented. Figure 4-17 shows a typical electrical/hydraulic system display. Notice the electrical system data is displayed at the top of the page and hydraulic data is presented in the lower half. At the very bottom of the display, the terms *MAN EVENT* or *AUTO EVENT* are shown when data displayed is from event memory. Also at the bottom of the display are the terms "L GEN DRIVE" (# 27) and "R HYD QTY" (#28); these are the specific events which triggered an auto event to be recorded. Each system has a limited number of items that can trigger the recording of auto event data. For example, there are five conditions which can cause the recording of an auto event for the electrical system; they are, "R GEN DRIVE", "L GEN DRIVE", "R IDG OIL TEMP", "L IDG OIL TEMP", and "IDG RISE TEMP".

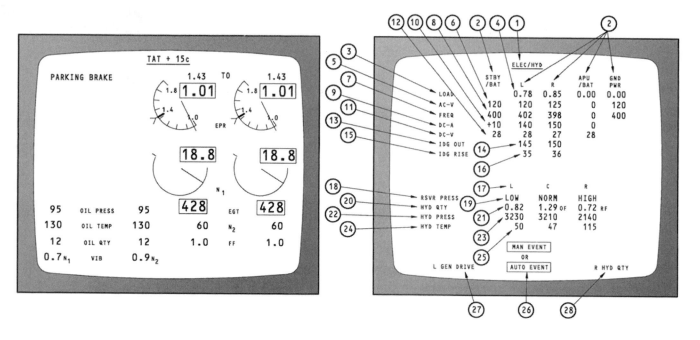

ELECTRICAL SYSTEM/ HYDRAULIC SYSTEM DISPLAY

No.	DISPLAY	RANGE	SOURCE	COLOR
1	PAGE FORMAT ELECTRICAL SYSTEM DISPLAY	N/A	EICAS COMPUTER	CYAN
2	POWER SOURCE	N/A	EICAS COMPUTER	CYAN
3	PARAMETER NAME AC-LOAD	N/A	EICAS COMPUTER	CYAN
4	LOAD READOUT	0 to 1.50 ******* (1.00 = 90KVA)	L, R, APU GCU, BPCU	WHITE
5	PARAMETER NAME AC-VOLTS	N/A	EICAS COMPUTER	CYAN
6	AC-VOLTS READOUT	0/100 TO 130V AC	*	WHITE
7	PARAMETER NAME FREQUENCY	N/A	EICAS COMPUTER	CYAN
8	FREQUENCY READOUT	0/380 TO 420 HZ	*	WHITE
9	PARAMETER NAME DC-AMPS	N/A	EICAS COMPUTER	CYAN
10	DC-AMPERES READOUT	0/2 TO 150 AMPS	**	WHITE
11	PARAMETER NAME DC-VOLTS	N/A	EICAS COMPUTER	CYAN
12	DC-VOLTS READOUT	0 TO 40V DC	***	WHITE
13	PARAMETER NAME IDG OUT	N/A	EICAS COMPUTER	CYAN
14	IDG OUT READOUT	0 TO 180° C	L, R, GCU	WHITE
15	PARAMETER NAME IDG RISE	N/A	EICAS COMPUTER	CYAN
16	IDG RISE READOUT HYDRAULIC SYSTEM DISPLAY	0 TO 180° C	L, R, GCU	WHITE
17	HYDRAULIC SYSTEM IDENTIFIER	N/A	EICAS COMPUTER	CYAN
18	HYDRAULIC RESERVOIR PRESSURE	N/A	EICAS COMPUTER	CYAN
19	HYDRAULIC RSVR PRESS READOUT	LOW, NORM, HIGH	HYD RSVR PRESS SENSORS	****
20	HYDRAULIC QUANTITY	N/A	EICAS COMPUTER	CYAN
21	HYDRAULIC QUANTITY READOUT	0.00 TO 1.50 (1.00 = 100%)	HYD QTY SENSORS	*****
22	HYDRAULIC PRESSURE	N/A	EICAS COMPUTER	CYAN
23	HYDRAULIC PRESSURE READOUT	0 TO 4000 PSI	HYD PRESS SENSORS	WHITE
24	HYDRAULIC RESERVOIR TEMP	N/A	EICAS COMPUTER	CYAN
25	HYDRAULIC TEMP READOUT AUTO/MANUAL EVENT DISPLAY	-60 TO +200° C	HYD TEMP SENSOR	WHITE
26	AUTO/MAN EVENT MODE DISPLAY	N/A	EICAS COMPUTER	CYAN
27	ELECT AUTO EVENT MESSAGE	******	REFER TO TEXT	WHITE
28	HYDRAULIC AUTO EVENT MESSAGE	******	REFER TO TEXT	WHITE

*STANDBY INVERTER, L & R GENERATOR, APU GENERATOR, GROUND POWER

**MAIN BATTERY, L & R TRU, APU BATTERY

***MAIN BATTERY, L & R DC BUS, APU BATTERY

****AUTO EVENT MESSAGES ARE NOT DISPLAYED IN MAN EVENT READ MODE

******* L.R. APU READOUT LESS THAN 0.05 SET TO 0.00 GND PWR READOUT LESS THAN 0.10 SET TO 0.00

****	RESERVOIR PRESSURE < 17 PSI	LOW	MAGENTA
	17PSI < RESERVOIR PRESSURE> 55 PSI	NORM	WHITE
	RESERVOIR PRESSURE > 55 PSI	HIGH	MAGENTA

*****	READOUT	WHITE
	OVERFILL (OF) QTY ≥ 1.22	MAGENTA
	REFILL (RF) QTY ≤ 0.75	MAGENTA

Fig. 4-17 Typical EICAS electrical/hydraulic system display. (Boeing Commerical Airplane Company)

Maintenance messages

The environmental control system/maintenance message display is used to access any maintenance messages recorded for all systems monitored by EICAS. Maintenance messages are the lowest priority advisory and are not accessible during flight. Maintenance messages would typically be accessed after the "write up" of a system malfunction or during routine scheduled maintenance. Figure 4-18 shows a typical ECS/MSG display. In the top right portion of the display are the maintenance messages (item #20). Up to eleven messages can be presented per page. If more than eleven messages are in memory the term "page 1" (item #21) is displayed. To view subsequent pages press the ECS/MSG switch. Once the maintenance messages have been retrieved, the information can be referenced to the aircraft's maintenance manual for details on system repairs.

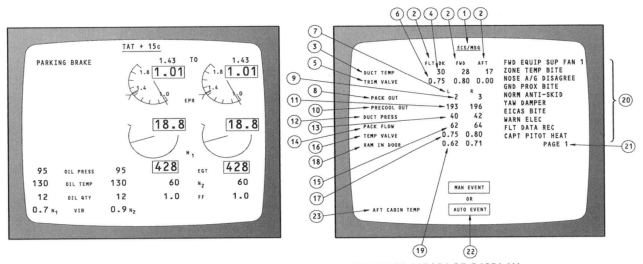

ENVIRONMENTAL CONTROL SYSTEM/ MAINTENANCE MESSAGE DISPLAY

No.	DISPLAY	RANGE	SOURCE	COLOR
	ENVIRONMENTAL CONTROL SYSTEM DISPLAY			
1	PAGE FORMAT NAME	N/A	EICAS COMPUTER	CYAN
2	FUNCTION LOCATION	N/A	EICAS COMPUTER	CYAN
3	FUNCTION NAME – DUCT TEMP	N/A	EICAS COMPUTER	CYAN
4	DUCT TEMPERATURE READOUT	-60 to +200° C	DUCT TEMP SENSORS	WHITE
5	FUNCTION NAME – TRIM VAVLE	N/A	EICAS COMPUTER	CYAN
6	TRIM VALVE POSITION READOUT	0.00 TO 1.00 (OPEN)	VALVE POSN SENSORS	WHITE
7	L, R PACK IDENTIFIER	N/A	EICAS COMPUTER	CYAN
8	FUNCTION NAME – PACK OUT	N/A	EICAS COMPUTER	CYAN
9	PACK OUT TEMP READOUT	-60 TO +200° C	PACK TEMP SENSORS	WHITE
10	FUNCTION NAME – PRECOOL OUT	N/A	EICAS COMPUTER	CYAN
11	PRECOOLER OUTLET TEMP READOUT	-60 TO +300° C	PRECOOL OUT TEMP SENSORS	WHITE
12	FUNCTION NAME – DUCT PRESS	N/A	EICAS COMPUTER	CYAN
13	DUCT PRESSURE READOUT	0 TO 100 PSI	DUCT PRESS SENSORS	WHITE
14	FUNCTION NAME - PACK FLOW	N/A	EICAS COMPUTER	CYAN
15	PACK AIR FLOW READOUT	0 TO 60 M³/MIN	PACK FLOW SENSORS	WHITE
16	FUNCTION NAME – TEMP VALVE	N/A	EICAS COMPUTER	CYAN
17	TEMP VALVE POSITION READOUT	0.00 TO 1.00 (OPEN)	VALVE POSN SENSORS	WHITE
18	FUNCTION NAME – RAM IN DOOR	N/A	EICAS COMPUTER	CYAN
19	RAM INLET DOOR POSITION READOUT	0.00 TO 1.00(CLOSE)	DOOR POSN SENSORS	WHITE
	MAINTENANCE MESSAGE DISPLAY			
20	MAINTENANCE MESSAGE READOUT	*	REFER TO TEXT	WHITE
21	OVERFLOW INDICATOR	*	EICAS COMPUTER	WHITE
	AUTOMATIC/MANUAL EVENT DISPLAY			
22	AUTO/MAN EVENT READ MODE DISPLAY	N/A	EICAS COMPUTER	WHITE
23	ECS AUTO EVENT MESSAGE	*	REFER TO TEXT	WHITE

* AUTO EVENT MESSAGE AND MAINTENANCE MESSSAGES ARE NOT DISPLAYED IN MANUAL EVENT READ MODE

Fig. 4-18 Typical EICAS environmental control system/maintenance message display. (Boeing Commerical Airplane Company)

System Troubleshooting Using EICAS

In most cases, the aircraft technician will have a copy of the pilot's squawk recorded in the aircraft's log, or the pilot may send a fault code to the technician while the aircraft is still in flight. This will give information as to what system(s) require attention. The pilot's log entry may state that a manual event record was made, or an auto event may have been recorded which relates to the squawk. In either case, the technician should access the EICAS memory for that system and study the parameters that occurred at the time the squawk was initiated. The system should also be operated and the current status observed using EICAS. If the fault has repaired itself, the technician should consider the defect may be intermittent.

It may be wise to access the data contained in the ECS/MSG page. If a maintenance message related to the problem is recorded in the nonvolatile memory, the solution may be easily found using the maintenance manuals. If no related maintenance messages are available, the next step would be to consult the troubleshooting or maintenance manuals for the system in question. In most cases, the manuals can point you in the proper direction for the repair.

Once the repair has been completed, the system should be operated in the same configuration as when the problem was first noticed. Access the real time information for the system through EICAS and look for any new auto events that may have been recorded. (Be sure previous auto events were already erased.) If the system operates within specified parameters, the repair was successful. If the system is still defective, continue the troubleshooting process.

Troubleshooting EICAS

Both EICAS computers continuously perform fault monitoring of all circuits that can affect the integrity of EICAS data. The fault monitoring system consists of software within the EICAS computers. Incoming digital data is monitored for validity using the ARINC 429 parity bit, and all input data (both analog and digital) is compared through the EICAS computers cross talk data bus. The health of each computer is monitored, and if one computer fails any necessary switching automatically occurs.

In the event of an EICAS malfunction, the status page will contain a message associated to the failure. For example the message *L EICAS CMPTR* will appear if a left computer fault or computer off is detected. If the EICAS monitoring system detects a fault, the maintenance message *EICAS BITE* will appear on the ECS/MSG maintenance page. The ECS/MSG page is accessed as previously described. If this message is present, the EICAS test should be run.

EICAS BITE tests

On the B-757, the **EICAS BITE** test is used to perform testing of the EICAS computers, upper and lower display units, the master caution and warning displays and various EICAS interfaces. The BITE test is initiated by pressing the test switch on the MCP (figure 4-15). The aircraft must be on the ground and the parking brake set for the test to begin. Each EICAS computer is tested individually. The EICAS test can also be initiated automatically due to a pin programming error. In this case, the EICAS test will be initiated at start up of the system.

At the beginning of each test the EICAS message "L EICAS TEST" and "TEST IN PROGRESS" will appear on the upper display. The message "R EICAS TEST" will be presented during the test of the right computer. The message "PROG PIN ER-ROR" will appear if the test was initiated automatically due to program pin error. During the EICAS test, the master warning and caution lights are activated, the related audio tones will sound, and the standby engine indicators are energized.

After the test is complete, the EICAS will present a display similar to figure 4-19 on both the upper and lower EICAS CRT. The following items 1-10 describe the test display format.

1) States why the test was initiated (automatically due to program pin error, or manually for L/R EICAS test).

2) Test results (test fail or test OK) are displayed.

3) Listing of any computer failures are displayed.

4) Listing of any display unit failures are displayed.

5) The *KEY* message allows for the operator to test EICAS inputs from the DSP, MCP and cancel/recall switches. After the test is complete the operator can press any of the associated switches. The switch name will appear next to the *KEY* message. In this example the *erase* button has been pressed.

6) This display indicates the current program pins status (in a hexadecimal format) for the left and right computers.

7) The L/R computer test results are displayed in a hexadecimal format. All zeros indicates no computer system failures. **NOTE**: If the test shows *L/R COMPTR FAIL*, be sure to check the *CU TEST* code shown in the upper right quadrant of the test display (figure 4-19). The *CU TEST* code is used to identify the specific problem with the L/R computer. The code displayed can be found in the ATA chapter 31 Fault Isolation Manual (FIM). The FIM will provide further instructions for system repair using the various CU TEST codes. Remember do not simply replace the EICAS computer if a failure is listed on the test display; verify the specific fault using the CU TEST code.

8) The results of the L/R computer input tests are displayed adjacent to *CU MONITOR*. The results are displayed in a hexadecimal format, where a
9) Any disagreements between engine, subsystem, and discrete data transmitted to EICAS will be displayed here. A readout of all zeros indicates no disagreements. Once again, the aircraft manuals will provide specific information on decoding any engine disagree message.

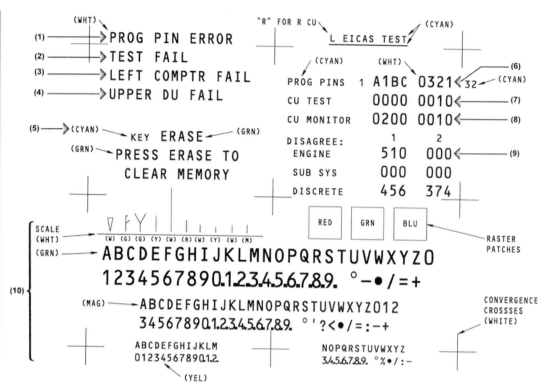

Fig. 4-19 EICAS test display. (Boeing Commerical Airplane Company)

9) Any disagreements between engine, subsystem, and discrete data transmitted to EICAS will be displayed here. A readout of all zeros indicates no disagreement. Once again, the aircraaft manuals will provide spacific information on decoding any engine disagree message.

10) Any special symbols, which are typically displayed by EICAS, are shown on the lower portion of the display during an EICAS test.

To clear the CU test, CU monitor and parameter disagree codes, the erase key must be pressed twice. With the first press of the erase key, the message *PRESS ERASE TO CLEAR MEMORY* will appear. Press the erase key a second time and the codes will revert to all zeroes.

As mentioned above, if the EICAS test results show a CU test, CU monitor, or parameters disagreement, the associated codes can be used to further troubleshooting efforts using the appropriate section of the aircraft manuals. The code(s) will direct the technician to a specific system or LRU suspected of creating the problem. These codes are the key to accurate troubleshooting using the EICAS test function; be sure they are not ignored. After the repair has been completed, the EICAS test should be run again to verify correct system operation.

Configuration/maintenance page

The configuration/maintenance page is also a valuable EICAS troubleshooting aid. As discussed earlier, this page is accessed through the maintenance control panel. A typical configuration/maintenance page is shown in figure 4-20. From this example it can be seen that a variety of EICAS computer configuration information is listed

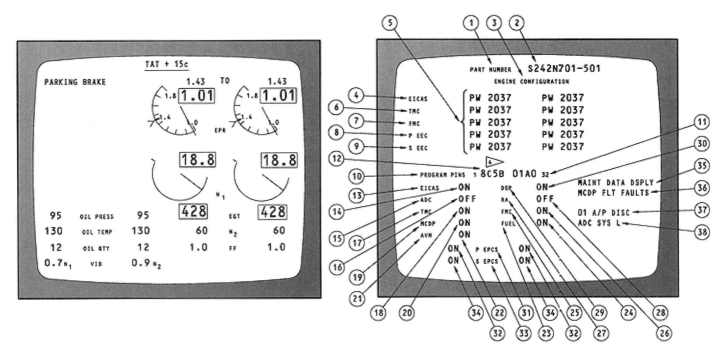

CONFIGURATION/MCDP DISPLAY

No.	DISPLAY	RANGE	SOURCE	COLOR
	ENGINE CONFIGURATION DIPLAY			
1	PARAMETER NAME – PART NUMBER	N/A	EICAS COMPUTER	CYAN
2	EICAS COMPUTER PART NUMBER	N/A	EICAS COMPUTER	WHITE
3	PARAMETER NAME – ENG CONFIGURATION	N/A	EICAS COMPUTER	CYAN
4	COMPUTER NAME – EICAS	N/A	EICAS COMPUTER	CYAN
5	ENGINE TYPE/MODEL	N/A	*	WHITE
6	COMPUTER NAME - TMC	N/A	EICAS COMPUTER	CYAN
7	COMPUTER NAME - FMC	N/A	EICAS COMPUTER	CYAN
8	COMPUTER NAME – P EEC	N/A	EICAS COMPUTER	CYAN
9	COMPUTER NAME – S EEC	N/A	EICAS COMPUTER	CYAN
	PROGRAM PIN CONFIGURATION			
10	PARAMETER NAME – PROGRAM PIN	N/A	EICAS COMPUTER	CYAN
11	BYTE START/STOP REFERENCE	1,32	EICAS COMPUTER	CYAN
12	PROGRAM PIN READOUT (HEXIDECIMAL)	0000 TO FFFF	EICAS COMPUTER	WHITE
	INPUT BUS ACTIVITY			
13	INPUT BUS NAME – EICAS	N/A	EICAS COMPUTER	CYAN
14	ACTIVITY/NO ACTIVITY	**	EICAS CU (INTER-CONNECT)	WHITE
15	INPUT BUS NAME – ADC	N/A	EICAS COMPUTER	CYAN
16	ACTIVITY/NO ACTIVITY	**	ADC (TAT)	WHITE
17	INPUT BUS NAME – TMC	N/A	EICAS COMPUTER	CYAN
18	ACTIVITY/NO ACTIVITY	**	TMC (TAT)	WHITE
19	INPUT BUS NAME – MCDP	N/A	EICAS COMPUTER	CYAN
20	ACTIVITY/NO ACTIVITY	**	MCDP (DSPLY DATA LABEL 357)	WHITE
21	INPUT BUS NAME – AVM	N/A	EICAS COMPUTER	CYAN
22	ACTIVITY/NO ACTIVITY	**	AVM (L ENG VIBRATION)	WHITE
23	INPUT BUS NAME – FUEL	N/A	EICAS COMPUTER	CYAN
24	ACTIVITY/NO ACTIVITY	**	FQIS (L TANK FUEL QTY)	WHITE
25	INPUT BUS NAME – AVM	N/A	EICAS COMPUTER	CYAN
26	ACTIVITY/NO ACTIVITY	**	FMC (GROSS WT)	WHITE
27	INPUT BUS NAME – RA	N/A	EICAS COMPUTER	CYAN
28	ACTIVITY/NO ACTIVITY	**	RA (RADIO ALTITUDE)	WHITE
29	INPUT BUS NAME – P EPCS	N/A	EICAS COMPUTER	CYAN
30	ACTIVITY/NO ACTIVITY	**	DSP (DISCRETE WORD #1)	WHITE
31	INPUT BUS NAME – P EPCS	N/A	EICAS COMPUTER	CYAN
32	ACTIVITY/NO ACTIVITY	**	L/R EEC (L/R P2)	WHITE
33	INPUT BUS NAME – S EPCS	N/A	EICAS COMPUTER	CYAN
34	ACTIVITY/NO ACTIVITY	**	L/R EEC (L, R Ps)	WHITE
	MAINTENANCE CONTROL & DISPLAY PANEL			
35	ANNUNCIATE MESSAGE***	N/A	MCDP LABEL 357	YEL/WHT
36	MODE MESSAGE	N/A	MCDP LABEL 357	YELLOW
37	MCDP TOP LINE DATA MESSAGE	N/A	MCDP LABEL 357	WHITE
38	MCDP BOTTOM LINE DATA MESSAGE	N/A	MCDP LABEL 357	WHITE

* SOURCE	LABEL	SOURCE	LABEL
EICAS	N/A	FMC	271
TMC	270	EEC	271

** ON, TEST, HCD, FAIL/OFF

***IF MCDP IS OFF, WHITE "MCDP OFF" MESSAGE REPLACES ALL OTHER MCDP MESSAGES

****SIA – LEFT COMPUTER SHOWN RIGHT COMPUTER $_1$ 845B D1A1$_{32}$

Fig. 4-20   EICAS configuration/maintenance display format. (Boeing Commerical Airplane Company)

on this page. During troubleshooting item 13 through 34 can be used to verify input bus activity for the EICAS computer currently being displayed. For each bus monitored, one of the five messages will appear: *ON, OFF, TEST, NCD,* or *FAIL.* No activity on the bus will generate the OFF message. Defective bus wiring or no data being transmitted will generate the OFF message. An ON message indicates that the bus is operating normally. If TEST is displayed, the transmitting system or LRU is currently undergoing an internal BITE test. The message NCD (no computed data) means the transmitter is operating correctly but there is no data to be transmitted. The message FAIL means that the system or LRU in question has failed its own internal BITE test.

Reversionary switching

Another quick EICAS test often used by line technicians is done through a simple switch of the EICAS computers. This is often done if normal operation of EICAS shows missing data for one of the engine (or other system) parameters. In this scenario the missing data is most likely the result of a defective input to one of the computers. To verify which computer is receiving the invalid or lost data, simply designate a specific computer using the display select panel. This technique is sometimes referred to as **reversionary switching.** For example, for an EGT fault the technician would select the *L* (left) to operate the display directly from the left computer. If the display still has missing EGT data, the EGT system feeding the left computer or its associated wiring is defective. If moving to the left computer solves the missing EGT data problem, the problem lies in the right side system. The same test could be run for the right side computer by turning the computer select switch to *R.*

Once a technician has become familiar with the aircraft's systems, this type of troubleshooting becomes second nature. If the EGT probe feeding the left computer was suspected as defective, the appropriate manuals would be reviewed and the repair initiated. The test executed in the previous paragraph would be performed and the appropriate paperwork competed. The aircraft would once again be equipped with a fully operational EICAS.

It was noted earlier that for some systems pin programming is used to determine what information is displayed by EICAS. On the B-757, if pin 27 is grounded, hydraulic system pressure is displayed. If pin 27 is open, hydraulic pressure is not presented on the display. This type of pin programming can sometimes cause difficulty when troubleshooting the system. For example, if pin number 27 should become loose on the left EICAS computer; the connection to ground would be eliminated or intermittent. Whenever operating on the left EICAS computer, the pilot would see the hydraulic pressure blink on and off the display as pin 27 connected/disconnected to ground (pin 27 intermittent). Or the hydraulic pressure information would be lost completely (pin 27 open). The technician would most likely suspect a faulty hydraulic pressure input to the left computer. A quick troubleshooting technique would be to manually select the opposite (right) EICAS computer to drive the displays. This switch would indeed fix the problem, leading the technician to assume the left EICAS computer input was faulty, or the left computer itself had a problem. Replacing the left computer would not solve the problem. A quick test of the hydraulic system feeding the left computer would show that system was operational. The cause of the defect is an intermittent or an open connection to ground on pin 27. In this case, an auto EICAS BITE should be initiated. The results of this test would show *Prog Pin Error* (program pin error). This should lead the technician to verify all program pins connections related to the hydraulic system.

A typical troubleshooting sequence

The following example illustrates the basic series of events that would take place when an EICAS system failure occurs. In this example the upper EICAS display unit has failed during flight. The following would occur:

1) The EICAS would automatically revert to a compact mode and display all necessary information.

2) As standard procedures, the flight crew would switch to the alternate EICAS computer to potentially solve the problem. In this example, the display operates normally on the alternate computer.

3) The first officer would reference a fault codes manual to determine the correct fault code for this failure. Fault codes adhere to ATA chapter/section sequencing; this fault is 31-41-07-04.

4) The first officer would record the fault in the appropriate log book either by code number or by description. Exactly what information gets recorded is a function of specific airline operating procedures. The maintenance code(s) could also be sent to the ground facility using ACARS. ACARS will be discussed latter in this text.

5) Once the aircraft has landed the maintenance crew would check the flight logs during aircraft turnaround. The technician would find a log entry and/or code 31-41-07-04 recorded along with the time at which the system failed. The flight crew might also discuss the fault with the technician.

6) The technician would find the fault code in the fault isolation manual. Figure 4-21 shows 31-41-07 and states that the *EICAS display is (blank, out of focus, distorted, or wrong color). The 04 defines the upper display.* The fault isolation reference tells the technician where to begin troubleshooting. In this case, the fault isolation manual chapter/section 31-41-00, figure 104, block 1 is referenced.

7) The technician would follow the troubleshooting procedures listed in 31-41-00, figure 104, block 1 (see figure 4-22). Note the prerequisites for the tests are listed at the top of the page. In this case, electrical power must be supplied to the system, and the listed circuit breakers must be in and operable. The maintenance manual (MM) 24-22-00 would be used as a reference for the electrical power procedures.

8) For our test, lets assume the technician followed the fault tree and each time a question was asked the answer was *NO*. This would bring the solution to block 25, *Replace right EICAS computer (MM 31-41-02). If fault persists, replace upper EICAS display switching module (MM 31-41-04).* The maintenance manual reference is given for each suggested repair.

9) The suggested repair would be done in accordance with the maintenance manual procedures. Once the fault has been repaired, the appropriate operational test(s) would be done and the aircraft would be returned to service.

BOEING 757
FAULT ISOLATION/
MAINTENANCE MANUAL

FAULT CODE	1. LOG BOOK REPORT 2. FAULT ISOLATION REFERENCE
31 41 04 00	1. Master caution lights did not illluminate when level B caution condition existed. (State conditon existing). 2. Examine and repair the circuit from left (right) EICAS computer connector D319A pin F8 (connector D321A pin F8) to captainís (F/Oís) master caution lighted switch (pin 16) (WM31-41-14, -24).
31 41 06 00	1. EICAS (ENGINE, STATUS) select switch does not (select, deselect) (secondary engine, status) format. 2. Replace the EICAS select panel, M10195 (MM 31-41-03).
31 41 07 --	1. (03=upper, 04=lower) EICAS display is (blank, out of focus, distorted, wrong color: describe fault). Operation norm on alternate computer 2. 31-41-00 Fig. 104 Block 1
31 41 08 --	1. (03=upper 04=lower) EICAS display is (blank, out of focus, distorted, wrong color: describe fault). Fault remains on alternate computer 2. 31-41-00 Fig. 105 Block 1
31 41 14 00	1. EICAS msg EICAS CONT PNL displayed. 2. 31-41-00 Fig. 106 Block 1
31 41 15 00	1. Max ind reset switch will not reset overlimit readout. 2. Replace the EICAS select panel, M10195 (MM 31-41-03).

EFFECTIVITY

31 FAULT CODE INDEX

ALL   04   Page 5
Jun 20/92

Fig. 4-21   EICAS Fault isolation manual/Fault code index. (Boeing Commerical Airplane Company)

Testing the standby engine indicator

A built-in test equipment (BITE) circuit is contained in the standby engine indicator and accessed using the self test switch located on the face of the LRU (see figure 4-23). Turning the switch clockwise runs the self test using the number 2 power supply. If the switch is turned counterclockwise the self test is run using number 1 power supply. This enables the BITE to test the operation of each independent power supply. The test switch is accessed using a screwdriver. During each test the displays should be viewed and compared to the appropriate test values (EGT, 885; EPR, 1.88; N_1, 88.8; N_2, 88).

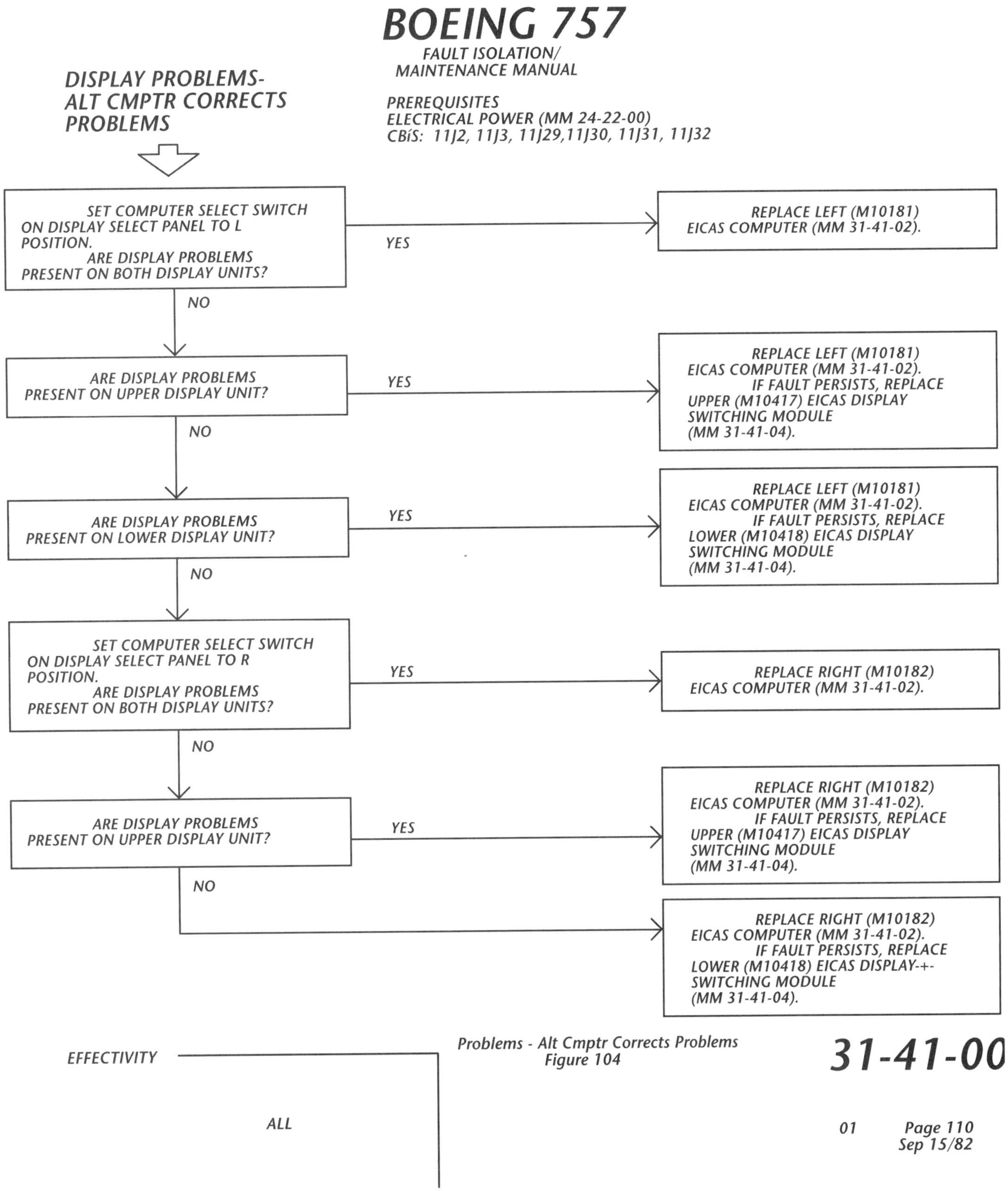

BOEING 757

*FAULT ISOLATION/
MAINTENANCE MANUAL*

*DISPLAY PROBLEMS-
ALT CMPTR CORRECTS
PROBLEMS*

PREREQUISITES
ELECTRICAL POWER (MM 24-22-00)
CBíS: 11J2, 11J3, 11J29,11J30, 11J31, 11J32

SET COMPUTER SELECT SWITCH ON DISPLAY SELECT PANEL TO L POSITION.
ARE DISPLAY PROBLEMS PRESENT ON BOTH DISPLAY UNITS?
— YES → REPLACE LEFT (M10181) EICAS COMPUTER (MM 31-41-02).
— NO ↓

ARE DISPLAY PROBLEMS PRESENT ON UPPER DISPLAY UNIT?
— YES → REPLACE LEFT (M10181) EICAS COMPUTER (MM 31-41-02). IF FAULT PERSISTS, REPLACE UPPER (M10417) EICAS DISPLAY SWITCHING MODULE (MM 31-41-04).
— NO ↓

ARE DISPLAY PROBLEMS PRESENT ON LOWER DISPLAY UNIT?
— YES → REPLACE LEFT (M10181) EICAS COMPUTER (MM 31-41-02). IF FAULT PERSISTS, REPLACE LOWER (M10418) EICAS DISPLAY SWITCHING MODULE (MM 31-41-04).
— NO ↓

SET COMPUTER SELECT SWITCH ON DISPLAY SELECT PANEL TO R POSITION.
ARE DISPLAY PROBLEMS PRESENT ON BOTH DISPLAY UNITS?
— YES → REPLACE RIGHT (M10182) EICAS COMPUTER (MM 31-41-02).
— NO ↓

ARE DISPLAY PROBLEMS PRESENT ON UPPER DISPLAY UNIT?
— YES → REPLACE RIGHT (M10182) EICAS COMPUTER (MM 31-41-02). IF FAULT PERSISTS, REPLACE UPPER (M10417) EICAS DISPLAY SWITCHING MODULE (MM 31-41-04).
— NO → REPLACE RIGHT (M10182) EICAS COMPUTER (MM 31-41-02). IF FAULT PERSISTS, REPLACE LOWER (M10418) EICAS DISPLAY-+- SWITCHING MODULE (MM 31-41-04).

EFFECTIVITY

ALL

Problems - Alt Cmptr Corrects Problems
Figure 104

31-41-00

01   Page 110
Sep 15/82

Fig. 4-22 EICAS Fault isolation manual/troubleshooting procedures. (Boeing Commerical Airplane Company)

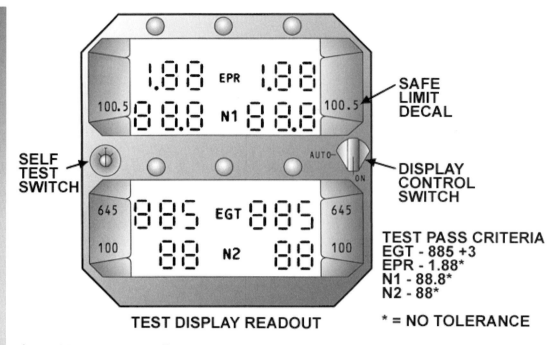

SAFE LIMIT DECAL

DISPLAY CONTROL SWITCH

SELF TEST SWITCH

TEST PASS CRITERIA
EGT - 885 +3
EPR - 1.88*
N1 - 88.8*
N2 - 88*

TEST DISPLAY READOUT   * = NO TOLERANCE

Fig. 4-23 EICAS standby engine indicators. (Boeing Commerical Airplane Company)

SECOND GENERATION ENGINE INDICATING AND CREW ALERTING SYSTEMS

For the most part, the first generation EICAS was designed in the 1970's and put in production in the 80's. During that time period improvements in microprocessor and computer technologies made great strides. Also, with the first generation systems in production it was easy to see what needed improvement and what portions of the system should be carried forward. Design of the second generation system began about the time first generation started flying. In the late 1980's, the second generation systems were introduced on the Boeing 747-400.

There were two significant changes made for the second generation engine indicating and crew alerting systems: 1) the new EICAS was part of an integrated system, and 2) the systems were better at self diagnostics. Integration of the B-747-400 EICAS made the system more compact through the use of shared components. Integrating the system with other flight deck displays made the EICAS more reliable by providing extended back up capabilities. Improved diagnostics also resulted from integration, in this case, integration with the ship's central maintenance system. The B-747-400 EICAS is capable of more accurate troubleshooting and can present information in a more easily understood format compared to previous systems.

The Boeing 747-400 is typical of aircraft that contain a second generation EICAS. The Boeing 747-400 EICAS will be presented during the following discussion.

System Architecture

The B-747-400 EICAS is part of a complete aircraft monitoring system called the **Integrated Display System (IDS)**. The integrated display system monitors a variety of aircraft and engine systems as well as various flight parameters using three **Electronic Interface Units (EIU)**. As shown in figure 4-24, the three EIUs manipulate the incoming data and send outputs to six display units. The display units on the B-747-400 are called **Integrated Display Units (IDU)** since the data on each display is interchangeable during operation. This system allows for a greater flexibility of display formats in the event of a single or multiple display failure. The EIUs also talk directly to the central maintenance computers. The **Central Maintenance Computer System (CMCS)** is used to monitor the health of various aircraft systems.

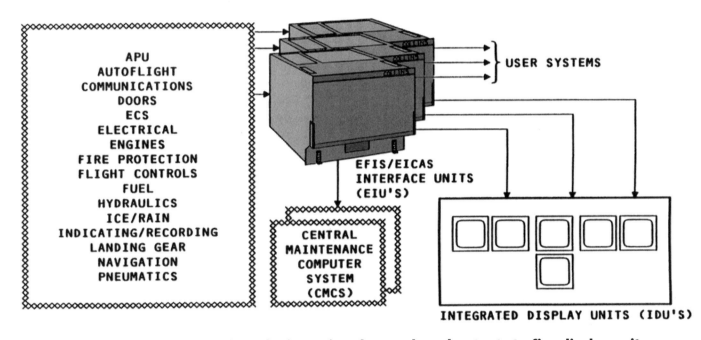

Fig. 4-24   The three EIUs manipulate the incoming data and send outputs to five display units. (Boeing Commerical Airplane Company)

The three EIUs are located in the main equipment center as shown in figure 4-25. These computers are static discharge sensitive and should be handled only when using proper precautions. The electrical leads of the EIUs are made through an 800 pin connector mounted on the rear of the unit. Take special care not to damage the pins or sockets during removal and replacement of the computers.

The 6 IDS display units are located as shown in figure 4-26. Shown in this figure are the normal display configurations. The **Primary Flight Displays (PFDs)** are used to display flight data such as airspeed, pitch attitude, and altitude. The **Navigational Displays (NDs)** are used for the display of navigational data, such as the compass rose, course map, and weather radar information. The **Main EICAS** display will usually show primary engine data, flight crew alert messages and total fuel flow (see figure 4-27). The **Auxiliary EICAS** display is located beneath the main EICAS display and is used to show one of the following formats: secondary engine data, status, synoptic, or maintenance pages.

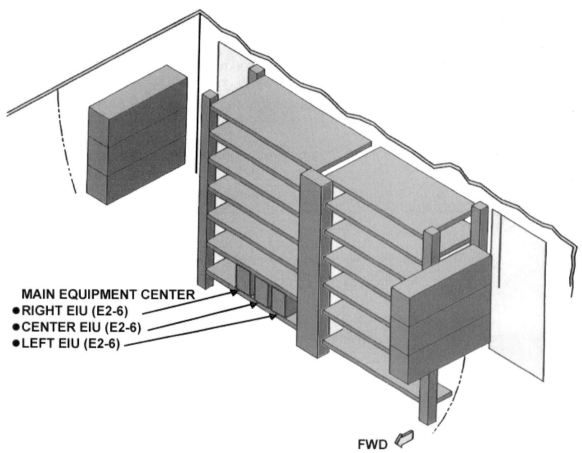

MAIN EQUIPMENT CENTER
- RIGHT EIU (E2-6)
- CENTER EIU (E2-6)
- LEFT EIU (E2-6)

FWD

Fig. 4-25 Location of the EFIS/EICAS interface units. (Northwest Airlines, Inc.)

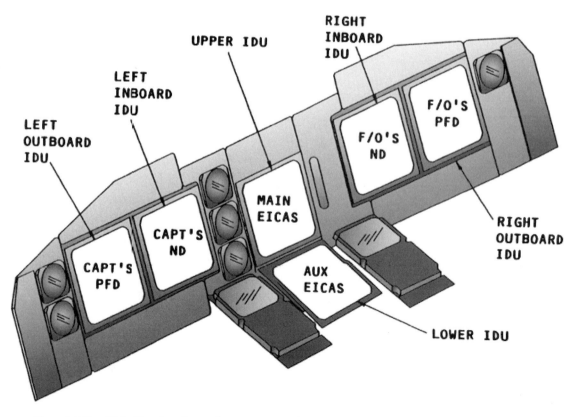

Fig. 4-26 IDS display locations, normal mode. (Northwest Airlines, Inc.)

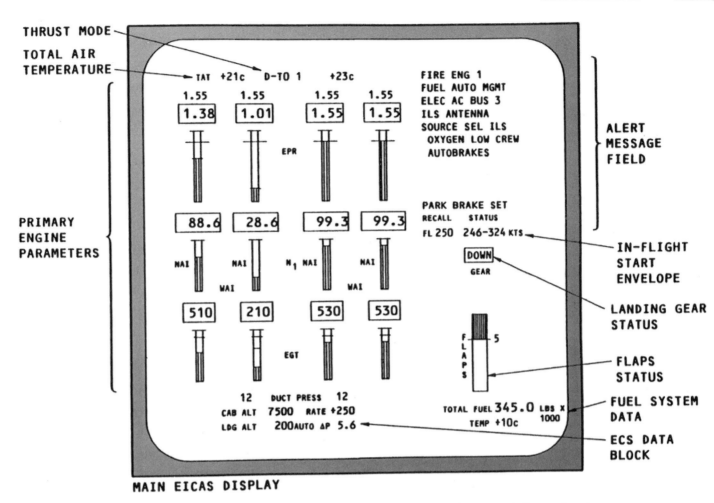

Fig. 4-27   Main EICAS display , primary format. (Northwest Airlines, Inc.)

Figure 4-28 shows a closer look at the extremely complex EIU input/output structure. The EIUs receive ARINC 429 digital data in both low- and high-speed formats, digital discrete signals, analog information, and analog discrete signals. Up to 108 digital input busses can transmit to each EIU. Of these 108 busses, a maximum of 8 are high-speed and the remainder are low-speed busses. Lightning protection is provided for up to 22 sensitive input busses. Each EIU transmits nine digital bus outputs, one low-speed, and 8 high-speed formats.

The EIUs each receive up to 450 discrete analog inputs from various aircraft systems, LRUs, and control switches. There are also 19-DC analog, 24-AC analog, 4-pulsed, 4-frequency modulated and 12-tachometer signals transmitted to each EIU.

The EIUs each transmit nine discrete analog outputs and 24 AC signals. On the left side of figure 4-28 are the EICAS control panels. The EICAS control panel is connected to the EIUs using a discrete analog connection. A close up view of the display select panel bus structure is shown in figure 4-29. The EICAS display select panel (DSP) talks directly to the EIUs for cancel/recall functions, all other DSP functions are sent to the EIUs through the L/R EFIS (not EICAS) control panels. The EFIS control panels communicate to the EIUs on a 429 bus through the L/R control display units (CDUs). The CDUs can act as a backup control panel (as discussed later in this chapter).

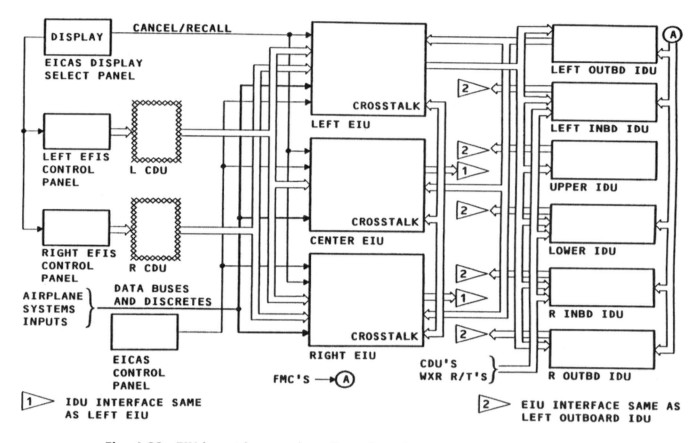

Fig. 4-28 EIU input/output interface. (Northwest Airlines, Inc.)

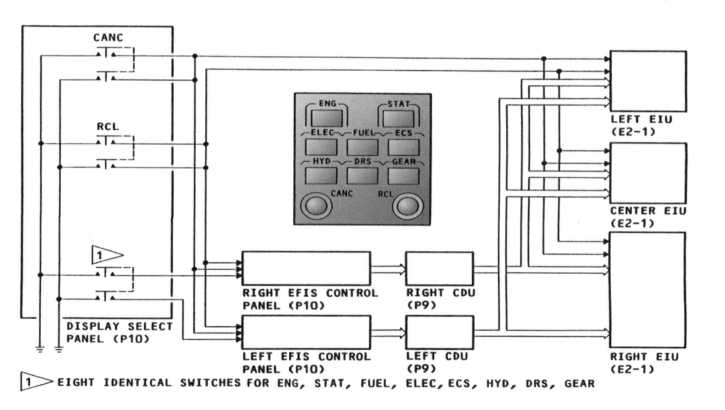

Fig. 4-29 EICAS display select panel interface. (Northwest Airlines, Inc.)

The integrated display units receive inputs from the EIUs on a high speed ARINC 429 bus. This bus carries the majority of all EICAS information. The IDUs used for electronic flight instruments also receive digital data from the weather radar receiver transmitters (WXR R/T's) on an ARINC 453 data bus and ARINC 429 data from the flight management computers (FMCs).

Each EIU has three functional sections; the input/output (I/O) interface, the processing section, and the power supply (see figure 4-30). The power supply receives 115V AC and converts that signal into various voltages for operation of the EIU. The input and output signals are interfaced through 4-ARINC 429, 3-discrete, 1-analog, and 2-frequency cards. The signal processing takes place with help of the controller and processor/memory cards. The memory is used to store any latched messages or system data, which may be accessed later.

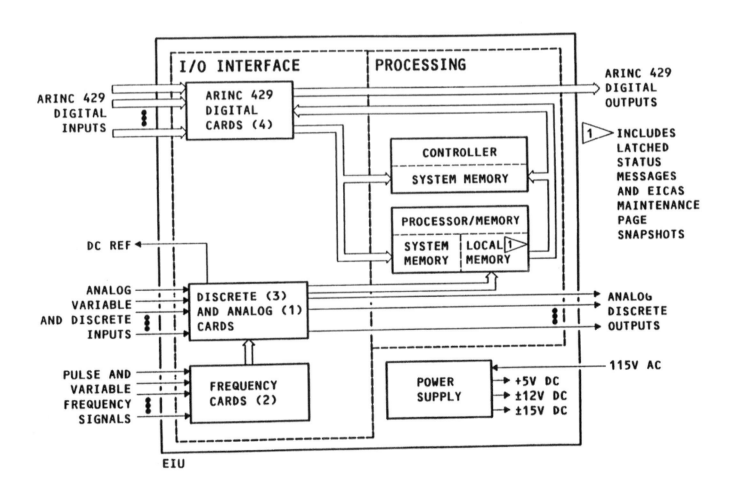

Fig. 4-30 EIU 3 subsections, I/O interface, processing, and power supply. (Northwest Airlines, Inc.)

EIU interface

Each EIU has the capability to "crosstalk" using a dedicated 429 bus to receive and transmit between each EIU (see figure 4-31). The following digital information is transmitted between EIUs:

> 1) BITE data consisting of internal fault information and LRU fault data received by the EIU. 2) EIU status data, and 3) EICAS message data which is transmitted from one EIU to the other; hence, causing each EIU to generate the same message.

Each EIU also has a discrete signal transmitted to the other two EIUs, which is used to communicate a complete EIU failure. In the event of a total EIU failure (or loss of input power), the discrete signal opens. The remaining operable EIU(s) will respond accordingly.

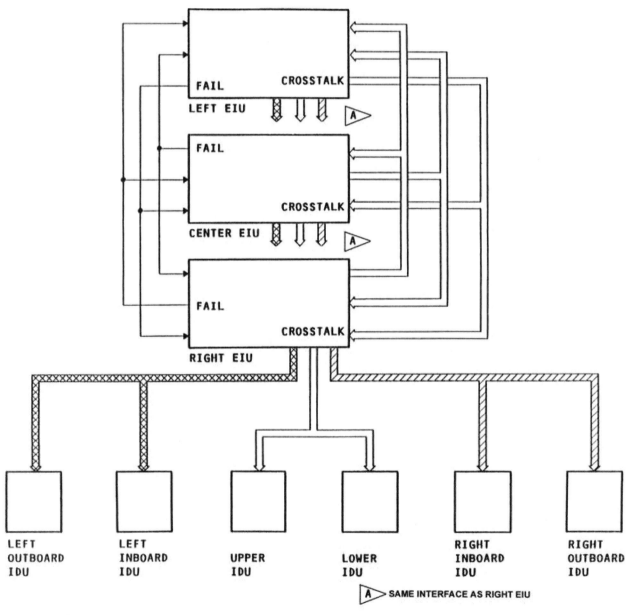

Fig. 4-31 EIU crosstalk and interface diagram. (Northwest Airlines, Inc.)

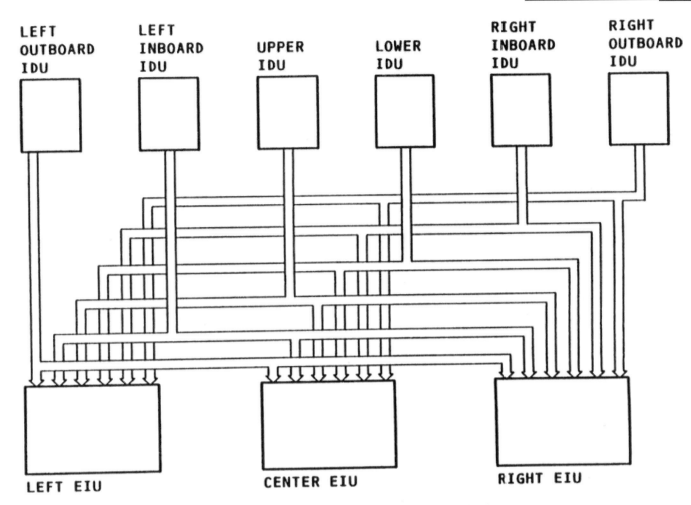

NOTE: ALL INTERCONNECTIONS SHOWN ARE ARINC 429 BUSES

Fig. 4-32 IDU/EIU data interfaces. (Northwest Airlines, Inc.)

Each EIU uses three separate busses to transmit data to six CRT displays (see figure 4-31). The left bus sends data to the left outboard and left inboard IDUs. The center data bus feeds the upper and lower center IDUs. The right bus transmits to the right inboard and outboard IDUs. The display formats on these displays are interchangeable to provide system redundancy. Normal and backup formats will be discussed later in the chapter.

Each IDU transmits ARINC 429 low-speed data to the three EIUs as a feedback signal. This "feedback" circuit is used to monitor health of the six IDUs. As shown in figure 4-32, one bus leaves each IDU and transmits the health status of all three EIUs.

IDS power is supplied through 3 different AC busses and 11 different circuit breakers. In order to provide input power redundancy, the captain's flight instrument transfer bus (115VAC), the first officer's flight instrument transfer bus (115VAC), and the standby AC bus (115VAC) each power one EIU and one or more of the IDUs. The IDS power input diagram is shown in figure 4-33.

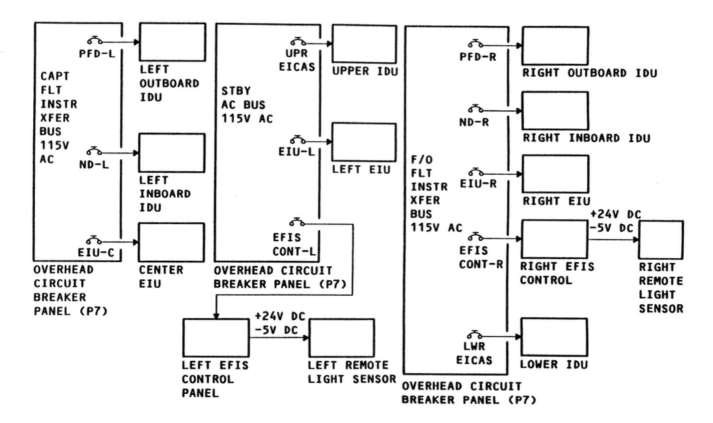

Fig. 4-33 Integrated display system power inputs. (Northwest Airlines, Inc.)

Pin Programming

The electronic interface units of the B-747-400 utilize hard-wired discrete analog inputs for **pin programming.** Pin programming is used to configure EICAS specifically for a given aircraft and it's related systems. Each EIU has 57 connections (on the 800 pin connector) dedicated to pin programming. One of those 57 pins is used as a parity bit. During aircraft manufacturing or during a system modification, the program pins should be connected/disconnected (ground or open) as specified by the manufactures data. A ground represents logic 1, open equals logic 0. Program pins are also used by the display units to identify what position (inboard/outboard, left/right, upper/lower, or center) the IDU is installed.

The 57 pin programming connections are assigned the following categories: 1) Pins 1-22 are EICAS program pins. These pins determine some of the items to be displayed, such as, tire pressure and fuel quantity units. 2) Pins 23-32 are unique option pins. These pins are dedicated for use by any software specifically requested by an airline. 3) Pins 33-56 are DISPLAYED pins; used to determine what information or format is displayed by the PFDs and NDs. 4) Pin number 57 is the parity bit. Note: this parity pin is specifically set according to the number of program pins set to logic 1. Odd parity is always maintained. For more information on odd parity see chapter 2.

EICAS Controls

The diagram of figure 4-34 shows the locations of various IDS control panels. The four panels used to control EICAS functions are: 1) the EICAS control panel (located above the upper EICAS display), 2) the EICAS display select panel (located on the glare shield just right of center), and 3) the captain's and first officer's display transfer modules (located above their respective right and left displays).

The EICAS control panel

The EICAS control panel has three controls: brightness, event record and EIU selection. As seen in figure 4-35 there are two brightness (*BRT*) controls. The *UPR* knob is used to adjust the brightness of the upper EICAS display. The *LWR* control has an inner and outer knob. The outer knob adjusts the brightness of the lower display. The inner adjustment is used for raster brightness on the lower IDU when it is used as an ND. The raster brightness will determine the intensity of the background information on the display, such as the radar display or attitude sphere. These controls consist of a simple potentiometer with +5V DC and ground references (see figure 4-35). The *EVENT RCD* (event record) switch is used to record manual events in the EICAS nonvolatile memory. Pressing this button will record the parameters of all EICAS systems simultaneously.

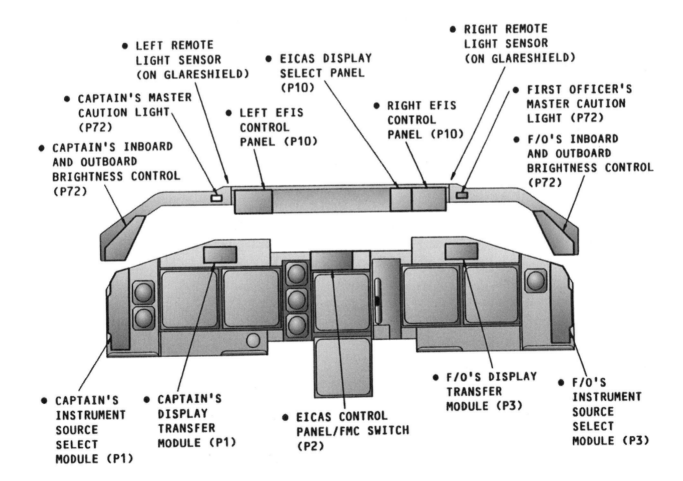

Fig. 4-34 IDS control panel locations. (Northwest Airlines, Inc.)

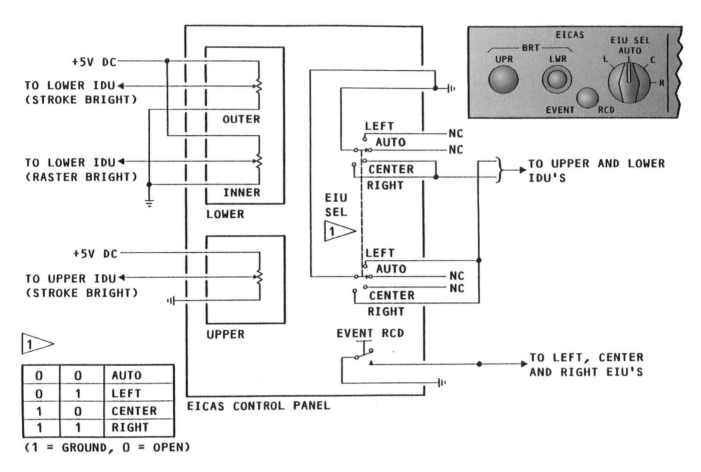

Fig. 4-35 EICAS control panel and related circuitry. (Northwest Airlines, Inc.)

The source selection switch is used to determine which EIU will drive the EICAS displays. The pilot can manually select *L*, *C*, or *R* to choose a specific computer. The *AUTO* position selects the left, center, and right computers respectively. If one EIU fails, the system will automatically select the next priority EIU. The EIU selection switch sends a two bit binary code to each EICAS display; the display's software then selects the correct EIU. It should be noted that data from all three EIUs is sent to each display unit and the DU simply "listens" to the selected EIU. As shown in the lower left portion of figure 4-35 the binary code *00* is an AUTO selection, *01* means L, *10* = C, and *11* means R.

The display select panel

The B-747-400 display select panel (DSP) combines the engine (ENG) and status (STAT) switches with the synoptic page switches (ELEC, FUEL, ECS, HYD, DRS, and GEAR). The cancel and recall switches are also part of the DSP. As presented earlier on the B-757, the synoptic pages are found on the EICAS maintenance panel. Figure 4-36 shows the layout of a B-747 DSP. The *ENG* (engine) push-button is used to call the secondary engine data to the lower EICAS display. The *STAT* (status) switch, when pressed, will present the status display on the lower EICAS CRT.

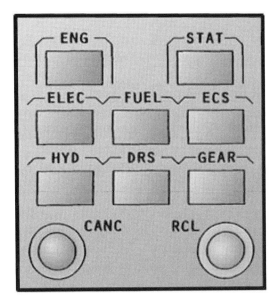

Fig. 4-36 EICAS display select panel.

Six synoptic pages are available with this EICAS: *ELEC* (electrical), *FUEL* (fuel), *ECS* (environmental control systems), *HYD* (hydraulic), *DRS* (doors), *and GEAR* (landing gear). The *CANC* (cancel) switch is used to cancel any level B. C, or D EICAS message. The *RCL* (recall) switch is used to recall any messages that are still active. Engine exceedance data is also accessed using the CANC/RCL switches.

The B-747-400 allows for EICAS control functions of the DSP to be transferred to the **control display unit (CDU).** The 747 has three CDUs located in the center console of the flight deck. The CDUs are used to input flight data, access the central maintenance computers, and as backup controls for the IDS. Figure 4-37 shows the layout of a typical CDU. They are enabled by failure of the EICAS DSP, or through failure of the EFIS (not EICAS) control panels.

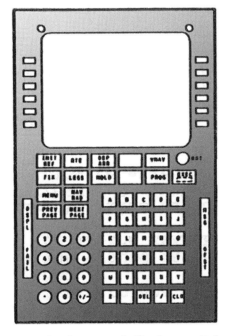

Fig. 4-37 Typical control display unit layout.

To operate the EICAS using the CDU, simply press the *MENU* key on the CDU (see figure 4-38). The CDU will present the menu allowing the operator to choose the "EICAS CP SELECT" key. This will activate the next display where the operator can select a portion of the DSP functions. The remainder of the DSP functions are available by pressing the synoptics page button. This function is available only if the EICAS control panel (or related circuitry) fails.

Display transfer modules

The main and auxiliary EICAS displays can be transferred to either the pilot's or first officer's inboard IDU. Likewise, the pilot's or first officer's PFD or ND can be transferred to the lower EICAS display. This transfer is controlled through the **display transfer modules (DTMs)** and is typically done in the event of a display failure. Figure 4-39 shows the pilots DTM. The first officer's DTM operates identically; however, the DTMs are not interchangeable due to switch position.

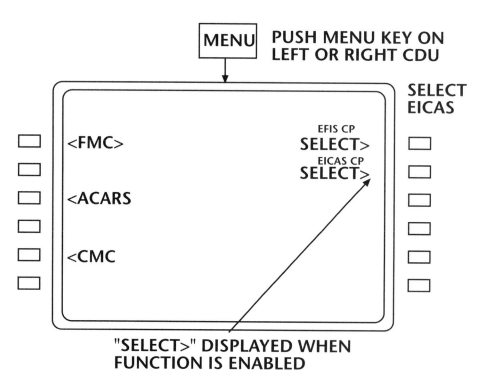

Fig. 4-38 Using the CDU to access EICAS functions. (Northwest Airlines, Inc.)

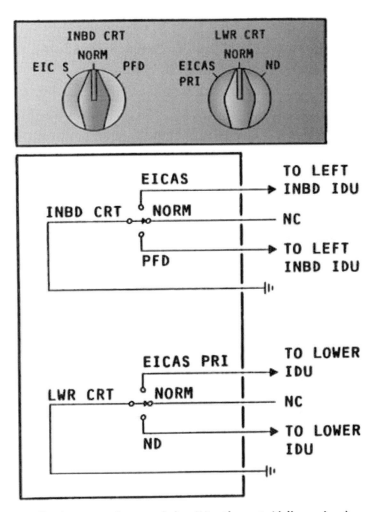

Fig. 4-39 Captain's display transfer module. (Northwest Airlines, Inc.)

The DTMs each have two switches, the inboard CRT (*INBD CRT*) and the lower CRT (*LWR CRT*). The term CRT is interchangable with IDU. The inboard CRT control is used to choose what information will be displayed on the pilot's or first officer's inboard CRT. The inboard CRT can display EICAS or PFD data if manually selected. The normal position will allow the IDU circuitry to switch display formats in the event of an IDU failure. The lower CRT switch can be set in the EICAS primary, or the navigational display position. The normal position for the lower CRT control will also allow the IDU to perform automatic switching. The DTM sends discrete signals of ground or open to the IDU for controlling switch selection.

With all DTM switches in the normal position, if the main EICAS display fails, the main EICAS data is automatically transferred to the lower CRT. EICAS is now operated in a degraded mode. As shown in figure 4-40 full EICAS capacity can be regained by selecting EICAS for display on the inboard CRT. Likewise, if the lower CRT switch is moved to the *ND* position, navigational data will be substituted for secondary EICAS information and EICAS will be operational in a degraded condition.

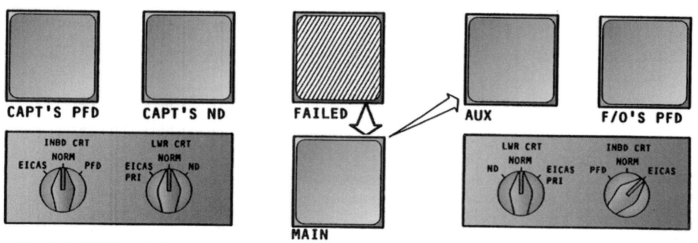

Fig. 4-40 EICAS switching; automatic switching moves data from upper display to lower display. Manual switching moves data from lower display to first officer's ND display position. (Northwest Airlines, Inc.)

Normal Operation Of The B-747-400 EICAS

During normal operations the B-747-400 EICAS utilizes two IDUs, the **Main EICAS** and **Auxiliary EICAS** displays. The main EICAS can display the primary engine, compacted, compacted-partial, and mini-synoptic formats. The auxiliary EICAS can display secondary engine, secondary-partial, status, synoptic, and maintenance pages formats. The compacted or partial formats will be discussed under abnormal operation.

Display formats

When the AC busses receive power, the EICAS automatically display the primary and secondary formats (see figure 4-41). The primary format consists of the primary engine data, fuel data, landing gear position (when down), flap position, environmental control system (ECS) data, and the requested thrust modes which are displayed on the upper EICAS display. Any active alert message would also be displayed. The lower EICAS display would show the secondary engine parameters including fuel flow, oil system data, and engine vibration. Pressing the ENG switch on the DSP can blank the lower display. During normal flight, the crew would most likely blank the lower display.

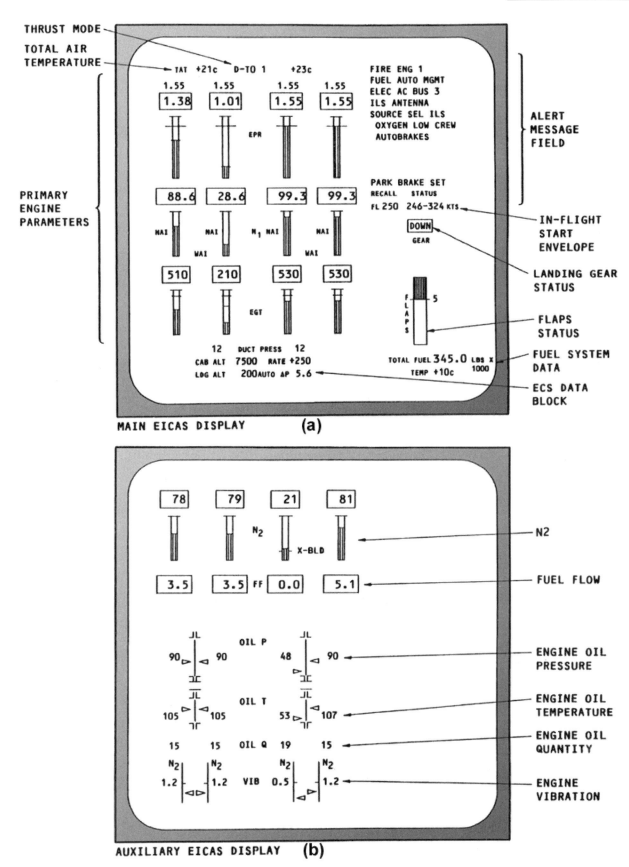

Fig. 4-41 EICAS displays: (a) primary format, (b) secondary format. (Northwest Airlines, Inc.)

The display format for the second generation EICAS is similar to the first generation systems. Both digital data and analog displays are used to provide engine parameter information. The B-747-400 analog representations are done using a vertical-scale type display. The B-757 utilizes a round-dial style of analog display.

In the event of a primary engine parameter exceedance, the displayed data turns red and the appropriate advisory message is displayed. With the B-747-400 EICAS in the event of a secondary engine parameter exceedance, the particular system effected will be displayed in the **partial format.** One of three parameters, N_2 speed, oil, or vibration are available in secondary-partial display format (see figure 4-42). It should be noted that the parameters for all four engines will always be displayed. During an exceedance, the secondary-partial format will automatically be displayed regardless of the current auxiliary EICAS display format.

Alerting and memo messages

The B-747-400 EICAS utilizes four levels of alerting and memo messages: warnings (level A), cautions (level B), advisories (level C), and memos (level D). Figure 4-43 show the various messages, as they would appear on the upper EICAS display. Their associated discrete audio and visual annunciators are also shown. It should be noted that all messages except level A can be canceled to "clean up" the EICAS display using the cancel button located on the DSP. If more than one message page is available, the cancel button is used to scroll through the pages. The recall button will bring back any previously canceled messages.

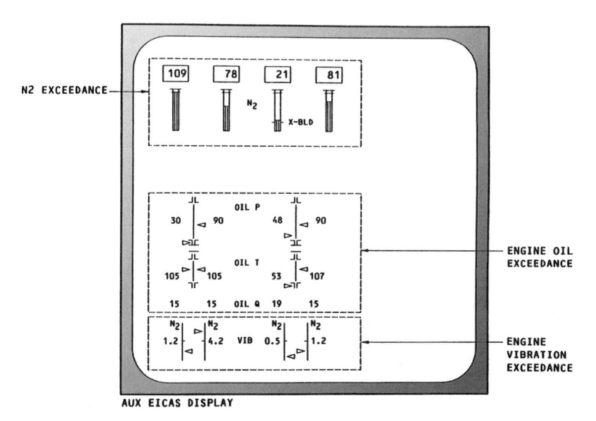

Fig. 4-42 EICAS secondary partial format. (Northwest Airlines, Inc.)

Status messages

Pressing the STAT button on the DSP will access the EICAS status page. The status page, containing APU, hydraulic and oxygen system data will appear on the auxiliary EICAS display. Status messages and a dynamic flight control surface display will also be shown in the lower portion of the status page (see figure 4-44). If more than one status page is available, pressing the *STAT* will cycle through all available pages and eventually remove the status page data. The status page data is typically used by technicians and the flight crew to determine the "status" of the aircraft prior to dispatch. If the status page shows a given system as inoperative, the technician would consult the minimum equipment list (MEL) to determine the airworthiness of the aircraft. If approved by the MEL, the repair would most likely be deferred until the next maintenance opportunity.

Status pages are initiated automatically by EICAS whenever a malfunction occurs in a system monitored by the EIUs. The EIUs can generate two types of status messages, **latched** and **non-latched**. Non-latched messages are "real time" information. If the problem corrects itself, the non-latched message will automatically remove itself.

On the B-747-400 there are three types of latched status messages: **ground only, air only**, and **unconditional.** Their names imply when the messages can be stored (latched) in the EICAS nonvolatile memory. Unconditional messages can be stored

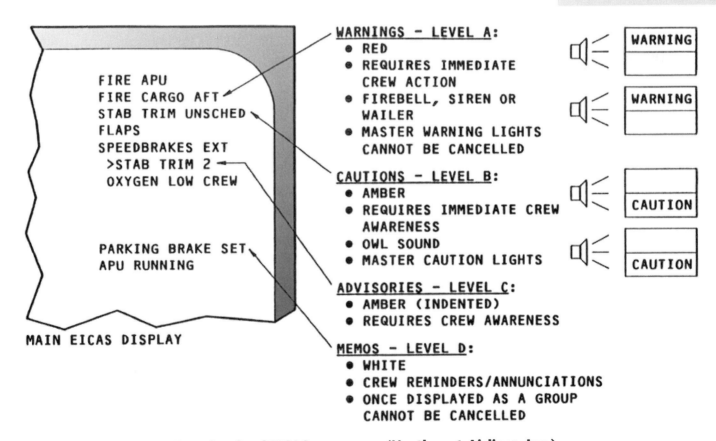

Fig. 4-43 Four levels of EICAS messages. (Northwest Airlines, Inc.)

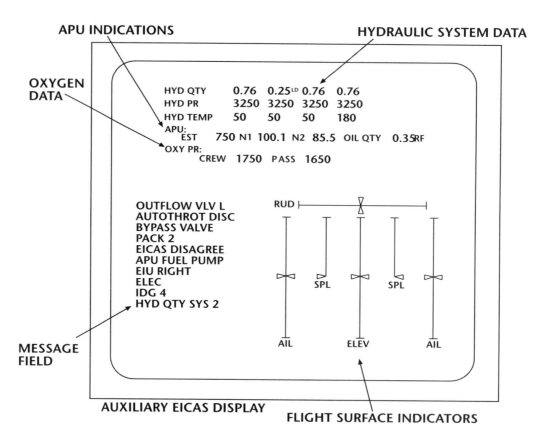

APU INDICATIONS   HYDRAULIC SYSTEM DATA

OXYGEN
DATA

HYD QTY   0.76 0.25ᴸᴰ 0.76 0.76
HYD PR   3250 3250 3250 3250
HYD TEMP   50   50   50   180
APU:
   EST   750 N1 100.1 N2 85.5 OIL QTY 0.35RF
OXY PR:
   CREW 1750 PASS 1650

OUTFLOW VLV L   RUD
AUTOTHROT DISC
BYPASS VALVE
PACK 2
EICAS DISAGREE
APU FUEL PUMP
EIU RIGHT
ELEC   SPL   SPL
IDG 4
HYD QTY SYS 2

MESSAGE
FIELD   AIL   ELEV   AIL

AUXILIARY EICAS DISPLAY
                    FLIGHT SURFACE INDICATORS

Fig. 4-44   EICAS status page. (Northwest Airlines, Inc.)

with the aircraft in flight or on the ground. The latched messages are extremely handy for maintenance of intermittent problems. For example, if a poor electrical connection causes a temporary flight management system malfunction, the maintenance crew may not be able to duplicate the problem. However, the associated latched status message will provide an excellent reference for troubleshooting. If we now assume the correction is made by replacing an LRU, the latched message for the next flight can be used to verify if the defect was properly repaired.

There are two methods available to erase the status messages. The primary method uses the control display unit to erase each message individually. This method will be discussed in chapter five. The alternate method of erasure employs the DSP. When the status page is on display, simultaneously pressing both the cancel and recall buttons (on the DSP) will erase all status messages from the latched memory. Any messages that are still active will automatically reset themselves and once again be stored in status memory.

Synoptic pages

The synoptic pages for the B-747-400 EICAS are accessed the same as the synoptic pages for the B-757. The major differences are that on the B-747-400, only one system is shown per page and each page has a more graphic format (see figure 4-45).

Abnormal Operation Of The B-747-400 EICAS

The B-747-400 EICAS can fail in four basic ways and still be operational: 1) loss of one or more parameters, 2) loss of the main or auxiliary displays, 3) loss of one or both EICAS control panels, and 4) loss of up to two system computers. In each of these four situations, the EICAS will be operational in some form. The redundancy of these systems makes it extremely unlikely that a complete EICAS failure will occur.

The loss of one parameter to EICAS will cause that particular portion of the display to go blank and the associated message to be displayed. The loss of the main or auxiliary displays will cause the EICAS to go into a compact format. The EICAS **compact-full** format is used when both primary and secondary engine data is requested and only one display is operable (see Figure 4-46). In compact-full format only EPR is shown in both digital and analog forms.

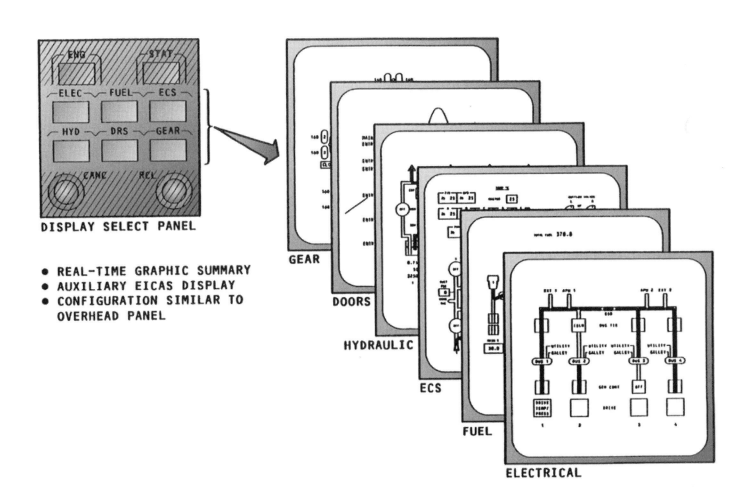

DISPLAY SELECT PANEL

- REAL-TIME GRAPHIC SUMMARY
- AUXILIARY EICAS DISPLAY
- CONFIGURATION SIMILAR TO
  OVERHEAD PANEL

GEAR

DOORS

HYDRAULIC

ECS

FUEL

ELECTRICAL

Fig. 4-45 EICAS synoptic pages. (Northwest Airlines, Inc.)

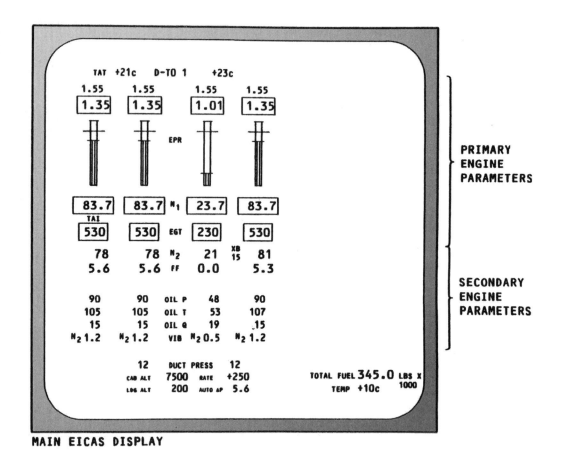

MAIN EICAS DISPLAY

Fig. 4-46 EICAS compacted mode. (Northwest Airlines, Inc.)

A synoptic **mini-format** is available on the second generation EICAS to display certain status data when only one EICAS display is operational. The fuel and gear status pages are the only systems with the mini-format option. In the mini-format, a compacted version of the standard synoptic page (fuel or gear) is displayed under the compacted primary engine data. The display is very similar to the compact-full format except the synoptic page replaces the secondary engine data.

In the event of a control panel failure, the EIUs will revert to a fail-safe mode and display necessary information. If the DSP fails the DCU can be used as a backup as previously discussed. If one or two EIUs fail, the third computer can be used as the sole provider of EICAS information. If all three EIUs fail, all EICAS functions are lost.

Maintenance pages

The second generation EICAS maintenance pages are accessed through the central maintenance computer system. There are 11 different maintenance pages available on the 747-400 system. These pages are each displayed on the auxiliary EICAS display during ground operations or during flight. In general, there are three types of information available for each maintenance page: real time, manual snapshot, and auto snapshot. The real time data is dynamic information and presents information as it currently exists. The snapshot data is information recalled from memory. EICAS maintenance pages will be studied in further detail in chapter five of this text.

Troubleshooting and maintaining EICAS

Many of the techniques discussed earlier in this chapter concerning troubleshooting the B-757 EICAS also apply to the B-747-400 EICAS. Some of the techniques, which are unique to the second generation systems, will be presented here.

Software data loader

One major change for the second generation EICAS is the ability to change portions of the programming through the use of a software data loader (SDL). The data loader can alter the software in all three EIUs, the upper and lower EICAS displays, and several other LRUs. The updated data and the program needed to access the LRU is contained on a standard 3.5 inch floppy diskette. Two types of data loaders are available for the B-747-400: a portable unit, and a permanently installed data loader located on the observers console of the flight deck. The data loader is also used to retrieve data from the CMCS. When troubleshooting a chronic problem, CMC data is often down loaded to a floppy disc. Engineering personnel can then use the data for in-depth system analysis.

According to the figure 4-47 the data loader sends a SDL enable signal to the SDL panel and the selected LRU (in this example the left outboard IDU). The LRU then returns a signal to the SDL when it is ready to accept the updated software. This system has made it possible to make modifications to the LRUs in a relatively easy fashion. In older systems, the same software change would require a LRU change and factory update, or a card replacement.

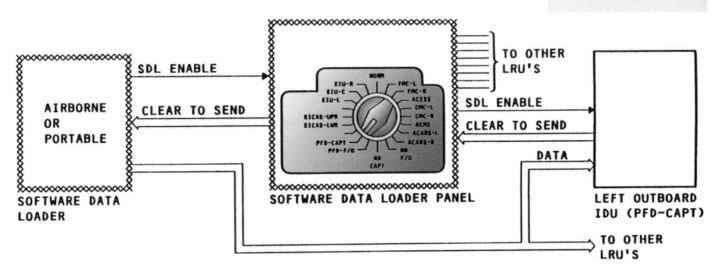

Fig. 4-47 Software data loader interface diagram. (Northwest Airlines, Inc.)

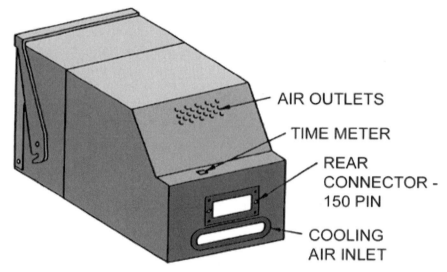

Fig. 4-48 Rear view of a typical EICAS display unit. (Northwest Airlines, Inc.)

CRT displays

The CRT displays on the second generation EICAS are slightly different than those found in the earlier systems. The display units are larger to improve readability, and incorporate an hour meter to track time in use. Cooling is still a critical item for the CRTs and these displays will automatically shut off in two stages when over heated. When the display temperature reaches 110° C, the display will shut down all graphic displays. If the display reaches 125° C the display will blank out all together. When the unit cools, the display will automatically return. Be sure the inlet and outlet air passages are kept clean during display replacement (see figure 4-48).

The B-747-400 provides a display unit protective cover located just below the lower EICAS IDU. This cover should be used during all IDU maintenance to protect the CRT screen. The screen of the B-747-400 CRT displays **CAN** be cleaned with standard glass cleaner. Since these displays are always on when their respective bus is energized, it is wise to turn down the brightness if the aircraft is to be powered for long periods during maintenance. This practice will help to lengthen life of the display. The display circuit breaker could also be opened to "turn off" the IDU. If this is done, take all necessary precautions to ensure that you reactivate the system.

BITE tests

The BITE test for the B-747-400 is activated through the central maintenance computers. The actual BITE tests and fault memory still occur in the individual LRU of the system. The CMC is simply used as an access point for all the aircraft's functional tests. A block diagram of the IDS BITE system is shown in figure 4-49. The CDU is used to access the EICAS EIUs and IDUs. The individual EIUs and IDUs monitor their respective health and input signals, then transmit that data to the CMCs. The CMCs display all BITE data through the CDU.

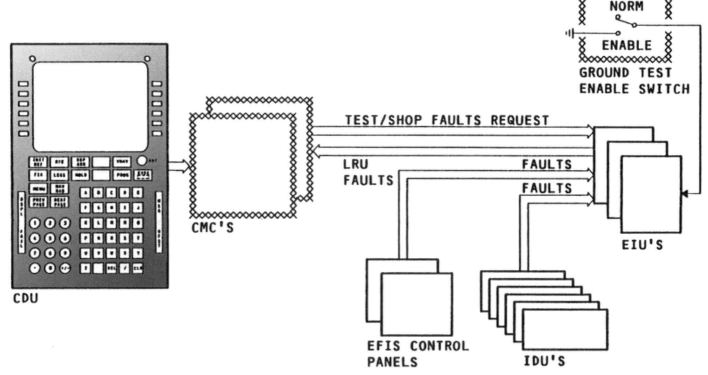

Fig. 4-49 Block diagram of the Integrated Display System BITE. (Northwest Airlines, Inc.)

To initiate an EICAS BITE test, the ground test enable switch must be placed in the *enable* position. The ground test enable switch is located on the overhead panel (P61). The next step would be to access the CMC menu using the CDU. From the menu, the ground tests menu would be selected, then the appropriate test (IDS) would be selected. During the test, the term *IDS* and *IN PROGRESS* will be displayed on the CDU. The test takes approximately 8 seconds; during that time the test format will be shown on all operable IDUs. The master warning and caution lights will illuminate and the audio tones will sound during the test. Any detected failures will be displayed on the CDU. There are 4 different fault messages associated with the EICAS; *IDU failed, EIU failed, EIU disagree, and EIU no test response*.

During the IDU BITE test, the IDUs will display a test format similar to figure 4-50. As shown in this figure, the system program pins are displayed in the upper left of the test format. If a program pin error exists the term *ERROR* will be seen adjacent to an error code. The error code can then be found in the aircraft's maintenance manuals.

In general, second generation EICAS offers improved troubleshooting over earlier systems. Use of the CMC has helped to simplify the troubleshooting process. The CMC will display fault codes and fault messages. Using the aircraft's fault isolation manual (FIM), these codes will direct the technician to the proper repair or further troubleshooting sequence.

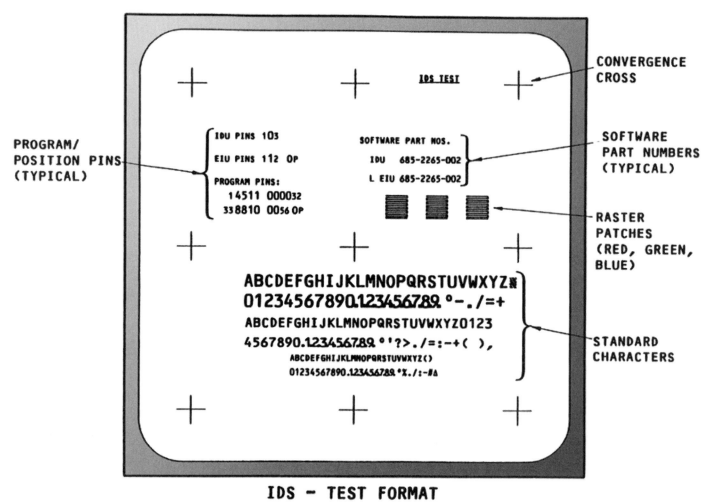

IDS — TEST FORMAT

Fig. 4-50 IDS test format. (Northwest Airlines, Inc.)

The CMC can also be used to monitor input data to many of the EICAS LRUs. The **input monitoring** function of the CMC is used to determine the status of ARINC 429 data to the EICAS electronic interface units (EIUs). In general, input monitoring is done through the CDU and compares system data received from the EIUs and system data sent directly to the CMCs (see figure 4-51). Complete details of the central maintenance computer system will be discussed later in this text.

ELECTRONIC CENTRALIZED AIRCRAFT MONITORING

Airbus transport category aircraft use a system called **electronic centralized aircraft monitor or ECAM** to monitor and display aircraft system and engine parameters. The system utilizes two CRT displays located in the center of the instrument panel, positioned either horizontally or vertically. The newest ECAM systems are integrated systems, which operate in conjunction with the aircraft's electronic flight instrument system. The integrated EFIS and ECAM system found on the A-320 is called the **Electronic Instrument System (EIS).** The following discussion on ECAM will be based on the A-320 aircraft.

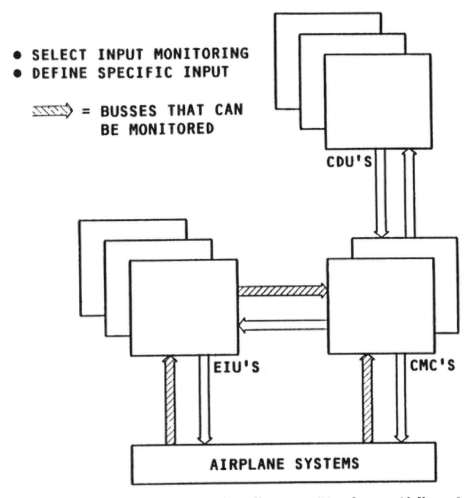

Fig. 4-51 CMC input monitoring diagram. (Northwest Airlines, Inc.)

System Description

The EIS consists of 6 CRT display units as seen in figure 4-52. Four displays are used primarily for flight data (EFIS). The remaining two displays are used primarily for engine and aircraft system data (ECAM). On the A-320, the two captain's side EFIS displays are called the **primary flight display (PFD 1)** and the **navigational display (ND 1).** The first officer's (copilot's) side consists of PFD 2 and ND 2. The two ECAM displays are referred to as the upper and lower ECAM displays. The upper ECAM display is used for **engine and warning** (E/W) information, the lower display shows **system and status** pages.

The major advantage of the integrated ECAM/EFIS system is the flexibility in the display of data. In the event of a display failure, one of the remaining displays can be used to show the missing data. Figure 4-53 shows the three means by which ECAM data is interchanged between displays. As indicated by the white arrows, the PFD and ND data can be switched manually by the flight crew. This is accomplished by the PFD/ND XFR switch located just outboard of the captain's (pilot's) and first officer's (copilot's) EFIS displays. The shaded arrows show that the data from the upper ECAM display is automatically transferred to the lower ECAM display in the event of a single ECAM display failure (figure 4-53). The black arrows show that the upper ECAM data moves to the EFIS displays in the event that both ECAM displays fail.

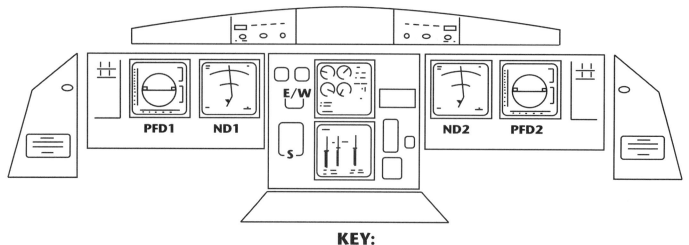

KEY:
PFD= Primary Flight Display
ND= Navigational Display
E/W= Engine Warning
S= System or Status

Fig. 4-52   A-320 Electronic instrument system display locations. (Airbus Industrie)

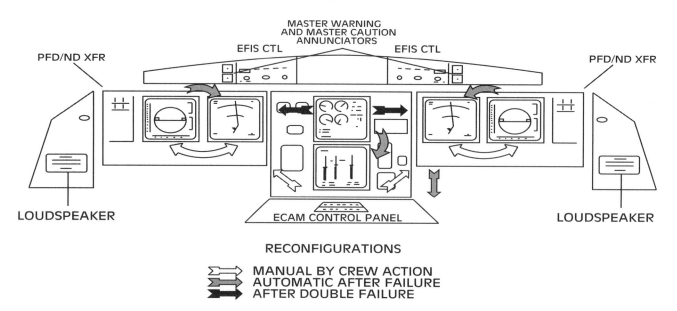

Fig. 4-53 ECAM data display switching. (Airbus Industrie)

Display formats

There are four basic types of information displayed by ECAM: **engine data, messages, system synoptic pages,** and **system status data.** Engine data is always displayed any time ECAM is operating. During normal operation, various memo messages are displayed on the upper ECAM display along with engine data as shown in figure 4-54. In the event of a system failure, a warning or caution message would be displayed on the upper ECAM along with the engine data.

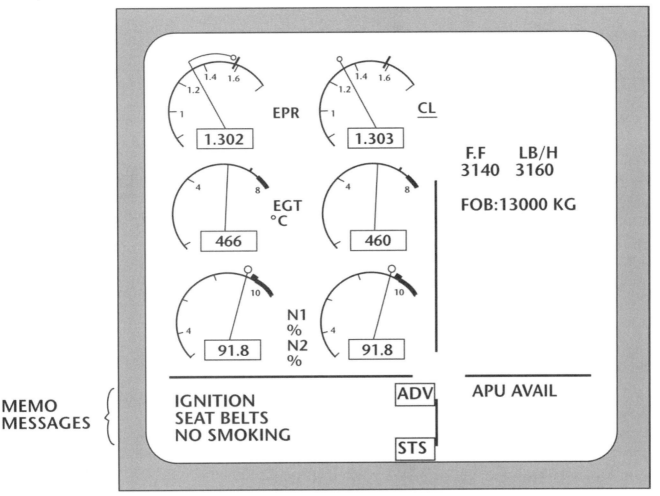

Fig. 4-54 Typical upper ECAM display, normal operation. (Airbus Industrie)

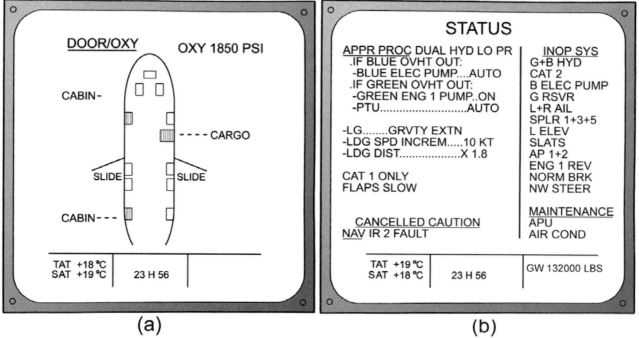

Fig. 4-55 Lower ECAM display, (a) System Synoptic Door/Oxygen Page; (b) A typical system status page. (Airbus Industrie)

The system synoptic pages and system status data is displayed on the lower ECAM display. The synoptic display gives a brief general description of the systems operation in a graphic format. The system data displayed is a function of aircraft configuration, pilot selection, or automatically displayed after a system failure. After a failure, status information is automatically recorded on the status page and can be displayed on the lower ECAM CRT when requested. The door/oxygen synoptic page and a typical status page are shown in figure 4-55.

All information displayed by ECAM is color coded to help the flight crew identify a system's status. The different colors also provide a quick reference to the technician during troubleshooting. In general, information in green indicates that all systems are operating normal. Magenta (red) indicates a critical system failure. Cyan (blue) indicates a system failure that is not critical, and white indicates general information.

Warning/caution annunciators

The ECAM system incorporates two discrete visual annunciators: the **master warning** and **master caution** lights. There is one each of these light assemblies on the pilots and copilots glare shield (see figure 4-53). To ensure visual warnings are displayed in the event of lamp failure, each annunciator contains 4 light bulbs. ECAM also uses audio tones to gain the pilot's attention during display of critical system failure. Two loudspeakers installed on the instrument panel are used for ECAM audio tones.

System Architecture

The electronic centralized aircraft monitoring system contains seven computers, which receive data from various engine and aircraft systems. The A-320 ECAM computers are classified into three general categories: **display management computers (DMC), flight warning computers (FWC),** and the **system data acquisition concentrators (SDAC).**

Figure 4-56 shows a block diagram of the computers and their relationship within the integrated EFIS/ECAM system. It can be seen from this diagram the system computers monitor data for both the EFIS and ECAM displays.

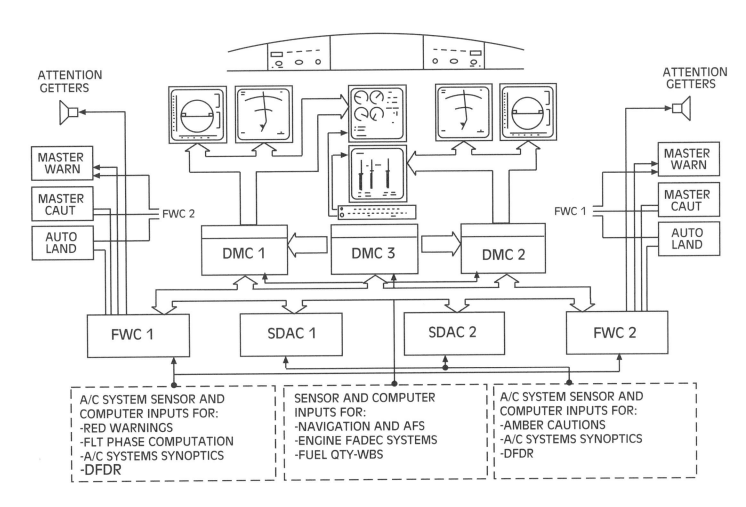

Fig. 4-56 EIS interface diagram. (Airbus Industrie)

Display management computer

The EIS contains three **display management computers (DMCs)** which process data from the SDACs, FWCs and certain discrete inputs. The discrete inputs include digital and analog signals for navigation and autoflight, digital engine control, fuel quantity, and weight and balance data (see figure 4-56). The data sent from the SDACs and FWCs is in digital format. A dual ARINC 429 data bus format is used for most systems to provide redundancy of transmitted data to the DMCs. If the data on one bus is invalid, the DMC will take the information from the other bus. If the data is invalid on both busses, the DMC will form an *XXXX* pattern on the display were the data would normally appear. The DMCs receive weather data through ARINC 453 busses and the FWC warning messages through RS 422 data busses (see figure 4-57).

The DMC sends a digital signal to a maximum of three display units. The output data contains the video information needed to produce the correct image on the EIS displays. The signals from the DMC to the display units are sent on a **dedicated serial data link** (DSDL), see figure 4-57. A feedback signal is returned from the display units to the DMC. The feedback signal provides information on the health of the display unit. The status of the display is monitored for automatic display transfer, built in test equipment (BITE) data, and various parameters used by the FWC. If any of the data returned by the DSDL does not match the data sent to the display, a system inoperative flag appears on the display.

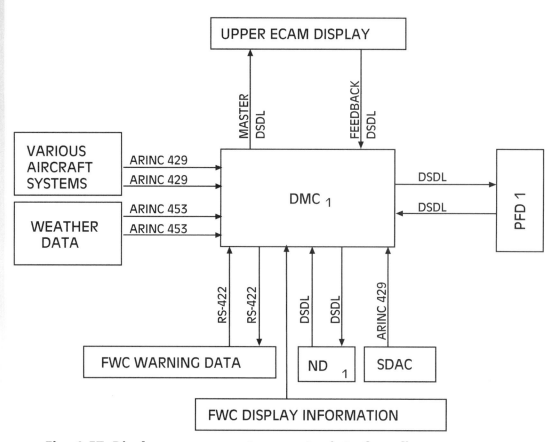

Fig. 4-57 Display management computer interface diagram.

During normal operation, DMC 1 supplies video information for PFD 1, ND 1 and the upper ECAM displays; DMC 2 supplies the video information for the PFD 2, ND 2, and the lower ECAM displays. The normal DMC output configuration is shown in figure 4-58. In the event DMC 1 or 2 fails, DMC 3 will supply data to the inoperative displays. During all operations the display management computers compares information to ensure the input and output data is valid. If a discrepancy is detected, the erroneous information is ignored and/or the system reconfigures itself to compensate. If any system fails, the appropriate message will be displayed by ECAM to alert the flight crew.

Flight warning computers

The two **flight warning computers (FWCs)** monitor the various systems necessary to generate all warnings on the ECAM displays. Other inputs include the data necessary for computation of flight phase information, and system synoptic pages. Warnings, flight phases, and synoptic pages will be discussed in greater detail later in this chapter. During normal operation, FWC 1 generates all output signals to the DMCs. The DMCs then create the video information for the display units. FWC 2 is always operational and will automatically take control of the output data if FWC 1 fails.

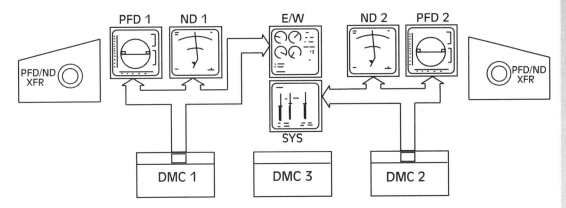

Fig. 4-58 Normal DMC output configuration. (Airbus Industrie)

The software of the FWCs perform the essential computations to generate the warning messages on the ECAM displays (via the DMCs). The software also controls the discrete aural and visual warning signals. The master caution and warning annunciators are each powered by both FWCs (see figure 4-56). There are four light bulbs in each annunciator as shown in figure 4-59. In this example, the FWC 1 controls bulbs A and D; FWC 2 controls bulbs B and C. This configuration ensures the master annunciators are operational in the event of a single FWC or multiple lamp failure.

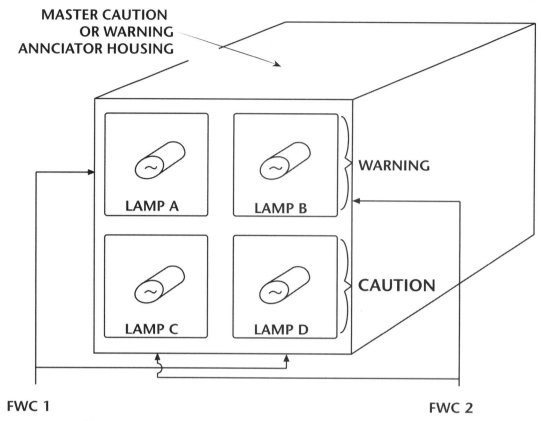

Fig. 4-59 Master warning and caution annunciator.

Figure 4-60 shows the input/output diagram for the flight warning computers. The FWC receives 6 synchro, 154 discrete, and 40 ARINC 429 data bus inputs for various aircraft and engine systems. The FWC output data is comprised of ARINC 429 data, RS 422 data, and discrete data. The discrete data is sent directly to both the captain's and first officer's caution and warning displays. Discrete data is also sent to both loudspeakers. The RS 422 data outputs are used to transmit warning messages to the DMCs. The ARINC 429 data busses transmit information to three DMCs, the opposite FWC, the Centralized Fault Display Interface Unit (CFDIU), and the Flight Data Interface Unit (FDIU). The CFDIU interfaces with the aircraft's built-in diagnostic system. The FDIU interfaces with the aircraft's flight data recorder.

System data acquisition concentrators

The **system data acquisition concentrators (SDACs)** monitor aircraft systems, perform the necessary computations, and transmit output data to the FWCs and DMCs. The systems monitored by the SDACs are less critical than those monitored by the FWCs. That is, the SDACs control the master caution annunciators; the FWC controls the warning annunciators. (Caution messages are considered a lower priority than warnings messages.) The SDAC output data is sent to the master caution annunciators via FWC 1 and 2. The synoptic page data is sent to the ECAM displays via the CMCs.

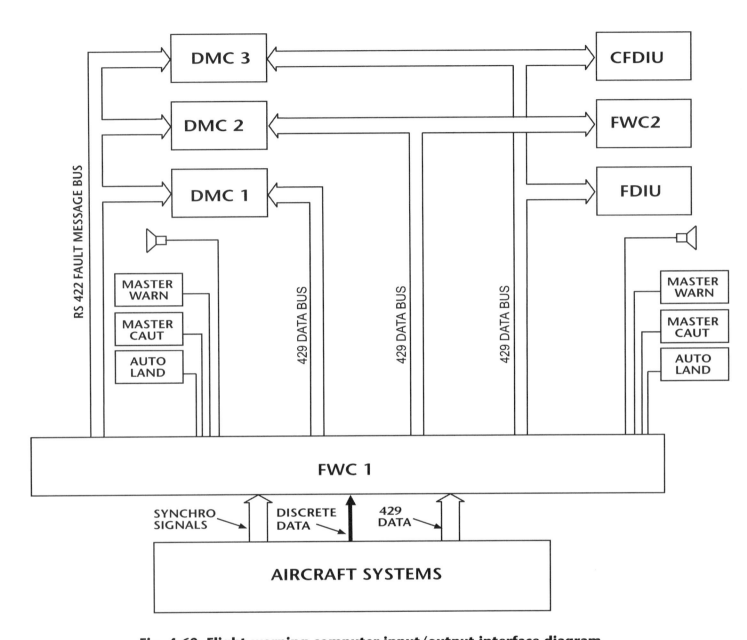

Fig. 4-60 Flight warning computer input/output interface diagram.

Figure 4-61 shows the input/output diagram for the SDACs. The SDAC's input data includes 388 discrete signals, 24 ARINC 429 data busses, and 12 synchro signals. Both SDACs receives identical inputs. Each SDAC transmits three ARINC 429 data bus outputs. One bus feeds DMC 1 and FWC 1. A second bus sends data to DMC 2 and FWC 2. A third bus transmits to DMC 3, the data management unit (DMU), central fault display interface unit (CFDIU), the data link, and flight data interface unit (FDIU).

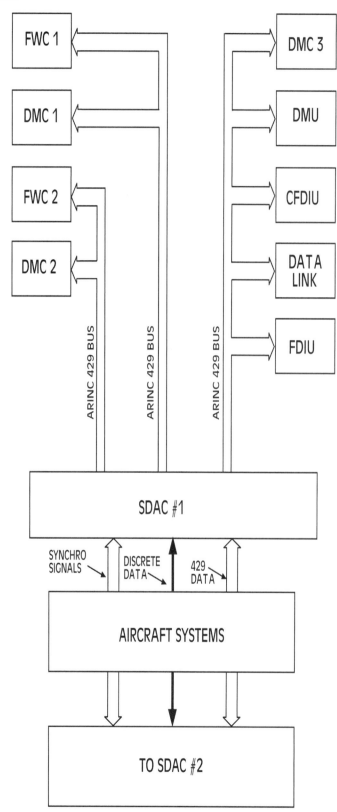

Fig. 4-61 System data acquisition concentrators (SDACs) input/output interface diagram.

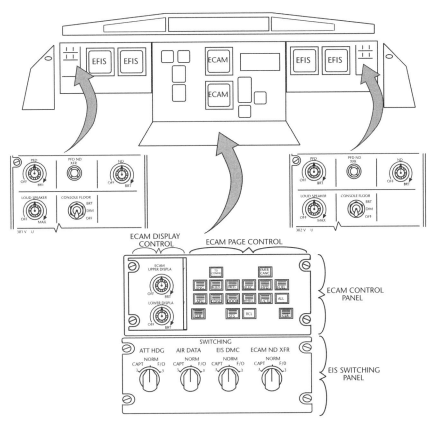

Fig. 4-62 EIS control panel locations. (Airbus Industrie)

In general, the LRUs of the ECAM system are interchangeable. That is, the DMC 1 can be installed in the DMC 2 or DMC 3 mounting racks and visa versa. The two FWCs are identical, as well as the two SDACs. All six display units are also interchangeable. This becomes very handy when performing troubleshooting. For example, if it is suspect that FWC 1 is defective; simply swap FWC 1 for FWC 2. If the fault now appears in FWC 2, the LRU, which was moved from the number 1 to number 2 FWC slot, must be defective. **Pin programming** of the FWC's wiring harnesses is used to identify the FWC 1 and FWC 2 installation racks. As the computer is installed into the equipment rack, a given pin is connected to ground (digital 0) to tell the LRU that it is installed in the FWC 2 rack. The same pin would be open (connected to digital 1) to identify the FWC 1 rack. Pin programming is used to identify the location of all interchangeable LRUs.

EIS Controls

The ECAM and EFIS control panels are shown on the component location diagram (figure 4-62). The ECAM displays are controlled by the ECAM LRU and the 8VU panel located just below the lower ECAM display. The brightness and transfer of the PFDs and NDs are controlled through the 301VU and 302VU panels located to the outboard side of the captain's and first officer's EFIS displays. Using these controls, the PFD and ND can be turned on/off separately as needed for maintenance. Located between the on/off/brightness controls is the PFD/ND XFR (transfer) control. This push-button switch is used to transfer the data shown on the PFD to the ND, and visa versa. The captain's and first officer's displays are controlled independently; therefore, identical control panels are accessible to both pilots.

Display controls

The ECAM control module is a separate LRU, which has two basic functions, **display control,** and **page control.** The **switching control** panel (8VU) is located directly below the ECAM control module. The display controls are used to turn on/off the individual ECAM displays. This control can be a useful troubleshooting tool for testing the mono mode of the ECAM system. Simply turn on the upper or lower ECAM display and the DMCs should automatically convert all display data to the mono format. It should also be noted, the average mean time before failure of these display units is relatively low. It is therefore very important that only the necessary EIS displays are operated during aircraft maintenance. This will help to "save" the displays and improve system reliability.

ECAM switching

The **ECAM switching** controls are used to select the source of information to be presented on the EIS displays. In other words, ECAM switching controls which DMC will control which display(s). As shown earlier in figure 4-58, during normal operation of the system, DMC 1 supplies data for the captain's EFIS and the upper ECAM displays. In the normal configuration, DMC 2 supplies the first officer's EFIS and lower ECAM displays.

The ATT HDG (attitude heading) switch controls the source of attitude heading information. If the ATT HDG switch is placed in the *CAPT 3* position, the captain's EFIS will receive attitude data from DMC 3. If switched to the *F/O 3* position, the first officer's attitude data will come from DMC 3. The air data switch operates in a like manner.

The EIS DMC switch determines which display management computer will drive the captain's and first officer's PFD, ND and the associated ECAM displays. Figure 4-63 shows the configuration if *CAPT 3* is selected on the EIS DMC switch. In this situation DMC 3 supplies data to PFD 1, ND 1, and the upper ECAM display. If *F/O 3* was selected, DMC 3 would be used to drive PFD 2, ND 2 and the lower ECAM displays.

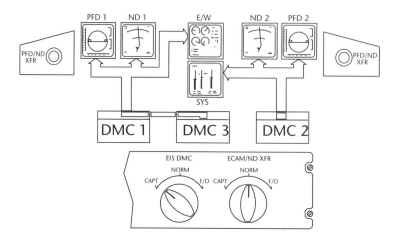

Fig. 4-63 EIS interface when captain's DMC 3 is selected. (Airbus Industrie)

The ECAM/ND XFR switch is used in the event one ECAM display fails. During a single ECAM display failure, the ECAM mono mode is automatically displayed on the operable CRT. If so desired, the flight crew can use ND 1 or 2 to replace the failed ECAM display using the ECAM/ND XFR switch. Figure 4-64 show the system with the lower ECAM display failed and the ECAM/ND XFR switch placed in the *F/O* position. In this situation, all ECAM data normally displayed or the lower ECAM is transferred to the first officer's navigational display (ND 2).

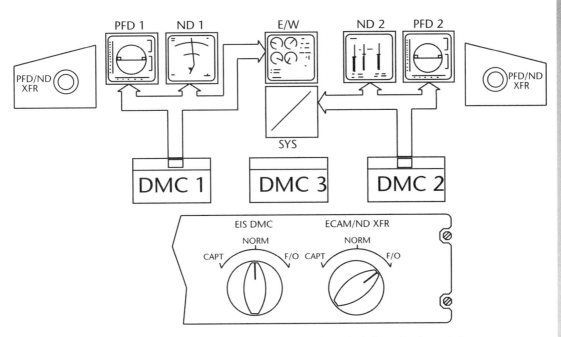

Fig. 4-64 Example of ECAM/ND transfer. (Airbus Industrie)

A simplified diagram of the EIS switching is shown in figure 4-65. In this diagram it can be seen that switching is done through software within the LRUs and through dedicated relays mounted in the equipment bay. Each display unit can be driven by its normal (N) or alternate (A) input. The switching for the PFD and ND displays are controlled via a 28VDC relay (see the relay box capt, and relay box F/O, figure 4-65). The pilot selects an EIS configuration using the PFD/ND control panel. The control panel commands a relay to send a given signal to the display unit software. The display unit then selects the requested input (alternate or normal). In like manner, the ECAM engine warning (E/W) and status (S) displays are also controlled via a 28VDC relay and display unit software.

Switching to determine which DMC will drive which display is done within the DMCs. Each DMC receives input signals from the ECAM switching panel. The DMC has the capability of moving the lower ECAM display to the captain's or first officer's ND. Control for the ECAM/ND transfer is provided by the ECAM switching panel through a 28VDC relay to the DMCs.

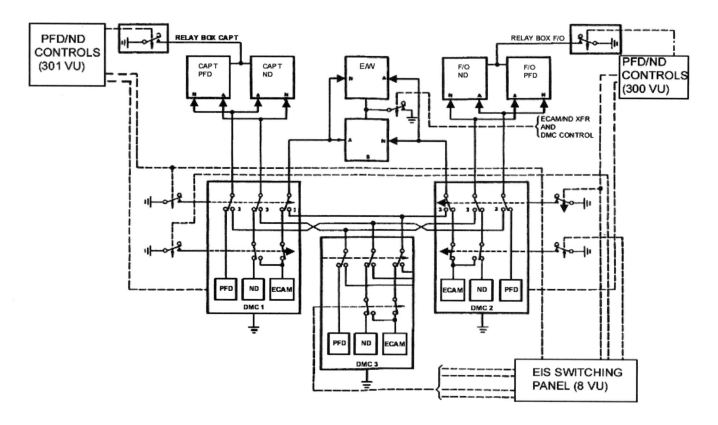

Fig. 4-65   A-320 EIS switching diagram. (Airbus Industrie)

Page control

The **page control section** of the ECAM control panel contains 18 push-button switches used to access different systems for display on the lower ECAM CRT (figure 4-62). It is important for the electronics/avionics technician to become familiar with this control panel. The various ECAM pages show real time status of all major aircraft systems. During troubleshooting, this information is extremely valuable. The following paragraphs contain a description of the ECAM page control switches and their associated ECAM displays.

The TO CONFIG (takeoff configuration) switch is used by the flight crew to determine if the aircraft is ready for takeoff. When this button is pressed, the message *TO CONFIG NORMAL* is displayed on the E/W ECAM display. If the system is not in the correct configuration for takeoff, the appropriate warning will be initiated by ECAM.

The EMER CANC (emergency cancel) will delete any ECAM aural warnings currently active; however, the master warning and associated ECAM message will remain on display. The EMER CANC switch will also cancel any caution messages and extinguish the master caution light. System cautions and warnings will be discussed later in this chapter.

There are 11 switches each labeled with a given system designator, such as ENG (engine), BLEED (bleed air), PRESS (pressurization), ELEC (electrical), etc. Pushing these switches will call the related system information to the lower ECAM display. On an ECAM system page, it is easy to determine the **real time** status of the system displayed. During troubleshooting, choosing a given system page will allow the technician to verify system operation.

The following is an example of a typical system page. Figure 4-66 shows the electrical system page with the APU generator operating at 94% output, producing 116 volts, at 400 Hz. All ac and dc busses are powered and the batteries are both receiving a 50 amp charge at 28 volts. Note that the power transfer line from the DC BAT bus contains an arrow pointing to the batteries. This indicates a current flow to the battery (i.e. the battery is being charged). The various busses, power transfer lines, and APU data are shown in green since all systems are normal. The GEN 1 and 2 are shown in amber since these systems are inactive.

The ALL button on the ECAM control panel is used to access any of the system pages. When the ALL switch is depressed and held down, the 11 system pages are displayed successively at one second intervals. If the button is released the system page currently on display will remain. It should be noted, the ALL button has dedicated wires to the DMCs; all other ECAM control switches communicate to the DMCs via the data bus. Since the ALL button is "hard wired," it provides a back up in the event the control panel or the associated data bus fails.

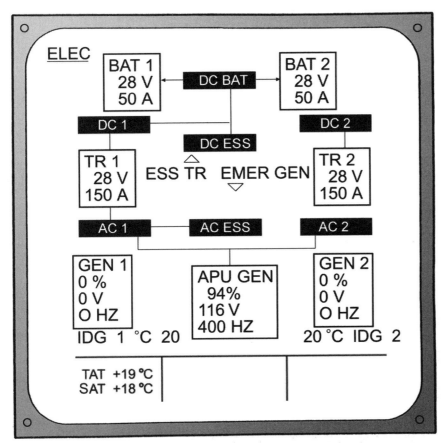

Fig. 4-66 A typical ECAM system page. (Airbus Industrie)

The **CLR** button is an illuminated switch, which turns on whenever there is caution/warning or status messages displayed on the ECAM. Depressing the CLR button (when lit) will clear the message(s) and eliminate the status page on the lower display. If more than one system or status page is present, the CLR button will cycle through all pages before clearing the display.

The **RCL** button will display any caution/warning messages, which have been cleared from the ECAM display. However, five minutes after engine shut down, stored ECAM messages are no longer kept in memory and cannot be recalled. In most cases, these messages are still available through the aircraft's central maintenance system.

The **STS** button will display the current status page for the aircraft. If no status messages are present, the term *NORMAL* will be displayed for 5 seconds. After 5 seconds the display will return to its previous configuration.

ECAM Normal Operation

The ECAM system can operate in two basic configurations **normal mode** and **mono mode.** In normal mode, the top ECAM display shows engine data and warning messages. This display is know as the E/W (engine/warning) display unit. The lower ECAM unit is used to indicate the configuration of systems and aircraft status data. The data indicated on these displays is generally consistent between aircraft. However, some displays show slight differences due to specific configurations requested by an airline.

The upper display

The upper ECAM display contains the E/W page which is divided in to four quadrants (see figure 4-67). The top left quadrant contains engine data, such as engine pressure ratio (EPR), exhaust gas temperature (EGT), turbine spool speeds (N_1), and fuel flow (FF). This data is displayed in both a digital and analog format. The analog display provides easy analysis of changing values, while the digital data offers precise reading of stable numbers.

The top right quadrant of the upper ECAM display typically shows **fuel on board** (FOB) in a digital format. A graphic representation of the slat and flap positions is shown below the FOB. The bottom two quadrants of the E/W display contains all warning, caution and memo messages recognized by ECAM. Primary and independent failure messages are shown in the lower left quadrant. Secondary failures are displayed in the lower right, and memo messages are displayed on either side. The various message types will be discussed in greater detail later in this chapter. The label STS will be displayed if ECAM status messages are available. The label ADV is shown in the mono mode to inform the flight crew of advisory messages not displayed.

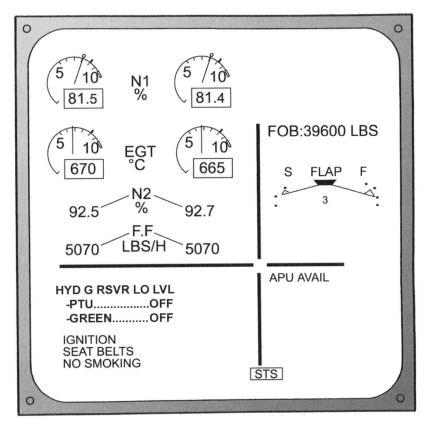

Fig. 4-67 ECAM Engine/Warning page. (Airbus Industrie)

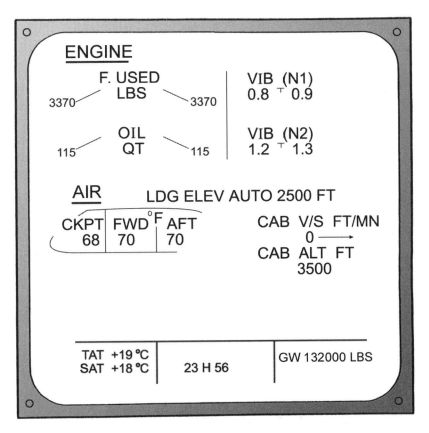

Fig. 4-68 ECAM Cruise page. (Airbus Industrie)

The lower display

The lower ECAM display will show one of three types of information, the **cruise page**, **system pages** or **status pages** during normal operation. As the name implies, the cruise page is displayed during normal cruise operations. As seen in figure 4-68, this page displays a combination of both engine and cabin data. There are 11 system pages which can be display by ECAM, the engine, bleed air, pressure, electrical, hydraulic, fuel, auxiliary power unit (APU), door, wheel, and flight controls. These pages are displayed under certain aircraft configurations or upon request through the ECAM control panel. Upon failure of any given system, ECAM will automatically call the related system page to the lower CRT. The engine page is shown in figure 4-69.

In the event of a system failure, ECAM will automatically display the message "STS" (status) on the E/W display (see figure 4-67). This indicates that aircraft status information is available on the lower display (status page). The status page contains information regarding the aircraft's condition after a system failure. This information tells the flight crew exactly how any given failure has effected the flight characteristics of the aircraft. The status page is displayed by depressing the STS button on the ECAM control panel. This button automatically illuminates after a system failure. The status page is also displayed any time the slats are lowered and the aircraft enters a landing configuration.

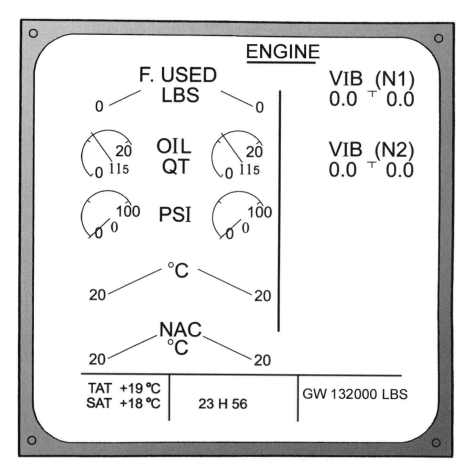

Fig. 4-69 ECAM Engine page. (Airbus Industrie)

As seen in figure 4-70, the status page contains procedures and limitations on the left side of the display. Procedures and limitations are used by the flight crew to determine the aircraft landing capabilities; hence, this page is automatically displayed during approach to land. Any cautions which were previously canceled by the crew are also displayed on the left side of the status page. Any inoperative systems are displayed on the right side of the screen in red or amber. Maintenance messages are displayed in white.

It should be noted that all ECAM messages are real time displays. In other words, all the data presented on the ECAM displays is a function of the aircraft's configuration at the time of the display. If a system's status changes, the ECAM display will respond immediately. Any warnings, cautions or memos are presented only if a system has failed and remains inoperative. The status page will also change if systems fail (or repair themselves) during flight. This real time ECAM presentations can be very helpful when troubleshooting various aircraft systems. ECAM can also be used for verification of a system's operation after a repair. In either case, simply operating the system while monitoring the aircraft's status on ECAM can provide a wealth of information.

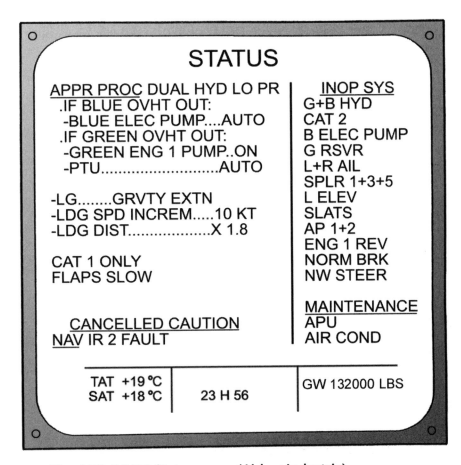

Fig. 4-70 ECAM Status page. (Airbus Industrie)

ECAM permanent data

On the lower portion of the ECAM system/status page five items known as **permanent data** are always displayed. Figure 4-71 shows permanent data displayed on the ECAM status page configuration. The permanent data items include: 1) true air temperature (TAT), 2) static air temperature (SAT), 3) gross weight (GW), 4) Greenwich mean time (GMT) in hours (h) and minutes, and above GMT is 5) the load factor (G load) or preselected altitude. The load factor is displayed in amber if the aircraft exceeds limits of greater than 1.4 g or below 0.7 g. If the load factor is within limits, the altitude selected for the flight guidance system is displayed above GMT.

ECAM mono operation

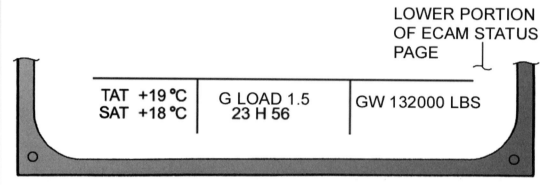

Fig. 4-71 ECAM status page permanent data. (Airbus Industrie)

The ECAM system automatically reverts to mono mode in the event of an ECAM display failure. The mono mode is also active whenever only one ECAM display is turned on. In the mono mode, E/W data has priority and will be displayed in its normal configuration. The ADV message will appear (on the E/W page) to identify when an advisory message is present. If the flight crew wishes to view the system/status page, the appropriate button is pressed on the ECAM control panel; as long as the switch is depressed, the system or status page will be displayed. Once the button is released the E/W page will reappear.

If the E/W display should fail, that information is automatically displayed on the lower ECAM display as seen in figure 4-72. If both ECAM displays fail, ECAM data is temporally lost. The flight crew will then select an ECAM/ND transfer and the E/W page will appear on the captain's and/or first officer's navigational display (ND).

Display priority

There are four basic modes of operation which determine the type of data displayed by ECAM. One of these modes is selected manually by the flight crew and three are displayed automatically. The automatic displays are a function of the current aircraft condition. For example, if a system failure had just occurred, the ECAM system would display a given system page as needed to inform the flight crew of the failure.

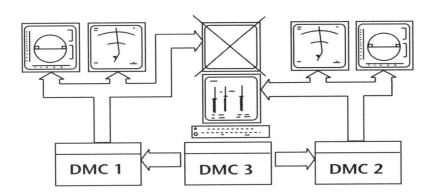

Fig. 4-72 Automatic transfer of Engine/Warning data to lower ECAM display. (Airbus Industrie)

The three automatic modes have a distinct priority to ensure the most important data is displayed first. Table 4-1 shows that the *automatic mode related to a failure* is the most important and will take priority over all other automatic display modes. After a critical system failure, ECAM will display any associated warnings or cautions on the E/W page and, in most cases, will automatically display an associated system page on the lower display. The *automatic advisory mode* is displayed any time a system parameter nears an out of limits (failure) condition. For example, if the engine oil temperature is higher than normal, but not too hot to trigger a warning, the *automatic advisory mode* will be active. In this mode the questionable parameter and title of the system page appear blinking on the display. In the advisory mode, the associated system page is also automatically displayed. The third priority is the manual mode. The lowest priority automatic mode is the *automatic mode related to a flight phase*. In this mode, certain data is displayed during different phases of the flight. Flight phases will be discussed later in this chapter.

The ECAM system pages are displayed with the following priorities :

1 - AUTOMATIC MODE RELATED TO A FAILURE

        An ECAM system page call is associated with most of the warnings.
        The page is automatically called and has priority over other display modes.

2 - AUTOMATIC ADVISORY MODE

        The advisory mode indicates a parameter which drifts out of its normal range before triggering a warning.
        The parameter and the title of the system page are displayed pulsing.
        In advisory mode, the ECAM system page is automatically displayed and the corresponding pushbutton comes on.

3 - MANUAL MODE

        When pressed in, an ECAM control panel pushbutton will come on and display the corresponding system page in normal mode.
        Pressing it again will return to the previous page.

4 - AUTOMATIC MODE RELATED TO THE FLIGHT PHASE

        If no other mode is selected, the ECAM system pages are automatically displayed according to the flight phase.

Table 4-1 ECAM mode priorities

The *manual mode* is activated by the flight crew to display a given system page. This mode can be activated at any time. However, if the manual mode is selected and a failure or abnormal system parameter occurs, the related automatic mode will take priority. As previously discussed, manual selection of system pages is done through the page control section of the ECAM control panel.

Failure categories

The failure of any system monitored by ECAM will have a distinct effect on the performance and/or capabilities of the aircraft. Since several hundred different failures can cause an ECAM message, a failure classification system was developed to help the flight crew and maintenance personnel identify the severity of each failure. Each system failure displayed by ECAM can be identified as one of three major categories, **independent failures, primary failures,** or **secondary failures.** Independent and primary failures are always displayed on the lower left portion of the E/W page. An **independent failure** will effect only that component and will not have adverse effects on other components or systems on the aircraft. A primary failure will cause a given component to fail as well as one or more other components or systems to fail.

Corrective action messages are displayed directly below primary and/or independent failure messages shown on the ECAM display. Corrective action messages are listed in blue and should be performed by the flight crew as soon as practical. Figure 4-73 shows a primary failure (*HYD G RSVR LO LVL*) and the corrective actions of (*PTU...OFF*) and (*GREEN ENG 1 PUMP...OFF*). In this situation, one of the flight crew members would turn off the PTU (power transfer unit) and the number 1 green engine driven hydraulic pump. It should be noted, the A-320 has three separate hydraulic systems (green, blue, and yellow).

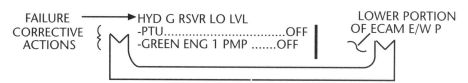

Fig. 4-73 Example of corrective actions during a system failure. (Airbus Industrie)

Secondary failures are defined as a loss of a component or system as a result of a primary failure. Secondary failures are always displayed on the lower right side of the E/W page (see figure 4-67). To learn more about the displayed secondary failure the flight crew would manually select the appropriate system page. In some cases the same system page is used to display the primary and associated secondary failures. An example of a primary and secondary failure would be Primary failure-*loss of hydraulic pressure in the green hydraulic system.* Secondary failures-*flight controls and nose wheel steering* would both be effected.

ECAM warning classes and levels

For each system or component failure monitored by ECAM, a warning message is displayed on the E/W page. These warnings are divided into three **classes** and three **levels.** The highest priority class is *class 1*. **Class 1** warnings are always displayed on ECAM and may also present a discrete aural and visual annunciator. Class 1 messages are divided into three levels (levels 3 through 1). Level 3 messages are displayed for the most significant failures. Class 1, level 3 warnings occur only for a handful of serious failures which require immediate crew action. For example, class 1, level 3 warning messages occur during cabin depressurization, engine fire, inappropriate slat or flap position during take off, low pressure in all three hydraulic systems, and dual engine failure.

Class 1, level 3 warnings always appear in red on the ECAM display, create an aural warning, and lights the red flashing master warning annunciators on the captain's and first officer's glare shield. The audio signal is a continuous repetitive chime broadcast by two flightdeck speakers. The master warning annunciators are actually momentary contact switches illuminated with the words "caution" or "warning" when appropriate. Depressing either of these annunciators will cancel the audio warning and turn off the annunciator lights. The ECAM message will remain on the CRT display as long as the fault exists. ECAM Class 1 level 3 messages are similar to the level A (warning) messages displayed by the Boeing EICAS. (EICAS was discussed earlier in this chapter.)

Class 1, level 2 ECAM warnings are generated by failures less serious than level 3 faults. Most of the ECAM warning messages are class 1, level 2. Level 2 messages occur from a fault that requires immediate attention and possible future action by the flight crew. For example, the loss of number 1 engine driven generator causes no threat to flight safety; but, typically requires the crew to start the auxiliary power unit (APU). For this situation ECAM would display the amber message *GEN 1 FAULT*. ECAM Class 1 level 2 messages are similar to the level B (caution) messages displayed by the Boeing EICAS.

Class 1, level 2 messages appear in amber on the ECAM display and illuminate the amber master caution annunciators. A single chime will sound for all class 1 level 2 faults. Pressing either the pilot or copilot's annunciator switch turns off the master caution annunciator. The ECAM message will remain as long as the fault exists.

Class 1, level 1 messages are the least critical class 1 message. Level 1 messages occur mainly on component failures which are used for redundancy or emergency backups. For example, failure of the static inverter produces a level 1 message. In this case, the ECAM would show the amber message *STATIC INV FAULT*. There is no related aural or discrete annunciator for level 1 messages. Class 1 level 1 faults are similar to the advisory messages found on the Boeing EICAS.

Class 2 messages are caused by faults which have no flight deck effect, but require flight crew attention prior to leaving the aircraft. These messages appear on the status page of ECAM. If class 2 messages are present, the *STS* (status) message flashes on the E/W display to alert the flight crew. These messages are present mainly to allow the flight crew the opportunity to record the failure in the aircraft log. The log "write-ups" are used by technicians to determine what maintenance is required on the aircraft.

Class 3 failure messages are recorded only by the aircraft's central maintenance system. The flight crew is totally unaware of these failures. Repair of class 3 faults becomes mandatory at the next heavy maintenance check performed approximately every 400 flight hours.

	MOST SERIOUS FAILURE ←		→		LEAST SERIOUS FAILURE
CLASS	CLASS 1			CLASS 2	CLASS 3
LEVEL	LEVEL 3	LEVEL 2	LEVEL 1		
VISUAL FLIGHT DECK EFFECT	MASTER WARNING	MASTER CAUTION		FAILURE INDICATED ON ECAM STATUS PAGE UNDER MAINTENANCE	
AUDIO FLIGHT DECK EFFECT	CONTINOUS REPETITIVE CHIME	SINGLE CHIME			
ECAM MESSAGE	ECAM YES	ECAM YES	ECAM YES		NO ECAM MESSAGE
	ALL FAILURE CLASSES RECORDED BY CENTERAL MAINTENANCE SYSTEM				

Fig. 4-74   Various classes, levels, and related flight deck effects reported by ECAM.

Figure 4-74 shows a graphic representation of the various classes, levels, and their related flight deck effects. It should be noted that all ECAM messages, regardless of class, are reported to the aircraft's central maintenance system. All ECAM messages are real time and are lost from memory after engine shut down. Keep in mind, actual fault data is available throught the aircraft's central maintenance system.

Local warnings

A **local warning** is considered any fault message presented on the specific control panel for a given system. For example, if the number 1 AC generator should fail, the number 1 generator control switch will illuminate with the word "FAULT." This switch is located on the electrical control panel on the overhead panel of the flight deck. The ECAM upper display will also show the message "GEN 1 FAULT" and the lower ECAM display will call the electrical status page. There are numerous local warnings associated with different ECAM messages. These warnings are activated at the same time the fault message appears on ECAM, but local warnings are not controlled by the flight warning computers. Each local warning is controlled by the computer for that particular system (i.e. electrical, APU, autoflight, etc.). It should be noted that local warnings remain lit even if ECAM master warning and caution annunciators are canceled.

Memo messages

Items which the flight crew should be aware of, but that are not related to a failure are displayed in the form of a **memo message.** These messages are displayed in the lower portion of the ECAM engine/warning page. Typical memo messages are items such as: *SEAT BELTS*, *NO SMOKING*, and *APU AVAIL*. These messages inform the flight crew that the seat belt and no smoking signs are illuminated, and the APU electrical power is currently available for use.

Flight phases

The various segments of a typical flight are broken down into ten segments known as **flight phases.** A breakdown of the ten flight phases is shown in figure 4-75. The flight phases are computed by the FWC and used by ECAM to determine system page presentations and inhibit warning messages. During each flight phase, the ECAM system determines which system page will be presented on the lower ECAM display. For example, during flight phase one, the door/oxygen page is displayed. It is important to remember that the flight phase system pages are the lowest priority. If a failure occurs or if a manual page selection is made, the flight phase system page would be replaced.

Under critical flight phases, such as landing and take off, it is best not to distract the flight crew with ECAM fault messages. Most of the ECAM messages do not require immediate attention and may divert the flight crew from more crucial tasks. The FWC uses the ten flight phases to determine which messages will be displayed and which will be inhibited. Figure 4-75 shows some of the major inhibits which occur during the various flight phases.

Current Aircraft Events which begin and end a flight phase	Flight Phase	ECAM System Page Displayed	ECAM Inhibits
Electrical Power applied to the Aircraft Busses	1	Door/Oxygen	None
1st Engine to Takeoff power	2	Wheel or Flight Control Page for 20 sec. when either side stick is moved or rudder deflected more than 22°	None
Aircraft reaches 80 Knots Airspeed	3	Engine Page	Most Warnings Inhibited
Lift of	4		
Slats are retracted and Engine thrust is reduced from takeoff power	5		
Aircraft Descends to 800 Ft.	6	Cruise Page (portion of Engine and Air Conditioning Page) or Wheel Page Displayed if landing gear are lowered	None
Touchdown	7	Wheel Page	Most Warnings Inhibited
Aircraft slows to 80 Knots Airspeed	8		
2nd engine to shut down	9		None
5 Minutes after 2nd engine shut down	10	Door Page	None

Fig. 4-75 Ten flight phases, Aircraft events, ECAM displays, and ECAM inhibits. (Airbus Industrie)

AVIONICS: SYSTEMS and TROUBLESHOOTING

ECAM system failures

Let's look at a typical system failure which is monitored by ECAM. Refer to figure 4-76 (ECAM displays labeled a-e) during the following discussion of the ECAM presentations. For this example, we will start with the aircraft in flight and the E/W page on the upper ECAM display; the lower ECAM will be showing the cruise page (see ECAM display a). If the green hydraulic system suffers a severe leak, the following scenario should occur:

a) Prior to the leak, the ECAM displays show all systems normal.

b) When the hydraulic fluid reaches a critical level, an amber caution message will appear on the E/W display. Beneath the caution message, the corrective actions will be displayed. At the same time, the lower display will be changed to the hydraulic page. A single chime will sound and the master caution annunciators will illuminate.

The flight crew will then depress the master caution annunciator to turn it off. Next, according to the corrective actions on the E/W page the PTU and green engine pump #1 will be turned off.

c) Once the corrective actions are completed, the primary and secondary failures will be shown on the bottom of the E/W display. The "*" next to the secondary failure indicates a system page available to display that failure.

d) The flight crew will then press the clear button on the ECAM control panel. This will replace the primary failure messages with the memo messages. The flight control page will appear on the lower CRT. If the clear button is pressed again the wheel page will appear on the lower ECAM display

e) The status page will be displayed when the clear button is pressed again. The E/W page also returns to its original pre-failure condition.

After the status page is reviewed by the flight crew, the clear button is pressed again. The lower ECAM reverts back to its flight phase mode (the cruise page), and *STS* will appear on the upper ECAM display to indicate the status page contains information regarding system failures. The recall button can be used to retrieve the status page.

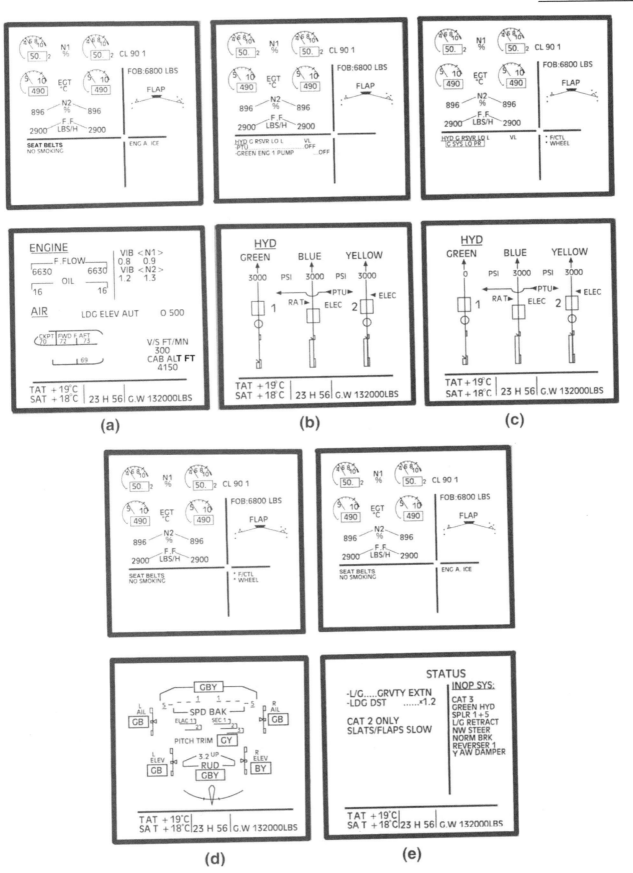

Fig. 4-76 Various ECAM displays; (a) all systems normal, (b) amber caution message, (c) primary and secondary failures, (d) memo messages returned and failed system displayed, (e) status page displayed and engine/ warning page returns to it's pre-failure condition. (Airbus Industrie)

Troubleshooting ECAM

The majority of ECAM troubleshooting is done through the aircraft's central maintenance system. The onboard diagnostics will pinpoint defective LRUs and many wiring problems. The defect should then be corrected and the system tested for proper operation. This troubleshooting sequence is relatively simple if the defect is a continuous fault. However, in cases where intermittent faults exist the troubleshooting process requires additional attention.

During flight, the various temperature extremes and vibrations often causes intermittent problems. Many of these faults repair themselves prior to landing which makes accurate troubleshooting difficult. In this type of situation, the central maintenance system will most likely recommend a LRU replacement, however, it is always wise to review the aircraft's history for this same defect. If the suggested repair has already been done, and the problem still exists, performing that same repair will **not** fix the system. Consider something new. Intermittent faults are often caused by loose connections. Perform a thorough inspection of the systems wiring and all related connector pins.

Whenever repairing ECAM or replacing components, be sure to remove power from the system and use electrostatic discharge sensitive (ESDS) precautions. ECAM computers can be damaged if removed during operation, and just as easily damaged by static electricity. Be sure to read the manual for all precautions.

AIRPLANE INFORMATION MANAGEMENT SYSTEMS

The **Airplane Information Management System (AIMS)** found on the Boeing 777 employs Line Replaceable Modules (LRMs) to integrate a multitude of electronic systems. The concept behind AIMS is simple; integrate various systems by sharing common functions and components. AIMS actually integrates data collection, computing functions, power supplies, and output functions for several subsystems. By sharing both hardware and software components AIMS enhances reliability, improves redundancy, and creates a substantial weight savings as compared to conventional systems.

Boeing and Honeywell designed AIMS to incorporate a fault tolerant software design. Fault-tolerance along with proper redundancy of critical systems hardware allows the AIMS software to detect a fault and reconfigure the system for uninterrupted operations. In many cases, the flight crew would not even be aware of the system failure. In theory this type of design will allow the B-777 to continue flying with the failed system until the next convenient maintenance opportunity. Typically, the repair would be performed during night maintenance, not during the short time period available between flight.

On the B-777, there are two AIMS cabinets, each with eight LRMs. Each cabinet contains four Input Output Modules (IOMs) and four Core Processing Modules (CPMs). The IOMs transfer data to and from the AIMS cabinet and within the four CPMs. The four CPMs perform the calculations for the various avionics systems serviced by AIMS. Each LRM is designed for quick removal and installation. Without the AIMS concept, several LRUs would be required to perform the same tasks.

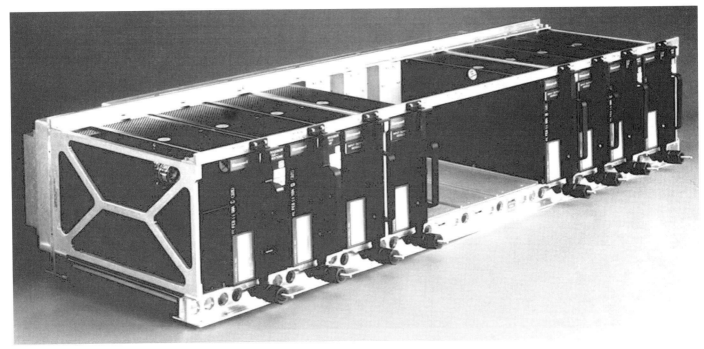

Fig. 4-77 AIMS interface cabinet. (Honeywell, Inc.)

The LRMs are smaller than the conventional LRUs since they share several functions with other LRMs within AIMS. Figure 4-77 shows a typical AIMS cabinet and LRMs.

As seen in figure 4-78, AIMS has the ability to transfer data in a variety of communication formats. To help simplify wiring and increase the speed of data transfer, AIMS uses the ARINC 629 data bus to communicate with a large number of the aircraft systems. There are three ARINC 629 data busses used for flight control, four ARINC 629 data busses for systems communications, and four ARINC 629 data busses used to connect the LRMS within the AIMS cabinets. As seen in figure 4-78, other data busses include: 1) ARINC 429 for communications with a variety of airframe, engine, and avionics systems. 2) ARINC 453 for weather radar information. 3) ARINC 717 and RS 422 used for communications with the flight data recorder, and the quick access recorder. AIMS also monitors a variety of analog signals to various components. The display units (PFD, ND, EICAS, and MFD) are connected to AIMS through display unit video busses. The display unit video busses are coaxial cables that transfer high speed digitized and compressed video signals to the six flat panel displays. AIMS also has a Local Area Network (LAN) to interface with the Maintenance Access Terminal (MAT) and Portable Maintenance Access Terminal (PMAT). This LAN is a fiber optics data transfer connection.

AIMS coordinates data input and outputs, and performs data processing calculations for a variety of systems. These systems include:
- Primary display system (PDS)
- Central maintenance computing system (CMCS)
- Airplane condition monitoring system (ACMS)
- Digital flight data recorder system (DFDRS)

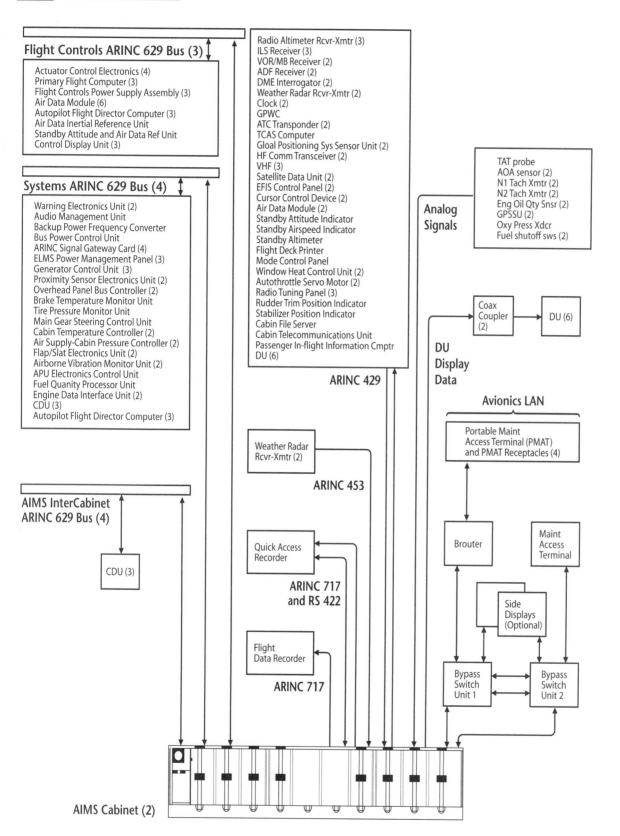

Fig. 4-78 Boeing 777 Airplane Information Management System (AIMS) interface diagram. (Boeing Commerical Airplane Company)

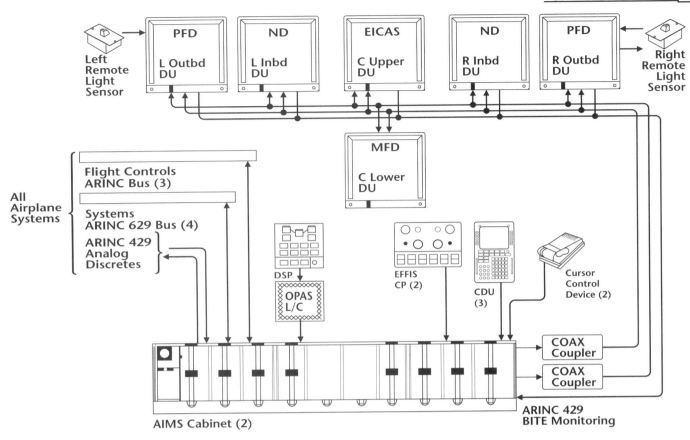

Fig. 4-79 Boeing 777 Primary display system. (Boeing Commerical Airplane Company)

- Data communication management system (DCMS)
- Flight management computing system (FMCS)
- Thrust management computing system (TMCS)

The primary display system (PDS) contains six flat panel display units. These full color LCD displays replace the conventional CRT; hence, saving both weight and space. The LCDs are arranged similar to the conventional CRTs found on other transport category aircraft. As seen in figure 4-79 the B-777 contains two PFDs, two NDs, one main EICAS display, and a Multifunction Display (MFD). As mentioned earlier, these displays interface with AIMS through dedicated display unit video busses.

Access to system fault data is provided by the central maintenance computing system (CMCS). The major computing functions for this system will be contained in the AIMS processors. On the B-777 the technicians will access the central maintenance system data using the **Maintenance Access Terminal (MAT)**. As seen in figure 4-80, five different functions will be accessed through MAT; *Line Maintenance*, *Extended Maintenance*, *Other Functions*, *Help*, and *Reports*. Each of these functions is used to access a specific operation of the CMCS. The MAT is located on the flight deck directly behind the copilot's seat. The terminal consists of an 8 x 10 inch LCD display, a keyboard which is stowed during flight, a disk drive and a cursor control device. The system is menu driven and designed to operate similar to a personal computer. The B-777 incorporates the most advanced central maintenance system ever used on aircraft.

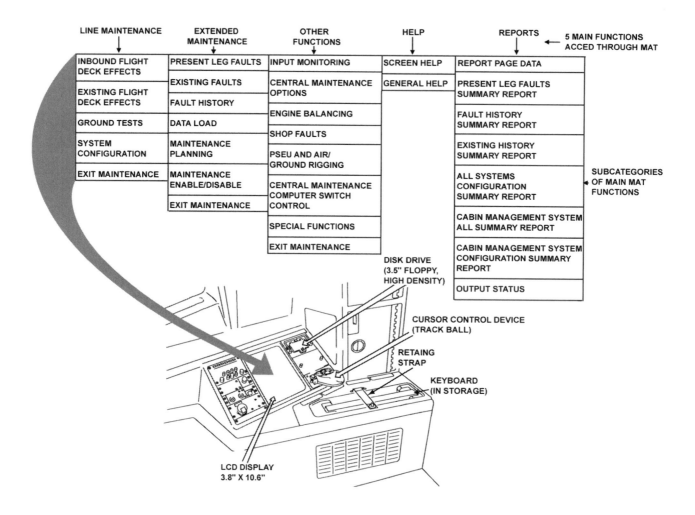

Fig. 4-80 Boeing 777 Maintenance access terminal. (Boeing Commerical Airplane Company)

CHAPTER 5
INTEGRATED TEST EQUIPMENT

INTRODUCTION

Aircraft designers have always been aware of the need for maintainability in the design of airplanes. In the past, maintainability was mainly a function of allowing technicians access to parts for replacement or inspection. Today's aircraft are so complex that design engineers must consider the ability to troubleshoot a system just as important as the ability to repair or inspect that system. Troubleshooting a complex digital aircraft would be nearly impossible without self-diagnostic systems. Technicians have always been accustomed to using various tools and test equipment to troubleshoot aircraft electrical systems. Simple items such as a test lamp or voltmeter could be used to diagnose many of the electrical problems on early aircraft. As radio equipment and autopilots became popular, more sophisticated test equipment became necessary. Many of the tests were too complex and required shop, or at least, hanger maintenance. Other problems could be repaired on the flightline given the proper carry on test equipment. To repair some systems, it required several pieces of test equipment and considerable time simply to troubleshoot the problem. As aircraft progressed, designers began to take advantage of computer technologies and added self-diagnostic circuits. Self-contained diagnostics used for electronic/avionics systems troubleshooting became known as **BITE (built in test equipment).**

BITE systems are typically found on transport category and many corporate type aircraft. BITE systems came into practical use with the invention of digital electronic systems. The Boeing 757 and 767 were the first civilian aircraft to make extensive use of BITE. As with most aircraft systems, an evolution process took place during the development of BITE. Early systems were dedicated to limited troubleshooting of a particular system. More modern BITE could monitor input or output signals for a given system. In general, early BITE systems were accessed through the specific LRU that housed the BITE circuitry. This meant early BITE systems were accessed in the electronics equipment bay or similar area.

The second generation of self-diagnostics equipment incorporates the use of a centralized monitoring system. Now, the faults detected through several BITE systems could be monitored in one location. The advanced built-in troubleshooting system found on Boeing aircraft is known as **CMCS (central maintenance computer system).** The advanced diagnostic system used by the Airbus Industries Corporation is called **CFDS (centralized fault display system).** Each of these advanced systems incorporate enhanced BITE features that aid in troubleshooting. In general, the advanced systems are more easily accessible and understood than older systems. The design and operation of individual and centralized diagnostic systems found on modern aircraft will be presented in this chapter. After studying this chapter the reader should be able to successfully use the majority of the built-in troubleshooting systems found on today's aircraft.

BUILT-IN TEST EQUIPMENT (BITE)

Built-in test equipment (BITE) systems are used in conjunction with many digital circuits to aid in system troubleshooting. BITE systems are designed to provide **fault detection, fault isolation,** and **operational verification after defect repair.** Fault detection is performed continuously during system operation. If a defect is sensed, the BITE initiates an appropriate control signal to isolate any defective component(s). In order to repair the defective system, the technician can utilize the BITE to identify faulty components or wiring. The majority of the aircraft digital systems contain several line replaceable units (LRUs). Defective LRUs may be quickly identified by the BITE system and exchanged during ground maintenance. Use of the LRU and BITE concepts greatly reduce aircraft maintenance down time. After the appropriate repairs have been made, the system should be run through a complete operational check. The BITE will once again monitor the system and verify correct operation if the system has been properly repaired.

A typical commercial airliner may contain several BITE units used to monitor a variety of systems. A Boeing 757 or 767 aircraft, for example, utilizes built-in test equipment systems on approximately 50 LRUs located throughout the aircraft. Seven separate BITE units located in the aircraft's electrical equipment bay or aft equipment center are used to monitor electrical power, environmental control, auxiliary power, and flight control systems. Each of these BITE boxes receives inputs from several individual components of the system being tested. Other individual systems also contain their own dedicated built-in test equipment. These BITE systems are relatively simple and are contained within the LRU being monitored. Systems that employ dedicated BITE on the B-757 include the following:

Engine indicating and crew alerting
VHF communication radios
HF communication radios
ARINC communication addressing and reporting
Selective calling
Passenger address
Weather radar
ATC transponder
Radio altimeter
Automatic direction finder
Inertial reference
Air data computer
Electronic flight instruments
Flight management computer
Radio distance magnetic indicator
Lighting
Fuel quantity
Fire and overheat
Antiskid-autobrake
Instrument landing
VHF omnirange receiver

Distance measuring equipment
Window heat
Proximity switch electronic unit
Hydraulic management
BITE test programs

A complex BITE system is capable of testing thousands of input parameters from several different LRUs. Typically a system performs two types of test programs: an **operational test** and a **maintenance test.** Normal operational checks start with initialization upon acquisition of system power (see figure 5-1). The operational BITE program is designed to check input signals, protection circuitry, control circuitry, output signals, and the operational BITE circuitry. During normal system operation, the built-in test equipment monitors a watchdog signal initiated by the BITE program. The watchdog routine detects any hardware failure or excessive signal distortion that may create an operational fault. If the BITE program detects either of these conditions, it automatically provides isolation of the necessary components, initiates warning, caution, or advisory data, and records the fault in a nonvolatile memory.

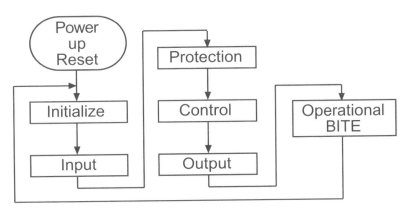

Fig. 5-1   BITE flow diagram showing the sequence of circuit tests.

The maintenance program of the built-in test equipment begins operation when the aircraft is on the ground and the maintenance test routine is requested. When initiated, the maintenance BITE will exercise all input circuitry and software routines of the system being checked. The corresponding output data are then monitored, and faults are recorded and displayed by the BITE system. An illustration of the Bus Power Control Unit (BPCU) maintenance BITE routine is shown in figure 5-2. This routine checks the input circuitry, voltage regulator circuitry, protection software, logic software and operational BITE circuitry. The test results are returned to the BPCU for storage and display. The software, or operating programs, of the system are tested through utilization. That is, input data is initiated by the BITE and manipulated by the software program. The BITE program evaluates the corresponding output data in order to determine the system's performance. If a discrepancy in the output data is detected, the BITE system considers the operational software faulty and provides the appropriate indication.

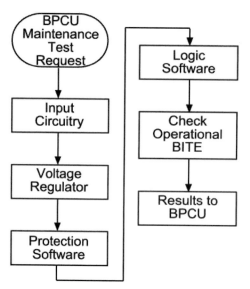

Fig. 5-2   Bus power control unit BITE flow diagram.

Discrete digital signals are used as the code language for BITE systems. Built-in test equipment interprets the various combinations of digital signals to determine a system's status. If an incorrect input value is detected, the BITE system records the fault and displays the information upon request. Figure 5-3 shows an LRU containing a 24-character Light-Emitting Diode (LED) display that can show a variety of fault information. The LED display is located on the face of the unit along with the appropriate test switches. This type of BITE is typically located on the face of the LRU, mounted in an equipment rack in the aircraft's electrical equipment bay.

Troubleshooting with Built-In Test Equipment (BITE)

There are several versions of built-in test equipment, which are in use today. Simple BITE systems typically incorporate a **go/no-go** red or green LED on the equipment black box or LRU. More complex systems use a multi-character display and monitor more than one LRU. Some BITE can also test the associated wiring. Even more advanced BITE systems incorporate displays which are activated from the flight deck, have paper printouts, and may have a means to transmit data from the aircraft to the maintenance facility during flight, known as ACARS (ARINC Communication Addressing and Reporting System). These systems will be discussed later in this chapter.

A bus power control unit BITE

The BITE system shown in figure 5-3 is incorporated into the B-757 Bus Power Control Unit (BPCU). This system monitors the entire electric power generation system, including the left, right, and APU generators, constant-speed drives, and their related control units. The BIT button is depressed on this system to activate the manual BITE test function. Typically, this type of BITE system will display fault information in a coded message. The technician then decodes this message through the use of the aircraft's maintenance manual. The appropriate manual will inform the technician of any LRU to be replaced or circuit to be repaired and their respective locations within the aircraft. The fault information on this system is displayed for two seconds and then automatically advances to the next fault (if any). This type of BITE system will make an appropriate indication when all fault data has been displayed.

After the system has been repaired, the BITE should be reset and an operational check performed. The repaired system should be run through a complete cycle of operation. In the case of the electric power generating system, the appropriate engine and AC generator should be subjected to a variety of operating parameters. The flight deck instruments and failure indicators are monitored during the test to detect any further problems.

After repair and operation of the system, the BITE fault display should be reactivated. This will initiate the readout of the nonvolatile memories and the BITE will display any remaining faults. If the system is found to be without fault, the BITE display will respond accordingly. Always be sure to reset the BITE prior to the verification test. The BITE fault messages are stored in memory and will remain unless erased by the reset function.

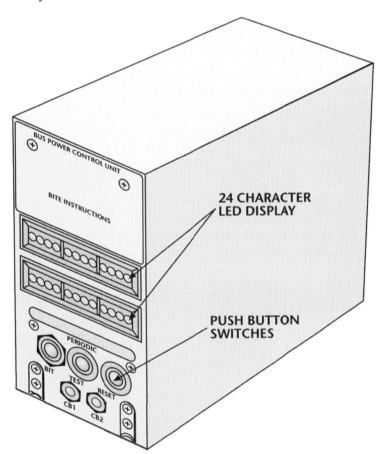

Fig. 5-3 Typical LRU with access to BITE controls on the face of the unit.

A Simple BITE circuit

Many aircraft systems built in the 1980's or later incorporated some type of BITE circuitry. Independent systems, such as the radio altimeter shown in figure 5-4, often contained BITE circuitry in the main LRU (Line Replaceable Unit) of the system. This radio altimeter is one of three installed on an early model Boeing 757. Each Radio Altimeter Receiver/Transmitter (R/T) contains its own BITE. To run the radio altimeter BITE, simply press the *TEST* button and monitor the LEDs on the face of the R/T. The procedures for the BITE test are shown in figure 5-4b. It should be noted that this BITE circuitry performs simple tests on the R/T, the antenna, and the radio altimeter display. Similar tests are extremely common on many LRUs found on transport category and corporate type aircraft.

Maintenance Control Display Units (MCDUs)

The Boeing 757 and 767 incorporate a **Maintenance Control Display Unit (MCDU),** which is used to monitor and test the flight control computers, flight management computers, and the thrust management computers. The MCDU was the beginning step in the development of the advanced integrated BITE systems found on today's state-of-the-art aircraft. MCDUs similar to the one presented here are very popular and found on hundreds of transport category aircraft. Through the MCDU, over 68 individual components, input signals, and systems are monitored.

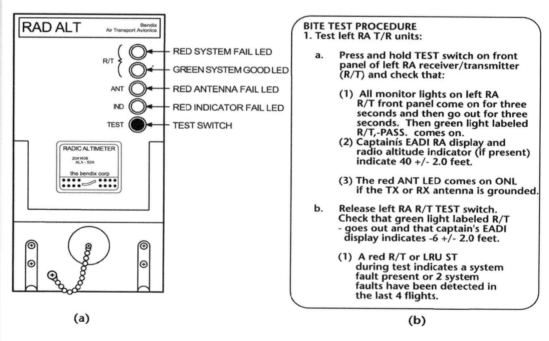

(a)   **(b)**

Fig. 5-4 Radio Altimeter BITE. (a) the LRU front panel; (b) the procedures for testing the left transmitter/receiver. (Boeing Commercial Airplane Company)

System data is accessed in the main electrical equipment bay via the control panel located on the front of the MCDU (see figure 5-5). On some aircraft an MCDU controller (a carry-on unit) is connected to the system on the flight deck. If the MCDU controller is used on the flight deck, the fault information is displayed on the lower EICAS CRT. During a complete MCDU ground test, certain controls must be operated and various indicators observed. Therefore, if the MCDU test mode is accessed from the equipment bay, two technicians will be required. Only one person can access fault data stored in memory from either location.

MCDU operation

The MCDU operation is similar to the previously described BITE system. The MCDU receives digital data in an ARINC 429 format transmitted from the thrust management, flight control, and flight management computers along with various other system inputs. The MCDU monitors in-flight faults and performs ground test functions. In-flight faults are directly correlated to the various flight deck effects associated with in-flight problems. **Flight deck effects** are considered to be any EICAS display or discrete annunciator used to inform the flight crew of an in-flight fault.

Upon landing, the MCDU automatically records any in-flight faults (from the last flight) in its nonvolatile memory. To access this memory, the technician must first cycle the MCDU off and on again. This action will perform an internal test of the MCDU. After the internal test is complete, the technician should select the in-flight mode of operation. This is done by pressing the *YES/ADV* button on the MCDU when the display reads *LAST FLT FAULTS* (see figure 5-5). The unit will respond accordingly with faults listed in order of occurrence. Press the *YES/ADV* switch to sequentially advance through all the faults. If no faults were found during the last flight, *NO LAST FLT FAULTS* will be displayed. Pressing the NO/SKIP button will skip to the next display.

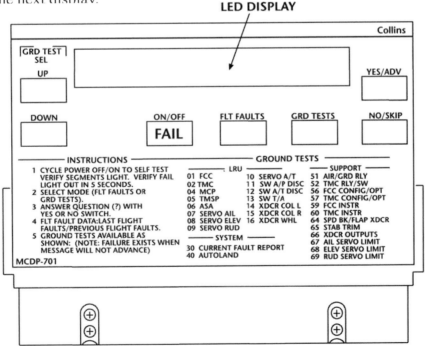

Fig. 5-5 Maintenance control display unit BITE controls and display located on the face of the MCDU. (Boeing Commercial Airplane Company)

At the end of the last flight fault data, the MCDU display will ask *PREV FLT FAULTS?* Answer *yes* to view faults which occurred during previous flights. To terminate the display of fault data press the *ON/OFF* or *GRD TEST* switch. The MCDU can store faults from a maximum of 10 previous flights.

MCDU ground tests

To run the ground tests function of the MCDU, the unit should be turned on and the self-test must be valid. The ground test will begin when the *GRD TEST* button is pressed. At that time, the MCDU will run an initialization test and display any associated fault messages. If no initialization faults are found, the MCDU will begin testing the Flight Control Computer (FCC) when the *YES/ADV* switch is pressed. The remaining ground tests are activated in like manner. If you wish to skip to a given test press the *NO/SKIP* button until the appropriate test appears in the MCDU display. At that time, press *YES/ADV*. The *UP* and *DOWN* ground test select switches can also be used to move through the menu to a given ground test. To terminate the ground tests function, turn the MCDU off or press *select* for the flight faults option.

Built-In Test Equipment For Light Aircraft

Light aircraft electronic/avionic systems have also moved into an era of self-diagnostics. Modern avionics equipment found on light aircraft are often digital systems that incorporate BITE software programs. Many of the communication and navigation systems, as well as electronic flight instrument systems (EFIS), utilize BITE for internal and limited external testing. BITE systems are most often found on individual avionics equipment for corporate type aircraft. However, the trend of increased BITE systems is continuing to move into even the smallest aircraft.

Single system BITE

The BITE systems found on single function avionics, like a VHF communications radio, typically employ only limited diagnostics. An LED indicator may be used to report the condition of the internal radio circuitry. If the technician encounters a radio problem, the test button on the transceiver would be pressed and the LED PASS/FAIL lights would be monitored. Obviously, if the unit fails, the transceiver would be removed for bench repair. If the transceiver passed its test, the fault lies in other portions of the system, such as the antenna or system wiring.

BITE systems for EFIS

Of all the BITE found on corporate type aircraft, the systems found on EFIS equipment are typically very advanced. Electronic flight instrument systems employ one or more CRTs which are used to display flight and navigational data. Therefore, EFIS is interconnected to a multitude of other avionics equipment needed to determine the information to be displayed on the CRTs. The EFIS built-in test equipment often tests both the EFIS circuitry and the incoming signals from the monitored systems.

The Bendix King EFIS 40/50 incorporates an extensive self-test during operation. The self-test begins with an analysis of internal circuitry and the data being sent to the EFIS processors. A more comprehensive test can also be initiated for maintenance purposes. If a failure is detected during operation, one or both of the EFIS CRTs will display a message related to the failure. Figure 5-6 shows an EFIS 40/50 with several fault messages shown on the display. Typically the most critical fault messages are displayed in red, less critical messages are shown in yellow.

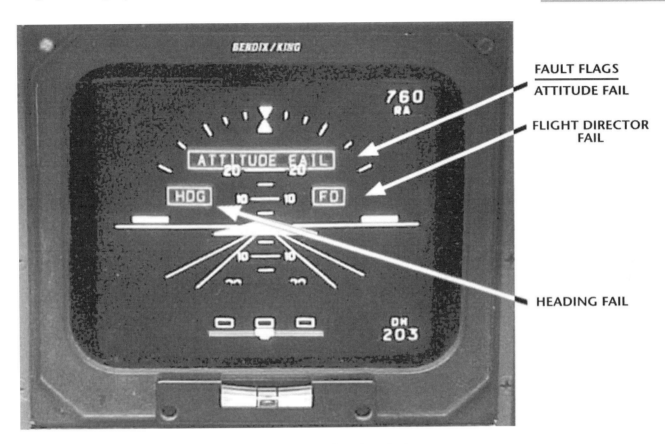

Fig. 5-6 Typical EFIS display showing fault messages.

Most electronic flight instrument systems for light aircraft allow the avionics technician to analyze input signals using the EFIS test functions. Input signal diagnostics help to troubleshoot the entire EFIS/autoflight system by providing a look at the data being analyzed by EFIS. In general, this type of data analysis requires reference to the system manuals. The Collins EFIS-85B & -86B utilize a **BITS Mode** to provide troubleshooting data stored in a random access memory. The BITS mode presents a display similar to figure 5-7. In this display the first two columns present the decimal (DEC) and hexadecimal (HEX) codes for the source of the data. The MPR (multiport RAM) data is shown in binary. The MPR can be decoded using the installation manual for the Collins EFIS. For a more detailed look at EFIS equipment and the related BITE systems, see chapter three of this text.

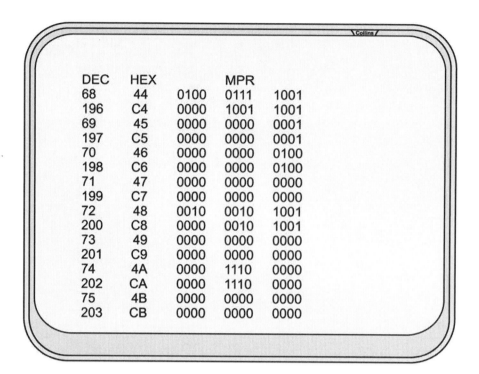

DEC	HEX		MPR	
68	44	0100	0111	1001
196	C4	0000	1001	1001
69	45	0000	0000	0001
197	C5	0000	0000	0001
70	46	0000	0000	0100
198	C6	0000	0000	0100
71	47	0000	0000	0000
199	C7	0000	0000	0000
72	48	0010	0010	1001
200	C8	0000	0010	1001
73	49	0000	0000	0000
201	C9	0000	0000	0000
74	4A	0000	1110	0000
202	CA	0000	1110	0000
75	4B	0000	0000	0000
203	CB	0000	0000	0000

Fig. 5-7 Typical BITS mode display. (Rockwell International, Collins Avionics Divisions)

CENTRAL MAINTENANCE COMPUTER SYSTEM (CMCS)

The **Central Maintenance Computer System (CMCS)** is found on a variety of modern Boeing aircraft. This system is designed to perform inflight and ground tests of virtually all aircraft systems, accessed from a central location. The CMCS incorporates greater memory capability and better information access compared to previous BITE systems. This leads to more reliable troubleshooting using CMCS. The Boeing 747-400 that incorporates state-of-the-art electronic systems will be presented in the following discussion on the central maintenance computer system.

System Description

The B-747-400 CMCS is accessed through one of four **control display units (CDUs).** There are three CDUs located on the center console of the flight deck and one located in the main equipment bay. Figure 5-8 shows a photograph of the center console and three CDUs. The CDUs contain an alphanumeric keyboard and a CRT display, which are used to access the CMCS data. This type display allows for a more descriptive message of faults that are directly correlated to flight deck effects created by the same fault. A **flight deck effect** is considered any failure condition that is displayed on EICAS or the EFIS. A CMCS printer is incorporated to provide a written report of the fault data, and a software data loader can be used to store faults on a computer disc. Aircraft equipped with ACARS are capable of transmitting fault data from the aircraft to a ground facility. ACARS will also answer all maintenance data requests from the ground facility.

The CDUs communicate directly with the **central maintenance computers (CMCs)** of the B-747-400 computerized maintenance system. The CMCs monitor virtually every electronic system on the aircraft. If a fault occurs, the CMCs record that information that can be recovered later through the CDUs. Figure 5-9 shows the location of the CMCs in the equipment bay. During flight, the CMCs receive fault data from the aircraft's EIUs (Electronic Interface Units) and other digital and discrete systems to record in-flight failures. The EIUs monitor system parameters and control the displays of the EICAS (engine indicating and crew alerting system) and EFIS (electronic flight instrument system). During flight, the CMCs also monitor the integrity of CMCS inputs and store any fault data.

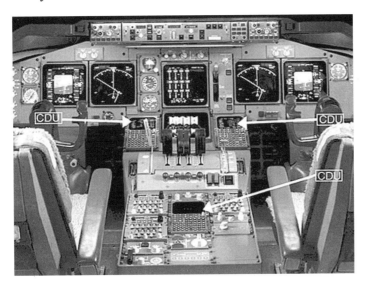

Fig. 5-8 Three central display units on a Boeing 747-400 aircraft.

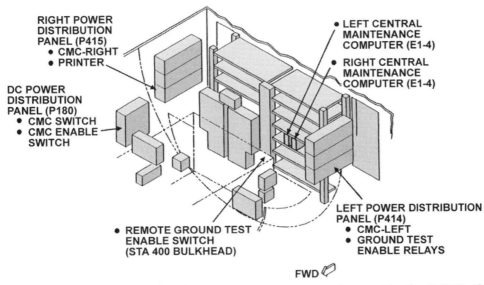

Fig. 5-9 Locations of central maintenance computer equipment in the B-747-400 equipment bay. (Northwest Airlines, Inc.)

Once on the ground, the CMC can be interrogated for fault history in its nonvolatile memory. Up to 500 faults can be stored in the CMC memory. From the failures detected by the CMC, one of 6,500 different fault messages can be displayed on the CDU. The CMC is also used to initiate various BITE tests of the aircraft's electronic LRUs. Each CMC weighs about 17 pounds and has a 300-pin connector on the rear of the unit. On the front of the CMC is an automatic test equipment (ATE) connector that is used to access CMC data during shop maintenance. The CMCs are electrostatic discharge sensitive so the technician must always take the necessary precautions during maintenance.

System architecture

Figure 5-10 shows a block diagram of the CMC data inputs and outputs. There are two central maintenance computers (CMCs) that receive up to 50 low speed and 6 high speed ARINC 429 data inputs. Most aircraft systems report to both the EIUs and the CMCs. The CMCs also receive information from the EIUs. The CMC compares fault information from the EIU and aircraft systems for fault verification. A list of ARINC inputs to the CMCs is shown in table 5-1.

Although there are only 10 output busses from each CMC, the information is shared by many aircraft systems. Table 5-2 shows the systems connected to ARINC busses 1-6. The remaining four busses are connected to the CDUs: the ACARS management unit, the printer, the airborne data loader, and the opposite side CMC.

Digital Inputs
These systems send ARINC 429 inputs to the CMCs:

-Left control display
-Center CDU
-Right CDU
-Left EIU (high speed)
-Center EIU (high speed)
-Right EIU (high speed)
-Left flight control computer (FCC)
-Center FCC
-Right FCC
-Left VHF transceiver
-Right VHF transceiver
-Left HF transceiver
-Right HF transceiver
-Left air traffic control (ATC)
 transponder

-Right ATC transponder
-MAWEA (master monitor card A)
-MAWEA (master monitor card B)
-MAWEA (crew alerting card)
-MAWEA (left aural synthesizer card)
-MAWEA (right aural synthesizer card)
-Lower yaw damper module (high speed)
-Upper yaw damper module (high speed)
-Audio management unit
-Left weather RADAR transceiver
-Right weather RADAR transceiver
-Digital flight data acquisition card (DFDAC)
-Captain's clock
-Software data loader panel (high speed)
-Multiple input printer
-Airborne communications addressing and reporting
 system(ACARS) management unit

Table 5-1 ARINC 429 inputs to the CMCS.

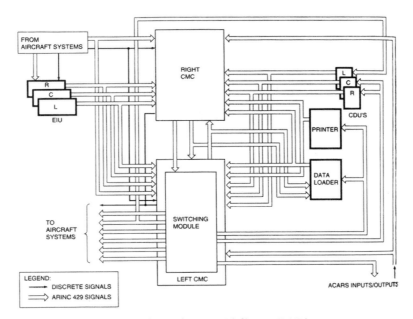

Fig. 5-10   CMC data bus structure. (Northwest Airlines, INC.)

The right CMC has ten ARINC 429 outputs; one is a cross-talk bus to the opposite side CMC. The outputs are sent to the various aircraft systems through the left CMC switch relays (see figure 5-11). The switch relays are normally closed to the left CMC outputs. If the left CMC fails, a ground will be applied to activate the switch relay; hence, the right CMC outputs will be sent to the airplane systems. In other words, if the left CMC detects internal faults, it automatically passes output data from the right CMC directly through the switch relay. It should be noted that the CMCs are identical. The switch relay is simply inactive in the right CMC. In some cases the aircraft can be operated with only one CMC. In this situation, the CMC must be installed in the left slot.

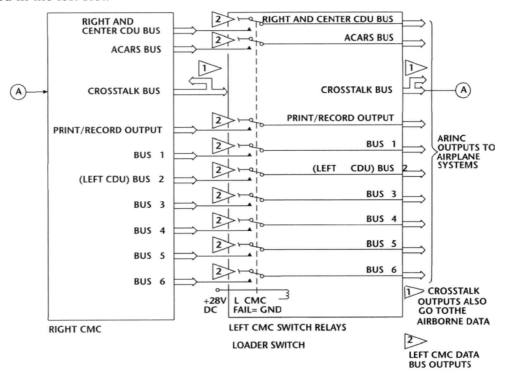

Fig. 5-11 Left CMC switching relay. (Northwest Airlines, Inc.)

ARINC output Bus 1 connects to the following systems:	ARINC output Bus 2 connects to the following systems:
-Pack temperature controller B -Cabin pressure controller A -Left flight control computer -Left stabilizer trim rudder ratio module -Upper yaw damper -Bus control unit 1 -Left flap control unit -Left window heat controller -Left VOR receiver -Left DME interrogator -Left ILS receiver -Left ADF receiver -Left range radio altimeter transceiver -Left ATC transponder -Right weather radar transceiver -Left EFIS/EICAS interface -Left flight management computer -Left air supply control/test unit -Electrical system card file	-Electrical system card file -Zone temperature controller -Tire pressure monitor unit -Center flight control computer -Center flap control unit -Ground flap control unit -Ground proximity warning computer -Center ILS receiver -Center radio altimeter transceiver -Hydraulic quantity monitor unit -Center EFIS/EICAS interface unit -Proximity switch electronic unit -Brake system control unit -APU control unit -Left control and display unit -Power supply - emergency lights
ARINC output Bus 3 connects to the following systems:	**ARINC output Bus 4 connects to the following systems:**
-Pack temperature controller A -Cabin pressure controller B -Right flight control computer -Right stabilizer trim rudder ratio module -Lower yaw damper -Bus power control unit 2 -Right flap control unit -Right window heat controller -Right VOR receiver -Right DME interrogator -Right ILS receiver -Right ADF receiver -Right radio altimeter transceiver -Right ATC transponder -Left weather radar transceiver -Right EFIS/EICAS interface unit -Right flight management computer -Right air supply control test unit -Electrical system card file	-Left VHF transceiver -Left HF transceiver
ARINC output Bus 5 connects to the following systems:	**ARINC output Bus 6 connects to the following systems:**
-Center VHF transceiver -Fuel quantity processor -Fuel system card file -Digital flight data acquisition card	-Right VHF transceiver -Right HF transceiver

Table 5-2   Connections for the CMC busses 1-6.

Each CMC will receive up to 22 discrete inputs. (Note: discrete inputs are typically analog singals.) All input data is wired in parallel and therefore shared between the two CMCs. Up to 41 discrete parallel outputs signals are available from each CMC. Four discrete signals are also used for pin programming of each CMC. There are two spare program pins, one used for a parity bit, and one to determine the left/right installation of the CMC. If the CMC is installed in the right-hand rack, the pin should be grounded. If the CMC is installed in the left-hand rack, the pin is open. The parity bit employs odd parity to ensure the correct signals are connected to the program pins. A list of the discrete inputs and outputs can be found in table 5-3.

Discrete Outputs

-Equipment cooling test relay through CMC ground test
 enable relay 3
-Left flight control computer (FCC)
-Center FCC
-Right FCC
-Center air data computer (ADC) through CMC ground
 test enable relay 1
-Right ADC through CMC ground test enable relay 1
-Left ADC through ground test enable relay 2
-R7421 left pitot probe heater test relay
-R7422 right pitot probe heater test relay
-Left inertial reference unit (IRU)
-Right IRU
-Center IRU
-R7683 wing thermal anti-ice system relay through
 CMC ground test enable relay 8
-MAWEA (left stall warning management card)
-MAWEA (right stall warning management card)
-MAWEA (configuration warning card)
-Right ozone valve, left ozone valve, and ozone switch
 (1 output)
-Fire test 1 through CMC ground test enable relay 3
-Fire test 2 through CMC ground test enable relay 4
-Fire test 3 through CMC ground test enable relay 5
-Fire test 4 through CMC ground test enable relay 6
-Fire test 5 through CMC ground test enable relay 7

Discrete Inputs

-Left radio communication panel
-Center radio communication panel
-Right radio communication panel
-R7421 left pitot probe heater test relay
-R7422 right pitot probe heter test relay
-V474 valve ozone catalytic converter bypass right
-V472 valve ozone catalytic converter bypass left
-V471 valve ozone catalytic converter left
-Flight control electronics (FCE) power supply module
 1 left (2 inputs)
-FCE power supply module 1 right (2 inputs)
-FCE power supply module 2 right (2 iputs)
-FCE power supply module 2 left (2 inputs)
-R7746 ozone converter command indication

Hardware Program Pins

-L/R CMC
-Spares (2)
-Parity

The L/R CMC pin defines the position of the CMC. The left CMC's pin is grounded. The right CMC's pin is an open.

The two spare pins are not connected.

The parity pin shows odd parity for the program pins. The left CMC's pin is an open. The right CMC's pin is grounded.

Table 5-3   CMCS discrete inputs and outputs.

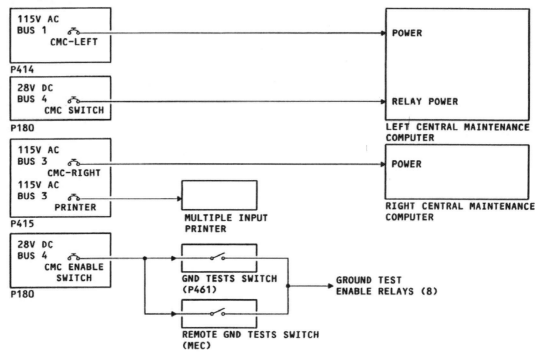

Fig. 5-12 CMC power inputs. (Northwest Airlines, Inc.)

The power to the CMCS is supplied through three different busses (see figure 5-12). The number 1 AC bus powers the left CMC. The number 3 AC bus powers the right CMC and the CMCS printer. The number 4 DC (28V) bus supplies the CMC switch relay and the ground test switches.

There are eight ground test enable relays that are used to ensure that certain LRU BITE tests are performed only on the ground and not during flight. The eight relays are activated by the ground test enable switch located on the flight deck or the remote ground test enable switch located in the main equipment bay. When energized, the relays send a discrete (ground), or discrete CMC signal to the affected LRUs. Figure 5-13 shows the two enable switches and four of the relays. Relay position is monitored by the EIUs, and whenever energized the message *GND TEST ENABLE* is displayed on the auxiliary EICAS panel. The operation of the CMC ground test will be discussed later in this chapter.

CMCS Operations

Any of the four control display units can be used to access the data contained in the CMC. In most cases, the technician will use the flight deck CDU since the EICAS displays can also be viewed from that location. The CDU contains a CRT display located at the top of the unit between twelve line select keys (see figure 5-14). The **line select keys (LSKs)** on the control display unit are used to select items for display or activate functions available on the CDU. Whenever a given line is active, a *caret symbol (<)* will be shown adjacent to the appropriate LSK. Other keys, which are used to access CMCS information, are the *MENU, PREVIOUS PAGE,* and *NEXT PAGE* keys.

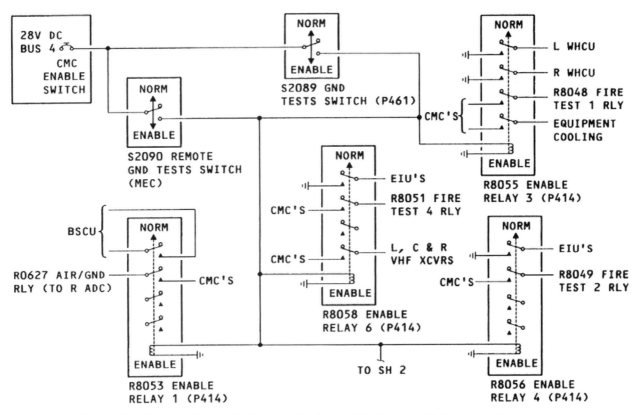

Fig. 5-13 CMC enable switches and relays. (Northwest Airlines, Inc.)

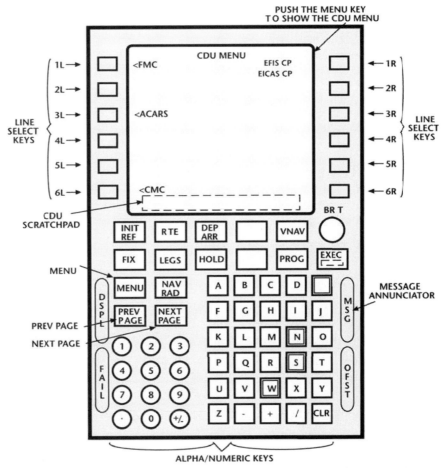

Fig. 5-14 A typical control display unit (CDU). (Northwest Airlines, Inc.)

To initiate CMC operations, the menu must be brought to the display screen by pressing the *MENU* button on the CDU (see figure 5-14). The line select key *6L* (which is adjacent to <*CMC* on the display) should be pressed to select the CMC menu. The two page CMC menu will then be available. (figure 5-15). Page one of the CMC menu will be the initial display. Pressing the *next page* button on the CDU accesses the second page. As seen on the CMC menu, there are a total of seven options available from the CMC, displayed on the two pages: 1) Present faults, 2) Confidence tests, 3) EICAS maintenance pages, 4) Ground tests, 5) Existing faults, 6) Fault history, 7) Other functions. To make a selection, press the LSK adjacent to any of the seven functions on the display.

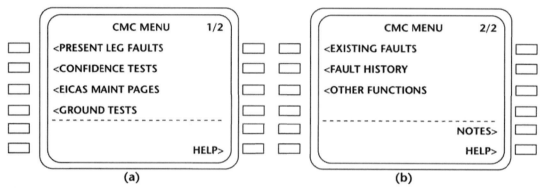

Fig. 5-15 CMC menu display. (a) page 1 of 2, mostly for line maintenance; (b) page 2 of 2 for extended maintenance troubleshooting.

Present Faults

The **present faults function** of the CMC is used to display any fault and related flight deck effect that occurred during the present leg. The present leg is defined as the elapsed time between first engine start and last engine shut down. When the first engine is started for the next flight the present faults data is moved to the fault history memory. There are two basic formats presented under the present faults function: **present leg faults**, and **present leg messages**. To view the present faults data, simply press the LSK 1L on the CMC menu page, the flight deck effect of the present leg faults will be displayed. For each flight deck effect, a related present leg message is available by pressing the appropriate LSK. A **flight deck effect (FDE)** is considered any EICAS message or parameter exceedance, any primary flight display (PFD) flag, or any navigational display (ND) flag. If no FDEs were reported during the present leg, the message *NO FLIGHT DECK EFFECTS REPORTED DURING THIS FLIGHT* will be displayed when the present leg faults page is accessed.

Present leg faults

Figure 5-16 shows the series of CDU displays for the present leg faults function of the CMCS. Refer to this figure during our discussion on present leg faults. The following information is presented on the present leg faults data page(s):

> 1) All present leg faults are displayed (listed by their FDE) in sequential order with the most recent fault at the top of the list.
> 2) The type of FDE (caution, memo, ND flag, etc.) is given along with a flight reporting manual (FRM) code for the fault. This infor-

mation is shown just above the FDE of each fault. The FRM code follows ATA specifications to establish a reference number for the fault. The FRM code can easily be referenced to the aircraft's maintenance and troubleshooting manuals.

3) The asterisk (*) shown next to the type FDE means that the fault is still active.

4) The number of pages of fault data is shown in the upper right corner of the display. To access additional pages simply press the *next page* button on the CDU.

5) A category for *NON-FDE FAULTS* is listed to allow access to all present leg faults that did not create a given FDE.

6) If the word *ERASE* is shown next to a latched status message, pressing that LSK will remove that message from the EICAS status page. This should be done after repairing the fault that caused the status message.

7) The *ERASE STATUS* LSK is used to remove all latched messages from the EICAS status page.

8) The *RETURN* LSK is used to change the display back to the previous menu.

9) The *REPORT* LSK is used to display the report menu.

10) The *HELP* LSK is available on most CMC page displays and is used to gain information concerning use of the CMC.

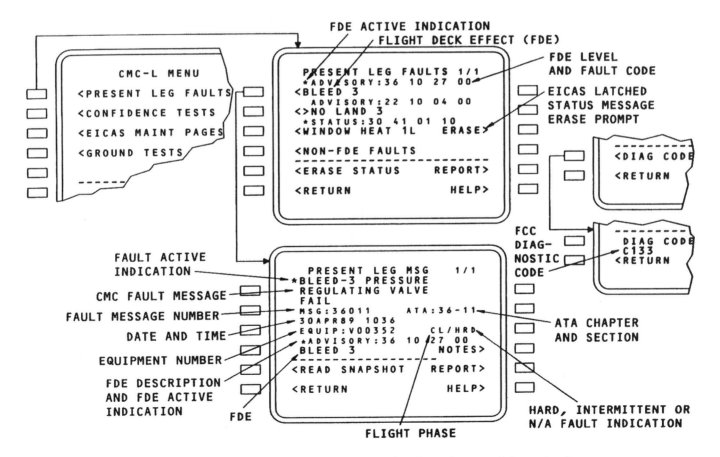

Fig. 5-16 CDU displays used for Present Leg Faults. (Northwest Airlines, Inc.)

Present leg messages

After a technician has viewed the list of present leg faults, he/she may select the associated message for that fault by pressing the adjacent LSK. The CMCS **present leg message** page provides a detailed look at the fault that caused the FDE selected from the present leg faults page. In our example, (figure 5-16) the FDE *BLEED 3* was selected. The present leg message page contains the following information:

1) The title of the FDE from the present leg faults page. The asterisk denotes the fault is still active.

2) The CMC fault message is listed beneath the FDE title.

3) The fault message number is used to find data about the fault in the aircraft's fault isolation manual (FIM).

4) The date and time at which the fault occurred is listed to help aid in troubleshooting. For example, if several messages occurred simultaneously they may be the result of the same defect. On the other hand, they may be nuisance messages caused by a normal operating procedure, such as, an electrical power transfer.

5) The equipment number to which the fault occurred.

6) The FDE description.

7) The flight phase in which the fault occurred and the type of failure (hard or intermittent). This information is extremely useful during troubleshooting.

8) The ATA chapter and section related to the failed system.

9) In most cases, an auto snapshot of the system is taken at the time of failure. Pressing the LSK adjacent to READ SNAPSHOT on the display accesses this information. As stated during the discussions on EICAS, auto snapshots will record various system parameters, which can be recalled during troubleshooting.

Nuisance messages

A **nuisance message** is considered any FDE or CMC message that is not caused by an actual fault. Nuisance messages may be caused by certain normal operating procedures, such as electrical power transfers, or abnormal operation of a system during preflight. To help eliminate nuisance messages the B-747-400, CMCS software incorporates **flight phase screening**. Flight phase screening is done by correlation of any fault with the flight phase at the time which the fault occurred. If the software determines that the failure occurred outside the normal operational flight region for that system, the CMC will ignore that failure. Keep in mind that if the fault still exists when that system enters its normal operating region, the CMC will then report the fault. Much of the flight phase screening is done while the aircraft is still on the ground.

Flight phases

Flight phases are recorded by the CMC to help the technician determine the airplane's configuration at the time of a fault. The CMC uses data from the EIUs to determine the flight phase. There are 14 different flight phases. The two-letter flight phase designators are listed below:

1) Power on (PO)
2) Preflight (PF)

3) Engine start (ES)

4) Taxi out (TA)

5) Takeoff (TO)

6) Initial climb (IC)

7) Climb (CL)

8) Enroute (ER)

9) Descent (DC)

10) Approach/Land (A/L)

11) Rollout (RO)

12) Taxi in (TI)

13) Go around (GA)

14) Engine shutdown (ES)

Printing and ACARS

The CMCS **report menu** contains two options: **printer** or **ACARS** (see figure 5-17). The ACARS function will transmit the present leg fault data to the airline's ground facility. The print option will print a detailed present leg faults summary as shown in figure 5-18. Printing the report is a handy way to document the fault information needed for troubleshooting, or warranty verification.

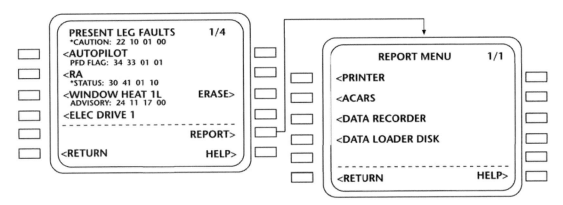

Fig. 5-17 CMC report menu. (Northwest Airlines, Inc.)

Existing faults

The **existing faults** consist of any real time faults monitored by the CMC that are present at the time of CMCS interrogation. Existing faults are often the only CMC function accessed during an aircraft turn around (the time between two flight legs, typically 30-60 minutes). Since existing faults show real time failures, this information is very helpful in determining aircraft departure status. Faults that are currently active can be identified via the existing faults function and correlated with the aircraft's minimum equipment list (MEL). If approved by the MEL, the aircraft may be dispatched without repairing the fault. The existing faults memory may contain some data which is not stored in the present leg faults memory. This occurs since flight phase screening restricts certain faults from entering the present leg faults category and existing faults are recorded regardless of flight phase. These restricted faults often occur during ground activities.

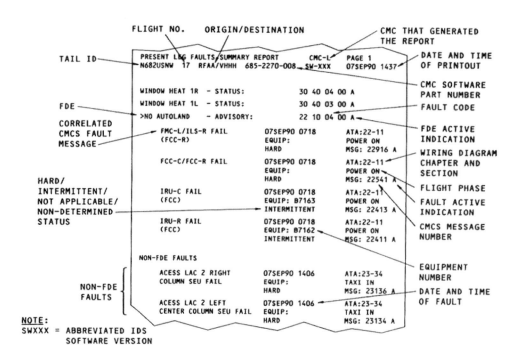

Fig. 5-18 Typical CMC report printout. (Northwest Airlines, Inc.)

To access the existing faults memory simply select the existing faults function from the CMC menu page on the CDU. As shown in figure 5-19 the existing faults menu consists of one or more pages that list the title of the failed system according to ATA chapter. (ATA chapters were discussed in chapter one of this text.) Lowest order chapters are listed first and only chapters that have a failed system will be listed. If no existing faults are present at the time of interrogation, the message *NO ACTIVE FAULTS* will be displayed.

To view the fault message(s) for a given system (ATA chapter) press the appropriate LSK. The first page of existing faults messages for that ATA chapter will be displayed on the CDU. Only one fault will be displayed per page. To access the next fault for that chapter press the *NEXT PAGE* key on the CDU keyboard. To access messages from other ATA chapters, press the LSK adjacent to *RETURN*. The information displayed on the existing faults message pages is nearly identical to the present leg faults messages (compare figure 5-19 and 5-16).

Fault history

To aid in the troubleshooting and repair of chronic problems, the CMCS **fault history** function should be accessed. There can be up to 500 faults from a maximum of 99 flight legs stored in the fault history nonvolatile memory. To access fault history, press the correct LSK on the CMC menu. All failure categories are listed according to ATA chapter as seen in figure 5-20. If the fault history memory is empty, the terms *NO FAULTS IN FAULT HISTORY* will be displayed.

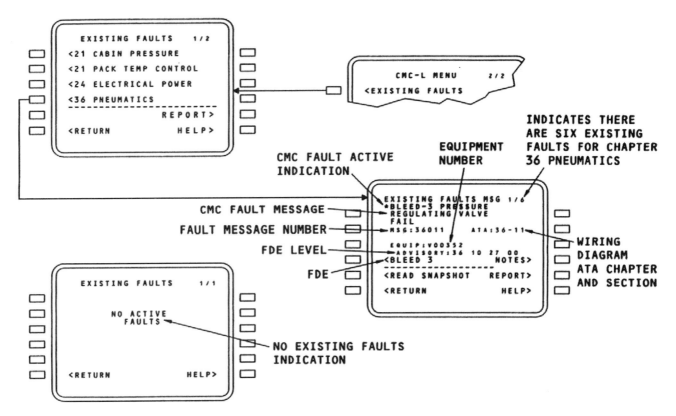

Fig. 5-19 CMC Existing Faults display. (Northwest Airlines, Inc.)

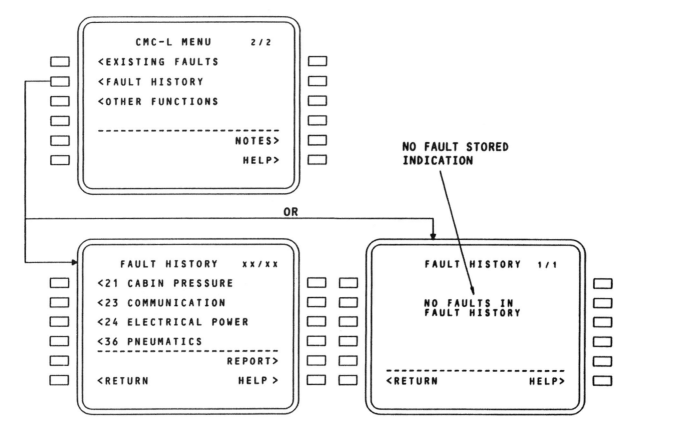

Fig. 5-20 CMC Fault History displays. (Northwest Airlines, Inc.)

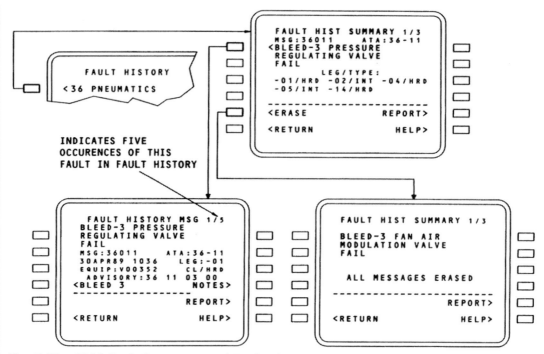

```
                                      FAULT HIST SUMMARY 1/3
                                      MSG:36011        ATA:36-11
                                      <BLEED-3 PRESSURE
                                       REGULATING VALVE
                                       FAIL
              FAULT HISTORY                     LEG/TYPE:
                                      -01/HRD  -02/INT  -04/HRD
              < 36 PNEUMATICS         -05/INT  -14/HRD
                                      ------------------------------
                                      <ERASE               REPORT>

             INDICATES FIVE           <RETURN              HELP>
             OCCURENCES OF THIS
             FAULT IN FAULT HISTORY


         FAULT HISTORY MSG 1/5                FAULT HIST SUMMARY 1/3
         BLEED-3 PRESSURE
         REGULATING VALVE                     BLEED-3 FAN AIR
         FAIL                                 MODULATION VALVE
         MSG:36011      ATA:36-11             FAIL
         30APR89 1036    LEG:-01
         EQUIP:V00352    CL/HRD               ALL MESSAGES ERASED
         ADVISORY:36 11 03 00
         <BLEED 3          NOTES>
         ------------------------            ------------------------
                           REPORT>                            REPORT>

         <RETURN           HELP>              <RETURN          HELP>
```

Fig. 5-21   CMC Fault Summary and Fault History Message pages. (Northwest Airlines, Inc.)

Pressing the appropriate ATA chapter LSK of the fault history menu page will display a **fault history summary** for each fault in that chapter. Each fault will be shown on a separate page. To access additional faults, simply press the *next page* key on the CDU. As seen in figure 5-21, the fault history summary page contains the following information: 1) the CMC fault message, 2) the leg(s) in which the failures occurred, and 3) the type of failure (hard or intermittent). At the top right corner of the fault history summary page is the page number and number of pages available in that summary.

The CMCS **fault history message page** gives a detailed summary for each occurrence of the fault selected from the fault history summary page. The fault history message page is activated by pressing the correct LSK on the fault history summary page (see figure 5-21). The data displayed on each page is listed in order of occurrence (by flight leg). The first page displayed shows the most recent flight leg in which the selected fault occurred. Subsequent pages display additional legs in which the fault occurred. The fault history message page contains information that is almost identical to a present leg message page. The exception being that the fault history message page also contains the leg number for the fault occurrence. Present legs are considered leg number 00, the fault previous to that is called flight leg -01, before that is -02, and so on.

EICAS MAINTENANCE PAGES - RECORDING SUMMARY

ATA	MAINTENANCE PAGE	MANUAL	AUTO
21	ECS	X	Ⓧ
24	ELECTRICAL	X	X
27	FLIGHT CONTROLS	X	NONE
28	FUEL	X	Ⓧ
29	HYDRAULIC	X	X
31	CONFIGURATION	N/A	N/A
32	GEAR	X	NONE
49	APU	X	X
73	EPCS	X	▷1
73	PERFORMANCE	X	X
73	ENGINE EXCEEDENCE	N/A	▷2

FUEL PAGES
MAIN1/MAIN 4
MAIN 2/MAIN 3
RESERVE 2/RESERVE 3
CENTER MAIN/STABILIZER

ECS PAGES
AIR CONDITIONING
AIR SUPPLY

▷1 STORE WHEN A PERFORMANCE
AUTO SNAPSHOT IS TAKEN
▷2 STORES AUTOMATICALLY
IN CUMULATIVE MANNER -
ERASED THROUGH CMC

Table 5-4 Manual and Automatic event recording related to various EICAS maintenance pages.

EICAS Maintenance Pages

The **EICAS maintenance page** function of the CMC is used to access real time data for 11 different aircraft systems. The EICAS maintenance pages are accessed through the CMC menu on the CDU; however, the data is displayed on the lower EICAS CRT. For more information on EICAS, see chapter three in this text. The EICAS maintenance pages present three types of information: **real time data, auto event data**, and **manual event data**. Real time data presents information about the selected system as it currently exists. Real time data is dynamic information and will change with changes in system parameters. Auto event data is information about the selected system, which was automatically recorded at the time of a system fault or exceedance. Auto event data is helpful in troubleshooting faults since it allows the technician to see the conditions of the system at the time of failure. Manual event data is a *snapshot* of the system parameters recorded by the flight crew or maintenance technician action. Table 5-4 shows a list of the 11 EICAS maintenance pages and the availability of manual and auto recorded events.

Accessing EICAS maintenance pages

Access of the EICAS maintenance pages is actually done through four different types of LRUs: the CDU, CMCs, EIUs, and EICAS displays. Therefore, each of these systems must be operable to access the EICAS maintenance pages. Pressing the correct LSK from the CMC menu will activate the choice of EICAS maintenance pages. Figure 5-22 shows the maintenance page menu and the components involved in retrieving maintenance pages.

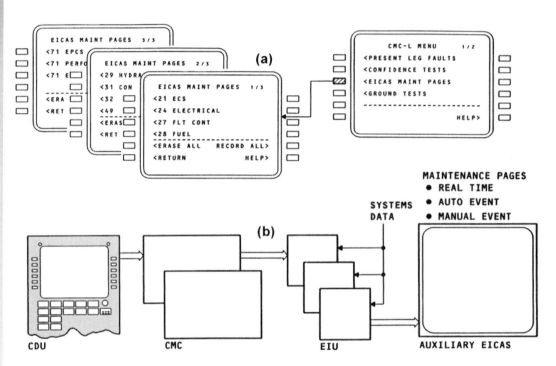

Fig. 5-22 EICAS maintenance pages. (a) typical EICAS maintenance menu pages; (b) components involved in the retrieval of EICAS maintenance pages. (Northwest Airlines, Inc.)

After a given system has been selected, the *EICAS page control* will be displayed (see figure 5-23). From this page the operator can select *display, record, manual snapshot erase system, auto snapshot, report, return,* or *help*. The display function will present the real time parameters for the previously selected system. The record function will take a manual snapshot of the parameters currently displayed on the EICAS CRT. (Keep in mind, since the EICAS maintenance pages are shown on the lower EICAS CRT, the EICAS page control will be displayed simultaneously on the CDU.)

The manual and auto snapshot functions will display a list of their respective snapshot data for the selected system (see figure 5-24). There are a maximum of five auto and five manual snapshots available for each system. The list of snapshots is displayed showing the flight leg, date, and time of the recording. Pressing the LSK adjacent to a given snapshot will display the system's parameters on the lower EICAS display. Auto snapshots are also available from five other CMC functions: present leg faults, confidence tests, ground tests, existing faults, and fault history. Manual snapshots are only available through the EICAS maintenance page function.

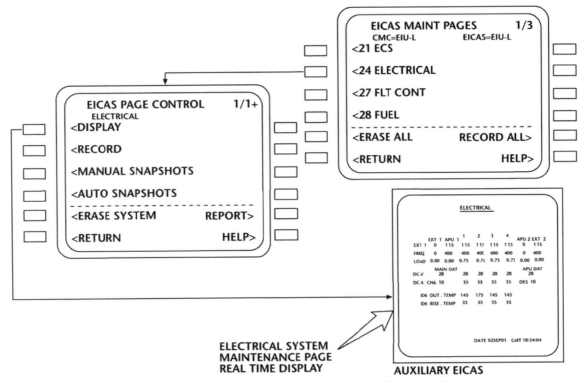

Fig. 5-23   EICAS Page Control display. (Northwest Airlines, Inc.)

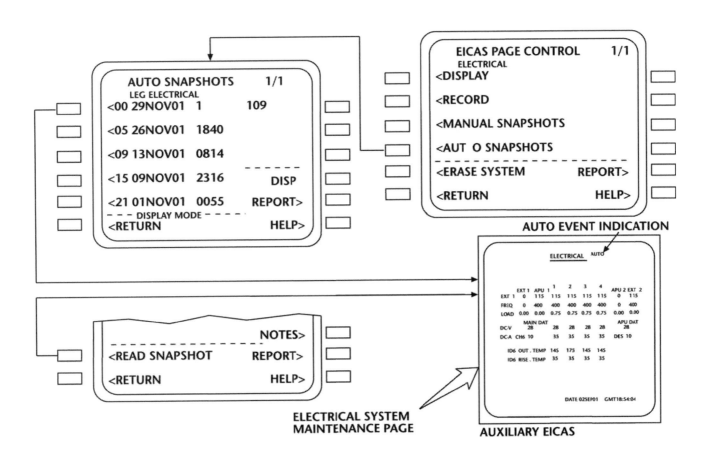

Fig. 5-24   EICAS maintenance page Auto Event Displays. (Northwest Airlines, Inc.)

Pressing the REPORT LSK on the EICAS page control menu will allow access to the *print on ACARS* function of the EICAS maintenance pages. The print function will send the currently displayed EICAS maintenance page to the CMC. The CMC will relay that data to the flight deck printer. This feature is handy to allow for retrieval of the parameter data, which can be studied at a more convenient time or place. The ACARS function will send the currently displayed data to the ACARS management unit for transmission to the airlines ground facility. The ACARS function is only used during flight.

Of the 11 systems available for display by EICAS maintenance pages, seven have one page formats such as the display in figure 5-25. Keep in mind that this one page can display real time, auto snapshot, or manual snapshot data. The four other systems available through EICAS maintenance pages (ECS, Fuel, Configuration, and Engine exceedance) have unusual page formats. The **ECS (environmental control system)** EICAS maintenance page format consists of two pages. One page displays conditioned air, the other shows supply air. The **fuel** system has four EICAS maintenance pages available for display. To allow access to each of these four pages an additional fuel menu page is displayed when requested using the CDU. The **Configuration** maintenance pages are used to access information on the integrated display system (IDS) software numbers and LRU pin programming. The configuration maintenance page does not have an auto or manual snapshot mode. The engine exceedance page consists of a single page format; however, not all engine parameters are always listed. Certain data is displayed as a function of an exceedance in that area. The engine exceedance page menu contains an erase function, which is used to clear the exceedance memory.

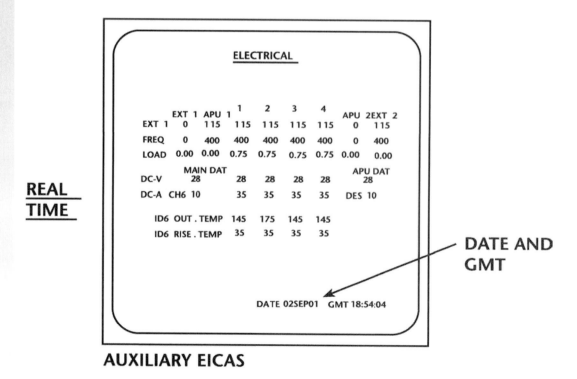

Fig. 5-25 EICAS maintenance page; ELECTRICAL, one page format. (Northwest Airlines, Inc.)

Confidence tests

The **confidence tests** function of the CMCS allows for preflight testing of three aircraft systems: the stall warning/stick shaker, the take off configuration of the aircraft, and the ground proximity warning computer (GPWC). In general, confidence tests provide a quick *GO/NO-GO* status of these systems. Pressing the appropriate LSK activates the various tests.

Ground tests

The **ground tests** function of the CMCS provides access to test functions of up to 80 different systems and LRUs. To initiate a ground test, the aircraft must be on the ground and the ground test enable switch must be placed in the *enable* position. When the ground test function is selected from the CMC menu, the list of ATA chapters is displayed on eight consecutive pages. The ATA chapters that have ground tests available are shown in table 5-5. Selection of a given ATA chapter will display a list of LRUs or systems for testing. Up to five tests are available for each page of the menu. Figure 5-26 shows the series of CMC menu pages used to access the DME ground tests.

```
ATA Chapters and Major Subsections

-Chapter 21                    -Chapter 31
AIR CONDITIONING               INDICATING/WARNING
CABIN PRESSURE                 RECORDING
EQUIPMENT COOLING
                               -Chapter 32
-Chapter 22                    BRAKE CONTROL
AUT OPILOT FLT DIR             PSEU SYSTEM
YAW DAMPER                     TIRE PRESSURE
                               BRAKE TEMPERATURE
-Chapter 23
COMMUNICATIONS
                               -Chapter 34
-Chapter 24                    AIR DATA
ELECTRICAL POWER               INERTIAL REFERENCE
                               NAVIGATION RADIOS
-Chapter 26                    FLIGHT MANAGEMENT
FIRE PROTECTION
                               -Chapter 36
-Chapter 27                    PNEUMATICS
FLAPS CONTROL
STALL WARNING                  -Chapter 45
                               CENTRAL MAINTENANCE
-Chapter 28
FUEL                           -Chapter 49
                               APU
-Chapter 29
HYDRAULIC POWER                -Chapter 73
                               ENGINE FUEL & CONTROL
-Chapter 30
ICE AND RAIN
```

Table 5-5   Ground test availability listed according to ATA chapter.

To start a given test, press the LSK adjacent to the LRU or system shown on the display. The CDU will initiate a digital signal to the CMC, which will in turn transmit a test signal to the appropriate LRU. The actual test will be performed within the LRU in question. During the test, the LRU will respond with a *test initiation* signal. Upon completion, the LRU will transmit the test results back to the CMC. The CMC will generate the appropriate message for display on the CDU.

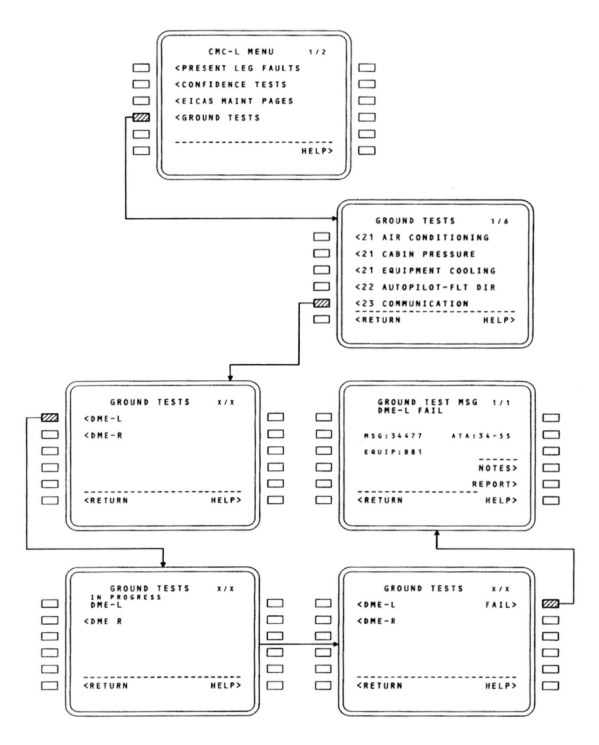

Fig. 5-26 CMCS Ground tests menu. (Northwest Airlines, Inc.)

At completion of a ground test, the CDU will display one of the following messages: *PASS*, *FAIL*, or *DONE*. Pass simply states that the component or system test found no faults. A test fail will record a CMC message. To access that message press the LSK adjacent to the *FAIL>* prompt on the CDU (see figure 5-26). The message *DONE* indicates the end of the test for systems that require the technician to determine the test results. For example, a test of the flight deck annunciators would require the technician to visually identify if the lights were illuminated at the appropriate time.

Shop faults

Shop faults are typically used to further define faults found during other CMC tests. The *other functions* LSK from the CMC menu is used to access shop faults (see figure 5-27). The shop faults option will present a list of ATA chapters from which to make a selection. When a given ATA chapter is selected, the LRUs which can be tested are listed on the CDU. When a given LRU is selected, the BITE for that LRU is activated and an internal test is performed. The results will be displayed as shown in figure 5-28. During a ground test for example, a fault in the cabin pressurization may be detected. The use of shop faults may isolate the given defect to a specific LRU, or within a given LRU.

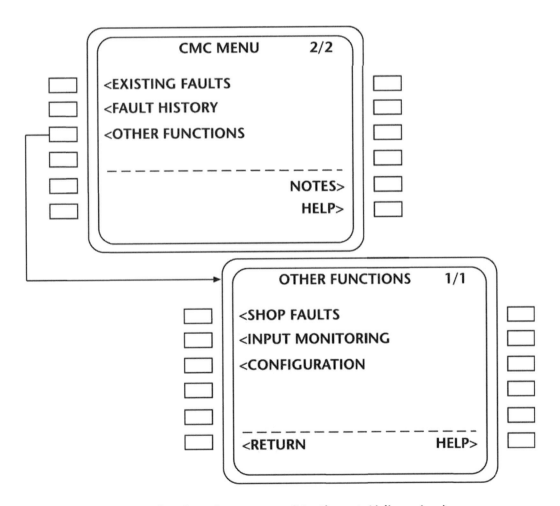

Fig. 5-27   CMCS Other functions menu. (Northwest Airlines, Inc.)

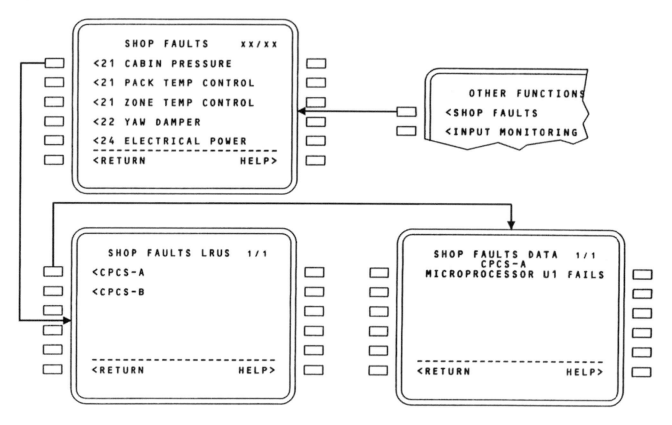

Fig. 5-28 CMCS typical Shop fault. (Northwest Airlines, Inc.)

Input monitoring

Input monitoring is accessible on the ground or in flight to display the ARINC 429 input data to various LRUs. Input monitoring data can be displayed in binary, hexadecimal, or engineering code formats. The CMC input monitoring function could be used to detect ARINC 429 bus problems similar to a carry-on data bus analyzer. As shown in figure 5-29, input monitoring is found through the *other functions* menu. Up to eight pages of input monitoring can be done at any given time. This is very handy when comparing the same data sent to several different LRUs.

To select a given system and bus to be monitored, enter the following four designators, using the CDU keypad:

1) The computer designator; enter either E for EIU, or C for CMC.
2) The specific port on the selected computer to be monitored; enter a three-digit identifier.
3) The label to be monitored; enter a three digit code. This label corresponds to the ARINC 429 specifications and identifies the information being transmitted to the selected CMC/EIU port.
4) The source destination indicator (SDI); enter a two digit code. The SDI is an ARINC specified number, which corresponds to the source transmitting the data.

The specific digits, which should be entered, are available from the aircraft's maintenance manuals, schematics manuals, or fault isolation manuals. When the code is initially entered, it appears on the scratch pad at the bottom left corner of the CDU (see figure 5-29). To select the number displayed in the scratch pad, press the top left LSK. To determine the format for the displayed data, press the LSK adjacent to the engineering (ENGR), binary (BIN), or hexadecimal (HEX) label.

The data displayed on the CDU is real time information for the system/bus selected. In example (figure 5-29), the data is displayed in binary language and each of the 32 bit characters is a binary 1 or 0. Remember, ARINC 429 data consists of a 32-bit word. On the input monitoring display, each data word consists of two lines of 16 bits each. The bits read from left to right, the top left is bit number 32 and the bottom right is bit number 1. The top two rows of data are the most recent sampling of data. The next two lines are from the previous sampling, and so on. Up to six lines (three samples) of data can be displayed at any given time. With each new sample of the data many of the binary digits may change from 0 to 1 and vise versa as the system being monitored changes parameters. The values displayed can be compared to the appropriate values (available from the aircraft manuals) to determine the validity of the data.

When the input monitoring function is updating data, the word *SAMPLING* is displayed on the CDU. During *sampling* the data will advance to the next sample approximately once a second. To study the data, press the LSK adjacent to the *FREEZE* prompt. When in the freeze mode, the data display remains stable and the message *FREEZE* replaces the word *SAMPLING* on the CDU display. The report function of the input sampling mode allows for printing of the data or transmission through ACARS to a ground facility.

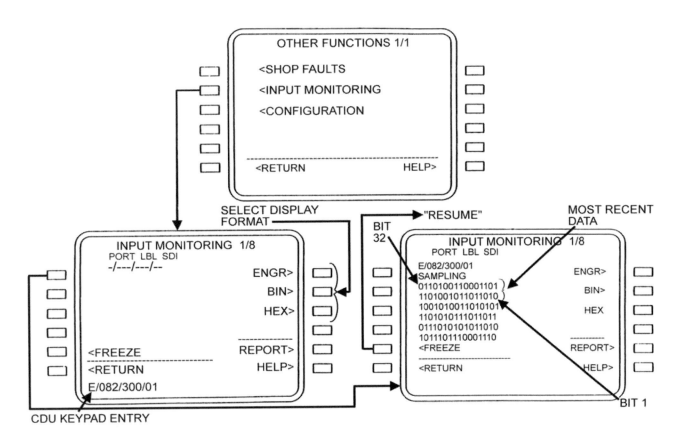

Fig. 5-29   CMCS example of input monitoring. (Northwest Airlines, Inc.)

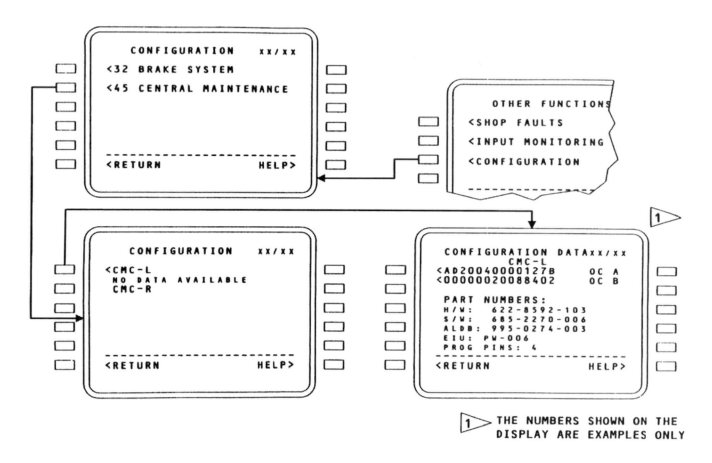

Fig. 5-30 CMCS Configuration page. (Northwest Airlines, Inc.)

Configuration

The **configuration** function of the CMC allows for verification of part numbers and programming options for the brake system control unit (BSCU) and the central maintenance computer (CMC). The configuration option is listed under other functions on page 2 of the CMC menu. As shown in figure 5-30, selection of the central maintenance computer (45) will activate the display of the left CMC data. As long as there is a left CMC installed, no data will be available from CMC-R. This is a normal condition because of the output switching relays (discussed earlier). To access data from the right computer, the left CMC must be deactivated. To select the left CMC, press the LSK adjacent to the CMC-L prompt.

A typical data display for the CMC configuration is shown in figure 5-30. This display shows the hardware (H/W) and software (S/W) part numbers, the airline database (ALDB) numbers, the EIU software configuration, and the CMC program pins. At the top of the display are the two airline **option codes** (OC A & OC B). To verify the CMC memory contains the correct option code, the hexadecimal form of the code is entered into the scratch pad. From the scratch pad, the code is moved to OC A or OC B by pressing the corresponding LSK. The CMC will then compare codes and display the appropriate response.

Options summary

The CMCS has a multitude of information available through the CDU. As a technician becomes familiar with the aircraft, it becomes second nature as to which function of the CMC will access the needed information. The CMC function accessed will be a direct result of the current needs of the technician and aircraft status. For example, if the aircraft is awaiting departure the existing faults function may be used; if the technician is troubleshooting a chronic fault, the fault history may be accessed. Table 5-6 shows the various functions of the CMC and what they are often used for during maintenance and troubleshooting.

Troubleshooting Using The CMCS

Whenever troubleshooting any system, the ultimate goal is to repair the aircraft. This should be accomplished in the least time possible without sacrificing safety. The CMCS is designed to help isolate faults (troubleshooting) and verify when the repair has been completed successfully (operational testing). For either of these operations, the CMCS cannot stand alone. The CMCS is always used in conjunction with the various aircraft maintenance manuals. For troubleshooting purposes the Fault Isolation Manual (FIM) portion of the maintenance manuals is extremely valuable. In many cases, the FIM is a completely separate manual or series of manuals. In some cases, the FIM data is included in the front of each chapter of the maintenance manuals.

The Fault Isolation Manual contains three sections, which are used extensively when isolating faults with the CMCS: the **Fault Code Index, EICAS Messages**, and **CMCS Message Index**. Each of these sections has a specific function, which may be used under different conditions to help isolate faults. These sections of the FIM will be presented in the upcoming examples of typical troubleshooting procedures. In all but a handful of situations, the fault isolation procedures begin with a write up from the flight crew. If the crew noticed a discrepancy during flight, a logbook entry would be made. Between flights a technician will examine the log to decide if any problems have occurred which require immediate attention. In some cases, the fault repair can be deferred.

VARIOUS CMC FUNCTIONS	
CMC FUNCTION	PURPOSE & TYPICAL USE
Present leg faults	Provides a list of faults and related fault data from the most recent flight leg. Typically used to confirm a fault reported by the flight crew during the last leg. Allows access of snapshots.
Existing faults	Provides data on all faults which currently exist (ie. real time data). Used most often to examine aircraft status prior to dispatch.
Fault history	Provides a list of fault data for failures within the past 99 flight legs. Used for troubleshooting reoccurring problems.
Confidence tests	Tests certain critical systems and provides data on the aircraft's configuration. Typically used by the flight crew for preflight testing.
EICAS maintenance pages	Presents the real time parameters for 11 different aircraft systems. Used for dynamic testing during troubleshooting or repair verification.
Ground tests	Displays test results for various systems and LRUs. Often used to verify a fault or as a test after repair
Shop faults	Displays data on specific LRUs or software failures. Typically accessed only for in-depth troubleshooting. Often used to pinpoint a defect after identifying a system fault through other tests.
Input monitoring	Functions as a data bus analyzer for ARINC 429 data transmitted to the CMCs or EIUs. Used during in-depth troubleshooting of a system.
Configuration	Presents specifics as to which hardware and software components are installed in the CMCS and BSCU. Used to verify the installation of the correct components.

Table 5-6 Various CMC functions

The log entry made by the flight crew typically consists of a description of the fault, the related EICAS message (if any), and the Fault Reporting Manual (FRM) fault code number. The FRM provides a fault code number for the various FDEs caused by system faults. The FRM fault code number is identical to the Fault Code Number, which is used to identify faults in the FIM. On some aircraft, ACARS is used to transmit fault code data to the airline ground facilities. This allows the technician the opportunity to review the fault information prior to aircraft landing. This creates an obvious advantage when maintenance takes place during a short turnaround period.

It should be noted, that although the CMCS is very effective at fault isolation, these systems monitor and test only electrical circuitry, not mechanical devices. In some situations, the mechanical device can be monitored using electrical/electronic means; in other situations it cannot. For example, limit-switches can be used to determine the position of a simple pressurization air valve; however, there is no simple electrical means to measure the integrity of the cargo door seal. If the CMCS detects cabin pressurization is too low it will look for several conditions to determine the fault. If the air valve motor has failed the CMCS message will report the mechanical failure of the valve assembly. However, if the cargo door seal has been damaged and too much air is leaking around the door, pressurization will be too low. In this situation, the CMCS will not be able to determine a defect or the CMC may fault a different portion of the system. Although these situations are rare, one must always consider that the mechanical defects are often missed by the CMCS.

System troubleshooting example #1

This example will provide a general guideline for troubleshooting a system fault. A specific detailed sequence will be provided in the second system troubleshooting example. The following steps should be used to isolate a typical system fault:

1) Access the flight log to determine the problem. The log should give a description of the problem, the fault code from the FRM, and the FDE. In some cases, a fault code number is not available for a given defect. For this situation, begin troubleshooting using the fault description.

In some cases, only the EICAS message is used to initiate troubleshooting. In this situation, the EICAS message portion of the FIM should be used to determine the fault code number.

2) Access the Fault Code Index found in the FIM or the appropriate chapter of the maintenance manual.

> NOTE: If the flight log entry did not include a fault code, use the general list at the beginning of the index. Figure 5-31 shows a portion of the Fault Code Index for a clock problem. In this case, the isolation procedures would be found in the maintenance manual (MM) 31-25-00/501.
>
> If the flight crew reported a specific fault code and FDE in the flight log, the fault code number should be found in the fault code index. As

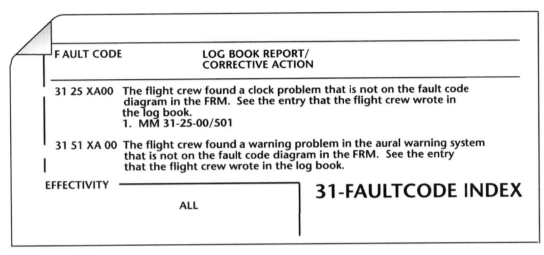

Fig. 5-31   Fault code data from the Boeing 747-400 Fault isolation manual.

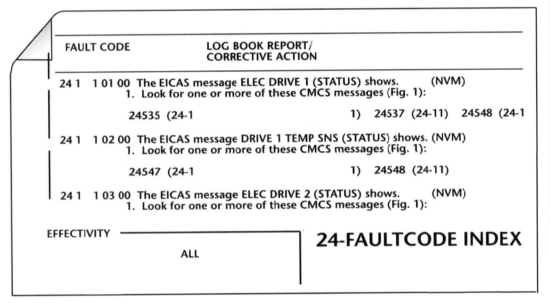

Fig. 5-32   Fault code index message 24 11 01 00, 24 11 02 00, and 24 11 03 00.

shown in figure 5-32, the fault codes are listed in numerical order in the Fault Code Index. For the fault code number 24 11 01 00, the EICAS message *ELEC DRIVE 1 (STATUS)* is given along with three possible fault message numbers: 24535, 24537, and 24548. Adjacent to the fault message numbers, listed in parenthesis, are the corresponding ATA chapters and sections.

3) If an EICAS message is being used to initiate the fault isolation procedures, the EICAS message pages from the FIM or MM must be used to find the corresponding fault code number. Then the Fault Code Index is used to find the fault message number as in step two above. An example of an EICAS message page is shown in figure 5-33. EICAS messages are arranged in alphabetical order and provide the message level (status, advisory, caution, or warning), description, and the fault code number.

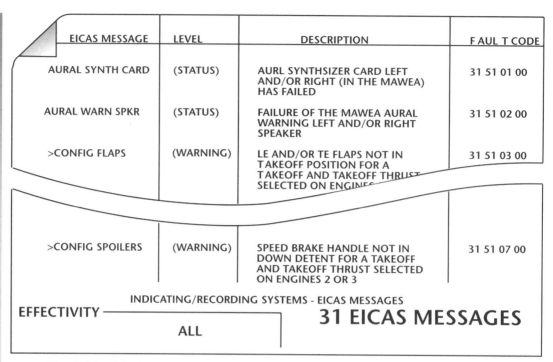

EICAS MESSAGE	LEVEL	DESCRIPTION	FAULT CODE
AURAL SYNTH CARD	(STATUS)	AURL SYNTHSIZER CARD LEFT AND/OR RIGHT (IN THE MAWEA) HAS FAILED	31 51 01 00
AURAL WARN SPKR	(STATUS)	FAILURE OF THE MAWEA AURAL WARNING LEFT AND/OR RIGHT SPEAKER	31 51 02 00
>CONFIG FLAPS	(WARNING)	LE AND/OR TE FLAPS NOT IN TAKEOFF POSITION FOR A TAKEOFF AND TAKEOFF THRUST SELECTED ON ENGINES	31 51 03 00
>CONFIG SPOILERS	(WARNING)	SPEED BRAKE HANDLE NOT IN DOWN DETENT FOR A TAKEOFF AND TAKEOFF THRUST SELECTED ON ENGINES 2 OR 3	31 51 07 00

INDICATING/RECORDING SYSTEMS - EICAS MESSAGES

EFFECTIVITY ———— **31 EICAS MESSAGES**

ALL

Fig. 5-33 EICAS messages form the Fault isolation manual.

4) Once the fault message number has been found through the aircraft's manuals, it must be correlated to the actual CMCS message. To do this, the technician must log on to the CMC using the CDU. A CMC message that related to the fault in question should be located in the CMC nonvolatile memory. In most cases, the search will begin in the Present Leg Faults memory. If the correct CMC message is not found there, the Existing Faults and/or Fault History function of the CMC should be accessed. Figure 5-34 shows a flow chart diagram from the B-747-400 Fault Isolation Manual, which provides guidance to CMC access.

A large portion of working your way through CMC data is simply knowing how to access different functions of the system. The CMC menu tree shown in figure 5-35 begins CMC access in the upper left corner of the diagram. The CDU menu is used to access the two-page CMC menu, from here the various fault nonvolatile memories and system tests are accessed. The bottom center portion of the menu tree shows that the help and report portion of the CMC program can be accessed from almost any function. These options become very handy if used properly.

5) The proper correlation is made between the CMC message and the flight log or EICAS message in order to find the correct fault code needed for the repair. Correlation helps to ensure the work performed is for the fault in question. After correlating the CMC message and the recorded fault, use the CMC Message Index portion of the FIM or MM. The CMCS Message Index will provide the necessary suggestions for repair and/or further fault isolation. As seen in figure 5-36 the fault codes are listed in numerical order in the CMCS Message Index. The index includes one or more CMCS messages for the related fault code, the EICAS message/type, and the corrective action. The corrective action portion of the CMCS Message Index identifies all necessary references, such as the maintenance manual (MM) or wiring diagram manual (WDM).

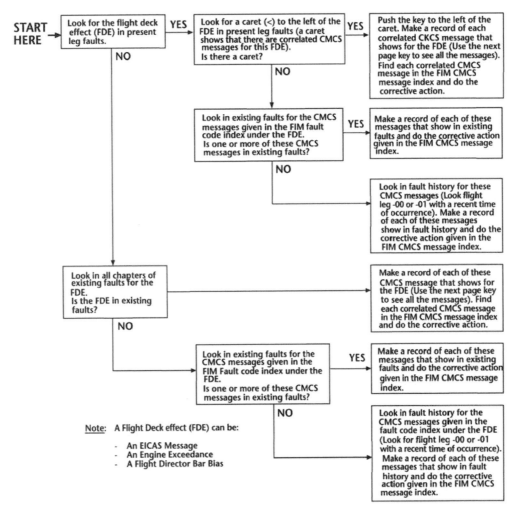

Fig. 5-34 CMCS access flow chart. (Boeing Commercial Airplane Group)

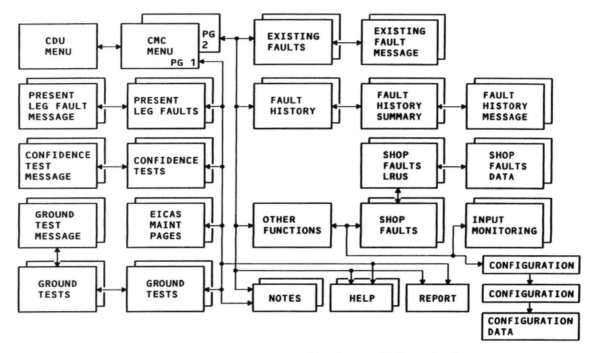

Fig. 5-35 CMCS menu tree. (Northwest Airlines, Inc.)

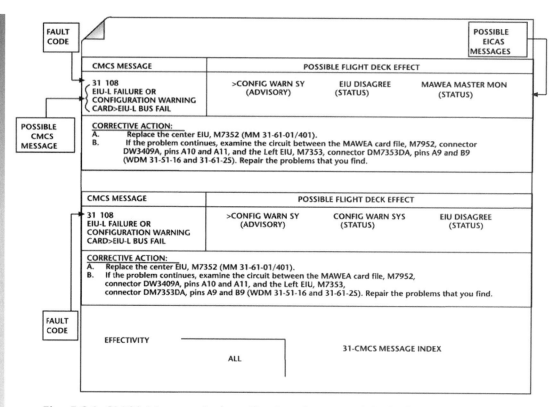

Fig. 5-36 CMCS Message index. (Boeing Commercial Airplane Group)

6) After finding the correct fault code, CMCS message, and related FDE, it would be wise to check the Boeing Service Tip information. The service tips provide helpful information to aid in the troubleshooting and repair of the aircraft. A typical service tip is shown in figure 5-37.

7) Access all necessary manuals as suggested in the CMCS Message Index and perform the necessary corrective actions. Be sure to always verify the repair through the correct testing procedures. In many cases, the CMCS can be used for ground test or review of existing faults of the repaired system. The correct testing procedures will be listed in the manuals under the appropriate page block. As discussed in a previous chapter, the page blocks of the maintenance manuals define the materials covered in that section. Page blocks typically used for troubleshooting and repair of a system include: 001-099 Description and Operation, 101-199 Troubleshooting (including electrical schematics), 201-299 Maintenance Practices, 401-499 Removal and Installation, and 501-599 Adjustment and Test.

In most cases the troubleshooting process begins with a reported system failure. The CMCS is then used for fault isolation of the reported problem. During the initial troubleshooting, always correlate reported faults to CMCS and EICAS messages. Correlation may require a thorough search of the CMCS information. Figure 5-38 provides a flow chart that can be used during investigation of the CMCS.

_____ SERVICE TIPS _____

1. **General**
    A. EICAS message ELEC TR UNIT (X) may be displayed if the Split
        System Breaker (SSB) is open and the right and left sync buses
        voltages differ by 2.5 volts AC.
    B. TRU currents may be at or near zero if the SSB is open.
    C. This may be a nuisance message.
    D. The 747-422 FIM 24- FAULT CODE INDEX will be updated with
        this information.
    E. A future BCU modification will reduce the occurrence
        of this nuisance message.

2. **Recommended Action**
    A. Disable power source to one SYNC bus half:
        (1) Set one APU (If on APU power) or one EXT (If on external
             power) switch to AVAIL position (the split system will close).

        _____
        NOTE: This may cause automatic load shedding

    B. If the ELEC TR UNIT (x) messages disappear and/or the TRU output
        current return to normal, the message is a nuisance.
    C. If the message still exists replace the TRU.

Fig. 5-37 Typical service tip.

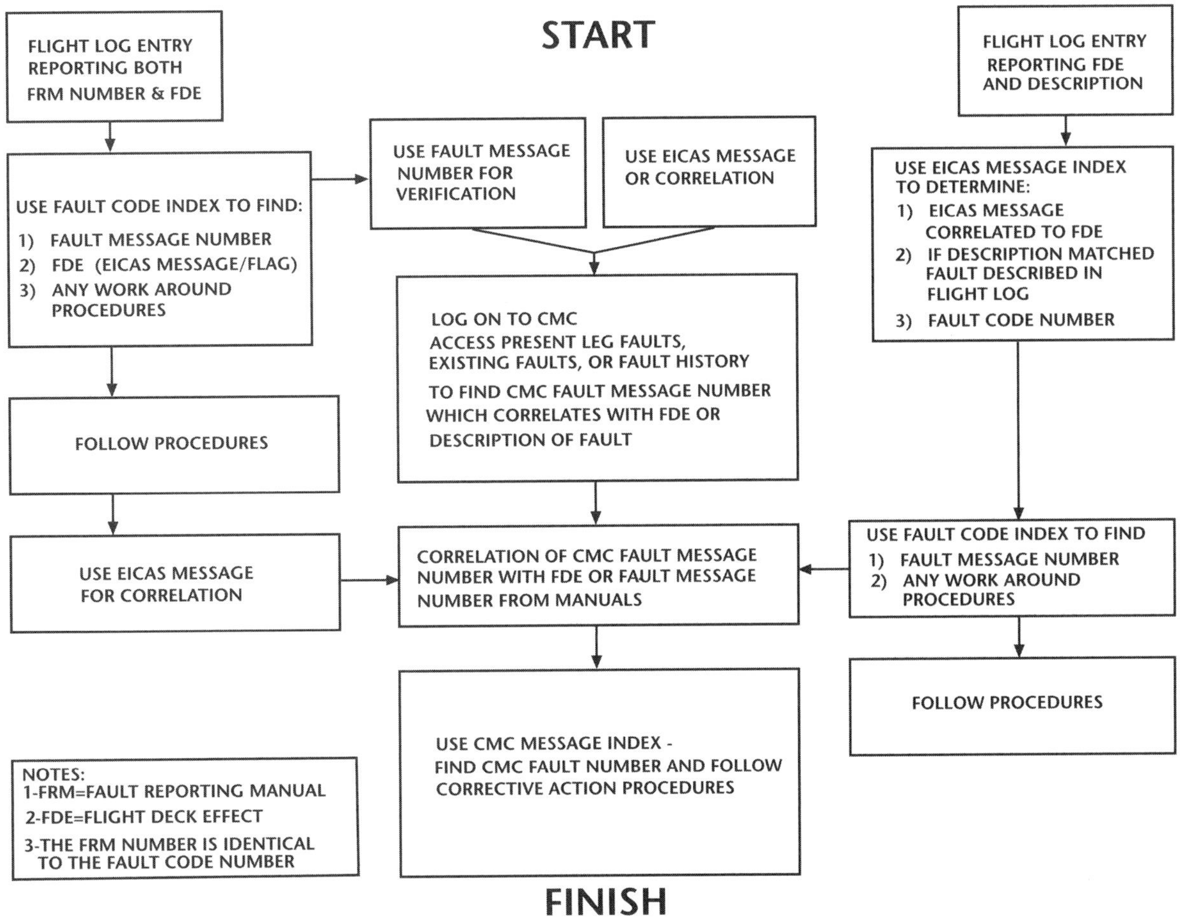

Fig. 5-38 CMCS flow chart.

If several improper FDEs are reported at once, suspect the same faulty condition may have caused several items to create an EICAS message. To determine if one failure is to blame for several FDEs, look for CMC messages that may have occurred at the same time. Also consider the time at which the faults occurred; did they correspond to a normal flight crew action? This normal action may have caused the FDEs. If so, the reported FDEs and related CMCS messages are nuisance messages and most likely require no further action.

System troubleshooting example # 2

In this example we will look at an actual fault and follow it through from logbook entry to repair and testing. The following steps should be taken to repair a problem with the APU battery system.

1) The technician at the gate received a flight logbook entry written as follows: *Fault # 24 30 01 00, EICAS status message BAT CHARGER APU appeared during cruise.* This message occurred during the last flight of the day; therefore, time was available for the repair. If the message appeared earlier in the day (between flights), the repair would most likely be deferred. To determine if the repair could be deferred the technician would have to consult the aircraft's minimum equipment list (MEL).

2) The troubleshooting process begins with the Fault Code Index. As seen in figure 5-39, the fault code number 24 30 01 00 (as recorded in the flight log) is used to locate the CMCS message number (fault message number). In this case, the fault message number is 24099.

Note: If the FRM code number was not reported in the flight log, the EICAS message pages of the maintenance manual could have been used to find the fault code number. The EICAS message page for this fault is shown in figure 5-40.

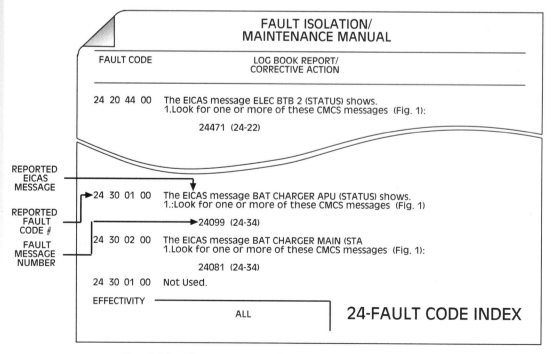

Fig. 5-39 Chapter 24 Fault code index.

FAULT ISOLATION/
MAINTENANCE MANUAL

EICAS MESSAGE	LEVEL	DESCRIPTION	FAULT CODE
BAT CHARGER APU	(STATUS)	ONE OR MORE OF THESE PROBLEMS OCCURRED: (1) THE APU BATTERY CHARGER HAS A FAILURE (2) THE INPUT POWER TO THE APU BATTERY CHARGER IS OFF OR HAS A FAILURE (3) THE APU BATTERY CHARGER INTERLOCK IS OPEN (4) THE APU BATTERY IS IN AN OVERHEATED CONDITION	24 30 01 00

Fig. 5-40 EICAS message for Fault code 24 30 01 00.

3) At this time, the technician should log onto the CMC to find the CMC message number for the fault in question. The first place to look would be in Present Leg Faults. If there were no faults that correlated to the one in question, move to the Existing Faults function. If that specific fault message number cannot be found in Existing Faults, access the Fault History function. When in Fault History, look at leg 00 or leg -01 for faults with a recent time of occurrence. When the correlation of the CMC message number and the fault message number (found in the Fault Code Index) is complete, move to the next step.

4) Use the CMCS Message Index to determine the corrective action (see figure 5-41). Since the message number 24099, EICAS message *BAT CHARGER APU*, the CMCS message, and the flight log entry all correlate, this is the correct location in the CMCS Message Index. In this example, the corrective action is a procedure that requires several tests to further identify the problem. Assume the following:

a) The resistance of the APU battery charger connector was measured. The reference for this test is given as WDM 24-31-21.

b) The measured value was 424 ohms; therefore, the instructions in part A apply. If after a one hour cooling period the second measurement was 424 ohms. The battery should be replaced.

5) Replacement of the battery would begin with reference to the maintenance manual (MM) 24-31-06, p. 401. This section of the MM includes general information, additional reference materials, locations, and procedures for complete removal and installation of the battery.

6) After completing the installation of the new battery, the system should be tested. Section 24-31-06 p. 501 of the MM should be referenced during testing (see figure 5-42).

7) If the APU operates correctly, any latched EICAS messages related to the fault should be erased. The proper maintenance log entry should be made and the aircraft returned to service.

CMCS MESSAGE	POSSIBLE FLIGHT DECK EFECT
24099 APU BATTERY CHARGER FAIL (BCU-2)	BAT CHARGER APU (STATUS)

CORRECTIVE ACTION:

NOTE: This CMCS message will also show if the APU battery becomes too hot. If you have used the APU battery heavily let at least 2 hours go by before you do more troubleshooting. Alternatively, you can replace the APU battery,M7432, (AMM 24-31-06/401) and continue trouble shooting.

NOTE: The APU battery charger is disabled wwhen the battery voltage drops below 4 volts. This is a protective function to prevent charger operation if a battery is not connected to the system. If the battery charger is faulty, and battery voltage has dropped below 4 volts, a new battery should be installed in addition to the battery charger.

(1) Measure the resistance between pins 11 and 12 of the APU battery charger, M7431, connector DM7431 (WDM 24-31-21).

A. If the resistance is less than 575 ohms, do the steps that follow:
(1) Let the battery cool for at least 1 more hour or replace the battery (AMM 24-31-06/401).
(2) If you have let the battery cool and the resistance is still less than 575 ohms, replace the APU battery, M7432 (MM 24-31-06/401).

B. If the resistance is more than 574 ohms, do the steps that follow:
(1) Measure the voltage at pins 4, 7, 10 of the APU battery charger, M7431, connector DM7431 (WDM 24-31-21).
(2) If the voltage on the pins is 115 volts ac, do the steps that follow:
(3) Replace the APU battery charger, M7431 (MM 24-31-07/401) and APU battery, M7432 (AMM 24-31-06/401).
(4) If the problem continues, examine the circuit between the APU battery charger, M7431, connector DM7431, pins 1, 3, 11, 12, and theAPU battery, M7432, connector DM7432B, pins 1, 3, 11, 12 (WDM 24-31-21). Repair the problems that you find.
(5) If the problem continues, examine the circuit between the APU battery charger, M7431, connector DM7431, pin 9, and the No. 2 BCU, G11, connector DG11CA, pin B-H6 (WDN 24-34-22). Repair the problems that you find.

C. If the voltage on the pins is not 115 volts ac, do the steps that follow:
(1) Make sure that the pins that follow are not shorted to electrical ground.
(2) The APU battery charger disable relay, R7218, connector DDR7218, pin X2 (WDM 24-31-21)
(3) Electrical system control module, M7307, connector DM7307F (WDM 24-31-21).
(4) APU battery, M7432, connector DM7432B, pin 8 (WDM 24-31-21).
(a) If the APU battery, M7432, connector DM7432B, pin 8 is shorted to electrical ground, replace the APU battery M7432 (MM 24-31-06/401 and WDM 24-31-21).
(b) If the problem continues, examine the circuit between the APU battery charger, M7431, connector DM7431, pins 4, 7, 10 and the APU BATTERY CHGR circuit breaker, C805 (WDM 24-31-21). Repair the problems that you find.

EFFECTIVITY

ALL

24-CMCS MESSAGE INDEX
01F.1

PAGE 33

Fig. 5-41 CMCS message number 24099. (Boeing Commercial Airplane Company)

MAINTENANCE MANUAL

APU BATTERY - ADJUSTMENT/TEST

1. General
   A. This procedure contains a task to do the operation test of the APU battery (M7432).
   B. The APU battery is installed on the E-33 equipment rack with the APU battery charger. The E-33 equipment rack is on the left side of the aft passenger cabin. Access to the E-33 equipment is through the E-33 access door .

TASK 24-31-06-705-001

2. Operational Test of the APU Battery
   A. References
      (1) 24-22-00/201, Manual Control
      (2) 45-24-00/201, MCS - Electrical Power
      (3) IPC 24-31-06 Fig. 1
      (4) WDM 24-31-21
      (5) SSM 24-31-02
   B. Access
      (1) Location Zone
          271 Passenger Cabin, LH
   C. Do a Test of the APU Battery

      S 865-002
      (1) Supply electrical power (AMM 24-22-00/201)

      S 715-003
      (2) Do a test of the APU battery:
          (a) Get access to the maintenance page for the electrical system
              (AMM 45-24-00/201).
          (b) Make sure the DC-Volts for the APU battery is more than 24 volts.
          (c) Make sure the DC-Amps is in the positive CHG mode.
          (d) Set the STANDBY POWER switch, on the P5 panel, to BAT.
          (e) Make sure the main EICAS display operates correctly
          (f) Make sure the advisory EICAS message BAT DISCH APU shows on the main EICAS screen.
          (g) Set the STANDBYPOWER switch, on the P5 panel, to AUTO.

      S 865-004
      (3) Remove electrical power, if it is not necessary (AMM 24-22-00/201.)

EFFECTIVITY ALL

24-31-06
PAGE 501

Fig. 5-42 APU battery adjustment and test procedures. (Boeing Commercial Airplane Company)

Troubleshooting the CMCS

The CMCS is an extremely reliable system which seldom requires maintenance. In most cases, when the CMCS does fail the CMC can pinpoint the defect and the repair can be made swiftly and accurately. The CMCS runs a self-test program whenever the system begins operation. If a defect is found, the appropriate message will be displayed on the CDU scratch pad. If the CMC is operable, the fault will be recorded in the CMCS nonvolatile memory.

If the left CMC fails, the right CMC will take control of the system. In this case, when you select the CMC from the CDU menu, the message *CMC-L FAIL* will appear on the CDU scratch pad. Also, the amber fail light (MSG) on the CDU keyboard will illuminate to indicate there is a message in the scratch pad. In most cases, the aircraft can be dispatched with one inoperative CMC. The repair can then be made at the next maintenance opportunity.

There are two types of failures that can be reported during the CMCS self-test: **priority one messages** and **priority two messages**. Priority one messages are the most important CMCS fault messages and have priority over priority two messages. Only one message can be displayed at a time in the CDU scratch pad. To cycle through all messages, press the clear (CLR) button on the CDU keyboard. There are a total of 11 priority one messages. Some of the priority one messages are: *CMC-L/R fail*, this indicates a CMC internal failure; *CMC-L/R program pin fail*, this indicates a program pin error detected by the parity program pin; *SW part number disagree*, this means there is different software installed in two CMCs. A part number error would only occur if one of the CMCs were recently changed or the software recently upgraded.

There are six priority two messages. For example, *fault history disagree* indicates that the fault history stored in the CMC memory differs between the left and right CMC. Both priority one and two fault messages are real time and are displayed unless canceled. If the fault clears itself and then fails again, the message will reappear. If either a priority one or two is displayed on the CDU, the aircraft's maintenance manual can be used to determine the correct actions to repair the fault.

There are several EICAS messages, which are used to show different types of CMC failures. If EICAS is active, the EIUs monitors the health of the CMCs. If a fault occurs, the message will be displayed using a standard fault message. A typical fault message might be *EIU-L-CMC-L BUS FAIL*, meaning there is a failure of the data bus between the left EIU and left CMC. The repair procedures would be outlined in the CMC message index. To isolate the fault, the technician should use standard procedures as demonstrated in this chapter.

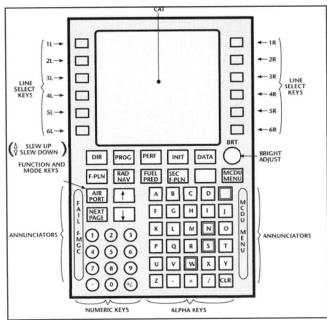

Fig. 5-43 An A-320 multipurpose control and display unit (MCDU).

CENTRALIZED FAULT DISPLAY SYSTEM

The Airbus Industries aircraft use the **centralized fault display system (CFDS)** for fault isolation and system analysis. The Airbus A-320 utilizes an advanced CFDS, which incorporates an integrated system used to access almost every computer on the aircraft. The A-320 CFDS will be presented in this portion of the chapter to describe a typical centralized fault display system.

System Description

As the name implies, the centralized fault display system can access various BITE systems throughout the aircraft and display the information in one central location. Previous generation aircraft required that each individual system be tested independently. The CFDS utilizes a **multipurpose control and display unit (MCDU)** to input commands and monitor replies of the CFDS. The MCDU used on the A-320 is nearly identical to the control display unit (CDU) used on the B-747-400. Figure 5-43 shows the alphanumeric control panel and CRT of an MCDU.

The CFDS offers greater standardization, better data interpretation and simplification of technical documentation over previous individual BITE systems. The CFDS also provides a more in-depth analysis of a system failure. This helps to reduce maintenance time and improve aircraft reliability. Documentation of data is simplified through the use of an onboard printer, which can be used to print most of the data that is displayed on the MCDU. The MCDU and printer are located on the flight deck in the center pedestal area (see figure 5-44). It should be noted that the MCDU gets the term multifunction since it is used by maintenance technicians for accessing CFDS data as well as the flight crew to access the flight management system.

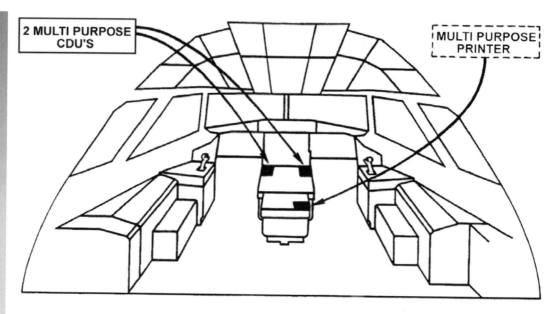

Fig. 5-44 Location of MCDU and multipurpose printer on an A-320. (Airbus Industrie)

Centralized Fault Display Interface Unit

The Centeralized Fault Display Interface Unit (CFDIU) is the main computer that manages the CFDS information. There is only one CFDIU on the A-320; however, a CFDIU backup channel is used to provide redundancy. As seen in figure 5-45, the CFDIU is in direct communications with the aircraft systems, flight warning computers (FWCs), MCDUs, and the airborne communication addressing and reporting system (ACARS) management unit. The ACARS system can be used to transmit information directly from the CFDIU to the airline ground facility. This capability allows the aircraft technician to begin troubleshooting the system even before the aircraft has landed.

Virtually every system which creates an ECAM message is also fed to the CFDIU. The failure or exceedance that caused the ECAM message is stored in memory and can be accessed on the ground through the MCDU. The various systems monitored by the CFDS are in table 5-7. It should be noted that not all LRUs within a system are directly connected to the CFDIU. In most cases, only one LRU reports BITE information for any given system.

System Architecture

The CFDS employs both discrete and digital data signals to communicate with the various aircraft LRUs. Figure 5-46 shows a generalized CFDIU interface diagram. On the A-320, 49 different systems report to the CFDIU with BITE information. It is important to remember that the CFDS is not a BITE system; the CFDS is a manager of the BITE system related to the specific LRUs. The actual built-in test circuitry is contained in the computer software for the individual systems. Most LRU BITE software can store failure data for up to 64 previous flight legs. This capacity for storage can be helpful when troubleshooting reoccurring problems.

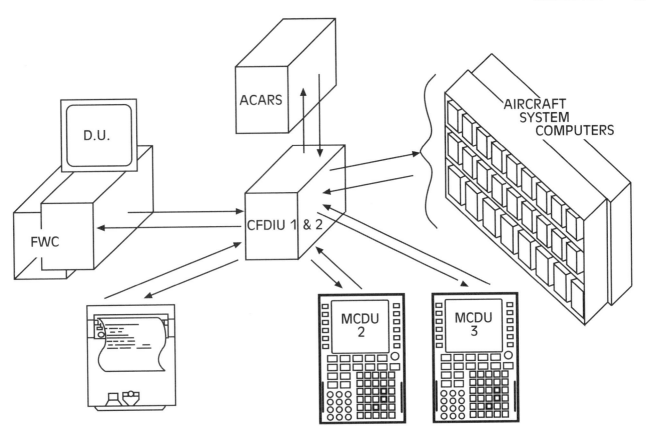

Fig. 5-45 Pictorial diagram of CFDIU communications interface. (Airbus Industrie)

TYPE OF DATA TRANSMITTED/RECEIVED TO/FROM THE CFDIU	SYSTEMS TRANSMITTING SPECIFIED DATA TO THE CFDIU	SYSTEMS RECEIVING SPECIFIED DATA FROM THE CFDIU
Flight number and city pair	Flight augmentation computer (FAC)	
Aircraft identification information and class 2 failures	Flight data interface units (FDIU)	
Flight phase and ECAM warning information	Flight warning computer (FWC)	
DMU class 2 failure informaion	Data management unit (DMU)	Flight warning computer (FWC)
Engine serial number	Display management computer (DMC)	Engine vibration monitoring unit (EVMU)
Time and date information	Aircraft clock	
Flight phase and time/date information		All type 1 systems
Aircraft identification and city pair information		ACARS management unit (MU)
City pair		Data management unit (DMU)

Table 5-7 Various systems monitored by ECAM during normal operation.

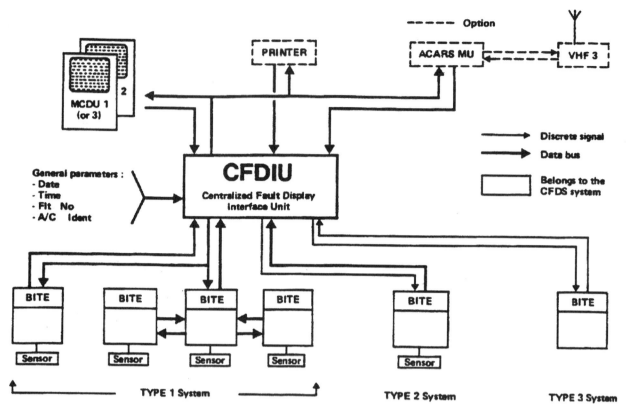

Fig. 5-46 CFDIU interface diagram. (Airbus Industrie)

System types

The CFDS architecture can be further divided according to the type of systems that report to the CFDIU. There are three system types, which encompass all LRUs monitored by CFDS: **type 1, type 2**, and **type 3**. The system type dictates the BITE memory and the interconnections between the CFDIU and the respective system LRU. Most of the LRUs monitored by the CFDS belong to type 1 systems (see figure 5-47). The type 1 systems transmit BITE data to the CFDIU through an ARINC 429 bus. The CFDIU also communicates to the type 1 LRUs using an ARINC 429 data bus. The data transmitted from the CFDIU is needed for BITE interrogation and to provide flight/ground information to the LRUs. Type 1 systems contain the most memory for storage of fault data, (typically up to 10 faults/flight for a maximum of 64 flight legs).

Type 1 system

In the centralized fault display system there are three categories of type 1 systems: **single computer, multi-computer**, and **duplicated systems** (see figure 5-47). A type 1, single computer system will always communicate directly to/from the CFDIU. Single computer systems are relatively simple LRUs, which perform a specific function. Examples of a type 1, single computer system are the VHF 1, 2, or 3 transceivers. Each of these units contains their own BITE that communicates directly to the CFDIU using a 429 data bus.

The multi-computer systems require more than one computer to perform their operations. These systems typically employ one designated computer to coordinate the BITE data and report to the CFDIU. More than one computer in the multi-computer system will most likely have BITE capabilities; however, only one computer will communicate directly to the CFDIU (see figure 5-47). Duplicated systems are LRUs that contain two different subsystems within one computer. For duplicate systems, each subsystem will communicate on a dedicated bus to/from the CFDIU.

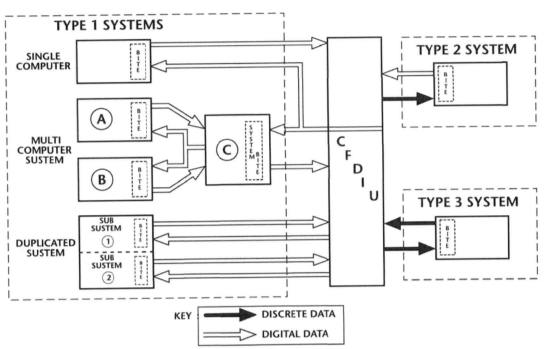

Fig. 5-47 CFDIU/BITE interface diagram. (Airbus Industrie)

Type 2 systems

Type 2 systems can only store fault data for the last leg; previous leg fault data is automatically erased. The BITE data for the type 2 LRUs is transmitted to the CFDIU through an ARINC 429 data bus. To initiate the BITE test, or retrieve the last failure, the CFDIU sends a discrete signal to the LRU. All memorized type 2 systems fault data is lost at the next engine start. (i.e., starting an engine is considered the start of a new flight leg.)

Type 3 systems

Type 3 systems have no capability of memorizing fault data. Therefore, type three data is considered *real time* information. Type 3 systems are typically LRUs, which perform switching or simple monitoring tasks. For example, transformer rectifier (TR) units and ice detection sensors are type 3 devices. Type 3 LRUs receive a discrete signal from the CFDIU to initiate a test. Type 3 LRUs transmit information to the CFDIU using a discrete *pass/fail* signal.

Internal and external failures

The CFDS has the capability to distinguish the difference between internal and external failures. An **internal failure** is defined as a failure that occurs to a component (LRU) that causes the system (or part of the system) to fail. For example, if the angle of attack sensor fails in the air data system, the air data system will report an internal failure (see figure 5-48). Other systems that use air data information will report an external failure. An **external failure** can be defined as a failure of a component outside of the failed system. In figure 5-48, system A, B, and C, will detect a component failure outside of their system. (Systems A, B, and C will report that the air data computer created an external failure in their system.) In this example, the CFDIU would receive four fault messages; one internal and three external. All four of these faults were caused by the failed angle of attack sensor. A faulty component can cause a failure of its own systems (internal failure) or failure of a related system (external failure). Any system reporting an external failure will also report what system caused the problem. The system that caused an external fault would be listed as an *identifier* of the fault.

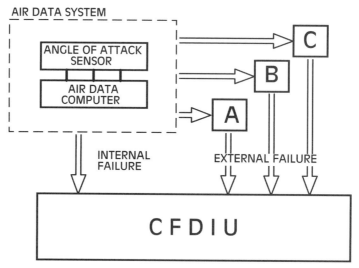

Fig. 5-48 Example of internal and external failures reported to the CFDIU. (Airbus Industrie)

CFDIU Architecture

There are five main functions of the CFDIU: **memory, management, correlation, monitoring**, and **detection**. The memory section of the CFDIU is used to store ECAM warning and system failure information for later display by the CFDS. The management section of the computer is used to add information, such as time and date, flight phases, and flight legs to any failures or ECAM reports, which are stored in the CFDIU memory. The correlation function of the CFDIU is used to relate one system failure to the failures of other systems. For example, if the number 1 **flight augmentation computer** (FAC 1) failed, that system would not be able to send data to several other systems. The correlation software of the CFDIU will record the FAC 1 failure; but, it will *ignore* messages that state *no data from FAC 1*. The CFDIU will present only the initial failure for the last leg report and the *IDENT* function will present the systems affected by the failure.

The monitoring function of the CFDIU is used to detect a total or intermittent input bus failure. As the name implies ,the monitoring software continuously monitors each input bus. The detection circuitry of the CFDIU is used to determine what type of failure occurred in the system. Different failure types include internal, external, intermittent, and class III.

CFDIU interfaces

The CFDIU interfaces with several different aircraft systems (see figure 5-49). The FWC (flight warning computer) is used to transmit flight phases and ECAM warnings to the CFDIU. The FAC transmits flight number and route to/from city pair to the CFDIU. This information is then sent to the **management unit (MU)** of the ACARS and the **data management unit (DMU)**. The clock is used as the central time/date base for the CFDIU management activities. The **flight data recorder interface unit (FDIU)** transmits aircraft identification to the CFDIU, which relays the data to the FWC. The CFDIU receives the engine serial numbers from the **display management computer (DMC)** and transmits that information to the **engine vibration monitoring unit (EVMU)**. The DMC receives the engine serial number from the **engine control unit (ECU)**.

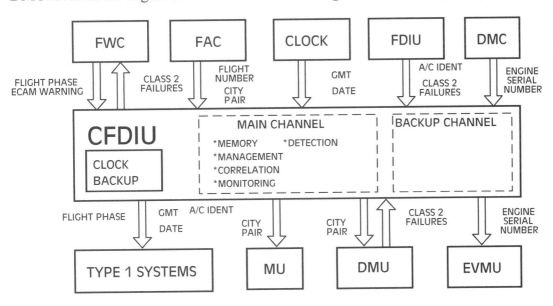

Fig. 5-49 Block diagram of CFDIU interface with other aircraft systems. (Airbus Industr*ie)*

CFDIU backup functions

The **backup** section of the CFDIU provides redundancy for two basic functions: backup of the system CFDS report/test function, and backup of the aircraft's clock. If the aircraft's clock fails while the CFDIU is powered, the CFDIU's backup clock software will provide time and date information to the CFDS and related systems. The backup channel of the CFDIU contains completely independent power supply and software for backup in order to provide the system report/test function in the event the CFDIU main channel fails.

CFDS Operations

To understand the CFDS operations, the system failure classification should be discussed. There are three **failure classifications** for the A-320 systems: Class I, Class II, and Class III. In general, systems which have the highest priority to flight safety, produce a Class I fault when they fail; lowest priority failures produce Class III faults.

Class I failures will have the following effects: 1) display a discrete caution or warning annunciator and a message on the ECAM display, 2) will have a direct consequence on the operation of the current flight, 3) may have dispatch consequences for the next flight, 4) must be reported by the pilots, 5) will provide maintenance information available through the CFDS at the end of each flight leg.

Class II failures have the following effects: 1) display a message on the ECAM status page only, 2) will have **no** direct consequence on the operation of the current flight, 3) have no dispatch consequences, 4) must be reported by the pilots, 5) maintenance information will be available through the CFDS at the end of each flight leg.

Class III failures will have the following effects: 1) have no message or status page information displayed by ECAM, 2) no operational consequence on the current flight, 3) no dispatch consequence, 4) maintenance information is available through the CFDS, 5) correction of these faults can be left until the next regularly scheduled maintenance opportunity.

CFDS Operating Modes

There are two modes in which the CFDS will display fault reports: **in flight** and **on ground**. The *in flight* mode of the CFDS allows the flight crew to access **current leg reports** and **current leg ECAM reports.**

Using the *on ground* mode of the CFDS, the following six reports can be assessed: 1) **last leg reports**, 2) **last leg ECAM reports**, 3) **previous legs reports**, 4) **avionics status**, 5) **system report/tests**, and 6) **post flight reports**. For both *in flight* and *on ground* modes the reports are activated using the MCDU (see figure 5-50).

MCDU Operation

The MCDU operation is relatively simple and typically requires operation of only a few select keys for CFDS access. Figure 5-51 shows the MCDU and highlights some of the CFDS controls. Keep in mind there are two MCDUs installed on the A-320; however, only one MCDU can be used to access CFDS information at any given time. The opposite side MCDU can be used at that time for functions other than accessing the CFDS. The MCDUs turns on automatically when aircraft electrical power is available. The first step to select any report on the MCDU is to press the *MCDU MENU* button.

The *current leg* and *last leg* reports are virtually identical. The difference is simply a function of the time at which the report is taken. The *current leg* report is available in the flight mode only, the current leg report data is moved to the *last leg* memory after landing. The same scenario just described is also true for the *current leg ECAM* and *last leg ECAM* reports.

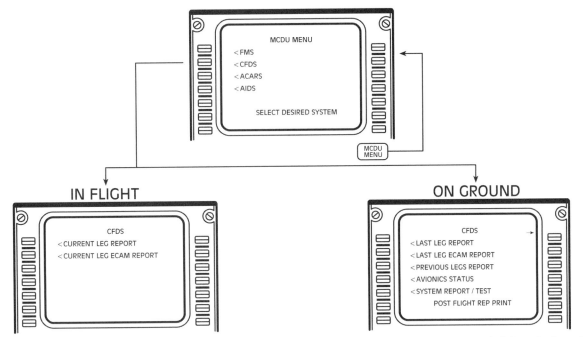

Fig. 5-50 Example of access to in flight and on ground reports using the MCDU. (Airbus Industrie)

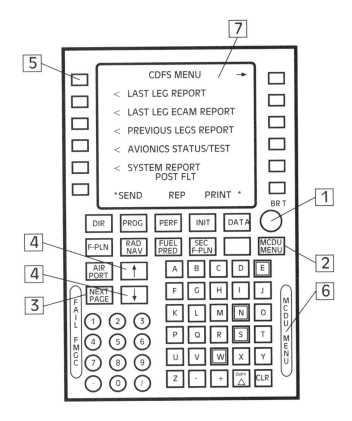

Fig. 5-51 MCDU control panel description. (Items descriptions: 1. The brightness knob, 2. MCDU MENU key is used to begin access of the MCDU functions, 3. The NEXT PAGE key is used to give access to various pages of a display sequence, 4. The scroll up/down keys are used to access data which requires more than one page for the entire display, 5. Twelve LINE SELECT keys provide access to the MCDU function adjacent to the selected key, 6. The MCDU MENU indicator illuminates when a system connected to the MCDU requests information to be displayed, 7. The CRT display contains a maximum of 14 lines each having 24 characters.) (Airbus Industrie)

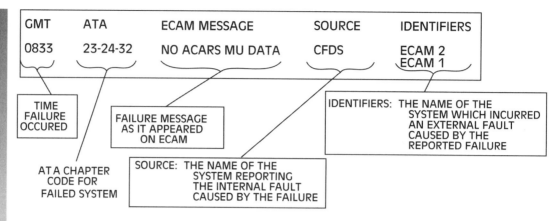

Fig. 5-52 CFDS Last Leg Report format.

Last (and current) leg report

The **last (and current) leg reports** can store up to 40 lines of fault information sent to the CFDIU from the individual BITE systems. These reports list only Class I and II failures. For each fault, the time at which the fault occurred is listed along with the associated ATA chapter for quick maintenance manual reference (see figure 5-52). It should be noted that the time could be displayed under UTC (Universal Time Constant) or GMT (Greenwich Mean Time). The last leg reports can also display the fault identifiers that incurred a related external failure (figure 5-52). If the report consists of more than one page an arrow is displayed in the upper right corner of the first page. Pressing the MCDU *next page* key will bring consecutive report pages to the display.

To access the current/last leg reports, press the appropriate line select key on the CFDS menu page (see figure 5-51). The title of the report (*LAST LEG REPORT*) and the date will appear at the top of the display. Figure 5-53 shows the sequence of displays accessible through the last (and current) leg reports. The report shows the LRU failure and its associated **functional identification number** (FIN) for each fault. The report can be printed if the print option is displayed in the lower left corner of the display. The < symbol displayed next to a fault indicates that source identifier information is available for that fault.

Last (and current) leg ECAM report

The **last (and current) leg ECAM report** is used to access the actual ECAM messages that have been displayed to the flight crew. This data is sent to the CFDIU from the flight warning computers for storage of the data. Up to 40 lines can be stored for each report in one or more pages. Figure 5-54 shows the typical page displays for the last (and current) leg ECAM report. The last (and current) leg ECAM report also displays the time that the ECAM message was initiated, the flight phase number (PN) at the time of the ECAM message, and the ATA chapter and section. Pressing the line select key labeled *print* * will print all pages of the ECAM report.

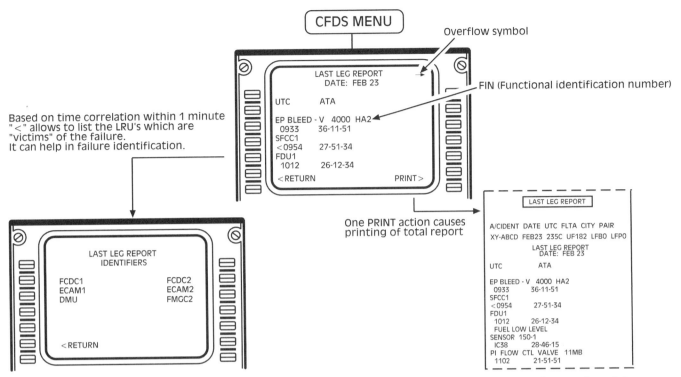

Fig. 5-53 Sequence of MCDU displays for the Last Leg (or Current Leg) Reports. (Airbus Industrie)

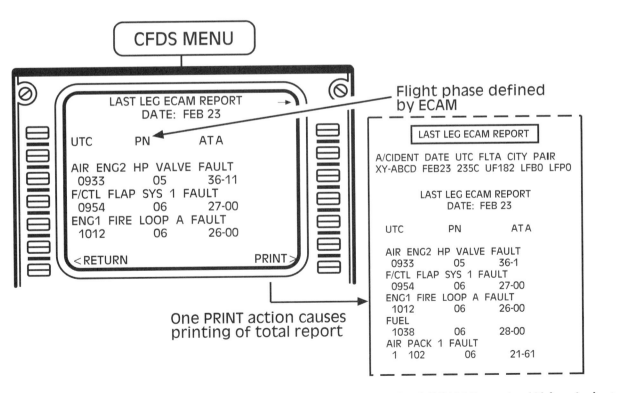

Fig. 5-54 Sequence of MCDU displays for the Last Leg (or Current Leg) ECAM Reports. (Airbus Industrie)

Previous legs report

At the beginning of each new flight (start of the first engine) the last leg report currently in memory is transferred to the **previous legs report** memory. The previous legs report is therefore comprised of a series of *past* last leg reports. A maximum of 63 flight legs and a total of 200 failures can be accessed through the previous legs report. The report contains the same information as the last leg report along with a two digit number used to show the flight leg when the fault was recorded (see figure 5-55). The designation (*INTM*) means the fault has occurred intermittently during that flight leg. The previous flight report is only available on the ground and the print feature will only print the page currently on the MCDU display. This printing arrangement is necessary since the previous legs report could be several pages long.

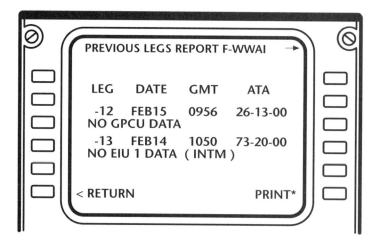

Fig. 5-55 Example of a Previous Legs Report. (Airbus Industrie)

Remember: when analyzing the previous legs report, leg *01* is actually two flight legs old. After engine shut down, the most recent flight (current leg) becomes the *last leg*, the flight before that is the *previous leg 01*. Each leg moves back in the memory sequence at the start of the first engine prior to a flight. If you are required to start an engine for troubleshooting purposes be sure to keep track of where the flight leg data is currently being stored.

Avionics status

The CFDS **avionics status** data presents a list of systems that are affected by a current failure. This list can be used during troubleshooting to help determine the departure status of the aircraft, or to determine the aircraft's current auto flight capabilities. The avionics status information is real time and is therefore continuously updated. During an avionics status inquiry many systems must be powered (and in some cases set to a normal operating condition) or they will be displayed as failed. As seen in figure 5-56, if the system is affected by a Class 3 failure it will be identified in the avionics status data. If the indication *Class 3* appears that system may be affected by a Class 2 or 1 failure also.

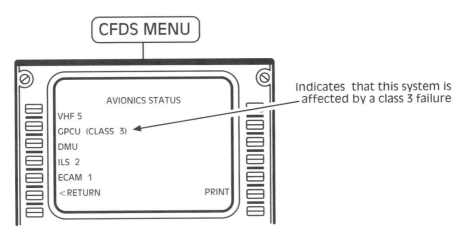

Fig. 5-56 Typical Avionics Status page. (Airbus Industrie)

System report/test

The CFDS **system report/test** function is used for an in-depth look at the faults that occurred in a particular system. System report tests are only available during ground operations. The system report/test will allow for access of fault and troubleshooting data for any system monitored by the CFDIU. As seen in figure 5-57, the main menu for the system report/test identifies the monitored systems listed in order of ATA chapter. The actual information available in the system report/test mode is a function of the type of system being monitored. Type 1 systems will provide the most in-depth analysis, type 3 systems offer the least information. (System types were discussed earlier in this chapter.) The data presented during system report/tests is retrieved directly from the individual system's BITE, not from the CFDIU memory. These reports/tests are therefore interactive between the CFDIU and the respective LRUs.

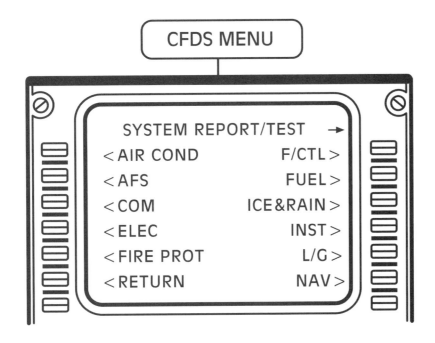

Fig. 5-57 Example of System Report/Test menu. (Airbus Industrie)

Type 1 system reports

Type 1 systems are connected to the CFDIU via an ARINC 429 data bus, and provide a menu specific to the selected system. Figure 5-58 shows the sequence of displays for a typical type 1 system. The displays are describes as follows:

a) The system report/test menu is retrieved from the main CFDS menu. From this display (or following pages), a given ATA chapter is selected. In this case, the technician presses the instrument (INST) key.

b) The CFDIU displays a list of the systems contained under this selection. The technician selects ECAM 1 for interrogation.

c) The CFDIU interrogates the ECAM 1 BITE, which sends a list of available reports back to the CFDIU. This display shows the type of reports and tests available for ECAM 1.

d) The last leg report shown here is similar to the last leg report previously discussed; however, in this case, the report only contains data on the ECAM 1 system.

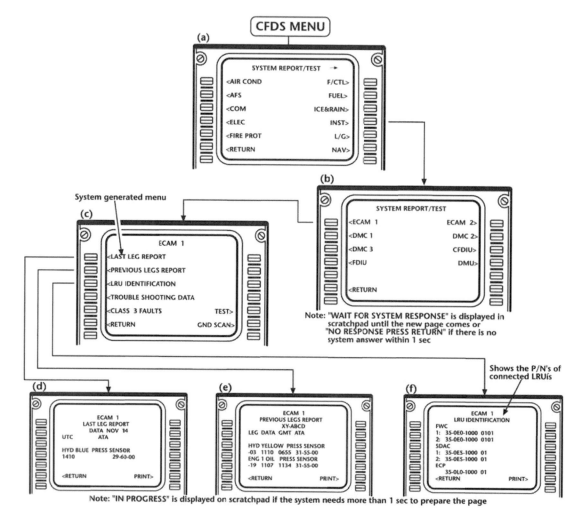

Fig. 5-58 Sequence of displays for a System Report/Test showing Last Leg Reports, Previous Legs Reports, and LRU Identification functions. (Airbus Industrie)

e) This previous legs report contains only the information that deals
with the ECAM type 1 system.

f) The LRU identification page is used to display the specific part
numbers of all electronic LRUs connected to the ECAM 1 system.

The remainder of the potential ECAM 1 report/tests are shown in figure 5-
59. The displays are described as follows:

a) This is the ECAM 1 system report/test menu. The technician will
make a selection from this display in accordance with his/her current requirements.

b) This page, *troubleshooting data*, provides access to the digital data
transmitted between the various LRUs in the system. The data is
displayed in a hexadecimal format and can provide an in-depth
look at the system operations. Typically this data is only accessed
when simpler troubleshooting methods fail. The hexadecimal codes
are typically sent to the airline engineering department to be analyzed. Engineering then relays information on a potential repair
back to the technician. Troubleshooting data will be discussed in
more detail later in this chapter.

c) Any Class 3 faults are listed in this display. Remember Class 3
faults are not critical repair items.

d) Pressing the *TEST* line select key will cause the ECAM 1 system
BITE to perform a real time system test. The results of that test
will be presented on this display. The associated ATA chapter will
be displayed adjacent to the failed component.

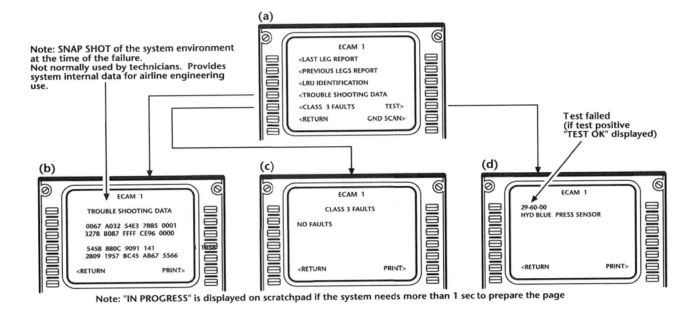

Fig. 5-59 Sequence of displays for a System Report/Test (type 1 systems) showing
Troubleshooting Data, Class 3 Faults, and Test functions. (Airbus Industrie)

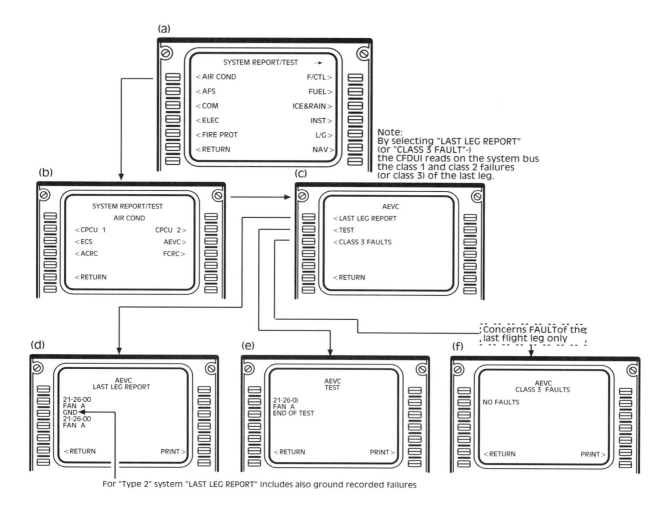

Fig. 5-60 Sequence of displays for a System Report/Test (type 2 systems). (Airbus Industrie)

Type 2 system reports

Type 2 systems are monitored by the CFDIU and allow a maximum of three options in the system report/test mode; last leg report, test, and Class 3 faults. The sequence of displays for a type 2 system is shown in figure 5-60. The displays are described as follows:

a) The air conditioning system was selected for this example.

b) From the air conditioning menu, a second subsystem selection is made. In this case, the avionics equipment ventilation computer (AEVC) was selected.

c) The options for the AEVC are displayed here. The last leg report is displayed as an option for all type 2 systems. A list of Class 3 faults and a real time test are also possible options on some type 2 systems.

d) The last leg report is similar to the last leg reports previously described; however, ground faults are also included in this display. If a ground fault is recorded, the term *GND* will be displayed adjacent to the fault.

e) The test feature will run a real time test for the displayed system. The test results will be shown at the end of the test.

f) Any Class 3 faults that occurred during the last flight leg will be displayed.

Type 3 system reports

Type 3 systems have only one function (test) and therefore have no associated menu. An example of a type 3 system display sequence is shown in figure 5-61.

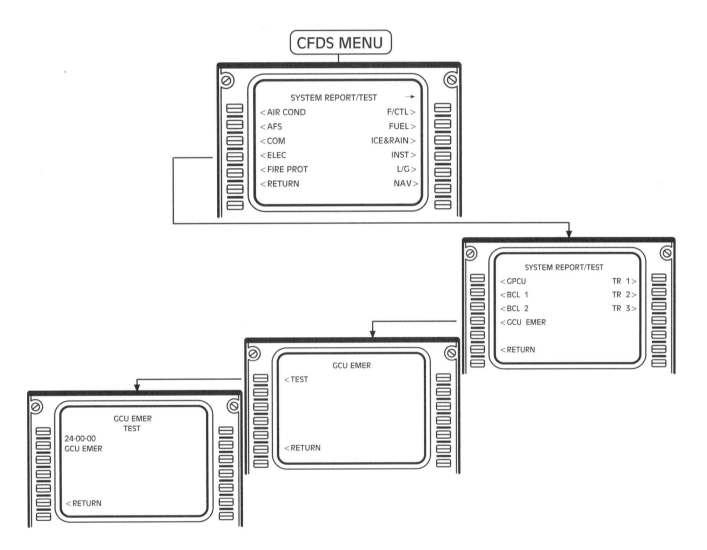

Fig. 5-61 Sequence of displays for a System Report/Test (type 3 systems). (Airbus Industrie)

Post flight reports

The CFDS **post flight report** is a combination of the last leg and last leg ECAM reports available from the CMCS printer. The post flight report is a handy means of accessing fault data from the last flight leg. This data must be accessed through the CFDS menu page and is available in print form only. An example of a typical post flight report is shown in figure 5-62. Be sure to print any desired post flight reports prior to the first engine start for the next flight. At that time the last leg and last leg ECAM reports move to the *previous flights* memory and are not available in the post flight report.

Clock initialization

In the event the aircraft's central clock fails, appropriate repairs must be made and the clock time/data must be reset. The second page of the CFDS menu allows access to the clock software. The clock option of the menu will be displayed only in the event of a clock failure and a CFDIU power interruption. In that case, the MCDU alphanumerical keyboard can be used to enter in the new time and date.

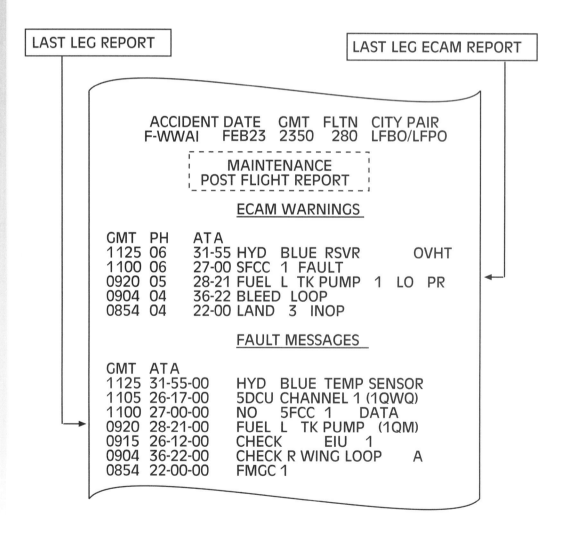

LAST LEG REPORT

LAST LEG ECAM REPORT

```
ACCIDENT  DATE    GMT   FLTN   CITY PAIR
F-WWAI    FEB23   2350   280   LFBO/LFPO

          MAINTENANCE
          POST FLIGHT REPORT

              ECAM WARNINGS

GMT   PH   ATA
1125  06   31-55 HYD  BLUE  RSVR          OVHT
1100  06   27-00 SFCC  1  FAULT
0920  05   28-21 FUEL  L  TK PUMP  1  LO  PR
0904  04   36-22 BLEED  LOOP
0854  04   22-00 LAND   3  INOP

              FAULT MESSAGES

GMT   ATA
1125  31-55-00    HYD   BLUE  TEMP SENSOR
1105  26-17-00    5DCU  CHANNEL 1 (1QWQ)
1100  27-00-00    NO    5FCC  1     DATA
0920  28-21-00    FUEL  L   TK PUMP   (1QM)
0915  26-12-00    CHECK           EIU  1
0904  36-22-00    CHECK R WING LOOP      A
0854  22-00-00    FMGC 1
```

Fig. 5-62 Typical Post Flight Report. (Airbus Industrie)

CFDIU backup mode

As discussed earlier, the CFDIU has a backup channel in the event of a primary system failure. The backup mode can be accessed during a partial CFDIU failure using the second page of the CFDS menu. The backup mode will be the only system operational in the event of a serious primary system failure. As seen in figure 5-63 the backup mode allows for access of the system report/test function only. The backup mode is operational on the ground, not in flight.

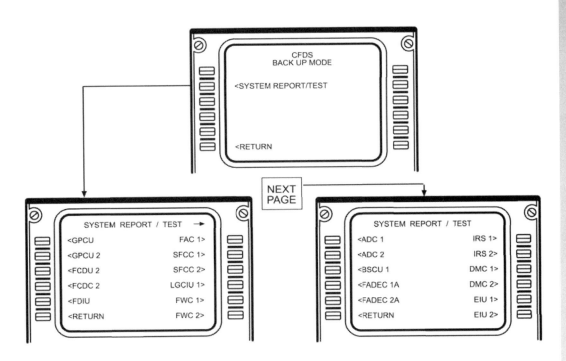

Fig. 5-63 CFDS Back Up mode allows access only to System Report/Test function. (Airbus Industrie)

ACARS/print program

Most modern Airbus aircraft are equipped with ACARS used to transmit CFDS information directly from the CFDIU to the ground. The ACARS/print program is used to determine what CFDS data will be sent via ACARS and what data will be printed onboard the aircraft. Dependent on the specific airline configuration, only certain data can be selected (or not) for ACARS transmission or onboard printing. ACARS can be an extremely valuable troubleshooting tool when linked to the aircraft's centralized fault display system. This combination can provide automatic transmission of system fault information to the airline ground facility. This allows the ground technician to prepare for aircraft repair even prior to aircraft landing. This can greatly reduce aircraft delays due to maintenance and enhance airline productivity. It should be noted, however many airlines limit the use of ACARS for transmission of aircraft systems fault data. Limits are imposed for maintenance information since ACARS frequencies are already overcrowded. In the future, it is anticipated that transmission of fault data via ACARS will become commonplace.

Figure 5-64 shows a typical display for the ACARS/print program. In the *SEND* column on the left side of the display, the selection of *yes* or *no* can be made to determine what information will be sent by ACARS. In the *PRINT* column (right side) the selection is made for the information to be printed. Simply pressing the corresponding line select key will change the label from *yes* to *no* and vise versa. It should be noted that if the label (yes or no) is displayed in blue, it could be changed manually. Labels displayed in green cannot be changed.

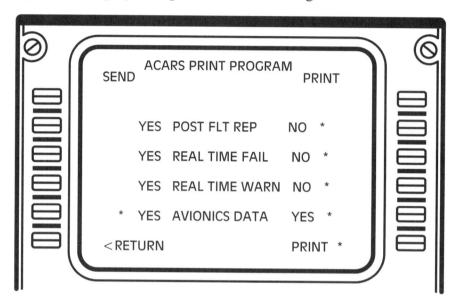

Fig. 5-64 Typical ACARS print program display. (Airbus Industrie)

Level 1 Troubleshooting

There are three levels of troubleshooting that can take place using the central fault display system. On the A-320, **level 1 troubleshooting** is performed at the gate during a quick turn around and is relatively limited in scope. Typically the technician will be under very tight time constraints to finish any necessary repairs and therefore will try to defer most maintenance. Level 1 troubleshooting will always begin with a flight crew log entry concerning a failed component or system. In many cases, the post flight report will allow the technician to access the correct repair using the aircraft maintenance manuals. An example of a CFDS message which would result in level 1 troubleshooting is *FAC 1 FAIL*. In this case the technician would quickly determine that the FAC (flight augmentation computer) #1 should be replaced. If the repair cannot be made quickly and the problem cannot be deferred, the technician will study the problem further and make the repair prior to dispatching the aircraft. This will require the use of level 2 or 3 troubleshooting techniques.

Level 2 Troubleshooting

The A-320 **level 2 troubleshooting** employs techniques of moderate complexity and depth, which often require several hours or longer to correct the fault. An example of a CFDS message requiring a level 2 troubleshooting procedure is *check pins FAC 1*. In this case the technician would be required to check the pin programming of the FAC (flight augmentation computer)1. This type of troubleshooting is always deferred if possible. If deferment is not possible because of other system failures, weather, or the system is simply to critical to fly in an inoperative condition, the aircraft must be repaired before takeoff.

Level 3 Troubleshooting

The A-320 **level 3 troubleshooting** is an in-depth study of the failed system, typically involving access of the binary or hexadecimal fault code data. As previously discussed, fault code data is found under *troubleshooting data* accessed through the *system report/test* of the CFDS. Fault code data is also available through some operational tests, and in some cases displayed in *last leg reports*. Level 3 troubleshooting is typically performed only when standard (level 1 & 2) troubleshooting fails to find the defect. In most cases, the information needed to analyze the fault code data is only available through the airline's engineering department. The specific fault codes can be printed using the CFDS printer, hand written, or down loaded to a floppy disc using the ship's data loader.

Figure 5-65 shows a CFDS page as displayed during an operational test of the engine bleed air system. In this example, the bleed-air monitor computer (BMC) current status fault code data is displayed. This information is displayed in binary language. Some codes are presented in hexadecimal numbers. The label (LAB) number 064, 065, 066, 067, or 055 bits and data bits (11-29) can be analyzed using the appropriate reference. Table 5-8 shows the decoding reference information for label 066. Comparing the binary bits to the appropriate columns of the reference information will provide valuable troubleshooting information.

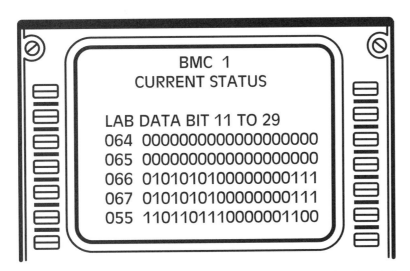

Fig. 5-65 Typical CFDS operational test display.   Data bits at the far right are bit #11; data bits at far left are bit #29.

Status Type	0	1
P+	Loss of Voltage	Applied Voltage
P-	Applied Ground	Loss of Ground

Label = 066				
BIT N°	**Parameter Definition**	**Status 0**	**Status 1**	**Comments**
11	Eng2 OPV Position	Fully Open	Not fully open	
12	Eng2 FAV Position	Fully Open	Not fully open	
13	Eng2 FAV Position	Fully Closed	Not fully closed	
14	Eng2 HPV Position	Fully Open	Not fully open	
15	Eng2 HPV Position	Fully Closed	Not fully closed	
16	Eng2 PRV Position	Fully Open	Not fully open	
17	Eng2 PRV Position	Fully Closed	Not fully closed	
18	Eng2 Bleed Push Button Position	Normal	Pushed	
19	Eng2 Fire Push Button Position	Normal	Pushed	
20	PRV2 Low regulation	No	Yes	
21	Starter Valve Eng2	Closed	Not closed	
22	Valve Closure (For Eng Start)	No	Yes	
23	CFM Validity	No	Yes	
24	IAE Validity	No	Yes	
25	Eng2 HPV Solenoid	Not Energized	Energized	Used by IAE only
26	Eng2 PRV Closure Control	No control	Closure Control	Output state
27	Eng2 Precooler Inlet Pressure Status	12PSIG<P< COPSIG	COPSIG<P< 12PSIG	
28	ENG2-BleedFault	Fault	No Fault	Self Maintained Signal Except Switch Upon OFF
29	ENG2 PRV PWR Supply	No Power	Power	

Table 5-8 Decoding information for label 066

In most cases, all three levels of troubleshooting begin with a log book report from the flight crew concerning a given fault. That information may lead to the correct troubleshooting manual or may require further interrogation of the CFDS. The flow chart in figure 5-66 shows the various reports and system tests available through the CFDS. In this chart, it is easy to see the depth of information available through the CFDS for troubleshooting purposes.

A Typical Troubleshooting Sequence

The troubleshooting sequence can involve many different steps which all lead to the common goal of repairing the aircraft. If the fault occurs during normal operations of the aircraft, the flight crew will initiate a log entry to inform ground crews of the need for maintenance. The fault information could have been transmitted (verbally or using ACARS) during flight to the airline ground facility. In either case, the troubleshooting would begin with referencing the aircraft's troubleshooting manuals. For the A-320, the troubleshooting manual has an **index of warnings and malfunctions** which can be directly referenced to the pilot's malfunction log entry. The index of warnings and malfunctions is divided into four categories: **ECAM** (ECAM warnings), **EFIS** (EFIS fault flags), **LOCAL** (local warnings), and **OBSV** (crew observations). The left hand column of each index is a list of the message, flight deck effect, or observation related to the fault. The right hand column shows ATA chapter and section of the troubleshooting manual (TSM) to be referenced next. All TSM references are to page block 101 (PB101).

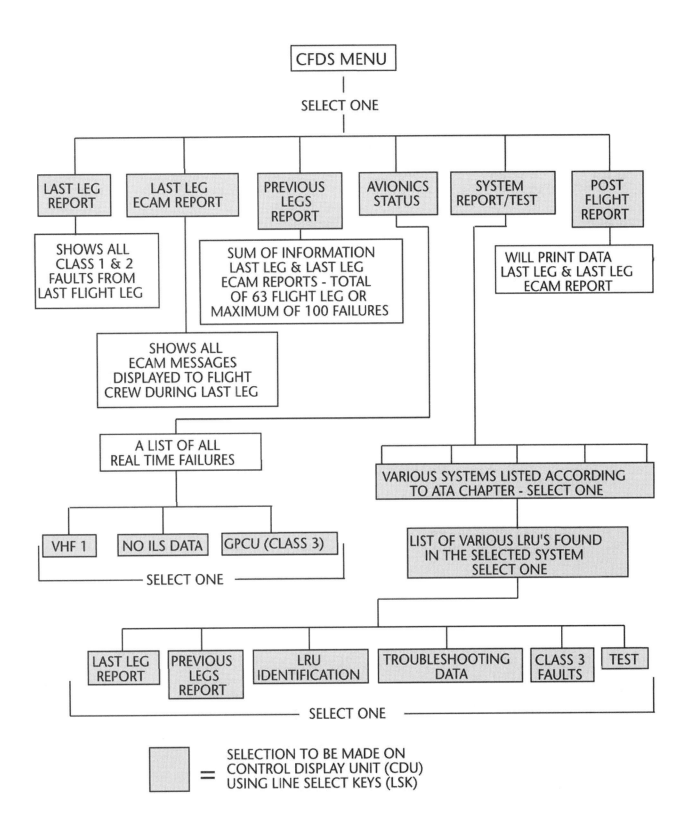

Fig. 5-66 CFDS access flow chart.

The TSM also contains an index of CFDS fault messages. If the exact CFDS message were known this index would be referenced (see figure 5-67). The CFDS message index provides the most information to aid the repair. This index should be accessed if the CFDS message is available. Any TSM index will lead the technician to the proper procedures to effectively troubleshoot and repair the system.

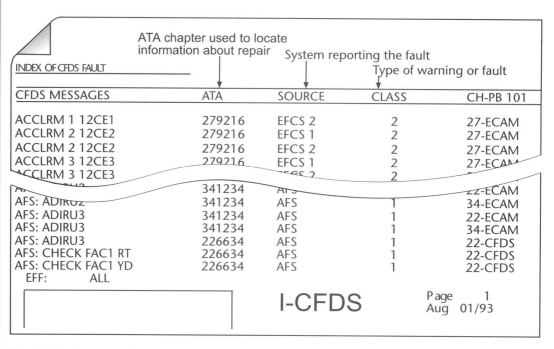

INDEX OF CFDS FAULT	ATA chapter used to locate information about repair	System reporting the fault	Type of warning or fault	
CFDS MESSAGES	ATA	SOURCE	CLASS	CH-PB 101
ACCLRM 1 12CE1	279216	EFCS 2	2	27-ECAM
ACCLRM 2 12CE2	279216	EFCS 1	2	27-ECAM
ACCLRM 2 12CE2	279216	EFCS 2	2	27-ECAM
ACCLRM 3 12CE3	279216	EFCS 1	2	27-ECAM
ACCLRM 3 12CE3		EFCS 2	2	
	341234	AFS		22-ECAM
AFS: ADIRU2	341234	AFS	1	34-ECAM
AFS: ADIRU3	341234	AFS	1	22-ECAM
AFS: ADIRU3	341234	AFS	1	34-ECAM
AFS: ADIRU3	226634	AFS	1	22-CFDS
AFS: CHECK FAC1 RT	226634	AFS	1	22-CFDS
AFS: CHECK FAC1 YD	226634	AFS	1	22-CFDS

EFF: ALL

I-CFDS   Page   1
                Aug 01/93

Fig. 5-67 An example of the Troubleshooting Manual index of CFDS Fault Messages.

In many cases, troubleshooting/repair of the system is deferred until the next maintenance opportunity. Deferring fault repair can be done if approved by the aircraft's Minimum Equipment List (MEL). The following paragraphs provide an example of the typical troubleshooting steps that follow the flow chart of figure 5-68.

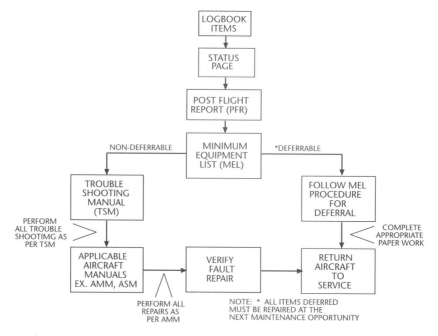

Fig. 5-68 Flow chart for aircraft fault troubleshooting.

AIRLINES NON ROUTINE MAINT.			ACFT	REF. NO.		FLIGHT	DATE	MONTH	LOG PAGE	STATION	CONT. NO.	383026
SFR- 4321B	SEE MM 7-0-3-1		4001			60	07	FEB		LGA		

ORIG	ATA	MB.	CTR	ENG POS.	LIMS. REQ.	IRQ. FLIGHT-DATE		IFS	MR NO. OR ENG	DASH OR S/N OFF	DASH OR S/N ON	FOB

METERS		INSP. FILE NO.		INIT	JOB NO	CROSS REF	

DEFECT: ENG 2 FADEC A Fault displayed on upper ECAM display during cruise.

MM CODE ☐☐☐☐☐☐☐

ACTION TAKEN:

DEF AUTH /		SPEC REF/						ENTRY BY/ 99999	
PARTS NEEDED				EMPL. FILE NO.	ON DAY	ON JOB		OFF JOB	INITIALS
PLACARD NO.		FORM RE-VIEW	EMPL. FILE NO.	INITIALS	EMPL. FILE NO.	ON DAY	ON JOB	OFF JOB	INITIALS OVR

Fig. 5-69 Example of a typical logbook entry.

1) The logbook entry received from the flight crew is reviewed by the technician (see figure 5-69). In this example, the entry states that the message *ENG 2 FADEC* was displayed by ECAM during flight.

2) The status page is accessed on the MCDU to verify that the fault is still active. If the fault no longer exists, it might be intermittent or related to a given flight phase. In this example, we will assume the fault is confirmed (see figure 5-70).

3) The post flight report should be reviewed to identify any CFDS message related to the fault (see figure 5-71). The appropriate fault message is found and a related ECAM message is generated.

4) A sample of the MEL is given in figure 5-72. In this example, the MEL states that the repair can be deferred if a placard is installed adjacent to the lower ECAM display. The required placard states *FADEC maintenance status message displayed on ECAM does not effect dispatch.*

5) The aircraft can be returned to service after the appropriate paperwork had been completed.

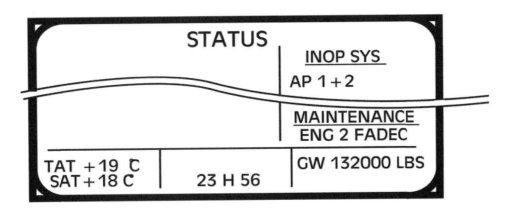

Fig. 5-70 Confirmation of a logbook entry using the CFDS Status Page.

357

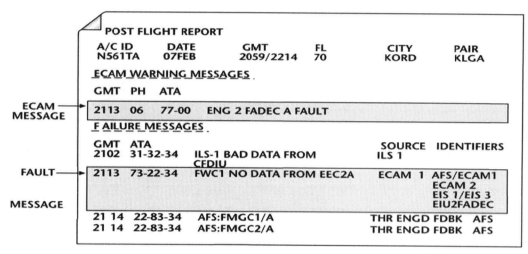

POST FLIGHT REPORT

A/C ID	DATE	GMT	FL	CITY	PAIR
N561TA	07FEB	2059/2214	70	KORD	KLGA

ECAM WARNING MESSAGES

GMT PH ATA

ECAM MESSAGE →

GMT	PH	ATA		
2113	06	77-00	ENG 2 FADEC A FAULT	

FAILURE MESSAGES

GMT	ATA		SOURCE	IDENTIFIERS
2102	31-32-34	ILS-1 BAD DATA FROM CFDIU	ILS 1	

FAULT MESSAGE →

GMT	ATA		SOURCE	IDENTIFIERS
2113	73-22-34	FWC1 NO DATA FROM EEC2A	ECAM 1	AFS/ECAM1 ECAM 2 EIS 1/EIS 3 EIU2FADEC
21 14	22-83-34	AFS:FMGC1/A	THR ENGD FDBK	AFS
21 14	22-83-34	AFS:FMGC2/A	THR ENGD FDBK	AFS

Fig. 5-71 Identify CFDS message using Post Flight Report.

MINIMUM EQUIPMENT LIST

Fault(s) indicated by FADEC Maintenance Status

ECAM message
  ENG 2 FADEC FAULT
Fault message
  FWC1 : NO DATA FROM EEC2 A

No.	DEFERRED POS.	PLCD#
7171B	ENG NO. 1......................	1485
7171A	ENG NO. 1 AND NO. 2...	1484
7171F	ENG NO. 2......................	1557

SPEC NOTES:
A. These maintenance status messages are displayed on lower ECAM.
B. CFDS interrogation not required.

MAINT :
A. Install DEFERRED. placard adjacent to lower ECAM.

OPS PLACARD: "__ FADEC MAINT STS MESSAGE DISPLAYED ON ECAM DOES NOT AFFECT DISPATCH."

Fig. 5-72 Example of a Minimum Equipment List Entry.

When time permits, the FADEC repair would be done in the following sequence:
To repair the FADEC may take several hours. In most cases, this would be too long for a gate turn around. Therefore, this type of repair would typically take place during an overnight stay or during the next scheduled maintenance check.

1) Reference the aircraft's TSM index as in figure 5-73. In this example the ECAM message from the post flight report was used along with the ECAM index of the TSM. The index states that the TSM chapter 77, ECAM messages should be used to find the potential fault. The warnings/ malfunctions, CFDS fault messages, and fault isolation procedures references are listed in the TSM (see figure 5-74).

2) The TSM task number 73-20-00-810-811, page 211 is now referenced (see figure 5-75). This portion of the TSM offers specific details on fault confirmation (par. 3) and fault isolation (par. 4). Once the fault has been repaired according to the aircraft maintenance manual (AMM) or aircraft schematics manual (ASM), the fault confirmation procedures should be repeated to verify that the repair was successful.

Fault Related ECAM Message	ATA Chapter & Section	Troubleshooting Manual Reference
ECAM	PFR ATA CH/SE	TSM ATA CH-PB 101
EDW WARNING(S)		
AUTO FLTA/THR OFF	22-00	22-ECAM
AUTO FLT A/THR OFF		77-ECAM
FLT AP OFF	22-00	
AUT O FLT A/THR FAULT	22-00	22-ECAM
ENG 1 FADEC FAULT	77-00	77-ECAM
ENG 2 BEARING 4 OIL SYS - HI PRESS	77-00	77-ECAM
ENG 2 BEARING 4 OIL SYS - SCAVENGE VALVE FAULT	77-00	77-ECAM
ENG 2 EIU FAULT	77-00	77-ECAM
ENG 2 FADEC A FAULT	77-00	22-ECAM
ENG 2 FADEC A FAULT	77-00	77-ECAM ←
ENG 2 FADEC B FAULT	77-00	22-ECAM
ENG 2 FADEC B FAULT	77-00	77-ECAM
ENG 2 FADEC FAULT7	77-00	77-ECAM

EFF: ALL

ECAM MESSAGE FROM POST FLIGHT REPORT

I-ECAM

Fig. 5-73 Example of a Trouble Shooting Manual index (I-ECAM). (Airbus Industrie)

False and Related Failure Messages

ECAM and the CFDS report all system malfunctions regardless of the cause. For example, during power transfer (from ground power to engine driven generator) the electrical system may go out of limits for an instant, which could trigger several CFDS messages. Many of these *false fault* messages may be totally unrelated to the electrical system; however, the power transfer caused a momentary power glitch and many systems incurred an instantaneous fault. That fault was then recorded by the CFDS. In this case, the problem is not aircraft related, but the CFDIU identified the fault and memorized the data anyway. The last leg, last leg ECAM, and post flight reports would all show these *false* fault messages.

To verify these faults as nuisance messages, check the time at which the fault occurred on the CFDS report. If the faults all happened at the same time, and if that time coincides with an engine start, the problem is most likely related to a power transfer. It can also be verified to see if the faults still exist. To do this, look at any real time display, such as ECAM.

To help eliminate the problem with false fault messages, troubleshooting procedures should be instigated by a pilot report of the problem. The pilots will not report nuisance messages that occur during power transfer or similar situations. A similar fault message can also be misleading if the time of occurrence is not considered. For example, if two CFDS messages occur at the same time it is likely they are caused by the same LRU failure.

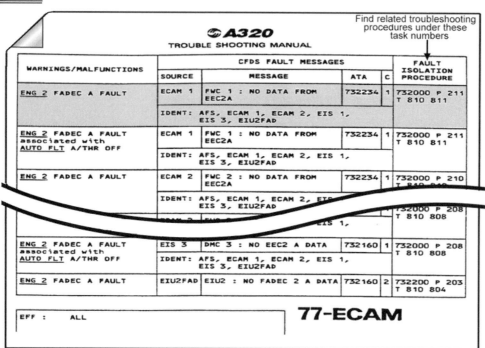

Fig. 5-74 Excerpt from Trouble Shooting Manual showing Warning/Malfunction, CFDS Fault Messages, and Fault Isolation Procedures. (Airbus Industrie)

A320
TROUBLE SHOOTING MANUAL

TASK 73-20-00-810-811

Loss of the Output 1 Bus on SEC 2 Channel A

1. Possible Causes

-EEC (4000KS)
-wiring of EEC A OUTPUT 1 signal from the EEC 2 (4000KS) to the first terminal block

2. Job Set-up Information

A. Referenced Information

REFERENCE	DESIGNATION

AMM 31-50-00-710-001	Operational Test of the Central Warning Systemsol (EEC)
AMM 73-22-34-000-010	Removal of the Electronic Engine Control (4000KS)
AMM 73-22-34-400-010	Installation of the Electronic Engine Contrl (4000KS)
ASM 73-25/10	

3. Fault Confirmation

A. Test

1) Do the operational test of the central warning systems (FWC) (Ref. AMM TASK 31-50-00-710-001).

4. Fault Isolation

A. If the test gives the maintenance message FWC1 : NO DATA FROM WEC 2A:
  - replace the EEC (4000KS), (Ref. AMM TASK 73-22-34-000-010) and (Ref. AMM TASK 73-22-34-400-010).

    (1) If the fault continues:
      - do a check and repair the wiring of EEC A OUTPUT 1 signal from the EEC 2 (4000KS) to the first terminal block, (Ref. ASM 73-25/10).

B. Do the test given in Para. 3.

EFF : ALL   **73-20-00**

Fig. 5-75 Excerpt from Trouble Shooting Manual showing specific possible causes (1), job setup procedures (2), fault confirmation (3), and fault isolation procedures (4). (Airbus Industrie)

CHAPTER 6
AUTOPILOT AND
AUTOFLIGHT SYSTEMS

INTRODUCTION

Autopilots have been installed on aircraft for several decades. The systems have been proven reliable and, for the most part, more accurate than human pilots. The variety of autopilot systems is almost as vast as the variety of airplanes. Simple autopilots are installed on many light aircraft. Transport category aircraft often incorporate complex systems with full autoland capabilities. Most twin-engine aircraft incorporate some type of autopilot system and many corporate aircraft employ complex systems similar to large passenger jets.

Autopilots were first developed to relieve the pilot/copilot from constantly having to "steer" the aircraft. On long flights this became especially important on older transport category aircraft like the Boeing 707. These aircraft were difficult to control and were physically exhausting, especially in bad weather. As technology improved, and systems became lighter and smaller, autopilots began to filter into the light aircraft market. In the early 1980's, autopilot technology advanced to the point where the machine became more efficient than the human; that is, flying with advanced autopilot systems would save both time and fuel. Today a wide range of autopilots are available for almost any type of aircraft.

Simple autopilots provide guidance along only the longitudinal axis of the aircraft. These systems, often found on light single engine aircraft, were called **wing levelers** because they were used to keep the wings level. More complex systems provide total control for attitude and navigation. Many modern transport category aircraft incorporate systems that provide aircraft control, yaw damping, navigational guidance, thrust control, and Category III landing capability. This chapter will examine the operational theory of autopilot systems, individual subsystems of the autopilot will be discussed, and two modern systems will be presented in depth.

BASIC AUTOPILOT THEORY

By definition, the autopilot is designed to perform the pilot duties automatically. The autopilot must first interpret the aircraft's current attitude, speed, and location. Second, if adjustments are necessary, the autopilot must move the appropriate control surface, (and throttles on advanced systems). Third, the autopilot must anticipate the aircraft's movement and reposition the control surfaces, and/or throttles, to prevent the aircraft from overshooting the desired course and attitude.

To perform the functions just described, the autopilot must monitor various aircraft parameters including airspeed, altitude, pitch, roll, and yaw (see figure 6-1). Navigational aides are also monitored to provide course data. Next, the autopilot will analyze the data to decide if attitude or course adjustments are needed. An autopilot computer is used to analyze the data and output the necessary control information. If adjustment is required, servos are used to move control surfaces and reposition the aircraft. **Servos** are devices used to move the flight control surface in accordance with autopilot command signals. Hence, older autopilot computers were sometimes referred to as **servo amplifiers**. A **follow-up system** is used to inform the autopilot computer that the control surface has changed position. The follow-up system allows the computer to anticipate when the control surfaces should be returned to the neutral position. Since there are three axes in which an aircraft can move (pitch, roll, and yaw), autopilot systems typically contain three channels or subsystems.

The autopilot must be capable of performing all of the necessary functions in a safe and reliable way. Two major safety considerations are given to every autopilot: 1) the system must be able to be quickly and positively disengaged, and 2) the system must be designed so the pilot can overpower the autopilot to manually control the aircraft if needed.

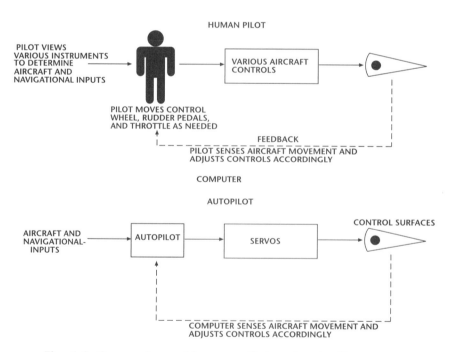

Fig. 6-1 Comparison of human pilot and autopilot.

The autopilot block diagram (figure 6-2) shows the inputs and outputs for a typical three axis autopilot. The autopilot computer receives aircraft inputs from: 1) air data sources; supplied by pitot static pressures, or electronic signals from an air data computer, 2) heading sources; provided by the aircraft's compass system, 3) navigational inputs, such as ILS, DME, or VOR, 4) attitude information from an inertial reference system (IRS) or attitude gyros. The pilot and copilot must have a means of quick disconnect for the autopilot, this is typically controlled through a disconnect push-button on the control yokes.

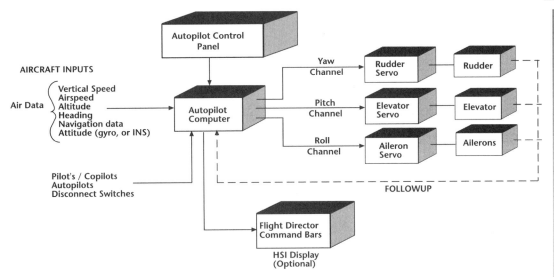

Fig. 6-2 Basic autopilot block diagram.

Each autopilot system will incorporate some means by which the flight crew can input commands. The two panels shown in figure 6-3 are used to control an autopilot for a typical corporate jet aircraft. The top panel is used to engage the autopilot and to control the manual pitch and roll functions. This panel also contains the *AP XFR*, for transfer of a dual autopilot system, and turbulence mode (*TURB*) pushbutton. The panel shown in figure 6-3(b) is used to select the different modes of the autopilot system. From this panel the pilot can select several different modes of operation including navigation (*NAV*) or vertical navigation (*VNAV*). The **navigation** mode allows the autopilot to fly the selected lateral navigational course (ie. control of North, South, East, and West directions). The **vertical navigation** mode allows the autopilot to fly the selected altitude or glide path.

The main autopilot outputs include the three signals used to control the pitch, roll, and yaw servos. As the respective control surfaces move according to autopilot signals, a followup signal is transmitted back to the autopilot computer. The autopilot computer may also have an output dedicated to the flight director control. The **flight director** presents a visual aid to the pilot that is used for manual control of the aircraft. As seen in figure 6-4, the flight director display is typically incorporated in the Attitude Director Indicator (ADI). Since the flight director utilizes many of the same inputs as the autopilot, the two systems often share components. In the example of figure 6-2, the autopilot computer drives the command bars for the flight director display.

Yaw Damping

Virtually all high-speed aircraft are designed with swept back wings. The aerodynamics of a swept back wing causes a stability problem known as **dutch roll**. Dutch roll is basically a slow oscillation of the aircraft about its vertical axis. Correct application of the rudder can prevent dutch roll; however, it requires constant repositioning of the rudder. This process becomes almost impossible for the pilot.

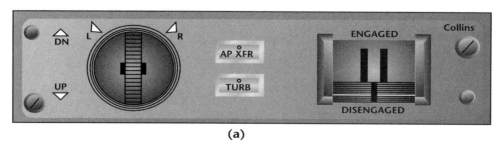

(a)

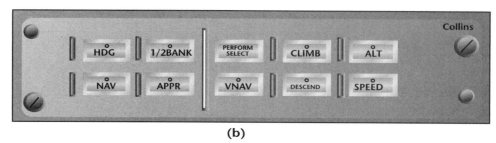

(b)

Fig. 6-3 Typical autopilot control panels from a corporate-type aircraft. (a) used to engage the autopilot and for manual pitch and roll functions; (b) used to select the different modes of the autopilot system. (Rockwell International, Collins Avionics Divisions)

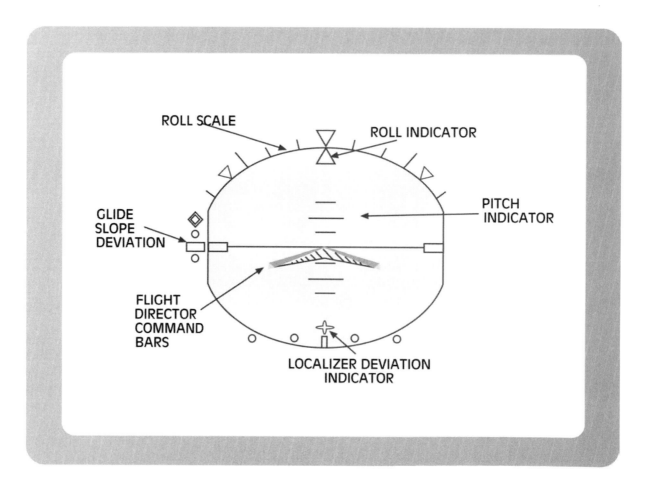

Fig. 6-4 A typical flight director display incorporated on an Attitude Director Indicator (ADI).

The **yaw damper** system is designed to control rudder and eliminate dutch roll. The yaw damper is basically an autopilot component dedicated to rudder control. If the system detects a slip or skid of the aircraft, the rudder is activated to correct this condition; hence, eliminating dutch roll. On most aircraft, the yaw damper is considered independent of the autopilot system, although they may share the same control panels, sensors, and computers. During flight, the autopilot and yaw damper may both be used to position the rudder. The autopilot positions the rudder to coordinate turn activity; the yaw damper positions the rudder to eliminate dutch roll. On most aircraft, the yaw damper can be engaged independent of the autopilot.

AUTOPILOT COMPONENTS

A variety of components are incorporated into every autopilot system. Many of these components or subsystems are not actually part of the autopilot, but are essential to autopilot operation. Older autopilots employ analog systems and mechanical sensors. Newer autopilot components are typically digital and communicate through data bus cable. The following discussion will present many of the basic elements of an autopilot system.

Gyroscopic sensors

There are two common types of gyroscopic sensors used in modern autopilots: **rotating mass gyros** and **ring laser gyros**. Most autopilots use gyro sensors to detect movement of the aircraft. Gyro outputs are also used for reference on certain navigation systems. Gyro systems are both fragile and expensive; it is very important the technician becomes familiar with the system before performing maintenance on gyros.

Rotating mass gyros

Rotating gyros have a tendency to stay stable in space. This effect allows a rotating mass gyro equipped with a rate sensor to detect aircraft motion (see figure 6-5). The output signal from the rate sensor can be sent to an autopilot computer and/or used to stabilize a gimbal platform.

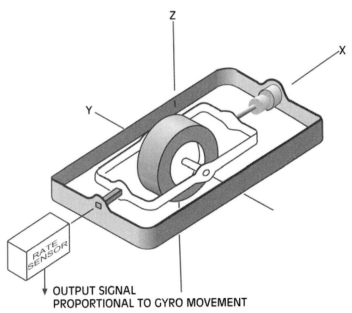

Fig. 6-5 Diagram of a basic rotating mass gyro.

Gimbal platforms

Rotating mass gyros are often used to stabilize acceleration sensors mounted on **gimbal platforms**. A gimbal platform is made rigid in space (parallel to the earth's surface) regardless of the aircraft's attitude. This is accomplished by mounting three rotating gyros to the platform as shown in figure 6-6. One gyro is needed for each axes of the gimbal. If the aircraft's attitude changes, the gyro rate sensor produces an output signal that is proportional to the amount of attitude change. This signal is monitored and amplified by the autopilot computer, then sent to the torquers on the stable platform. The torquer produces the counter-force (typically a magnetic field) needed to stabilize the platform. With the acceleration sensor mounted on a stable platform, an accurate acceleration can be monitored by the system. If the acceleration sensors were not mounted to a gimbal platform, the sensors would measure attitude changes as well as aircraft accelerations.

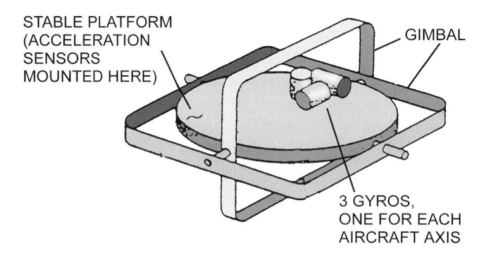

STABLE PLATFORM
(ACCELERATION
SENSORS
MOUNTED HERE)

GIMBAL

3 GYROS,
ONE FOR EACH
AIRCRAFT AXIS

Fig. 6-6 Diagram of a typical gimbal platform.

Laser gyros

The ring laser gyro (RLG) is actually an angular rate sensor and not a gyro in the true sense of the word. Conventional gyros generate a gyroscopic stability through the use of a spinning mass. The gyro's stability is then used to detect aircraft motion. The RLG uses changes in light frequency to measure angular displacement.

The term *laser* stands for "light amplification by stimulated emission of radiation." The RLG system utilizes a helium-neon laser; that is, the laser's light beam is produced through the ionization of a helium-neon gas combination. A typical RLG is shown in figure 6-7. This system produces two laser beams from the same source and circulates them in a contra-rotating triangular path. As shown in figure 6-8, the high-voltage potential (approximately 3,500 V) between the anodes and cathodes produce two light beams traveling in opposite directions. The laser is housed in a glass case, which is drilled with precise holes to allow travel of the light beam. Mirrors are used to reflect each beam around an enclosed triangular area. A prism and detector are installed in one corner of the triangle. The prism reflects the light to allow both laser beams to be measured by the detector.

Fig. 6-7 A ring laser gyro. (Honeywell Inc.)

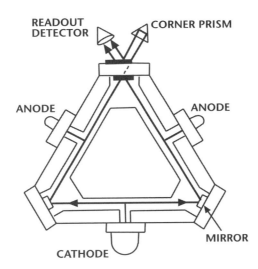

Fig. 6-8 Pictorial diagram of a ring laser gyro.

The resonant frequency of a contained laser is a function of its optical path length. When the RLG is at rest, the two beams have equal travel distances and identical frequencies. When the RLG is subjected to an angular displacement around an axis and perpendicular to the plane of the two beams, one beam has a greater optical path and the other has a shorter optical path. Therefore, the two resonant frequencies of the individual laser beams change. This change in frequency is measured by photosensors and converted to a digital signal. Since the frequency change is proportional to the angular displacement of the unit, the system's digital output signal is a direct function of the angular rate of rotation of the RLG.

The RLG system is typically coupled to a complete navigation system. The digital signals from the RLG can be used to control inertial reference and navigation systems and/or autopilot functions. Each inertial sensor assembly contains three triangular lasers. In figure 6-9, two of the lasers can be seen mounted within the sensor assembly. RGL technology provides a new era of aircraft safety. Since the RGL has no moving parts, it has a much greater reliability than the conventional rotating mass gyro system.

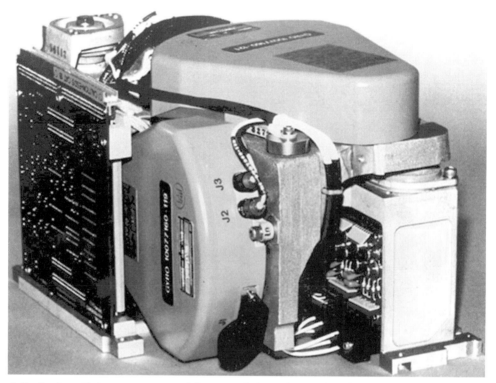

Fig. 6-9 An inertial sensor assembly containing three ring laser gyros. (Honeywell Inc.)

No matter which type of gyro is employed, all three axes of the aircraft must be monitored for a fully functional autopilot. Pitch, roll, and yaw can only be monitored by sensors aligned in the correct position. On most systems, alignment becomes very important during the installation, or removal and replacement of rate sensing LRUs. Each unit must be oriented in the correct position with respect to the rest of the aircraft. For example, if a unit is installed 180° out of alignment, the system could react exactly opposite of what is necessary. Be sure the unit is installed according to the manufacturer's instructions. Many gyro units are interchangeable between locations on the aircraft. Pin programming is used to identify the specific installation location.

Maintenance and Troubleshooting

In general, laser gyro systems are relatively maintenance free, and rotating mass gyros seem to be subject to frequent failures. Every gyro provides some type of electrical output signal. The simplest way to detect the operation of a gyro is to check the output signal. On the most modern systems, the output is probably a digital signal. On many rotating mass gyros, the output is a three-phase AC signal. Many rotating mass gyros are combined with rate sensors. The output signal from the rate sensor is typically a single AC voltage.

On any gyro system, the output signals are relatively low current; therefore, a poor connection can easily create a loss of output. If a gyro's output is inaccurate or missing, check the electrical connections to the unit. Rotating mass gyros all produce a humming noise as the gyro rotates. This noise should be present any time the gyro is active. If the gyro is completely quiet, the unit is defective or there is no power to the gyro assembly. If the gyro assembly makes an unusually loud rumble noise, the gyro bearings are probably worn beyond limits and the unit should be replaced.

All gyros require some warm-up period. Laser gyros must reach a given temperature to stabilize; rotating mass gyros must reach a certain RPM. In many gyro systems, it is important the aircraft remain stationary during this warm-up (initialization) period. Be sure to provide sufficient warm-up time to all gyro systems whenever performing maintenance.

Accelerometers

An **accelerometer** is a device that senses aircraft acceleration. It should be noted that acceleration is a vector force, and therefore is measured in both magnitude and direction. Since an aircraft can accelerate in three directions, a minimum of three accelerometers are used on most installations.

A basic accelerometer resembles a simple pendulum (see figure 6-10 (a)). The pendulum will swing to the right as the aircraft moves forward, and to the left as the aircraft decelerates. Figure 6-10(b) shows the arrangement of a simple aircraft accelerometer. This design incorporates two springs to center the pendulum. As the aircraft moves, the indicator moves in the opposite direction, relative to the aircraft.

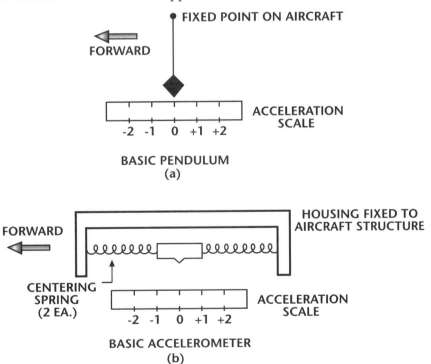

Fig. 6-10 Simple pendulum used to measure acceleration. (a) a basic pendulum; (b) a basic aircraft accelerometer.

An actual accelerometer incorporates a **pick-off** device to convert pendulum movement into an electric signal (see figure 6-11). The electric signal is amplified, and in some cases converted to a digital format. Some accelerometers use a secondary current flow to keep the armature centered over the pick-off coils. A torquer current is used to produce a magnetic field that centers the armature in the null position. This produces an armature that is very stable and increases accelerometer accuracy.

On many transport category aircraft, there are at least two accelerometers to monitor the acceleration of each aircraft axis. The outputs of both accelerometers are combined to provide an extremely accurate measurement. The accelerometers for each axis are usually mounted at opposite ends of the aircraft (ie., the tail and the nose). This also improves accuracy.

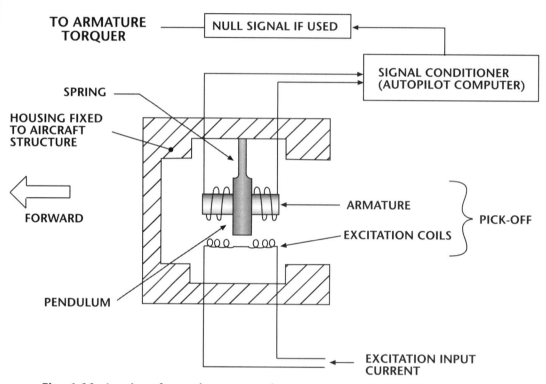

Fig. 6-11 An aircraft accelerometer showing the pick-off unit.

AIR DATA

To fly safely, it is essential to know the aircraft's speed (air speed) and the current elevation (altitude). These items are difficult for the pilot to sense and are best-determined using flight instruments. To measure these basic flight parameters, the air mass surrounding the aircraft must be monitored. The measurement of this air mass is known as **air data**. The three air data elements typically measured are **temperature, static pressure,** and **pitot pressure**.

There are two different temperatures typically measured by an air data system; **static air temperature (SAT)** and **true air temperature (TAT). Static air temperature** is the temperature of the undisturbed air surrounding the aircraft. **true air temperature** is a measure of the air as it is compressed by the moving aircraft. Temperatures are an important air data reference used to improve the accuracy of other parameters.

Static pressure

Static pressure is the absolute pressure of the air that surrounds the aircraft. Static pressure varies inversely with the altitude of the aircraft and also changes with the general atmospheric conditions of the area. On a standard day (59° F) at sea level the static pressure is 29.92 inHg (1013 mb). Static air pressure should be measured in undisturbed air, which is difficult to find near a moving aircraft; therefore, correction factors are often employed when determining static pressure. Static pressure is used to determine the aircraft's altitude and vertical speed.

Pitot pressure

Pitot pressure is an absolute pressure of the air that is "pressed" into the front of the aircraft. With the aircraft at rest the pitot pressure is equal to static pressure. As the aircraft increases speed, the pitot pressure will increase with respect to static pressure. The difference between pitot pressure and static pressure is often referred to as **dynamic pressure**. Dynamic pressure is used to determine the aircraft's airspeed. Figure 6-12 shows the installation locations of typical static ports and pitot probes. Both pilot and temperature probes are typically heated to prevent ice formation. Redundant probes are installed on the opposite side of the aircraft to avoid errors caused by aircraft yaw.

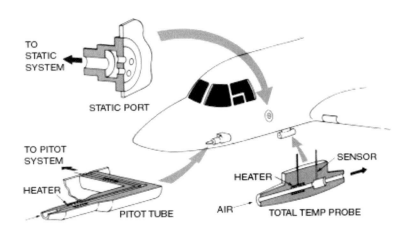

Fig. 6-12 Typical locations for installation of pitot tubes, static ports, and temperature probes. (Rockwell International, Collins Avionics Divisions)

Types of air data systems

There are basically three types of air data systems currently in use: **pneumatic, electro pneumatic**, and **electrical repeater**. Each of these three systems can be connected to an autopilot. A pure pneumatic system relies solely on the static and pitot pressures to drive the instruments (see figure 6-13). The electro pneumatic air data system employs electronic circuitry to provide signals for vertical speed calculations, altitude warnings, and temperature indicators (see figure 6-14). Electronic signals may also be transmitted to other aircraft systems, such as the flight data recorder. Electrical repeater is the newest of the systems and provides the most accuracy. Electrical repeater systems incorporate an air data computer (ADC) which converts all pressure data into electrical signals. The electrical signals are then distributed to electric/electronic instruments for display of the data (see figure 6-15).

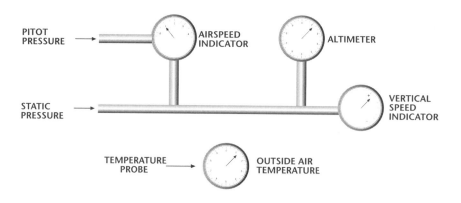

Fig. 6-13 A pneumatic pitot static air data system.

Air Data Computer Systems

An **air data computer (ADC)** system monitors pitot pressure, static pressure and air temperature to determine various parameters, such as airspeed, altitude, and vertical speed. The ADC converts air data information into electrical signals (analog or digital, depending on the system) and transmits the data to the autopilot, flight instruments, and other aircraft systems. Air data computer systems are relatively complex and are therefore typically found on newer corporate-type and transport category aircraft. Autopilots that do not rely on air data systems simply monitor the pitot-static pressures, and air temperature directly. The main advantage of a modern air data system is the increased accuracy, and the weight savings through the elimination of most of the pitot and static plumbing.

An air data computer receives air pressure inputs from both the pitot and static air source, as well as an air temperature input (see figure 6-15). The air temperature input is most likely an analog electric signal produced by a temperature transducer. For most systems, the ADC would receive pitot, static, and air temperature inputs from redundant sources. The ADC will transform the pressure inputs to an electrical signal using pressure transducers. The electrical signals are then sent to the processing circuitry where the data is manipulated into useful information, such as indicated air speed, vertical speed, altitude, etc. The ADC outputs are sent to the autopilot and other systems that require air data information. The components of a typical corporate-type jet aircraft's air data system are shown in figure 6-16.

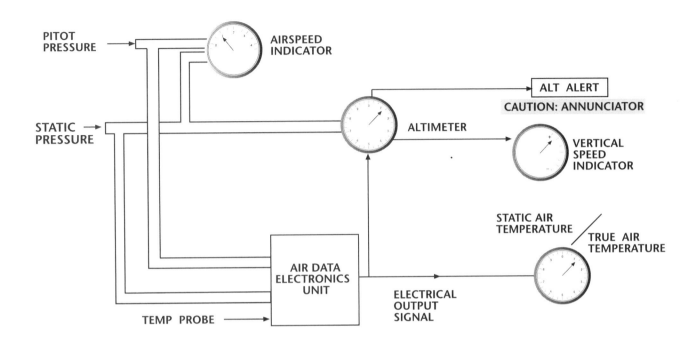

Fig. 6-14 Block diagram of a typical electro pneumatic air data system.

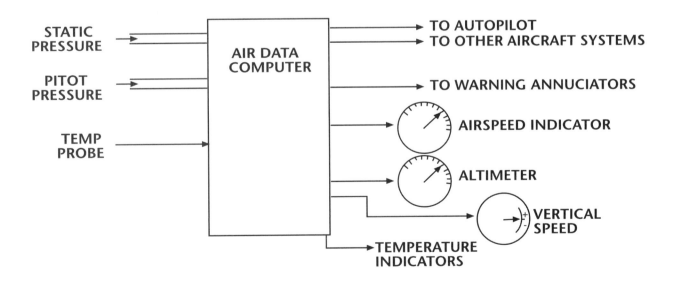

Fig. 6-15 Distribution of air data signals to the various aircraft systems.

ALI-80A
BAROMETRIC ALTIMETER

ASI - 80D
AIR SPEED INDICATOR

PRE - 80A
PRESELECTOR/ALERTER

MSI-80F
MACH SPEED INDICATOR

TAI - 80A
TEMP/TAS INDICATOR

VSI - 80A
VERTICAL SPEED INDICATOR

ADC - 82A
AIR DATA COMPUTER

Fig. 6-16 Components for a typical corporate aircraft air data system. (Rockwell International, Collins Avionics Divisions)

Maintenance & troubleshooting

Modern air data systems incorporate built-in diagnostics that can be accessed through the aircraft's central maintenance computer system or through a specific LRU. If electro mechanical instruments are used to display air data, they often employ a self-test button. Pressing the self-test will cause a specific indication on the instrument. If this indication is not displayed, the instrument should be replaced.

In many cases, built-in diagnostics can detect the fault on the first attempt. However, even on modern systems there are many mechanical faults to the system plumbing that cannot be detected by the diagnostics. Faults with partially clogged pitot tubes or static ports or leaking plumbing lines can give inaccurate reading on air data instruments. Static ports are especially prone to clogs since their openings are relatively small. Bent or misaligned pitot tubes will also create accuracy problems. On dual systems, if only one probe is damaged the ADC may report a disagree message on the fault diagnostics. Any time an air data system malfunctions, be sure to inspect the static and pitot probes for damage or clogs.

COMPASS SYSTEMS

Any autopilot that performs basic navigation functions must receive data from the ship's compass system. The aircraft's magnetic compass is too inaccurate; therefore, the autopilot system must rely on an electrical/electronic compass system. The **Flux Gate Compass** system employs one or more remote sensors to produce an electric signal that can be used to determine the aircraft's position relative to magnetic north.

The remote sensor used in a Flux Gate Compass system is called a **flux detector** or **flux valve** (see figure 6-17). The flux detector receives a constant input of 115 VAC, or 28 VAC, at 400 Hz. The output voltage is a function of the alignment of the detector with the earth's magnetic field. The sensing unit in the flux detector is the **flux valve**. The flux valve is comprised of a three-spoked frame with an output winding on each spoke (see figure 6-18). An excitation winding is located in the center of the flux frame. The frame is suspended in a sealed case on a universal joint, which allows it to pivot and remain relatively stable at different aircraft attitudes. The unit is surrounded by oil to dampen the flux frame movement.

Fig. 6-17 A flux detector typical of those found on corporate aircraft.

The operation of the flux detector relies on the interaction of the earth's magnetic field and the magnetic field induced in the flux frame by the 400 Hz excitation coil. Without the earth's magnetic field, each output coil of the flux frame would produce an equal voltage. As the aircraft moves with respect to magnetic north, different legs of the flux frame become saturated with magnetism. As the saturation of the frame changes, different voltages will be induced in the three output coils. Therefore, the output coils produce a three-phase AC voltage that changes characteristics relative to the aircraft's heading.

The output signals from the flux detector can be sent directly to a remote slaved compass such as those found in Horizontal Situation Indicators (HSI). The flux detector output signals may be sent to an electronic circuit where they are amplified and distributed to various systems that will use the data. Systems that monitor flux detector signals include the autopilot system, the flight data recorder, the Radio Magnetic Indicator (RMI), and/or the Electronic Flight Instrument Systems (EFIS). On some aircraft, such as the B-747-100, the output of the flux detector is sent to a compass coupler. The compass coupler contains a mechanical servo/synchro combination. The output form the compass coupler is directed through relays to the RMI or HSI. The relays are used for reversionary switching to select which source would be used to drive which RMI or HSI.

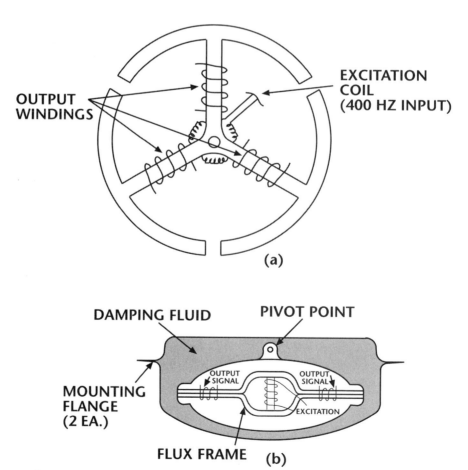

Fig. 6-18 Flux detector internal components. (a) the flux frame; (b) the flux frame and housing assembly

Maintenance and troubleshooting

In most installations, the flux detector is placed in the aircraft's wing to isolate it from magnetic interference caused by the other electrical systems (see figure 6-19). The mounting structure for the flux detector contains an adjustment that is used to ensure the unit is correctly aligned. This alignment becomes an important maintenance item when replacing the unit.

In general, alignment of the flux detector is done by placing the aircraft facing a known compass point and moving the flux detector until the remote compass (on the HSI or EFIS) reads the correct compass heading. This is a simplified explanation of the procedures since most systems contain two flux detectors. Always follow the manufacturers recommended procedures during flux detector alignment.

To isolate a defective component in the compass system, test all LRUs containing built-in test equipment. For example, the compass coupler on the B-747-100 contains a fault ball that is visible any time the unit detects an internal failure. Considering the entire slaved compass system, the compass coupler is probably the most likely to fail. If the aircraft contains mechanical switching relays in the compass system, be sure to include these items as possible fault suspects. The relay coils can fail or the contact points can become pitted and create excess signal loss. Troubleshooting the RMI or HSI typically becomes a remove and replace procedure. That is, the suspect component is swapped with a known operable unit and the system is tested.

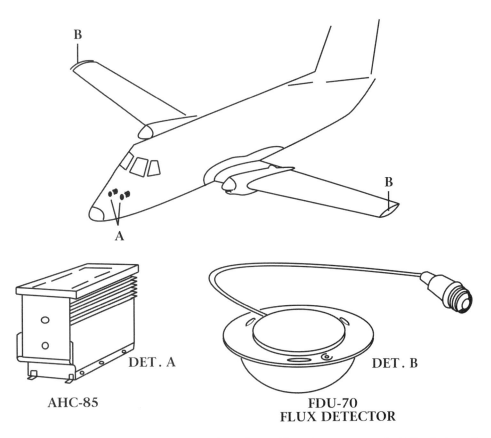

DET. A   DET. B

AHC-85   FDU-70
                         FLUX DETECTOR

Fig. 6-19   Typical installation locations for a flux detector and attitude heading computer. (Rockwell International, Collins Avionics Divisions)

CAUTION: Whenever testing any flux gate compass system, be sure the aircraft is away from items which may interfere with the earth's magnetic field. The test must be done outside the hangar and away from other aircraft, cars, railroad tracks, and power cables. Also be aware that many metal items are not visible, such as, buried power lines, fuel tanks, or concrete reinforcing rods. Placing the aircraft on a compass rose is the best way to test the system. These general practices must also be observed by the pilot during a preflight test. If a problem occurs with the flux gate system on the ground, be sure to retest the compass system in a known environment.

Compass system tests should be performed with various electrical equipment both on and off. If the slaved compass is affected by the operation of certain electrical equipment the problem must be fixed. First check to see if all wires near the flux detector are properly shielded. If proper shielding fails to produce the desired results, reroute electrical wires (or equipment) away from the flux detector to illuminate the error.

INERTIAL REFERENCE SYSTEMS

An **Inertial Reference System (IRS)** is a combination of laser gyros and accelerometers used to sense the aircraft's angular rates and accelerations. IRSs are relatively expensive and typically found only on corporate, transport or military type aircraft. The LRUs of a typical inertial reference system are shown in figure 6-20. The laser gyros and accelerometers are installed in the **Inertial Reference Unit (IRU)**, which is typically installed in the aircraft's equipment bay. The IRU also contains the computer circuitry for signal processing and system interfacing.

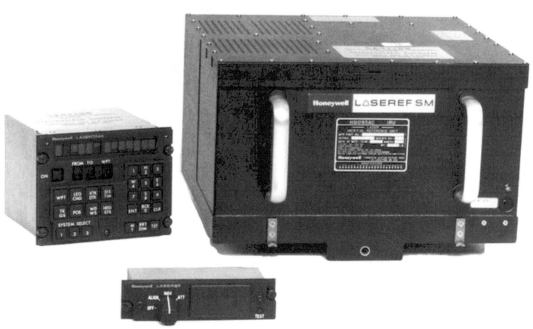

Fig. 6-20 Components of a typical laser inertial reference system (Honeywell Inc.)

The data produced by an IRS is used in conjunction with a total autoflight system. The IRS data is typically combined with air data outputs to compute:

1) attitude (pitch, roll, yaw), 2) angular rate changes (pitch, roll, yaw), 3) aircraft velocity, 4) course track angle, 5) inertial altitude, 6) linear accelerations, 7) magnetic heading, 8) true heading, 9) position (latitude and longitude), 10) vertical speed, 11) wind speed, and 12) wind direction.

The output data from the IRS is a primary input for a modern autoflight system. IRS outputs are also sent to electronic flight instrument systems for display of attitude and navigational data. IRS data is also sent to the flight data recorder along with other aircraft systems.

Many of the latest IRSs are so accurate, the need for a Flux Gate Compass is eliminated. For example, aircraft such as the B-757/767, the B-747-400, and the A-320 use the IRS for magnetic heading data. The IRS sends magnetic compass data to the RMI and/or EFIS for display to the flight crew.

Initialization

Since an IRS can only measure **changes** in position, the unit must be given a starting reference point. The procedure used to provide the IRS with an initial latitude and longitude is called **initialization**. Initialization typically occurs at the aircraft gate before the first flight of the day. If the aircraft has not been moved overnight, the position in memory can be used. If the aircraft has been towed to a new location, the crew must enter the correct latitude and longitude into the IRS. This is typically accomplished using a multifunction alphanumeric keyboard.

During initialization, the IRS accelerometers measure the direction of the earth's gravity force to determine the aircraft's **local vertical.** Local vertical is a direction perpendicular to the rotational axis of the earth that intersects the aircraft's position (see figure 6-21). During initialization, the IRS rate sensors measure the speed and direction (relative to the aircraft) of the earth's rotation. This, along with the latitude, longitude, and local vertical allow the system to determine true north. At the completion of the initialization process, the IRS computer contains the necessary data to compute the aircraft's current position and heading. It should be noted that initialization takes approximately five to ten minutes and the aircraft cannot be moved during this time.

Theory of operation

Each IRS unit is made up of three laser gyros and three accelerometers. One each of these units is aligned with the pitch, roll, and yaw axis of the aircraft (see figure 6-22). Figure 6-23 shows the three laser gyro assemblies and accelerometers within a typical IRU. The three gyros measure angular displacement about their respective axis (pitch, roll, and yaw). The accelerometers are used to measure the rate of acceleration about each axis. Each of the three axes must be monitored since the aircraft travels in three-dimensional space. Also, most aircraft will contain two or three IRUs, each with the capability to monitor all three axes of the aircraft. Using multiple IRUs provides the redundancy needed for safety and reliability.

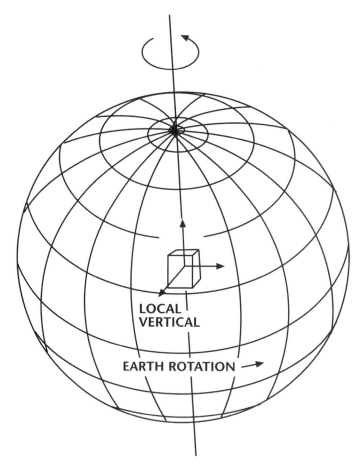

Fig. 6-21 Local vertical is measured between the aircraft's location and the rotational axis of the earth.

Once the IRU has been initialized, the system knows where it is located (in all three dimensions) and which direction the aircraft is facing (current heading). As the aircraft moves in any direction from its initial position, the IRS will sense the movement and compute the new location and heading using a high speed processor. For explanation purposes, let us assume the aircraft is stationary at a given point in space (see point A in figure 6-24). If the longitudinal accelerometer measures an acceleration of 2 ft/sec^2; this mean the aircraft is accelerating forward. After ten seconds the aircraft would be flying at a velocity of 20 ft/sec (10 s x 2 ft/sec^2 = 20 ft/sec). Assume the aircraft stops accelerating and the velocity remains constant at point B. If the aircraft continues to fly with a velocity of 20 ft/sec, for 100 seconds the aircraft's new location (point C) is 2,000 feet from point B. Distance equals velocity multiplied by time (20 ft/s x 100 s = 2,000 ft).

The IRS computer performs similar calculations for the angular rate changes measured by the laser gyros. Let us assume the IRS detects a yaw rate of 5° per second for 15 seconds. The computer would determine that the heading has changed 75° from the aircraft's original heading. The IRS computer continuously performs acceleration and angular rate calculations for all three axes. By measuring both accelerations and angular rates, the IRS can provide a constant update on the aircraft's location and heading. Of course, heading and location information are constantly being compared on multiple IRU systems to ensure accuracy.

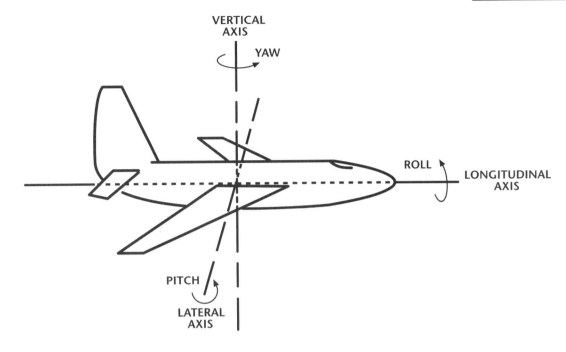

Fig. 6-22   The three axis of the aircraft; one IRS unit must be aligned with each axis.

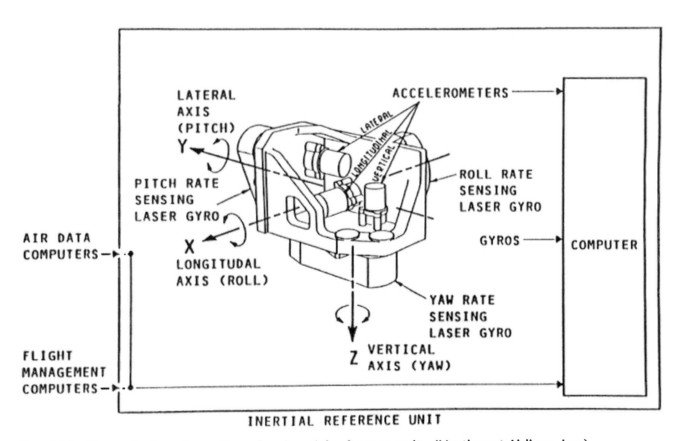

Fig. 6-23   The typical configuration of an inertial reference unit. (Northwest Airlines, Inc.)

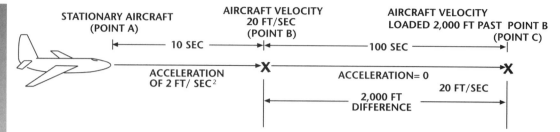

Fig. 6-24 The IRS measures acceleration and time to calculate the aircraft's change of position.

There are several other factors that can affect the accuracy of the IRS. For example: 1) the earth's rotation is approximately 15.04° per hour, 2) the spherical shape of the earth means aircraft do not travel in a straight line over the surface, 3) the laser gyro is subject to drift over time. This drift is much less than rotating mass gyros; however, it is still important to consider. To compensate for these inherent errors, the IRS software is programmed to make the necessary corrections while processing the data. On many systems, another means of ensuring accuracy is to periodically cross-reference location data with other navigational aids, such as, a GPS or DME signal.

Maintenance and troubleshooting

Most aircraft that employ an IRS also contain some type of Centralized Maintenance Computer System (CMCS). IRS troubleshooting is typically done using this system. Faults are stored in a nonvolatile memory, and displayed on a CRT when requested by the technician. In most cases, the aircraft will contain two or three IRUs. Each of these units is interchangeable and can be swapped to help identify a defective unit, or reversionary switching can be used to "swap" units from the flight deck. Whenever removing any IRU, make sure to handle the unit with care; the IRUs are fragile and can be severely damaged if dropped. Any unit that has been dropped is **Unairworthy**. Whenever shipping an IRU, be sure to use the appropriate shipping container to help protect the unit.

Whenever testing an IRS, be sure to allow proper time for the aircraft to initialize. The aircraft must remain stable during the initialization process. Large wind gusts or maintenance being performed on the aircraft may upset the initialization procedure. In this case, the procedure should be repeated on a "quiet" aircraft. Never condemn an IRS that will not align on the first attempt, it may be caused by a moving aircraft.

Installation of the IRS in the correct position with respect to aircraft axes is also an important consideration. For example, if the IRS unit was installed with the gyros and accelerometers out of alignment, the system could not produce accurate data. This is typically not a problem for removal and replacement since the LRU is installed in a rack that is permanently mounted to the aircraft. However, if the rack should become bent, cracked or somehow misaligned, the IRS will not work properly.

INERTIAL NAVIGATION SYSTEMS

A modern **inertial navigation system (INS)** uses airborne equipment for aircraft navigation without relying on external radio signals. Laser gyros and accelerometers provide three dimensional navigation capabilities. This system is employed mainly by the military. The major advantage of the system is that it requires no external navigation aids. All the equipment for worldwide navigation is contained in the INS.

Many older transport category aircraft, such as the DC-10, early B-747s, and the L-1011 employ an INS which utilizes rotating gyros, gimbal platforms, and accelerometers to sense aircraft position. This system is called INS since it is responsible for coordinating the aircraft's navigational parameters, including flight plan and waypoint selection. The INS can be interfaced with the autopilot or flight director system to steer the aircraft. The difference between and INS and IRS is that the INS provides waypoint and flight plan capabilities; a modern day IRS must work in conjunction with a Flight Management System (FMS) to provide these functions.

SERVOS

A **servo** is a device used to apply a force to the aircraft's control surface in response to an autopilot command. There are basically three types of servos: pneumatic, electrical, and hydraulic. Each servo must incorporate some type of clutch mechanism so the pilot can override the autopilot command. They may also contain a feedback system that provides a return signal to the autopilot computer.

Pneumatic Servos

Pneumatic servos are vacuum actuated units used on simple autopilots for light aircraft. As seen in figure 6-25, the pneumatic servo operates using a vacuum applied to the servo diaphragm. The autopilot computer controls the vacuum. Two servos are required for each control surface. A bridle cable is used to connect the servo to the control surface cable. Pneumatic servos offer limited range of travel and provide a relatively weak actuating force; therefore, pneumatic servos have limited use.

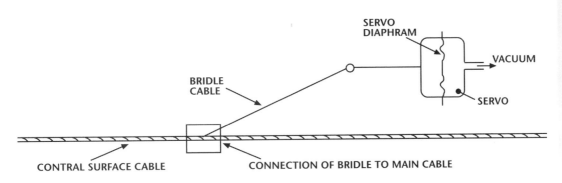

Fig. 6-25 Diagram of a simple pneumatic servo.

Electric Servos

Electric servos utilize an electric motor and clutch assembly to move the aircraft's control surface according to autopilot commands (see figure 6-26). Due to their reliability and excellent torque production, electric servos are commonly found on all types of general aviation aircraft, including corporate-type turbine and turboprop aircraft. A bridle cable is installed between the servo's capstan assembly and the control surface cable. The capstan is used to wind/unwind the bridle cable; hence, moving the control surface. Figure 6-27 shows a capstan with the test fixtures installed in preparation for adjustment of the slip clutch assembly. This figure also shows the torque adjustment nut for the slip clutch. The slip clutch gives the pilot the ability to overpower the unit in the event of a malfunction.

Fig. 6-26 A typical electric servo. (Rockwell International, Collins Avionics Divisions)

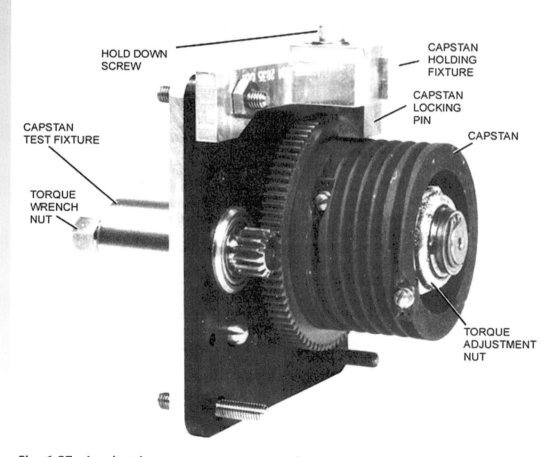

Fig. 6-27 An electric servo capstan mounted to the capstan test fixture. (Rockwell International, Collins Avionics Divisions)

Most electric servos use some type of clutch assembly to connect the servo motor to the capstan. During manual operation of the controls, the clutch is disengaged and the capstan moves freely. During autopilot operation, the clutch is engaged and the capstan is connected to the servo motor. The electric motor and clutch assembly typically operate on direct current; however, some transport category aircraft may employ AC motors. In order to keep capstan RPM relatively low while the motor operates at high speed, the unit contains a gear reduction assembly. The gear assembly is typically self-lubricating and requires no regular maintenance.

Hydraulic Servos

Hydraulic servos are the most powerful type of servo actuator; hence, these units are typically used on transport category aircraft. Ever since the B-727 and DC-9 took to the air, transport category aircraft have employed hydraulically operated control surfaces. In brief, the systems operate using an engine driven hydraulic pump and employ a control valve to route hydraulic fluid to a control surface actuator. The control surface actuator is mechanically linked to the control surface. On most transport category aircraft, the control surface actuator is linked to the control wheel and rudder pedals through control cables. On the newest aircraft, like the A-320 and B-777, the control wheel (or side stick controller) and rudder pedals are connected to the control surfaces via electrical wiring and computer circuits.

The Airbus A-320 hydraulic servos operate in two different modes: active and damping. As seen in figure 6-28, the active mode is employed when the servo valve is pressurized and the solenoid valve is energized by the Elevator Aileron Computer (ELAC). In the active mode, hydraulic fluid is controlled by the servo valve and directed through the mode selector valve to the aileron actuator.

In the damping mode, the actuator follows control surface movement as hydraulic fluid is allowed to flow through the restricting orifice. The servo is in damping mode whenever the solenoid is deenergized or hydraulic pressure is not supplied to the servo valve. In this situation, the mode selector valve moves to the right via spring force and connects the actuator hydraulic fluid to the restricting orifice. It should be noted that the mode selector valve and the main aileron actuator both produce a feedback signal to the computer (ELAC 1).

Although there are a large variety of hydraulic servos for transport category aircraft, they all operate in a similar fashion. Each hydraulic servo will contain some type of servo (or control) valve, typically actuated by an electric solenoid. The control valve will move hydraulic fluid to the main actuator in order to move the control surface. Every unit will also contain some type of bypass or damping mode in the event of system failure. On fly-by-wire aircraft, the flight deck controls are moved via an artificial force produced by the autoflight computer(s). On more traditional aircraft, the control wheel and rudder pedals move via a cable linked to the actuators.

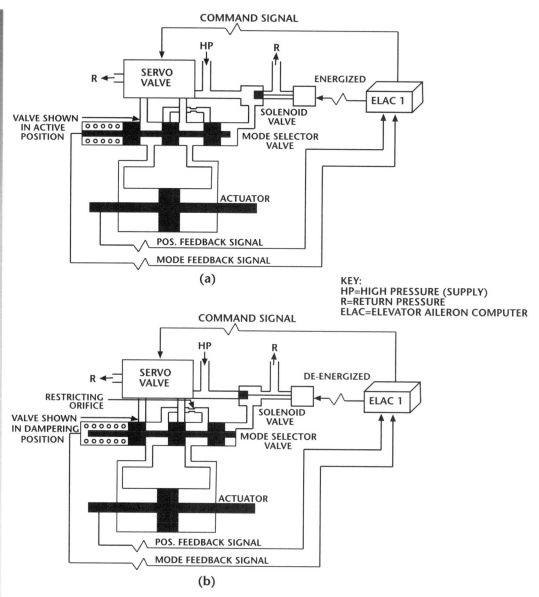

Fig. 6-28 Diagram of a typical hydraulic servo and related control circuitry; (a) active mode, (b) damping mode.

Servo Feedback Systems

As mentioned earlier, all servo systems must contain some type of feedback circuit to inform the autopilot computer that the servo has moved. The feedback system produces an electrical signal that is directly proportional to the movement of the servo actuator. There are two common devices used to generate the feedback signal: an **AC synchro**, and a **differential transducer**. Synchros are typically employed on electric servos and hydraulic servos used in conjunction with analog autopilot systems. State-of-the-art autopilot systems found on transport category aircraft often employ differential transducer feedback systems.

Synchro Systems

The most common **autopilot feedback synchro** is a transformer-like device that monitors angular displacement using a stationary primary winding and pivoting secondary winding. As shown in figure 6-29, the primary winding receives an input voltage of 26 V 400 Hz AC. The output voltage of the secondary is a function of the angular position of the secondary winding. Figure 6-30, position #1, shows the secondary winding in the null position, or perpendicular to the primary winding. In this position, no voltage is induced in the secondary. As the secondary rotates clockwise, the voltage induced in the secondary increases until the secondary is parallel to the primary (position #4). The voltage then decreases as the rotor continues to turn clockwise. A "second null" is reached when the rotor becomes horizontal to the primary once again (position #7). It should be noted that the secondary voltage is in phase with the primary voltage for rotor positions 2 through 6.

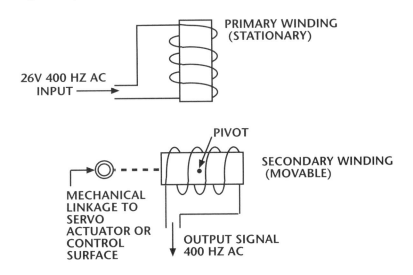

Fig. 6-29 Components of a typical autopilot feedback synchro.

As shown in figure 6-30, when the secondary winding rotates past the "second null" (position #7), the output voltage is 180° out of phase with the primary voltage. The voltage value continues to change as the secondary continues to rotate clockwise. The out of phase condition exists until the rotor reaches the "first null" position once again. The output voltage is 180° out of phase with respect to the input voltage in positions 8 through 12.

AC synchros provide excellent feedback signals for many autopilot systems. The phase shift principle, discussed above, allows for accurate measurements of even small control surface movement. When placed in the null position, any movement clockwise or counter clockwise is easy to measure due to the phase shift and voltage change. The most accurate measurements using a synchro are therefore obtained near the null positions. On most systems, the synchro rotor is connected to the servo output or control surface through a mechanical linkage, hence the synchro rotor moves in unison with the control surface.

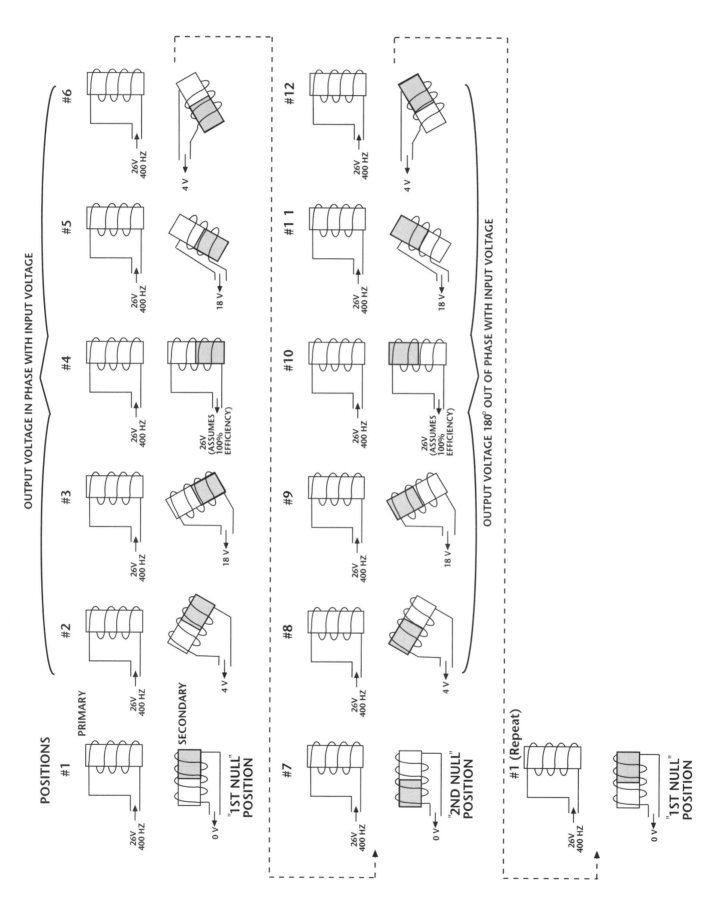

Fig. 6-30 Voltage and phase relationship of an autopilot feedback synchro as it rotates 360°.

Troubleshooting Synchro Systems

Most synchro systems are fairly reliable. The electrical components are simply wire coils and therefore seldom fail. The secondary pivot bearing can fail or become worn which causes inaccurate feedback signals. Likewise, if the mechanical linkage connecting the synchro becomes worn or binds during movement, inaccurate signals will result. The mechanical linkage and pivoting secondary coil are critical and must have free movement. On many systems, adjustment of the synchro to the null position is critical for proper operation. Many electric servos drive the synchro to the null position prior to engaging the servo clutch. This ensures the synchro "starts" in the null each time the autopilot is engaged.

The electrical system of a synchro can be tested for the proper input voltage to the primary. In most cases, the input is 26V 400Hz alternating current. The synchro primary and secondary coils can be tested for continuity and shorts to ground. When measuring continuity, it is critical that the coil resistance be within specifications. A change in resistance of just a few ohms can create inaccurate readings most likely resulting from a breakdown of the coil's insulation. If an insulation breakdown is suspected, be sure to monitor the system closely during the next several hours of operation.

Differential transducers

The differential transducer is typically used to provide a feedback signal from hydraulic servos. There are two common types of transducers used for autopilot feedback systems, the **Linear Voltage Differential Transducer (LVDT)** and the **Rotary Voltage Differential Transducer (RVDT)**. LVDTs and RVDTs produce a relatively weak electrical signal and are found on modern digital autoflight systems. Since they are used in conjunction with hydraulic servo systems, both LVDTs and RVDTs are typically found on transport category and high performance corporate-type aircraft. LVDTs and RVDTs are also found in other (non-autopilot) systems to measure position or rate of motion. For example, the Airbus A-320 employs an LVDT on the turbine engine to measure stator vain position. The nose wheel steering system of the A-320 employs RVDTs to measure nose wheel position.

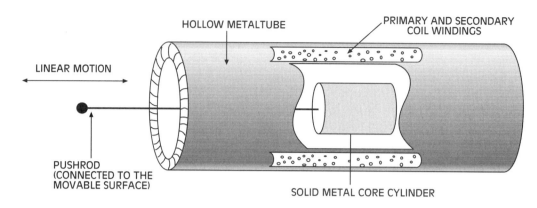

Fig. 6-31 Cut-a-way of a LVDT (Linear Voltage Differential Transducer).

The increasing popularity of LVDTs and RVDTs stems from the simplicity of their design. As seen in figure 6-31, the LVDT consists of a hollow metallic tube and a solid metal cylinder that is allowed to slide inside the tube. Around the tube are two electrical windings, a primary and secondary, similar to a transformer. A push rod is used to connect the solid metal cylinder to the movable object that is being monitored.

An LVDT (or RVDT) is a mutual inductive device. The primary winding is flanked by two secondary windings as shown in figure 6-32. The secondary windings are wired to form a series opposing circuit. The primary receives an alternating current. This AC will induce a voltage into both secondary windings. If the core material is exactly centered, the output signals from the secondary windings will cancel. As the core is displaced from center, the output signal increases in amplitude. The phase of the AC output signal (with respect to the input signal) is determined by the direction of core displacement. Hence, the transducer can measure both direction and magnitude of any movement. Figure 6-33 shows the relationship between core position and the secondary output signal.

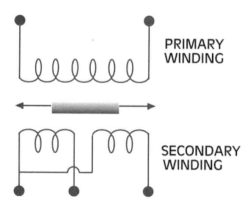

Fig. 6-32 Wiring diagram of a LVDT.

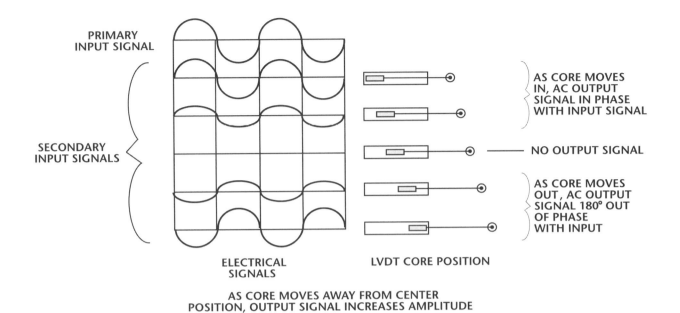

PRIMARY
INPUT SIGNAL

SECONDARY
INPUT SIGNALS

ELECTRICAL
SIGNALS

LVDT CORE POSITION

AS CORE MOVES
IN, AC OUTPUT
SIGNAL IN PHASE
WITH INPUT SIGNAL

NO OUTPUT SIGNAL

AS CORE MOVES
OUT, AC OUTPUT
SIGNAL 180° OUT
OF PHASE
WITH INPUT

AS CORE MOVES AWAY FROM CENTER
POSITION, OUTPUT SIGNAL INCREASES AMPLITUDE

Fig. 6-33   The relationship between core position and LVDT output.

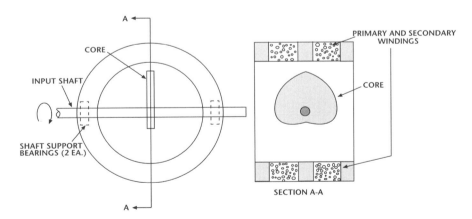

A

CORE

INPUT SHAFT

SHAFT SUPPORT
BEARINGS (2 EA.)

A

PRIMARY AND SECONDARY
WINDINGS

CORE

SECTION A-A

Fig. 6-34   Internal components of a typical RVDT.

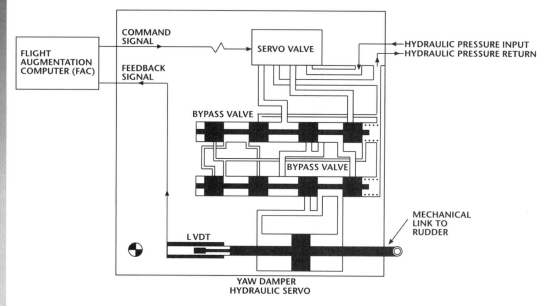

Fig. 6-35   An LVDT used in a Yaw Damper servo.

An RVDT operates similarly to a LVDT except it is designed to detect rotational movement. The RVDT contains a heart-shaped core material that rotates within a hollow tube (see figure 6-34). As the core is rotated, it changes the output voltage and phase of the secondary. The rotational movement measured by an RVDT is typically 120° or less, and the highest resolution is obtained in the first 40° of rotation. An RVDT contains two bearing assemblies, required to support the input shaft.

Both LVDTs and RVDTs are always used in conjunction with some type of electronics circuitry. The circuit is used to interpret the output signals of the LVDT or RVDT. On many aircraft, the secondary's output signal is sent to an LRU, which converts the AC voltage and phase relationships into usable data. Figure 6-35 shows a hydraulically operated yaw damper servo containing an LVDT. The output signal from the LVDT is sent directly to the Flight Augmentation Computer (FAC).

Troubleshooting LVDTs and RVDTs
Both LVDTs and RVDTs are relatively maintenance free. The LVDT contains only one moving part (the core), which is typically supported so there is no contact between the core and the coil housing. The RVDT core is supported by two bearings that have extremely long life due to the light loads on the input shaft. Both LVDTs and RVDTs therefore have virtually infinite mechanical life unless damaged by some external force.

Electrically, the primary or secondary winding of the transducer may fail due to an open or shorted circuit. In many cases, an ohmmeter can be used to detect these failures. Simply disconnect the transducer from the aircraft wiring and perform a continuity test of the coils. Opens most often occur due to induced stress on the windings caused by vibration, or a failed solder or crimped connection. However, even these failures are rare and the MTBF (mean time before failure) of a typical aircraft quality transducer is over one million hours.

Another means of troubleshooting an LVDT or RVDT is to measure the input and/or output voltage of the transducer. The input signal must be an AC voltage within the limits established by the manufacturer. A high impedance voltmeter must be used for this test to ensure the meter does not distort the signal being measured. The output signal from the secondary winding is determined by the position of the core. A dual channel oscilloscope can be used to show the **voltage** and **phase relationship** of the input and output signals. A voltmeter can be used to measure the transducer output voltage as the core changes position. This test is typically sufficient since it is virtually impossible to change the phase relationship of the output signal once the transducer is installed correctly.

As noted earlier, the output voltage of the secondary should be zero when the core is exactly centered (located at the null position). This condition rarely exists in the real world since the excitation voltage often contains high-level harmonics, which induce stray voltages into the secondary. Whenever measuring the output voltage at the null position, remember that a small AC signal (approximately 0.25% of maximum) may be acceptable. If the null position output voltage exceeds this amount, be sure to check the purity of the input voltage to the primary.

On transport category aircraft, the troubleshooting process for LVDTs and RVDTs often becomes simplified through the use of built in test equipment (BITE) or central maintenance systems. The integrated test equipment can be used to monitor the output of the transducer as the unit is moved through its operating range. On many aircraft, a comparison is made of input and output signals to verify correct operation. For example, on the A-320, the Centralized Fault Display System (CFDS) monitors engine stator vain position (see figure 6-36). The CFDS checks the input signal to the stator vane actuator and compares that to the output signal from the RVDT, which monitors the stator vane position. Using the CDFS, the technician can read command channel and monitor channel signals on a CRT display. The stator vane's position is measured in angular degrees.

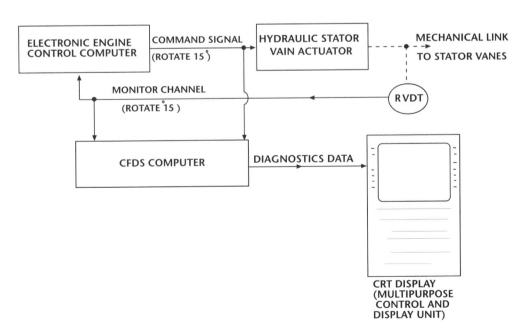

Fig. 6-36 RVDT output signals monitored by the centralized fault display system.

On this system, a 1° tolerance is allowable. It should be noted the CFDS also performs continuous fault monitoring of the stator vane positioning system. If a fault were detected during flight, the CFDS would take the necessary corrective actions and record the failure in memory for later recall. A transducer system found on the Boeing 767 is used to monitor aileron travel as specified by the autoflight computer. On this aircraft, the CMC (Central Maintenance Computer) monitors signals from three LVDTs. The technician can display transducer output on a CRT for comparison purposes.

Whenever troubleshooting an LVDT or RVDT, always remember that these units are extremely reliable. In most cases, the associated wiring or electrical connectors are more likely to fail than the transducer itself. However, if a transducer is to be replaced in the field, take caution to ensure the proper installation of the new unit. Note any markings on the core and/or housing assembly. Be sure the core is installed in the correct configuration. Some LVDTs and RVDTs can only be replaced at an overhaul facility. In this case, if the transducer has failed, the entire autopilot servo assembly must be replaced in the field. Whenever changing any transducer or servo assembly, always follow the manufacturer's instructions on installation and rigging very carefully. Some installations may also require a flight test of the autopilot system.

Tachometer Generators

Tachometer generators (or **tach generators**) are often used in electric servo systems as rate sensors. The tach generator measures the rotational speed of the electric motor and provides feedback to the servo amplifier (autopilot computer). This feedback signal is typically used to regulate and limit motor speed. The tach generator consists of a permanent magnet and armature assembly. The generator spins in direct relationship to the servo motor. Hence, the electrical signal produced by the generator is directly proportional to the motor's movement. A typical tach generator installation is shown in figure 6-37. Tach generators can easily be tested for correct operation by measuring the output voltage as the generator spins. If the generator produces inadequate voltage it must be replaced.

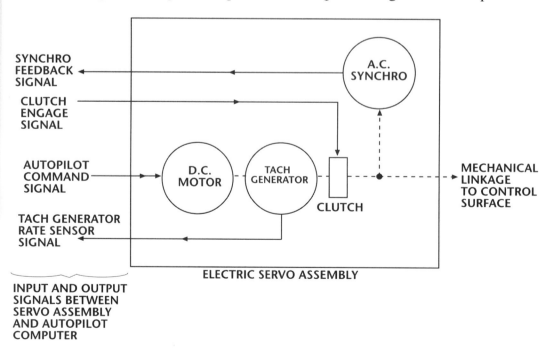

Fig. 6-37 An electric servo assembly containing a tach generator.

COLLINS APS-85 AUTOPILOT SYSTEM

General description and operation

The **APS-85** is a typical digital autopilot system found on high performance corporate-type aircraft. The system includes both autopilot and yaw dampening capabilities. The components include a mode select panel (two panels are used for a dual system), a flight control computer, an autopilot panel, three primary servos, and three servo mounts (see figure 6-38). The APS-85 **Flight Control Computer (FCC)** is a dual channel system that provides redundancy for the autopilot. The FCC, located in the equipment rack, is cooled with forced-air. The three servos are mounted throughout the aircraft in appropriate locations for their respective control surfaces. The control panels are located on the flight deck and are accessible to both pilots.

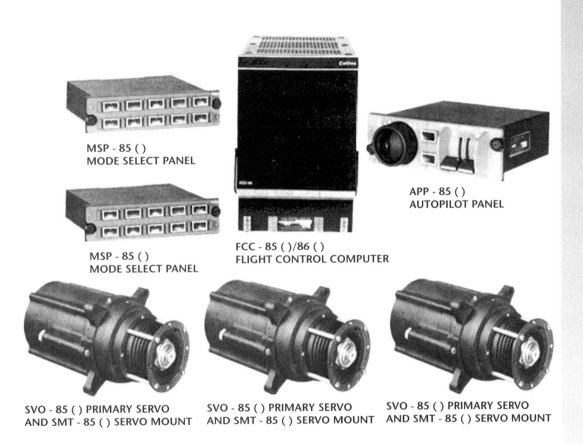

MSP - 85 ( )
MODE SELECT PANEL

MSP - 85 ( )
MODE SELECT PANEL

FCC - 85 ( )/86 ( )
FLIGHT CONTROL COMPUTER

APP - 85 ( )
AUTOPILOT PANEL

SVO - 85 ( ) PRIMARY SERVO
AND SMT - 85 ( ) SERVO MOUNT

SVO - 85 ( ) PRIMARY SERVO
AND SMT - 85 ( ) SERVO MOUNT

SVO - 85 ( ) PRIMARY SERVO
AND SMT - 85 ( ) SERVO MOUNT

Fig. 6-38 System components of the APS-85 autopilot. (Rockwell International, Collins Avionics Divisions)

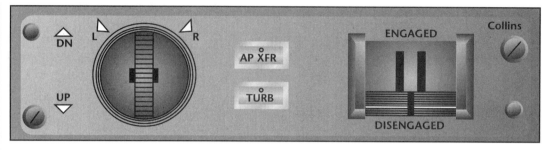

Fig. 6-39 **The APS-85 autopilot panel. (Rockwell International, Collins Avionics Divisions)**

Autopilot panel

The autopilot panel (APP) shown in figure 6-39 contains the main controls for the system. The autopilot and yaw damper switches are guarded levers and must be raised to engage the respective systems. On some aircraft, the autopilot may be engaged independent of the yaw damper; on other aircraft, both systems must be engaged simultaneously. Prior to activating the autopilot/yaw damper, the FCC monitors the system for faults. If a fault is detected, the FCC will not activate the autopilot/yaw damper. Whenever the autopilot/yaw damper is engaged, the appropriate message is displayed by EFIS.

The APP **pitch wheel** is a spring-loaded rotary switch (see figure 6-39). Moving the pitch wheel up or down is used to modify the vertical reference being flown by the autopilot. As a new vertical reference is entered, the value is displayed by EFIS. The **turn knob** is a bidirectional switch used to initiate a roll mode and define a given roll rate. The turn knob is inoperative while the autopilot is in the approach mode. The autopilot transfer (*AP XFR*) switch is used on systems that employ a dual flight guidance system. The transfer switch is used to shift from the left to the right FCC. The turbulence (*TURB*) push-button switch is used to "soften" the ride when flying through rough air. When in the turbulence mode, the FCC lowers autopilot gain signals. This degrades the intensity of control surface movement.

Mode select panel

The mode select panel (MSP) consists of ten push-button switches used to control the various autopilot modes (see figure 6-40). The various mode select push-buttons are somewhat self descriptive. The *HDG* (heading) switch commands the autopilot to steer a given heading. The *1/2 BANK* mode reduces all bank angles to approximately 13.5 degrees. The 1/2 bank mode is inoperative during approach to land. The *NAV* (navigation) mode causes the autopilot to follow the navigation source currently displayed on EFIS. The *APPR* (approach) mode is used when the pilot wishes to navigate via a localizer and glide slope signal during an approach to land. It should be noted: the NAV and APPR modes can be armed without actually "capturing" the mode. Capture of the mode occurs only when valid navigation signals are available. If no valid navigation signal (such as the localizer) is received the approach mode will be armed (not captured). The autopilot will fly the aircraft's last heading until capture occurs.

Fig. 6-40 The APS-85 mode select panel. (Rockwell International, Collins Avionics Divisions)

The vertical mode switches are described as follows: The *CLIMB* button is used to activate a given climb rate (IAS or MACH) according to FCC software. Three different rates can be selected (low, medium, or high speed) using the *Performance Select* button. The *ALT* (altitude) button is used to maintain the current barometric altitude of the aircraft. The *VNAV* (vertical navigation) mode is used to fly a vertical profile established by the flight management system. The *DESCEND* mode commands the autopilot to fly a preprogrammed descent rate. The APP pitch wheel can be used to increase or decrease that rate. The *SPEED* mode will cause the autopilot to fly a given speed by adjusting aircraft pitch accordingly.

Flight control computer

The flight control computer (FCC) is a dual channel unit designed to receive input data, process the information, and send the appropriate outputs to the autopilot servos and the electronic flight instrument system. While the autopilot is engaged, the FCC controls aircraft attitude through the control surface servos. With the autopilot engaged **or** disengaged, the FCC controls the **V-bar** position on the EADI. As mentioned earlier, the V-bars are part of the flight director system that provides visual reference to the pilot.

Servos

The APS-85 uses three electrically actuated servos. Each servo is equipped with an engage/disengage clutch that allows for quick response time of the servo mechanism. The servos also employ a slip clutch that is used as a backup for manual override. The APS-85 is designed to interface with the aircraft's trim motor assembly and therefore the aircraft controls do not require additional trim servos.

Theory of Operation

Refer to the block diagram of (figure 6-41) the APS-85 system during the following discussion on theory of operation. The FCC contains two channels (A and B), which receive identical inputs for data processing. The dual channels share the same FCC housing, yet operate completely independent. The FCCs perform system monitoring to ensure autopilot reliability, and present all diagnostic data through the aircraft's EFIS displays. The FCC outputs control servo operation and flight director displays on the EADI.

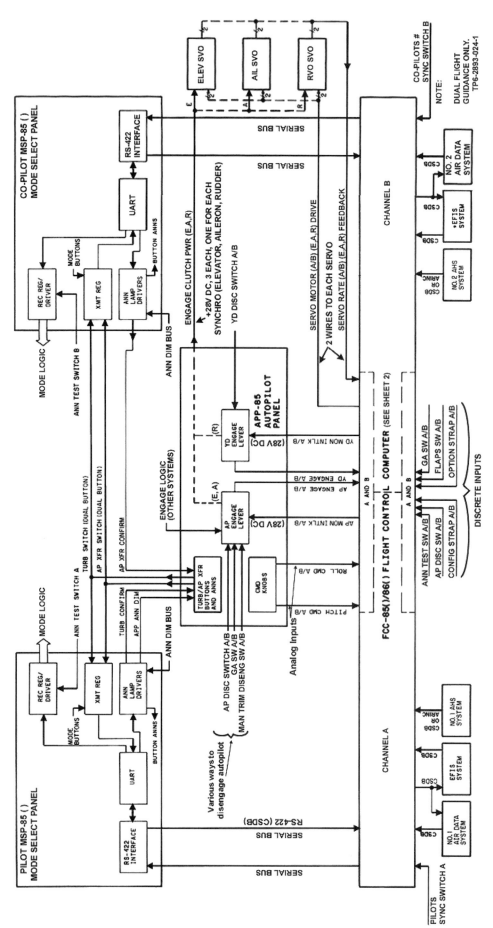

Fig. 6-41 Block diagram of the APS-85 autopilot system. (Rockwell International, Collins Avionics Divisions)

A voter circuit is contained within each channel of the FCC (see figure 6-42). The voter circuit determines which channel calls for the least servo movement and sends that signal to the motor. A torque limiter is used as a current limiting device, which allows the pilot to manually overpower the motor in the event of a servo runaway.

Flight control computer interface
Inputs to the FCC include:

1) Digital data in CSDB format from the two MSPs, the air data system, the Attitude Heading System (AHS), and EFIS. EFIS supplies all navigation inputs to the FCC. The FCC will also accept AHS data in ARINC 429 format.

2) Discrete inputs are received from the annunciator test switch, the autopilot disconnect switch, the go-around switch, the flaps switch, configuration strapping, options strapping (see lower/center of figure 6-41), as well as the pilot's and copilot's sync switches.

3) The APP sends analog data to the FCC for pitch and roll commands and autopilot/yaw damper engage commands.

4) The three servo units send analog servo rate data to both channels of the FCC (see figure 6-42).

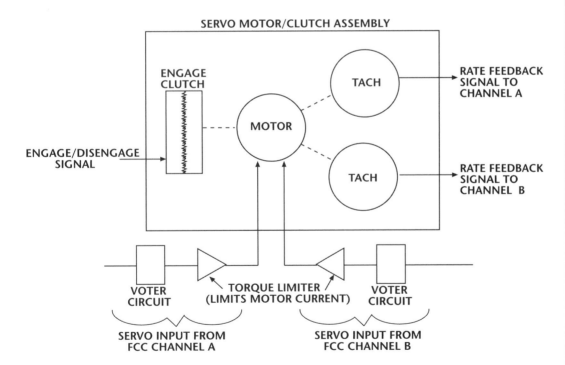

Fig. 6-42 Diagram showing the connections between a two-channel FCC and a control surface servo.

The FCC outputs include:

1) A two wire 28 Vdc analog signal to each servo motor.

2) A CSDB output is sent to EFIS with flight director information. The same data is sent to the air data system.

3) A CSDB output is sent to both MSPs. This signal informs the MSP of the current FCC operating mode(s).

Attitude heading system

The APS-85 is designed to interface with a unique attitude heading reference system, the **AHS-85**. The AHS-85 measures angular rates and accelerations along all three aircraft axes using piezoelectric sensors (see figure 6-43). The AHS-85 system employs an **Attitude Heading Computer (AHC)** containing the piezoelectric sensors. The piezoelectric sensors replace the accelerometers, gyros, and rate sensors found on conventional attitude heading systems (see figure 6-44).

Fig. 6-43 A piezoelectric sensor from the AHS-85 attitude heading system. (Rockwell International, Collins Avionics Divisions)

The AHS-85 AHC uses a conventional flux detector. The flux detector is needed to provide a magnetic heading reference to the system. Since each aircraft has slightly different magnetic characteristics, a **compensator unit** must be used to correct for magnetic errors and flux detector misalignment. Remember, the compensator unit corrects for a specific aircrafts magnetic error, and therefore must stay with the aircraft when changing other attitude heading system components.

The AHC contains a **dual sensor assembly** that houses two rotating wheels mounted at 90° angles from each other (see figure 6-45). The spinning wheels, which rotate at a constant 2,500 RPM, contain four piezoelectric crystals called **benders**. One pair of benders measures acceleration, the other pair of benders measures rate changes (see figure 6-46). It should be noted, a conventional rotating mass gyro spins at 20,000 RPM. The slower rotational speed of the AHC sensor assembly (2,500 rpm) makes this unit much more reliable than a conventional gyro.

Fig. 6-44 Attitude/heading system components. (a) early version systems with several independent components: gyros, flux detector rate sensors, and accelerometers; (b) newer version system containing all the necessary components in two units, the attitude/heading computer and the flux detector. (Rockwell International, Collins Avionics Divisions)

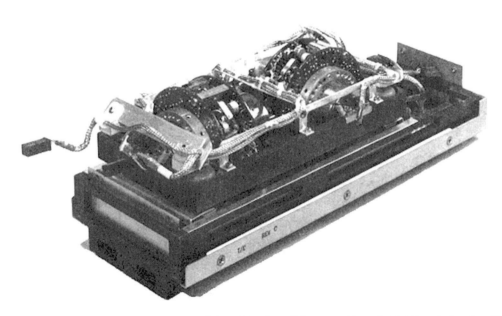

Fig. 6-45 A dual sensor assembly. (Rockwell International, Collins Avionics Divisions)

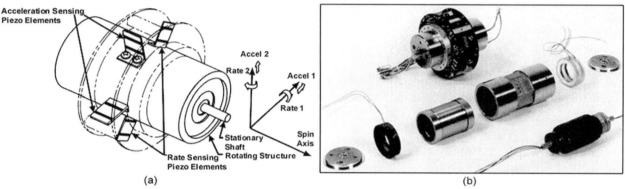

Fig. 6-46 A sensor wheel assembly. (a) diagram showing piezoelectric crystals (benders); (b) an expanded view of the sensor wheel. (Rockwell International, Collins Avionics Divisions)

When pressure is applied to a piezoelectric material, a voltage is produced. As the rotating piezoelectric crystals are subject to an acceleration or rate change, the materials bend. Bending the material applies a pressure to the crystals; therefore, the benders produce a voltage. The direction in which the crystal bends determines the polarity of the output voltage (see figure 6-47). The rotating wheel contains a timing mark. This mark provides a reference point for AHC. The rotating wheels also contain two transformer primary coils. These transformers are used to induce the signal from the benders to the stationary portion of the sensor assembly.

It is important that the Collins AHC-85 be oriented in the correct position for the system to operate properly. AHC configuration strapping is used to tell the computer which direction within the aircraft the AHC is facing. Once the computer knows the magnetic heading (determined by the flux detector) and the gravitational reference (determined by the rotating sensors), any changes in angular rate or acceleration can be easily converted into changes in aircraft position. These changes in position are transmitted to the autopilot FCC.

Like the inertial reference system discussed earlier, the AHS must be initialized prior to use. This process requires that the aircraft be parked in an area free of magnetic interference. This allows for proper flux detector operation. The aircraft must also remain still during the initialization process. This system also has the capability to initialize during smooth, straight and level flight.

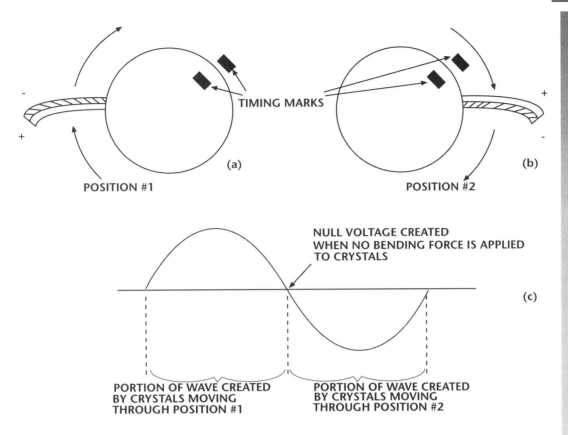

Fig. 6-47 Diagram showing the voltage produced by bending crystal sensors. (a) as the crystal sensors rotate from straight down through position #1 to straight up the positive portion of the sine wave is produced; (b) as the crystal sensors rotate from full up through position #2 to straight down the negative portion of the sine wave is produced; (c) the sine wave is produced as the crystal rotates and the aircraft is under an acceleration force. The amplitude of the sine wave changes proportionally to the acceleration force on the aircraft.

Inspection and maintenance

The APS-85 requires very little routine maintenance. Operational tests should be performed in accordance with the aircraft's approved maintenance schedule. The servos are time-limited components and require regular maintenance. Every major aircraft overhaul or every 10,000 flight hours, the main control surface servos and servo mounts should be inspected by an authorized repair facility. It is recommended that any trim servos be inspected in as little as 1,000 flight hours.

On the aircraft, the servo capstan should be inspected for wear, security, and proper cable alignment. The servo unit should be operated through its entire range and observed. If a cable binding or fatigue is present, the problem must be corrected. If the servo unit makes a grinding or rubbing sound, the servo and servo mount should be removed, inspected, and repaired. At regular intervals, the servo slip clutch must be tested. The test procedures are outlined in the service manual.

Troubleshooting procedures

The APS-85 is designed to operate in conjunction with the Collins electronic flight instrument system (EFIS). This allows the autopilot diagnostics to be displayed on the EFIS primary flight display or multifunction display. The APS-85 diagnostics operate in three modes: input, report, and output mode. The **input mode** displays various input parameters that report to the FCC. The input mode can therefore be used to determine the operational outputs of systems, such as AHS, air data, and navigational aids. The **report mode** presents data on various systems that are monitored by the FCC. If a fault flag is displayed during flight, the report mode should be accessed prior to turning off electrical power. The **output mode** is used to display FCC software outputs. In general, the report mode is used for primary troubleshooting, and the input/output modes are used for detailed fault isolation.

The pilot(s) of any aircraft using an APS-85 should be made aware of the autopilot diagnostics. In the event of an autopilot problem, the pilot should enter the diagnostics mode on EFIS and record all fault codes prior to power shut down. This will help the troubleshooting process since the Report Codes give technicians important diagnostics data. The Report Codes are decoded using the autopilot maintenance manual. Report mode data includes categories such as *REPAIR CODE* (general fault data), *AP DIS CODE* (faults causing an autopilot disengage), *STEER CODE* (flight director steering faults), and *RAM ERRORS* (FCC memory faults). Diagnostic codes of 000000 indicate a system with no faults detected. Other fault codes must be decoded using the autopilot maintenance manual. The EFIS chapter of this text explains diagnostic procedures using the Collins electronic flight instrument systems.

Autopilot and flight director problems

Whenever troubleshooting any autopilot system, it is sometimes difficult to determine if the fault lies in the autopilot or flight director. Keep in mind: 1) the autopilot computer typically controls the flight director indications, and 2) the autopilot and flight director typically receive different inputs from the autoflight computer.

If the flight director indicator is an electromechanical type, test the indicator if possible, using the appropriate test switch. If an EFIS is used, check for an EFIS fault using the built in diagnostics. If either the electromechanical instrument or EFIS show faults, the defect lies in the indicator and the autopilot is most likely OK. If the indicator tests OK, suspect an autopilot problem.

In general, when the autopilot is engaged, the flight director and autopilot functions are isolated. This is the best time to troubleshoot the two subsystems. If the pilot commands a left turn and the flight director responds, but the autopilot does not, the fault most likely lies in an autopilot component (perhaps a defective aileron servo). If the pilot commands the same left turn and the autopilot "flies" the aircraft into the turn, but the flight director does not respond, the fault lies in the flight director (perhaps a fault in the ADI or EFIS interface). If both the flight director and autopilot fail to respond, the problem lies in a component common to both subsystems (perhaps a defective APP or FCC).

Troubleshooting: Helpful Hints

The following are several troubleshooting techniques that may help isolate faults on the APS-85, as well as other autopilot systems:

1) Many autopilot problems come from the subsystems that feed the autopilot computer. One of the most complex and frequently failed subsystems is the attitude heading system. This is especially true for attitude heading systems that employ rotating mass gyros. The AHS-85 contains both the flux detector (for heading information) and the attitude sensors for attitude data. As shown in figure 6-48, the attitude sensors are part of the Attitude Heading Computer (AHC). Attitude and heading data are each processed and distributed through the AHC. This means: 1) if EFIS displays a heading flag (*HDG*), the flux detector is most likely at fault; 2) if EFIS displays an attitude flag (*ATT*), the attitude sensors are at fault and the AHC must be replaced; 3) if both flags (*ATT* & *HDG*) are displayed, the AHC processor software is faulty and the AHC should be replaced. Always become familiar with any subsystems that feed the autopilot.

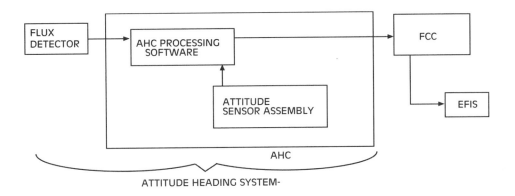

Fig. 6-48 Block diagram of an attitude heading system.

2) If a YELLOW message such as *HDG* (heading) appears on EFIS, this normally means a dual system disagreement. That is, two redundant subsystems that feed the autopilot are not transmitting the same data. In this case, determine which unit is faulty by operating the subsystems independently. Independent operation can be done through reversionary switching, or by opening circuit breaker(s) to one of the subsystems.

3) Any time a heading inaccuracy problem occurs, consider one of the flux gates may be too close to a metal object. Move the aircraft and see if the problem corrects itself. If a heading disagree fault is stored in the diagnostics memory, check the time of occurrence. If the fault occurred shortly after starting the engines, the fault is most likely caused by the aircraft taxiing too close to a metal structure.

4) While operating in autopilot mode, if the aircraft consistently changes altitude while in a banked turn, the fault is most likely a misalignment of the attitude heading system. It is very important that the AHS-85 mounting tray be aligned correctly. An alignment fixture can be used to verify alignment of the AHS mounting tray. The tray can be shimmed to adjust the alignment if necessary. On any attitude heading system, if a misalignment occurs, inspect the mounting structure for cracks, bends or looseness.

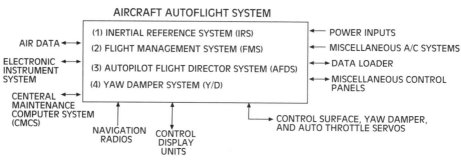

Fig. 6-49   The four subsystems of the B-747-400 Autoflight system.

5) Any autopilot system is only as good as the related control surface elements. If the control surfaces are improperly installed, loose or poorly balanced, the autopilot will most likely be unable to hold a steady attitude. If the control surface cables are too loose, the aircraft will oscillate (or porpoise) while in the autopilot mode. This will be especially evident at capture of a given attitude or heading.

THE BOEING 747-400 AUTOFLIGHT SYSTEM

A modern autoflight system operates using microprocessor technologies, and communicates with a variety of aircraft systems via digital data busses. The Boeing 747-400 autoflight system is typical of a fully integrated digital system found on modern transport category aircraft. As seen in figure 6-49, the B-747-400 autoflight system is comprised of four major subsystems: 1) inertial reference system, 2) flight management system, 3) autopilot flight director system, and 4) the yaw damper system. The autoflight system also receives data from and communicates with a variety of other aircraft systems. The following discussion on the B-747-400 autoflight system will include an overview of the four major systems listed above.

Inertial Reference System

The B-747-400 inertial reference system (IRS) is used to provide both vertical and horizontal navigation, attitude information, acceleration, and speed data to a variety of aircraft systems. The IRS consists of one **mode select unit** located on the flight deck and three **inertial reference units (IRUs)** located in the main equipment center. As seen in figure 6-50, each IRU interfaces with the air data computers, control display unit/flight management system, central maintenance computer system (CMCS), integrated display system (IDS), and the various IRS data users.

Controls

The IRS **mode select unit (MSU)** is used to select one of the three IRS operating modes: **1) align, 2) navigation, or 3) attitude.** The MSU can also be used to turn off each individual IRU (see figure 6-50). While in the **align** mode, the IRU performs the alignment procedures to determine the aircraft's local vertical, heading, and present position. While in the **navigation** mode, the IRU will provide: attitude data, accelerations, heading data, horizontal and vertical velocities, wind speed and direction, latitude and longitude, ground speed, and inertial altitude. If the mode selection switch is moved from *OFF* to the *NAV,* the IRU will perform the alignment procedures before operating in the navigation mode. The navigation mode is used during normal flight configurations. The IRS **attitude** mode is a backup mode used in the event the navigation mode fails. The attitude mode provides attitude data, heading information, accelerations, and vertical speed.

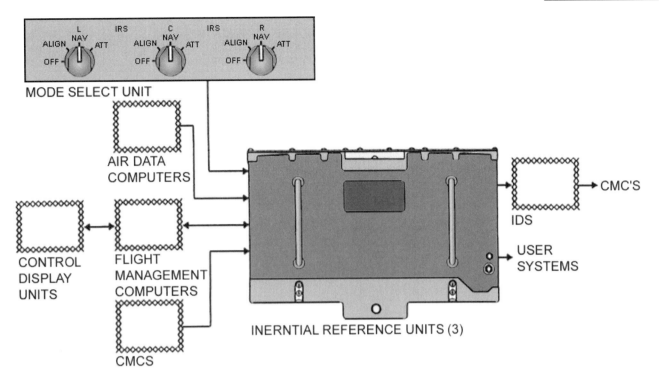

Fig. 6-50 IRS interface diagram. (Northwest Airlines, Inc.)

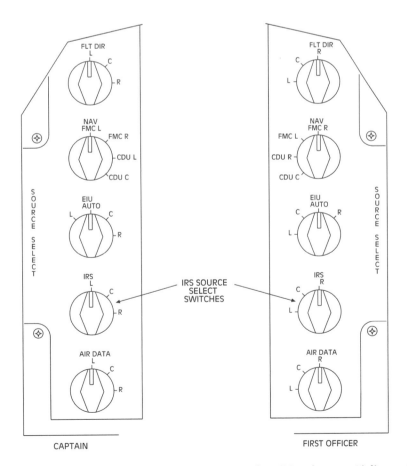

Fig. 6-51 B-747-400 IRS source select panels. (Northwest Airlines, Inc.)

There are two source select panels located on the flight deck that are used to choose which IRS will provide data to the electronic flight instruments (see figure 6-51). The captain and first officer can each select a different IRS source for their respective electronic flight displays. The first officer's IRS select switch also controls the source for the standby RMI.

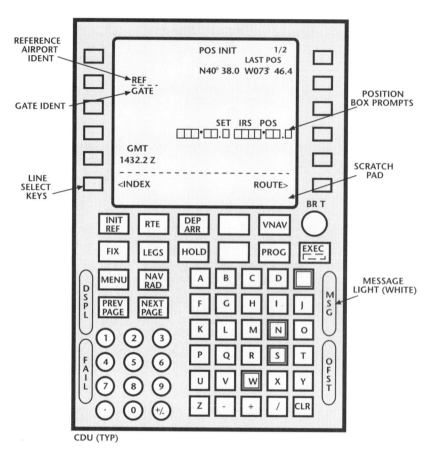

Fig. 6-52 Control display unit showing the IRS display during initialization procedures. (Northwest Airlines, Inc.)

Operation

At power-up, the IRU computer performs a BITE test that verifies the health of the system. The test monitors internal circuitry and the power supply switch-over capabilities. Each IRS has the ability to automatically switch from 115 VAC to 28 VDC power in the event of a bus failure. If the IRU passes the power-up test, the unit moves into the eight-minute initialization process. At this time the pilot must enter the aircraft position (latitude and longitude) using the control display unit. Figure 6-52 shows the typical display during latitude/longitude data entry. During initialization, the white memo message *IRS ALIGN MODE L/C/R* will be displayed on the main EICAS display.

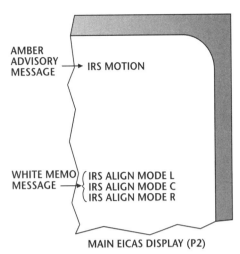

Fig. 6-53 IRS display caused by excessive motion during alignment. (Northwest Airlines, Inc.)

If the alignment process is disturbed by excessive aircraft motion, the main EICAS display will show an amber advisory *IRS MOTION* (see figure 6-53). This message will be displayed until 30 seconds after the motion stops. At this time, the corresponding white memo message will be removed from the display. The IRU will automatically continue the alignment procedure when the motion stops. After alignment is complete, the three IRUs compare position data to ensure accuracy. If a miscompare message is displayed by EICAS, the alignment procedure should be repeated. Once alignment is complete, the IRS is ready for operation.

Maintenance and troubleshooting

Whenever troubleshooting the IRSs, remember each IRU (left/right/center) receives power from different sources. Also, each IRU must be supplied with 115V AC power and DC power supplied from the APU hot battery bus. If the APU battery is below 18 volts, or is removed from the aircraft, the IRUs will fail the power-up test and will not function.

The three IRSs are each monitored by the aircraft's central maintenance computer system (CMCS). If a fault occurs during operation, the CMCS will cause the appropriate message to be displayed by EICAS. To verify the fault and determine the suggested repair, the CMCS can be accessed through the control display unit.

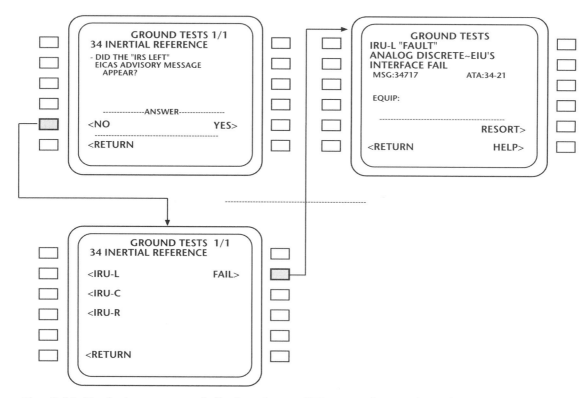

Fig. 6-54 Typical sequence of displays for an IRS ground test. (Northwest Airlines Inc.)

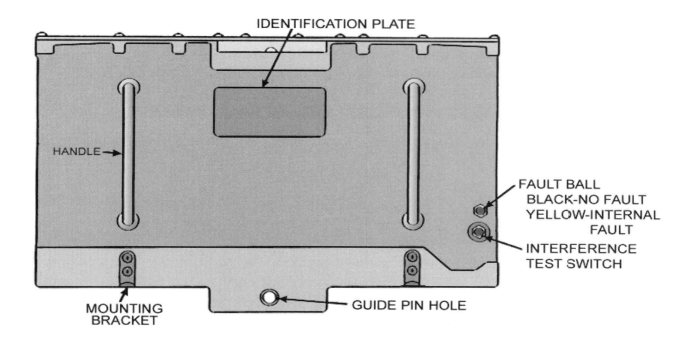

Fig. 6-55 Diagram of the IRS inertial reference unit. (Northwest Airlines, Inc.)

A ground test can be performed on each IRU using the CMCS. The test is accessed through the ground test menu of the CMCS. The technician should select the appropriate IRU for testing and follow the test preconditions. Pressing the line select key adjacent to *START TEST>* initiates the test process. If the ground test is passed, the CMCS displays a question asking if the advisory message "*IRS LEFT*" (for the left IRS) was displayed by EICAS during the test. The technician should answer the question accordingly. If "yes" is selected, the message *PASS* appears on the CMCS display. If the technician selects "no", the appropriate ground test message is displayed as shown in figure 6-54.

As mentioned earlier, during power-up of the IRS, an IRU BITE test is automatically conducted. The IRU BITE can also be activated by the **interface test switch** located on the face of the IRU. If the IRU fails its internal BITE test, the fault ball will be visible on the face of the IRU (see figure 6-55).

During removal and installation of the IRU, be sure to handle the unit gently. Also, the IRU must be installed with precise alignment if accurate output data is to be obtained. Each IRU mounting rack contains an alignment pin, which must fit accurately into the IRU alignment hole for proper installation.

Flight Management System

The **Flight Management System (FMS)** is a computer-based system that reduces pilot workload by providing automatic radio tuning, lateral and vertical navigation, thrust management, and the display of flight plan maps. Automatic radio tuning is performed by the FMS for all navigational aids used during normal flight. Radio tuning includes selection of the appropriate radio, tuning to the correct frequency, and selection of the correct course bearing. Vertical and lateral navigational parameters are computed by the FMS and sent to the flight director and autopilot systems. Thrust management automatically controls engine thrust as needed for a given flight condition. The flight plan map displays are constantly updated by the IRU computer according to the programmed flight plan. The flight plan maps are displayed on the EFIS CRTs. The **Flight Management Computer (FMC)** is the main element that provides interfacing and data processing for the FMS

Controls

The B-747-400 contains three **Control Display Units (CDUs)**, which are used to enter data into the FMS and interface the FMS with other aircraft sensors and systems. The CDU keyboard contains four types of switches: alphanumeric keys, mode keys, line select keys, and function keys (see figure 6-56). The line select keys perform a specific function according to the items displayed on the CRT, all other keys perform a given function as labeled on the key.

There are four annunciators, which illuminate to show specific messages related to the CDU. The display (*DSPY*) annunciator illuminates whenever the currently displayed page is not related to the active flight plan. The *FAIL* annunciator illuminates if the selected FMC fails. The message (*MSG*) illuminates if a message appears in the FMC scratch pad. The offset (*OFST*) light illuminates when navigating using an offset route. The annunciator lamps are accessed by removing two screws that hold the annunciator assembly to the CDU face plate (see figure 6-56).

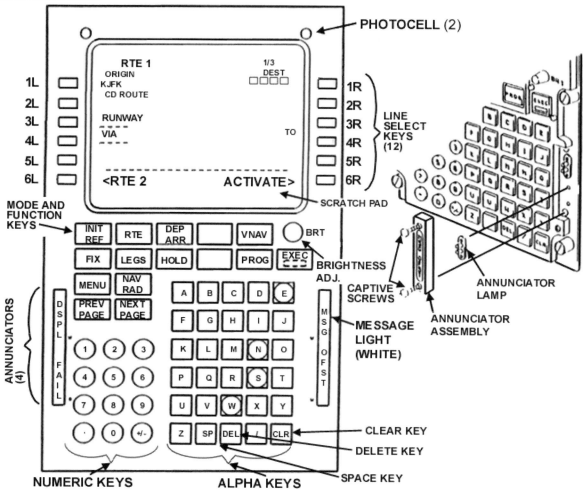

Fig. 6-56 Typical control display unit (CDU). (Northwest Airlines, Inc.)

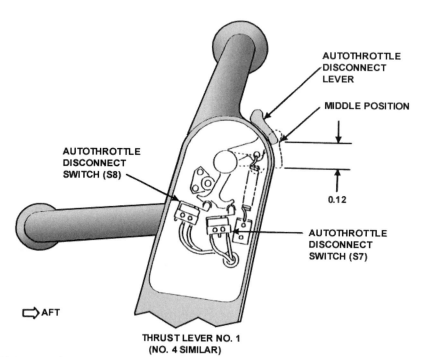

Fig. 6-57 Diagram showing the autothrottle disconnect switches located in the number one and four thrust levers. (Northwest Airlines, Inc.)

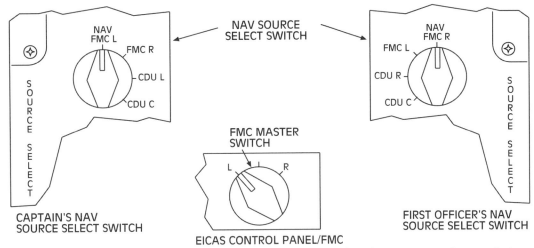

Fig. 6-58 The flight management control system navigation source select switches. (Northwest Airlines, Inc.)

The autothrottle (A/T) system is activated through the mode control panel (to be discussed later). The autothrottle disconnect switches are located in the number 1 and 4 thrust levers as shown in figure 6-57. Two switches are activated by each disconnect lever to provide redundancy.

The FMC master switch is located on the EICAS control panel (see figure 6-58). The master switch selects which FMC (left or right) will control commands for: autopilot, autothrottle, and radio tuning. The navigation source select switches control which FMC is used to drive EFIS displays. The navigation source select switches and EICAS/FMC control panel are located on the instrument panel.

Architecture

The B-747-400 contains 2 complete flight management systems (right and left). The **Flight Management Computers (FMCs)** are located in the main equipment bay and perform all the necessary interface and data processing functions for the FMS. The flight crew selects one FMC as "master". The opposite side FMC operates in hot standby in the event that the master FMC fails.

A FMS interface diagram is shown in figure 6-59. The FMC receives data from the control display unit and a variety of aircraft systems and sensors. A data loader is used to input preprogrammed navigational parameters, such as flight routes, way points, and airport data. Some of the FMC outputs are sent directly to the user, while some are sent via the FMC master relays. The use of relays allows either FMC (right or left) to send critical output information to four systems: the EECs, FCCs, MCP, and the navigation radios. The FMC sends output data to the **Electronic Engine Controls (EECs)**, which provide the control signals to the autothrottle servo motor. The servo motor generators send a feedback signal to each FMC. The **Flight Control Computers (FCCs)** receive FMC data for control of autopilot and flight director functions.

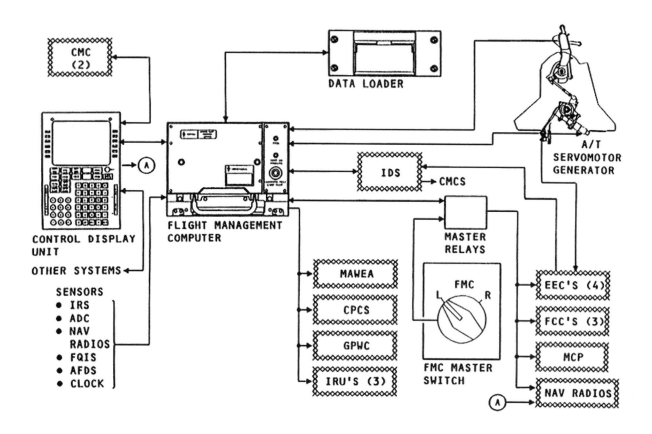

Fig. 6-59 Block diagram of the flight management system. (Northwest Airlines, Inc.)

The FMC sends output data to the Integrated Display System (IDS). This data is used to display FMS information on the PFD, ND, and EICAS. The data sent to the IDS is also used to communicate with the Central Maintenance Computer System (CMCS). The CMCS stores all fault data that can later be retrieved by the technician for analysis and fault isolation.

Autothrottle architecture

The autothrottle function of the FMS is regulated through the **Mode Control Panel (MCP)**, which is located in the center of the flight deck glare shield. As seen in figure 6-60, the MCP communicates directly to the right/left FMC. The FMC then transmits control signals to the autothrottle servo. Whether the throttles are moved manually or by autothrottle, a feedback signal is sent from the Throttle Resolver Angle (TRA) transducers to the EECs. The EECs send data to the engine Fuel Control Units (FCUs), which provide *coarse* adjustments of engine thrust. The FMC provides *fine* adjustment of engine thrust.

The FMC provides engine trimming commands. **Trimming** the engines is simply a fine thrust adjustment in order to precisely equalize the thrust of all four engines. The FMC receives engine thrust data from the **EFIS/EICAS Interface Units (EIUs)** and calculates the trim commands. The trim commands, along with Air Data Computer (ADC) information, are sent to the EECs.

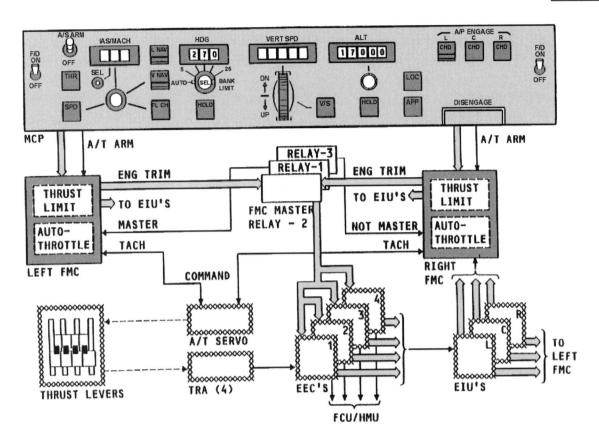

Fig. 6-60 Thrust management system interface diagram. (Northwest Airlines, Inc.)

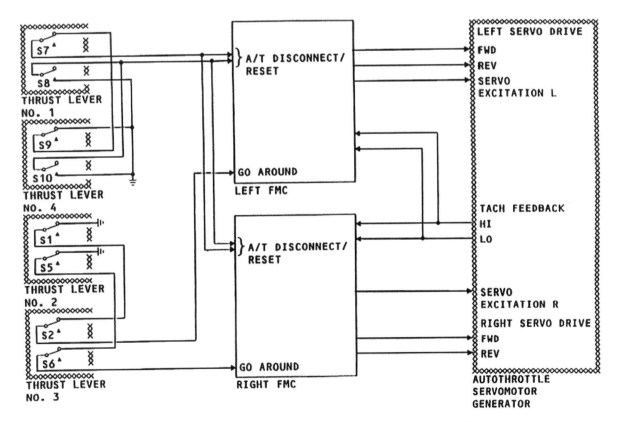

Fig. 6-61 Autothrottle/FMC interface diagram. (Northwest Airlines, Inc.)

An autothrottle/FMC interface diagram is shown in figure 6-61. The autothrottle assembly contains both a servo motor and a tachometer generator. The tachometer generator sends a feedback signal to each FMC. The FMCs send servo commands to the autothrottle servo motor consisting of a 115 VAC excitation voltage and a 28 VDC forward and reverse signal. The autothrottle disconnect and go-around signals are sent from the throttle lever switches to each FMC. The FMCs then send a discrete signal to the autothrottle servo motor generator assembly to command go-around or disconnect.

FMS power inputs

Power inputs to the FMC come from six different circuit breakers and five different power distribution busses. The 28 VDC busses 1 and 2 supply power for the autothrottle servos and master relays 1 and 3. The captain's 115 VAC transfer bus supplies power to the left side FMC for internal FMC functions. The first officer's 115 VAC transfer bus supplies autothrottle servo excitation power along with tachometer generator excitation. The 28 VDC battery bus powers the FMS warning circuits. The first officer's (F/O) 115 VAC transfer bus powers the internal functions of the right FMC.

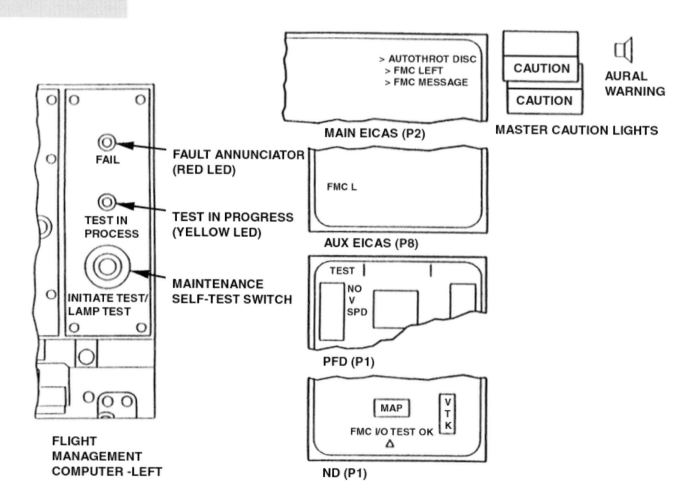

Fig. 6-62 Flight management control system BITE test can be controlled by the maintenance self-test switch found on the FMC. (Northwest Airlines, Inc.)

Maintenance and Troubleshooting

The FMS continually monitors itself using BITE systems programmed into the FMC software. The BITE is initiated at every power-up of the FMC. The BITE can also be initiated through the central maintenance computer system or using the *INITIATE TEST/LAMP TEST* switch on the front of the FMC (see figure 6-62). During this 15 second test, the main and auxiliary EICAS, the PFD, and ND each present specific test messages. During the test, the master caution and warning lights and aural tones sound for a short period. On the FMC, the red "fail" lamp illuminates while the FMC test switch is held in, or at the end of the test if the FMC BITE fails. The "test in process" light illuminates any time the test is in progress.

Two major subsystems of the FMS can be accessed through the Central Maintenance Computer System (CMCS): the FMC and the FMC servo loop. Both of these systems can be accessed for either the right or left FMS. The CMCS tests for the FMS can only be performed on the ground since the FMS is inoperative during CMC interrogation. FMS fault data stored in the CMCS memory can be accessed through the CMC *existing faults* or *present leg faults* page.

Autopilot Flight Director System

The B-747-400 **autopilot flight director system (AFDS)** receives inputs from various systems and sensors throughout the aircraft, and provides steering commands for automatic and/or manual control. For manual steering, the flight director provides the interface between the AFCS (Automatic Flight Control System) and the pilots. During automatic steering, the aileron, elevator, and rudder servos provide an interface between the AFDS and the control surfaces. The autopilot is capable of pitch control to maintain a given airspeed, altitude, vertical speed, or vertical navigation (including glide slope). Roll commands can maintain a given heading, track, lateral navigation, or attitude (including localizer). The autopilot yaw function provides control for adverse yaw, and crab angle.

Controls

The **mode control panel (MCP)** is the main interface between the flight crew and the AFDS. The mode control panel is located on the glare shield, cooled by forced air, and connected to the system through three connector plugs located on the rear of the unit. As seen in figure 6-63, a lighted push button assembly is removed from the face of the unit for lamp replacement. Each lamp assembly contains four bulbs, two powered by 5 VAC, and two powered by 28 VDC.

Refer to figure 6-63 during the MCP control explanation in this paragraph. The captain's flight director is activated by the toggle switch on the far left of the MCP, the first officers flight director toggle switch is located on the right of the panel. The autothrottle engage switch is located just right of the captain's flight director switch. Indicated air speed (IAS)/mach speed can be selected from the speed mode of the autothrottle function. Lateral navigation (*L NAV*) or vertical navigation (*V NAV*) can be selected using the appropriate lighted push button switch. Pressing the **flight level change (FL CH)** switch will engage both vertical and lateral navigation. The HDG control can be used to select a given heading for the autopilot or flight director. Vertical speed is entered

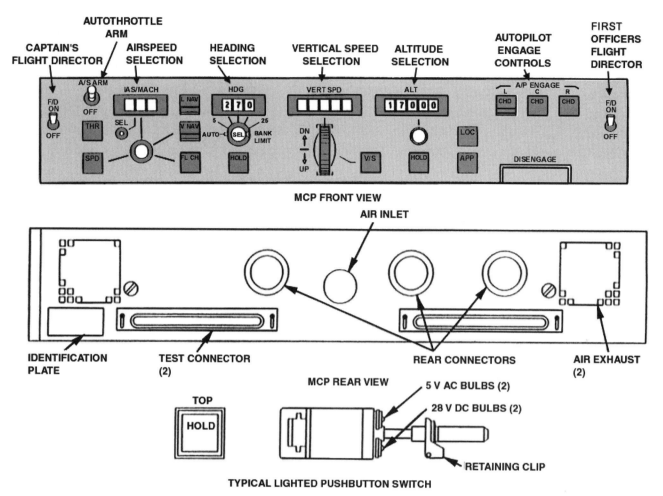

Fig. 6-63 Boeing 747-400 Autopilot Flight Director System (AFDS) mode control panel. (Northwest Airlines, Inc.)

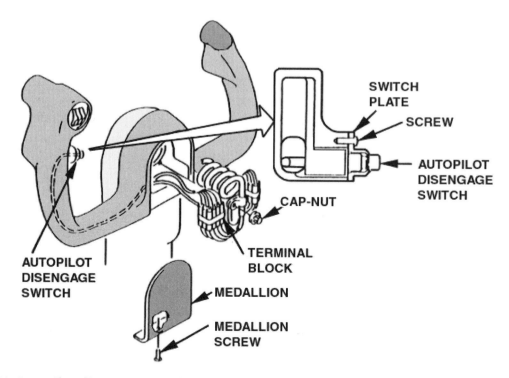

Fig. 6-64 Autopilot disengage switch. (Northwest Airlines, Inc.)

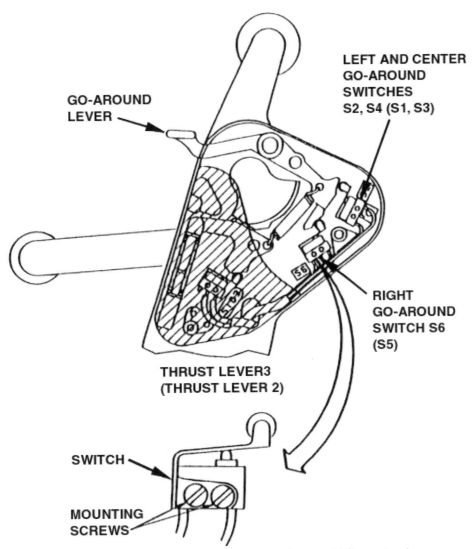

GO-AROUND
LEVER

LEFT AND CENTER
GO-AROUND
SWITCHES
S2, S4 (S1, S3)

RIGHT
GO-AROUND
SWITCH S6
(S5)

THRUST LEVER3
(THRUST LEVER 2)

SWITCH

MOUNTING
SCREWS

Fig. 6-65 Autopilot go-around switches. (Northwest Airlines, Inc.)

into the MCP using the vertical speed thumb wheel. A given altitude can be selected and displayed in the *ALT* window. The autopilot engage push buttons allow the pilot to select the left, center, or right flight control computer for command of autopilot/ flight director functions.

An autopilot disengage switch is located on both the captain's and first officer's control wheel. These switches are removed by a screw located on the front of the switch plate (see figure 6-64). The switch wiring is fed through the control wheel to a terminal block. The autopilot go-around switches are located on the number 2 and 3 thrust levers (see figure 6-65).

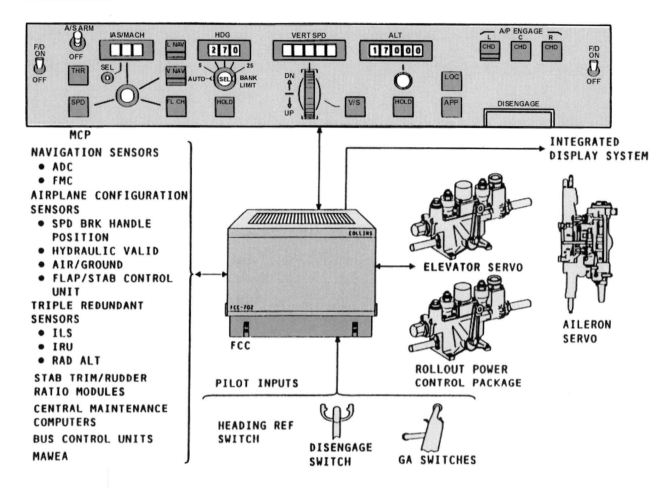

Fig. 6-66 Interface diagram of the Flight Control Computer (FCC) and various aircraft systems. (Northwest Airlines, *Inc.*)

Architecture

There are three **Flight Control Computers (FCCs)** which interpret data and provide the necessary calculations for the autopilot and flight director functions. The pilot selects inputs to the FCC through the MCP, the heading reference switches, the disengage switches, and go-around switches (see figure 6-66). The FCC receives three types of system inputs: navigational, airplane configuration, and triple redundant sensors. Navigational inputs are provided by the FMC and ADC. Airplane configuration sensors monitor items necessary for autoflight, such as hydraulic status and flap position. The triple redundant inputs are those needed for autoland functions. Triple redundant sensors include: ILS, IRU, and radio altimeter data.

The three FCCs each control a separate servo, one each for the ailerons, elevator, and rudder. The servos use electrical signals from the FCCs to control the flow of hydraulic fluid which in turn controls the position of the related control surfaces. The FCC outputs display data to the **EFIS/EICAS interface units (EIUs)**. As seen in figure 6-67, all three FCCs send a parallel data signal to each of the EIUs. The FCCs send a discrete warning signal to the **Modularized Avionics and Warning Electronic Assembly (MAWEA)** for annunciation of *Warning* data. *Caution* information is sent from the FCCs to the three EIUs.

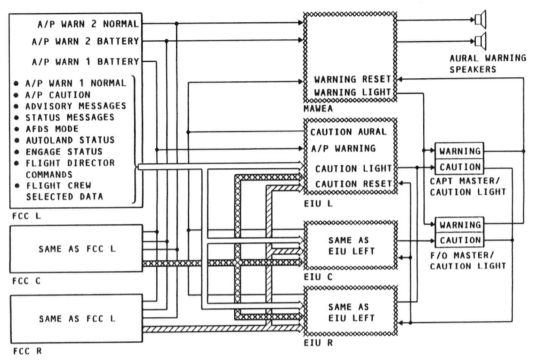

Fig. 6-67 FCC/EIU/MAWEA interface diagram. (Northwest Airlines, Inc.)

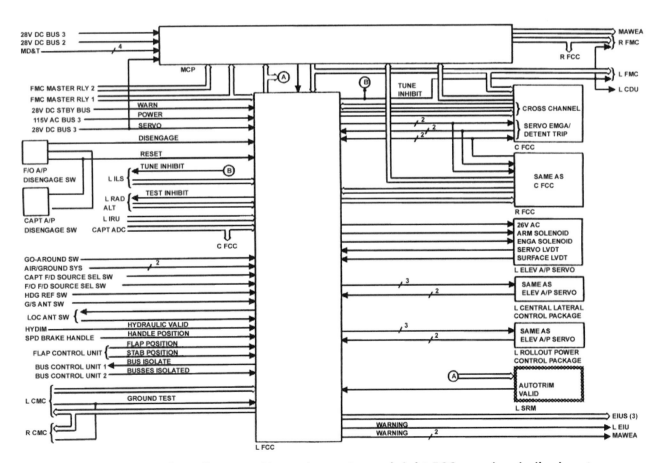

Fig. 6-68 Left FCC interface diagram. Note: the center and right FCCs receive similar inputs.
(Northwest Airlines, Inc.)

The FCCs communicate to each other via a cross channel data bus for exchange of health monitoring, and to provide redundancy for servo engage data. The ability to cross talk between FCCs improves system safety by allowing the comparison of information between computers. If any FCC detects a failed FCC or critical system out of tolerance, the autoland capability will not be available.

Figure 6-68 shows the interface of the FCC and various aircraft systems. In the upper-right portion of the diagram are the cross channel busses for communication between the left/right/center FCCs. The data busses to the Central Maintenance Computers (CMCs) are shown in the lower left portion of the diagram. Discrete data, represented by a single line on the interface diagram, comes from a variety of other aircraft systems to the FCCs. In the top left portion of the interface diagram are the power inputs to the FCC and MCP. To operate this autoflight system, there are a total of nine different circuit breakers fed from seven different power distribution busses. Whenever troubleshooting the system, be sure power is available to all necessary circuits.

Maintenance and Troubleshooting

The B-747-400 Autopilot Flight Director System (AFDS) contains BITE circuits that continuously monitor the health of the FCCs and related systems. The BITE circuits are located within each FCC and report all autoflight failures to the central maintenance computer. The flight crew is made aware of failures by a flag on the PFD or ND, an EICAS message, and/or a discrete annunciator and audio tone. EICAS will always display a warning, caution, advisory, or status message for the various AFDS faults. The *A/P DISCONNECT* message is the only EICAS **warning** applicable to AFDS. (Remember, warnings are the most serious EICAS message and require immediate crew action.) This message will display on EICAS for either a manual or automatic disconnect. In the event of a manual (pilot activated) disconnect, the CMC will not store the message as a fault.

Any fault that is sensed by the BITE circuitry is automatically recorded in the CMC nonvolatile memory. The technician can access current or previous failures through the *Existing Faults* or *Fault History* pages of the CMCS. Ground tests can also be performed using the CMCS. To access AFDS test functions, go to the **Ground Tests** page of the CMC menu (see figure 6-69). Next, select chapter 22 (*AUTOPILOT FLT DIR*) and choose the appropriate test from the menu. Table 6-1 is a list of the tests available through the AFDS ground tests menu.

It should be noted that several of the tests have preconditions that must be met before the tests can take place. Preconditions are listed on the control display unit after the test selection has been made (see figure 6-69).

> **CAUTION:** Whenever performing operational tests on any autopilot, be sure the aircraft is clear of personnel and machinery. Many of the autopilot tests will operate various control surfaces and/or thrust reversers. These control surfaces could cause damage to the aircraft or bodily injury to unsuspecting individuals. Also, be sure that other maintenance being performed on the aircraft will not adversely affect the autopilot tests and create a potential hazard. For example, if another technician was servicing the hydraulic system, the autopilot functional test should not be performed.

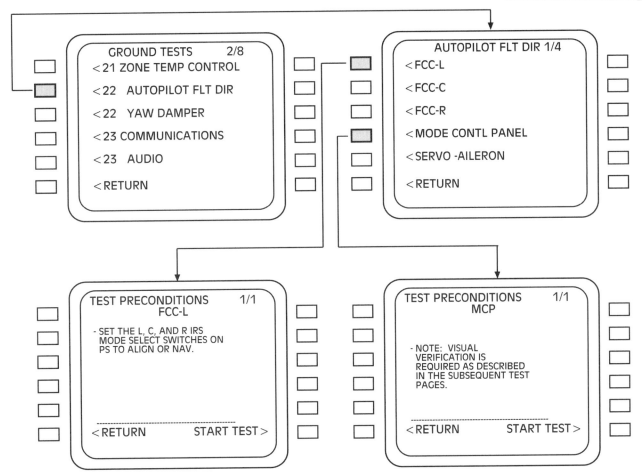

Fig. 6-69 Sequence of AFDC ground test displays showing test preconditions. (Northwest Airlines, Inc.)

List of the tests available through the AFDS ground tests menu

1) L/R/C FCC	Tests the FCC's and the systems/sensors that interface with the FCC's
2) MCP test	Tests the displays, switches, and control of the MCP
3) Aileron servo	Command and engage signal are sent to the aileron servo, the FCC monitors the response
4) Elevator servo	The elevator servo is tested same as aileron servo test
5) Rudder servo	The rudder servo is tested same as the aileron servo test
6) Autopilot disconnect switches	Tests function of A/P disconnect switch
7) Go-around switches	Tests function of G/A switches
8) Autoland unique test	Tests the operation of several functions critical to the autoland function
9) Air ground relay	Tests that all three FCC receive air/ground data
10) FCC configuration	Shows pin configuration of FCC's
11) FCC instrument	Monitors the interface between the FCC's and the integrated display system
12) Speed brake transducer	Monitors function and interface of S/B transducers
13) flap transducer	Monitors function and interface of flap transducers
14) Stabilizer trim	The autopilot sends a given signal to the trim system and the FCC's monitor the response
15) Surface limit	Test to ensure each FCC has the same control surface travel limits (separate test conducted for aileron, redder, elevator)
16) Transducer output	Tests the stabilizer, aileron, rudder, elevator, speed brake, and flap transducer outputs

Table 6-1 Autopilot Flight Director System (AFDS) ground tests.

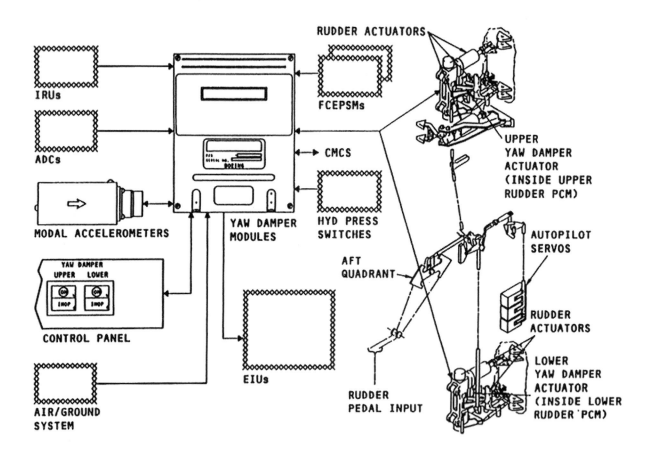

Fig. 6-70 Block diagram of the yaw damper system. (Northwest Airlines, Inc.)

Yaw Damper

The B-747-400 **yaw damper** system provides dampening for dutch roll prevention, turn coordination, and suppression of structural modal oscillations. **Structural modal oscillations** are an undesired effect created by turbulence, which causes bending of the fuselage around the wing area. There are two redundant yaw damper systems, each containing a yaw damper module powered by the FCEPSM (Flight Control Electronics Power Supply Modules). As seen in figure 6-70, each yaw damper module receives inputs from the IRUs, ADC, dedicated modal accelerometers, the yaw damper control panel, air/ground systems, the CMCS, hydraulic pressure switches, and a feedback signal from the yaw damper actuators.

The main yaw damper module outputs go to the yaw damper actuators. Output signals also go to the EFIS/EICAS interface units (EIUs), and the central maintenance computer system. The yaw damper control panel also receives an output from the yaw damper modules to verify the current operation of the yaw damper system.

CHAPTER 7
CREW COMMUNICATIONS & PASSENGER ENTERTAINMENT SYSTEMS

INTRODUCTION

Even during the early days of aviation, communication between flight crew and ground crew personnel was extremely important. The first aircraft were started by hand as the pilot simply yelled to his ground crew. Hand signals were often used for communication during times of adverse noise conditions. As aircraft began to carry passengers, they too needed to be informed of certain flight details. Early crew to passenger communications were simply a matter of loud conversations. As aircraft grew in size and complexity, it became evident that better communications between flight and ground crews, and between crewmembers and passengers were necessary.

Today's transport category aircraft contain a variety of systems, all dedicated to communications. For example, the flight crew can communicate with passengers, ground crew, air traffic control, and flight attendants. Some systems allow communication between airline operations or maintenance facilities and the aircraft central maintenance system. Gate changes, passenger lists or other pertinent data can be transmitted and printed using an onboard printer.

Many modern transport category aircraft also include an extensive passenger entertainment system. Entertainment systems include multichannel audio and video programs. All of these passenger systems must be linked to the flight crew in the event the pilot (or flight attendant) needs to make an announcement.

Today's transport category aircraft employ a sophisticated audio system to coordinate communications. This chapter will present an overview of these information and entertainment systems by examining the B-747-400 aircraft. Most of the systems found on this aircraft are similar to those of other large aircraft. Keep in mind, this chapter will present general concepts; the specific information for each aircraft must be accessed prior to performing any maintenance activities.

AIRBORNE COMMUNICATIONS ADDRESSING AND REPORTING SYSTEM

Airborne Communications Addressing and Reporting System (ACARS) is a digital air/ground communications service designed to reduce the amount of voice communications on the increasingly crowded VHF frequencies. ACARS allows ground to aircraft communications (in a digital format) for operational flight information, such as fuel status, flight delays, gate changes, departure times, and arrival times. ACARS can also be used to monitor certain engine and system parameters and downlink relevant maintenance data to the aircraft operator. Prior to the use of ACARS, this information was transmitted using voice communications. ACARS can be thought of as e-mail for the aircraft. Since the message is transmitted in digital format via ACARS, it occupies much less time on a given frequency than conventional voice communications, and since ACARS is an automatic system, transfer of information requires virtually no flight crew efforts.

There are two major corporations that provide ACARS services worldwide: ARINC Incorporated and a French organization known as SITA. As stated earlier in this text, ARINC Incorporated is a global corporation based in the United States with primary stockholders consisting of various U.S. and international airlines and aircraft operators. ARINC provides services related to a variety of aviation communication and navigation systems. One of the services available through ARINC is called **GLOBALink**. GLOBALink is the name of the ACARS service provided by ARINC Incorporated. It should be noted there are other providers of ACARS in various parts of the world; however, ARINC and SITA provide the majority of the ACARS services. In general, airborne equipment designed to operate using GLOBALink will operate with other ACARS services throughout the world.

ACARS Theory of Operation

The airborne components of ACARS connect to various sensors throughout the aircraft. These sensors are used to detect various parameters to be transmitted by ACARS. For example, virtually all transport category aircraft transmit OOOI (Out, Off, On, and In) data. *Out* stands for out-of-the-gate. ACARS uses a parking brake (or similar) sensor to determine out-of-the-gate time. *Off* means aircraft off-the-ground; this can be determined by a landing gear sensor. *On* refers to the aircraft touchdown (on-the-ground). Once again, a landing gear sensor can detect this condition and send the information to ACARS. *In* is determined when the aircraft is in-the-gate. ACARS could transmit *In* data when the parking brake is set. Remember, ACARS operates by transmitting digital data. A code of 1s and 0s is transmitted to deliver all information. If one were to listen to ACARS, it might sound similar to a modem connecting to an Internet web site.

ACARS transmits all information using VHF frequencies. The airborne equipment transmits through a VHF transceiver (typically located in the aircraft's equipment bay). At the time this text was written, new standards were being developed to link ACARS through satellite transmissions. The future use of satellites will improve worldwide coverage and enhance ACARS performance. In the future, more data will be transmitted through some form of ACARS and it will most likely be broadcast through satellites.

ACARS is a digital system that initiates all U.S. transmissions on the VHF frequency 131.55 MHz. The ACARS airborne equipment contains a control unit, typically located on the flight deck, a Management Unit (MU) and the necessary VHF transceiver located in an electronics equipment bay. A typical ACARS control unit is shown in figure 7-1. The ground-based equipment contains antennas and VHS transceivers located at various sites, a data link via telephone lines to one or more ACARS control facilities, and a data link to the various airlines (see figure 7-2). In some cases, the communications between ACARS ground facilities and the airlines are accomplished through microwave transmitters or satellite links. Of course, the airline element of ACARS is an elaborate system of command and control subsystems, which are used for maintenance, crew scheduling, gate assignments and other day-to-day operations.

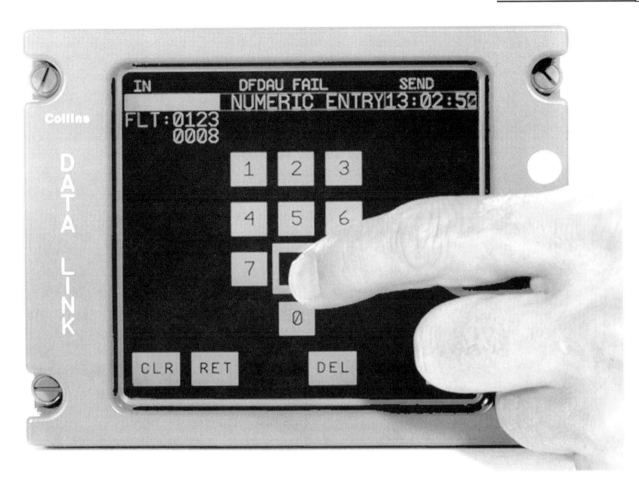

Fig. 7-1 A typical ACARS control unit. (Collins Divisions, Rockwell International)

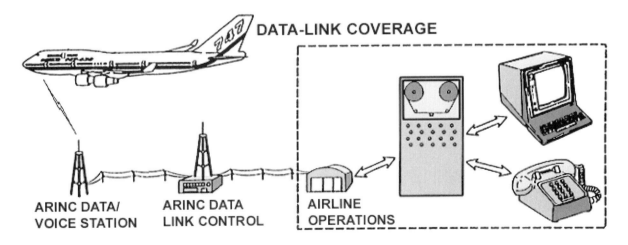

Fig. 7-2 Boeing 747-400 ACARS interface diagram. (Northwest Airlines, Inc.)

The ACARS system is designed to use a variety of VHF frequencies from 129.00 MHz to 137.00 MHz. In North America, all ACARS transmissions begin on 131.55 MHz. The ACARS ground facilities may then assign a different VHF frequency (between 129.00 and 137.00 MHz) to the aircraft. The assigned frequency will be a function of the aircraft location and the particular VHF frequencies used in that area. The ACARS ground facilities will then reassign new VHF frequencies to the aircraft ACARS as needed. Any frequency changes are totally automatic and therefore unknown to the flight crew. It is very likely that ACARS will change frequencies several times during a given flight.

Each aircraft using the ACARS system is given a specific address code. This code is used by the ground base facility whenever calling the aircraft. The airborne equipment will monitor all ACARS data transmissions on their assigned frequency and accept only those with the correct address code.

Since the VHF frequencies assigned to aircraft are relatively crowded, all ACARS messages must be as short as possible. To achieve a short message, special digitized code blocks using a maximum of 220 characters are transmitted in a digital format. If a longer message is needed, more than one block will be transmitted.

In general, the ACARS system operates in two modes for data communications: the **demand mode** and the **polled mode**. The **demand mode** will allow the flight crew or airborne equipment to initiate communications. To transmit, the airborne Management Unit (MU) determines if the ACARS channel is free from other communications. If the frequency is clear, the message is transmitted; if the frequency is busy, the MU will wait until the frequency is available. The ground station will send a reply to the message transmitted from the aircraft. If an error message, or no reply is received, the MU will continue to transmit the message at the next opportunity. After six unsuccessful attempts, the airborne equipment will notify the flight crew.

In the **polled mode**, ACARS operates only when interrogated by the ground facility. The ground facility will routinely uplink "questions" to the aircraft equipment and when a channel is clear, the MU will respond with a transmitted message. The MU organizes and formats all data prior to transmission. Upon request, the flight information is transmitted to the ground facility. Information for ACARS is collected from several aircraft systems, including the flight management system (FMS), and the central maintenance computer system (CMCS).

Architecture

The various systems that interface with the MU communicate via an ARINC 429 data bus. As shown in figure 7-3, the Boeing 747-400 ACARS management unit interfaces with:
1) Modularized Avionics and Warning Electronics Assembly (MAWEA), for chime tones needed to announce a call.
2) Control Display Units (CDUs), for basic operational controls.
3) EFIS/EICAS Interface Units (EIUs), for airplane identification and current flight status.
4) Airborne Data Loader (ADL), for the loading of ACARS MU software.

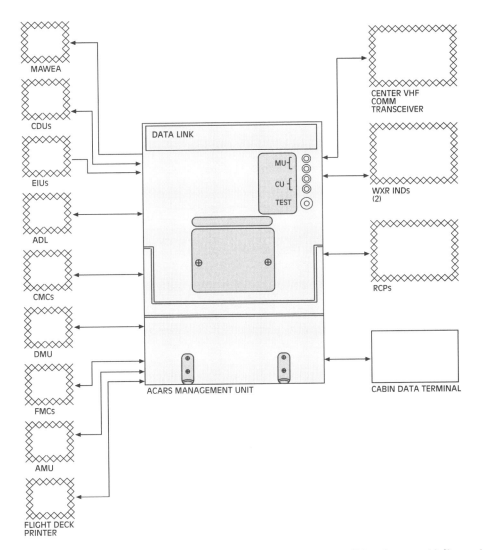

Fig. 7-3 Boeing 747-400 ACARS interface diagram. (Northwest Airlines, Inc.)

5) Central Maintenance Computer (CMC), for transmission of systems fault data.
6) Data Management Unit (DMU), for transmission of aircraft condition monitoring parameters.
7) Flight Management Computer (FMC), for transmission of various flight plan parameters.
8) Audio Management Unit (AMU), coordinating voice call lights.
9) Flight deck printer, for the reception of printed data.
10) The Center VHF Communications Transceiver, for transmission and reception of ACARS radio signals.
11) Weather Radar Indicators (WXR INDs), for display of received messages.
12) Radio Control Panels (RPCs), used for voice or data mode selection.
13) Cabin Data Terminal, for downlink or uplink of nonverbal data.

On the B-747-400, there is typically only one ACARS MU, which is located in the main equipment center. There are two power sources required by the ACARS MU: **115 VAC**, from **AC bus 3** and **28 VDC** from the **hot battery bus**. The power from the hot battery bus is used by the MU's internal clock. The MU performs all of the coordination functions of the airborne ACARS. The MU performs the following functions: 1) control of operational modes, 2) stores and decodes messages, 3) encodes messages and coordinates data transmissions, 4) provides timing using an internal clock, 5) provides all interface activities between airborne ACARS subsystems, and 6) monitors ACARS signals and accepts data transmitted to the specific aircraft.

Controls and displays

The ACARS control is managed through any one of the three Control Display Units (CDUs) located on the flight deck. ACARS is accessed through the CDU menu where a selection is made from the ACARS menu (see figure 7-4). Pressing the appropriate line select key (LSK) will access the different functions for ACARS. The flight crew can then input data as needed. All data entered will be displayed on the CDU display for visual reference. The message can then be sent automatically or manually using the ACARS menu. For example, pressing the LSK adjacent to *RECEIVE MESSAGES>* will allow ACARS to display any received message(s).

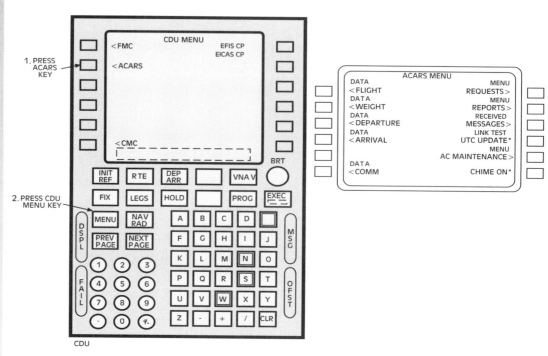

Fig. 7-4   A control display units (CDU) showing access to the ACARS menu. (Northwest Airlines, Inc.)

To activate the ACARS voice mode using the CDU, press the LSK adjacent to < *COMM*, then follow the prompts on the display. A request for voice communications can also be made from the ACARS ground network. The request will contain an assigned frequency. When the network is ready for voice communications, a call light on the audio control panel will illuminate and a soft chime will sound.

The B-747-400 **Radio Communications Panel (RCP)**, located on the flight deck, is used for radio selection and tuning of the HF and VHF radios. As seen in figure 7-5, the RCP displays the active and standby VHF/HF frequencies. If the center VHF radio is selected and ACARS is in use, the active communications frequency is replaced by the term *ACARS*. This indicates that ACARS is transmitting. The active frequency automatically returns when ACARS transmission is complete.

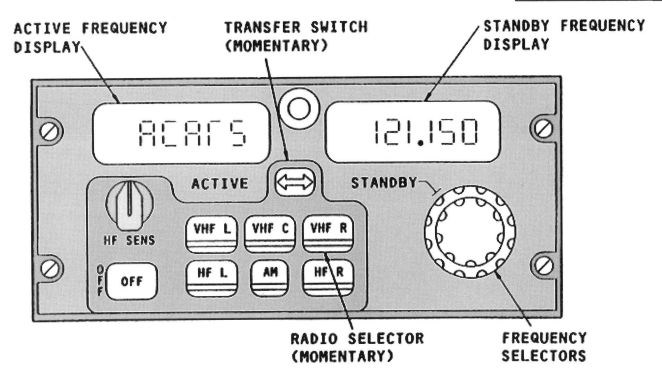

ACTIVE FREQUENCY DISPLAY **TRANSFER SWITCH (MOMENTARY)** **STANDBY FREQUENCY DISPLAY**

RADIO SELECTOR (MOMENTARY) **FREQUENCY SELECTORS**

Fig. 7-5 A radio communication panel showing the display of ACARS in the active frequency selection and 121.150 MHz in the standby frequency selection. (Northwest Airlines, Inc.)

Various EICAS maintenance pages and CMC fault reports can be sent to ground facilities using ACARS. This process becomes very helpful to maintenance personnel at line facilities. When aircraft have short turn around times, it is essential that maintenance is performed as quickly as possible. If the technician has received a copy of the CMC fault data, the troubleshooting process can begin while the aircraft is still in flight. In many cases, the CMC fault data can provide the technician with enough information so replacement LRUs, the necessary paperwork, and the appropriate tools can all be waiting for the aircraft's arrival.

To downlink a CMC fault report or EICAS maintenance page, first the crew would access the desired fault data on the CDU. In the example of figure 7-6 the pilot wishes to send the present leg faults page. To send the report, press the *REPORTS>* line select key, then select *< ACARS*. If all systems are operational, the message *IN PROGRESS* will be displayed. The message should change to *COMPLETE TO MU* and then, *COMPLETE TO GROUND*. If a failure has occurred or ACARS is processing an uplink message, one of the following messages will be displayed: *NO RESPONSE, NO COM, MU FAIL,* or *SERVICING UPLINK*.

On the B-747-400, the ACARS data terminal is typically located at the cabin video control center in the main cabin of the aircraft (see figure 7-7). The unit is installed in the wall and folds down for use. The terminal contains a standard alphanumeric key board and liquid crystal display, resembling a typical lap-top computer (see figure 7-8). This unit is typically used by the flight attendants to access ACARS data transmission/receive modes.

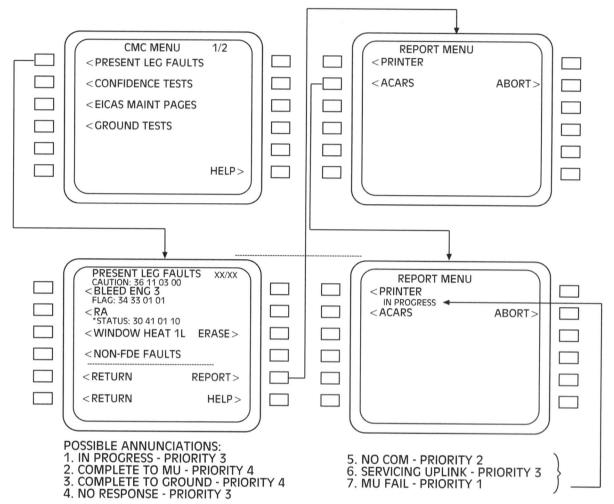

POSSIBLE ANNUNCIATIONS:
1. IN PROGRESS - PRIORITY 3
2. COMPLETE TO MU - PRIORITY 4
3. COMPLETE TO GROUND - PRIORITY 4
4. NO RESPONSE - PRIORITY 3

5. NO COM - PRIORITY 2
6. SERVICING UPLINK - PRIORITY 3
7. MU FAIL - PRIORITY 1

Fig. 7-6 Sequence of CMC displays used to send the Present Leg Faults through the ACARS transmitter. (Northwest Airlines, Inc.)

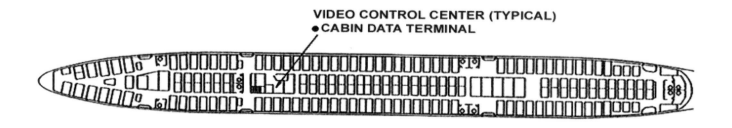

Fig. 7-7 Location of the cabin ACARS access unit on the B 747-400. (Northwest Airlines, Inc.)

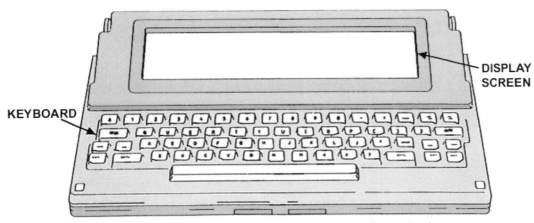

Fig. 7-8 ACARS cabin access terminal. (Northwest Airlines, Inc.)

ACARS data can also be displayed on either of the two weather radar indicators located on the lower right and left sides of the instrument panel. Keep in mind weather radar data is typically displayed on the navigational display; therefore, accessing ACARS data does not inhibit the viewing of weather information. To display ACARS data, the WXR indicator control switch must be in the auxiliary (AUX) position. ACARS data is displayed in up to 12 lines of 32 characters each. Data can also be printed using an airborne printer.

Maintenance and troubleshooting

For ACARS operations on the B-747-400, the aircraft battery power must be available to the Hot Battery Bus. If the battery is removed from the aircraft during maintenance, the ACARS unit cannot be tested and the ACARS MU internal clock must be reset. The clock reset is accessed through the ACARS menu on the CDU.

The ACARS MU and related airborne subsystems are monitored by a BITE circuit within the MU. The BITE circuit continuously monitors the health of ACARS components and reports all failures to the central maintenance computer system. The MU test switch, located on the face of the MU, can also be used to initiate an ACARS test (see figure 7-9). During the test, all four lights should illuminate for three seconds (to test lamp operation), all lamps should extinguish for the next three seconds. After six seconds, the appropriate lamp should illuminate; green means 'OK', red means failed system. Red and green indicators are available for both the MU and CU.

To test the ACARS link between airborne equipment and ground based equipment, the system must be configured for a **link test**. The link test is performed by accessing the ACARS menu on the CDU and pressing the LSK adjacent to *UTC UPDATE* (see figure 7-4). The link test will activate the MU transmitting, receiving, and system interfacing capabilities. For this test to take place, the aircraft must be within range of an ACARS ground station. Keep in mind, VHF transmissions with the aircraft on the ground are very limited.

Another system to consider when troubleshooting ACARS is the center VHF transceiver. Since ACARS will only transmit using the center VHF radio, that system must be working for ACARS operations. Of course, different aircraft may use a different VHF transceiver for ACARS.

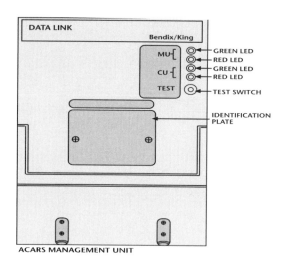

Fig. 7-9 ACARS management unit showing the MU test switch. (Northwest Airlines, Inc.)

INTERPHONE SYSTEMS

The **interphone system** is found on transport category aircraft to provide communications between flight crew, ground crew, flight attendants, and maintenance personnel. Due to the size of transport category aircraft, communications between different areas of the aircraft, or inside and outside of the plane are nearly impossible without the aid of the interphone system. Considering that most airports are noisy places to begin with, the interphone system becomes nearly essential to perform any maintenance or ground service activities.

The B-747-400 interphone system is divided into four basic subsystems: 1) **flight interphone**, 2) **service interphone**, 3) **cabin interphone**, and 4) the **crew call system**. The flight and service interphones, as well as the crew call system, all operate in conjunction with the audio management unit, and will be discussed here. The cabin interphone system is not controlled directly by the audio management unit and will be discussed in the next section of this chapter.

Flight Interphone

The B-747-400 **flight interphone** system is used to provide communications between flight crewmembers and/or other aircraft operations personnel. The flight interphone system is typically used for communications between flight crewmembers; however, the system may be interconnected to the service and cabin interphones when needed.

Flight Interphone Architecture

Refer to the flight interphone system interface diagram in figure 7-10 during the following discussion. The **Audio Management Unit (AMU)**, located in the main equipment bay, is used to process and control all audio signals required by the flight crew. The AMU receives input from the captain's and/or first officer's Audio Control Panel (ACP) which receives commands from the control wheel and glareshield Press-To-Talk (PTT) switches. The audio signals to the AMU can be sent from the captain's or first officer's hand held microphone (MIC), the oxygen mask MIC, or the headset boom MIC. The audio outputs can be sent to the headset, headphones, or cockpit speakers. The miscellaneous switch control module contains the switch to connect the service interphone with the flight interphone system. If the captain's or first officer's audio system fails, the observer audio system switch allows for transfer of the failed system to the observers station. Locations of the flight interphone system components are shown in figure 7-11.

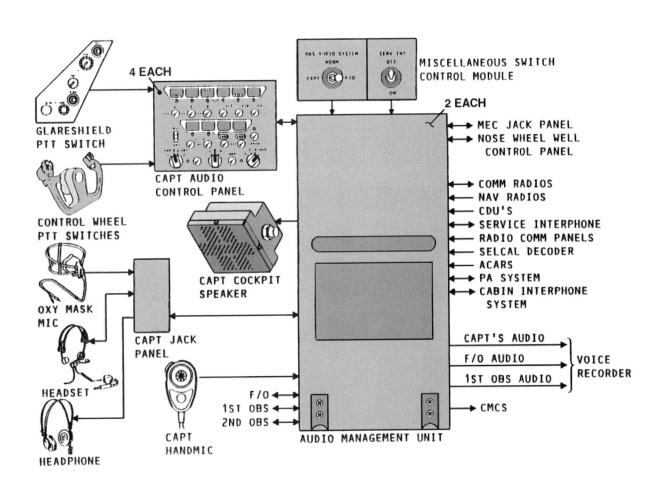

Fig. 7-10   Flight interphone interface diagram. (Northwest Airlines, Inc.)

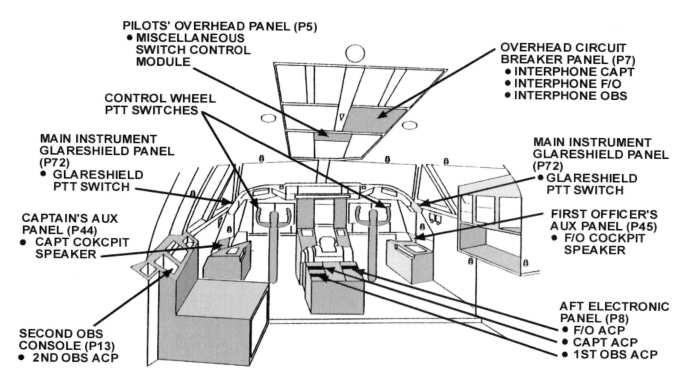

PILOTS' OVERHEAD PANEL (P5)
- **MISCELLANEOUS SWITCH CONTROL MODULE**

CONTROL WHEEL PTT SWITCHES

MAIN INSTRUMENT GLARESHIELD PANEL (P72)
- **GLARESHIELD PTT SWITCH**

CAPTAIN'S AUX PANEL (P44)
- **CAPT COKCPIT SPEAKER**

SECOND OBS CONSOLE (P13)
- **2ND OBS ACP**

OVERHEAD CIRCUIT BREAKER PANEL (P7)
- **INTERPHONE CAPT**
- **INTERPHONE F/O**
- **INTERPHONE OBS**

MAIN INSTRUMENT GLARESHIELD PANEL (P72)
- **GLARESHIELD PTT SWITCH**

FIRST OFFICER'S AUX PANEL (P45)
- **F/O COCKPIT SPEAKER**

AFT ELECTRONIC PANEL (P8)
- **F/O ACP**
- **CAPT ACP**
- **1ST OBS ACP**

Fig. 7-11 Location of flight interphone system components. (Northwest Airlines, Inc.)

Since the AMU controls all voice transmissions to/from the flight crew, the AMU must interface with the various radio systems on the aircraft, the passenger address system and the cabin interphone system. The AMU also sends all audio signals to the aircraft's voice recorder. The AMU is comprised of four separate sections to provide redundancy: the captain's, first officer's, first observer's, and second observer's cards (see figure 7-12). There are three separate circuit breakers that feed the AMU and each card is internally fused. Diodes inside the AMU ensure that a shorted card will not draw power from an operational portion of the AMU. As seen in figure 7-13, each card of the AMU is dedicated to a specific group of inputs and outputs.

Microphones and headsets

There are several options for each flight crewmember regarding communication microphones (MICs) and speakers/headsets. The captain and first officer can choose from the following: 1) headset (MIC/earphone combination), 2) hand held MIC, 3) oxygen mask MIC, 4) headphones, or 5) cabin speaker (see figure 7-14). The headset contains an acoustic tube MIC that carries sound waves to the transducer. The transducer converts the sound waves into an electrical signal, which is sent to the preamplifier located in the headset cord. The headset MIC is keyed by one of three Press-To-Talk (PTT) switches. The headset earphone is a standard earphone assembly that allows the crewmember to monitor communications.

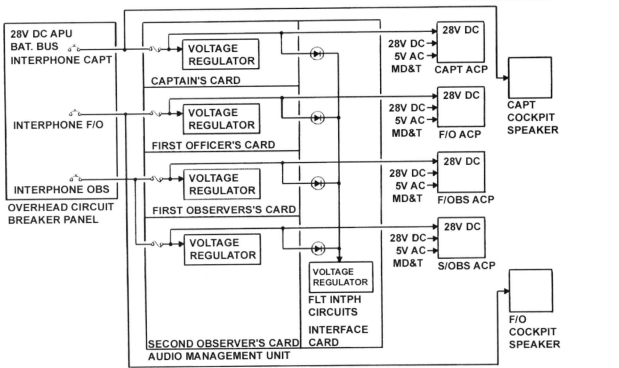

Fig. 7-12 Flight interphone system input power diagram; Note: the audio management unit is subdivided into different cards to provide redundancy. (Northwest Airlines, Inc.)

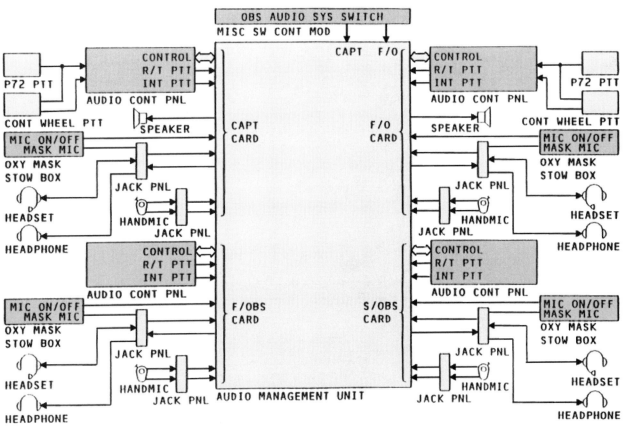

Fig. 7-13 Audio management unit interface diagram. (Northwest Airlines, Inc.)

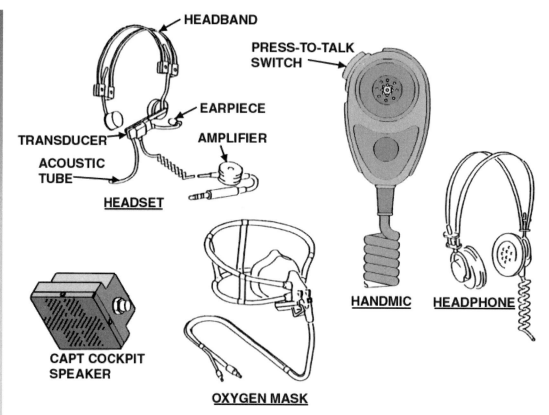

Fig. 7-14   Various microphones/earphone combinations available to the flight crew.

The oxygen mask contains a carbon-type MIC used for communications when the flight crew requires oxygen. The oxygen mask MIC is keyed by any of the PTTs. The hand held MIC assembly contains a preamplifier and microphone assembly. The hand held MIC is keyed by a switch on the side of the unit. The crewmember can select the headphones or cockpit speaker to monitor communications when using the oxygen mask or hand held MIC.

Controls

The **Audio Control Panels (ACPs)** provide command signals to the AMU for control of the communication and navigation audio signals. Each of the four ACPs (captain, first officer, first observer, and second observer) is operated independently and connects to the AMU via an ARINC 429 data bus. The Receive/Transmit (R/T) and Interphone (INT) PTT switches send a discrete signal to the AMU.

There are 13 audio receiver controls located on the ACP. Each receiver control is dedicated to a specific communication system. The receiver controls are a push on/ off switch-volume control combination. When a receiver control is in the *on* position, the associated green indicator illuminates.

As seen in figure 7-15, there are ten transmit switches located on each of the four identical ACPs. Each of the transmit switches are dedicated to one of the following systems: left/center/right VHF, flight interphone (FLT), cabin interphone (CAB), passenger address (PA), and the left/right HF, left/right satellite communications (SAT). The transmit switches are lighted push-button switches divided into two sections: top and bottom. The bottom section displays a *CALL* light when the related system has an incoming message for the flight crew. When the transmit switch is pressed, the *CALL* light extinguishes and the *MIC* light illuminates on the top half of the switch. If a given system is already selected for transmission, the *CALL* light extinguishes when a PTT switch is pressed.

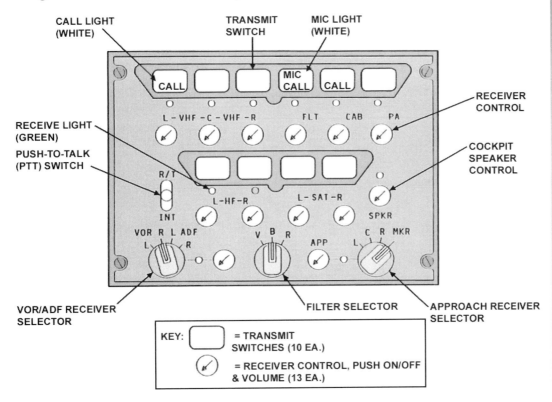

Fig. 7-15 Boeing 747-400 audio control panel. (Northwest Airlines, Inc.)

The ACP controls related to the interphone system are labeled **FLT, CAB,** and **INT.** The ACP flight interphone (FLT) call light illuminates whenever a request for communications is made by the ground crew via the flight interphone connection in the nose wheel well (see figure 7-16). Pressing the transmit switch allows the flight crew member to talk on the service interphone system via the hand held MIC. Pressing the INT PTT switch will key the headset or oxygen mask MIC to the interphone system. The FLT volume control adjusts the volume for all flight interphone messages received at that station. The cabin interphone (CAB) call light illuminates when a communication request is made by the flight attendants via the cabin interphone system.

The lower controls on the ACP are used for navigation radio audio reception (see figure 7-15). The flight crew can choose to listen to the **L/R VOR, L/R ADF,** or the **L/C/R marker beacon (MKR).** The filter selector is used to choose what type of VOR audio will be transmitted to the crew member. In the *V* position, only voice frequencies will be passed. In the *R* position only the station range code identification will be available. In the *B* position, both the VOR range and voice frequencies will be sent to the flight crew.

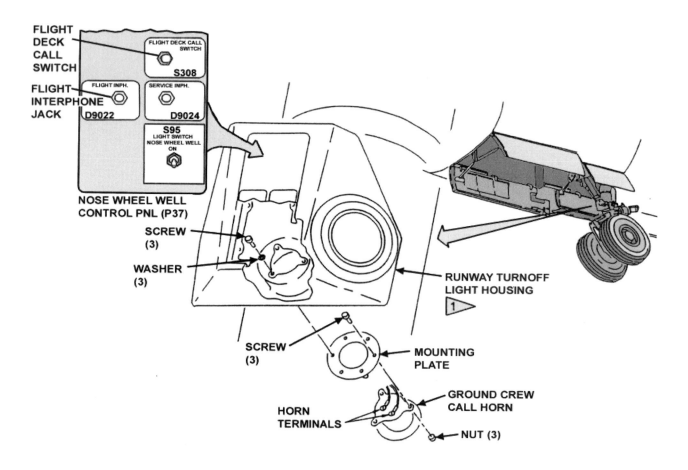

Fig. 7-16 Wheel well flight interphone connection.

Both the captain's and first officer's control wheel contain a two-position press-to-talk switch. As seen in figure 7-17, the switch control is a rocker assembly (trigger). Pressing the upper portion of the trigger keys the oxygen or headset MIC to the selected communication system. Pressing the lower portion of the trigger will activate the interphone system regardless of the communication system selected on the ACP.

Service interphone

The **service interphone** system permits ground crew communications through various access points located throughout the aircraft. Technicians often use this portion of the interphone system during various maintenance operations. The service interphone may also be used during ground service operations. As seen in figure 7-18, on the B-747-400 there are 19 different service interphone connection points. The interphone connection points are combined into three groups: **forward, mid,** and **aft**. Each input connection of the group is paralleled and sent to the audio management unit.

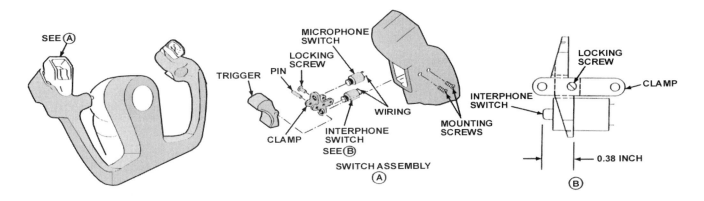

Fig. 7-17 Control wheel press-to-talk switches. (Northwest Airlines, Inc.)

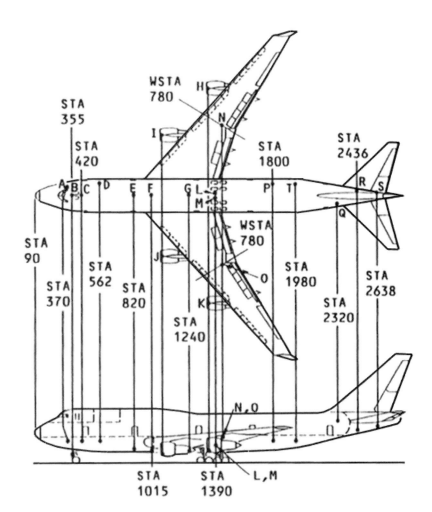

Fig. 7-18 Boeing 747-400 service interphone locations. A - Forward equipment center (internal), B - Nose wheel well panel (external), C - Main equipment center (internal), D - Forward cargo compartment (internal), E - Center equipment center (internal), F - Air conditioner equipment bay (external), G - Ground air inlet (external), H - Engine 4 nacelle (external), I - Engine 3 nacelle (external), J - Engine 2 nacelle (external), K - Engine 1 nacelle (external), L - right wheel well (external), M - Left wheel well (external), N - Right refueling station (external), O - Left refueling station (external), P - Aft cargo equipment panel (internal), Q - Aft equipment center (internal), R - Tail cone (internal), S - APU T - Bulk cargo door (internal) (Northwest Airlines, Inc.)

441

The **Audio Management Unit (AMU)** receives input signals from the service interphone jacks. As discussed earlier, the AMU coordinates interphone system communications (see figure 7-19). The service interphone inputs are amplified by the AMU and sent back to the service interphone jacks or flight interphone system (if the service interphone switch is on).

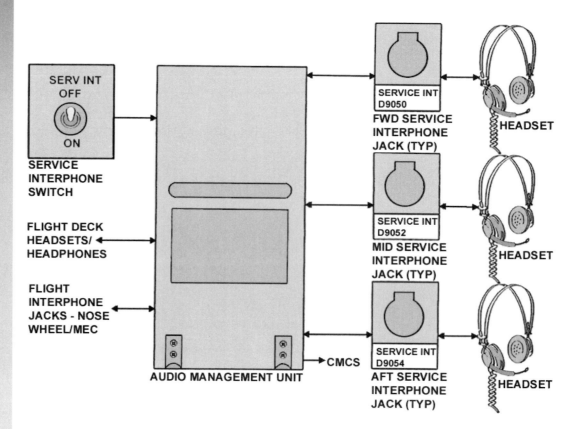

Fig. 7-19   Block diagram of the service interphone system. (Northwest Airlines, Inc.)

Crew call system

The **crew call system** is used to alert ground crew personnel of an interphone message from the flight crew, and to alert the flight crew of a message from the ground crew. A horn, along with the interphone control panel is located in the nose wheel well (see figure 7-16). Pressing the flight deck call switch sends a signal to the Modularized Avionics and Warning Electronics Unit (MAWEA). MAWEA activates a chime through the flight deck speakers and illuminates a light on the audio control panel. The **Cabin Interphone Controller (CIC)** is used to interface the various subsystems of the crew call system. The CIC will be discussed later in this chapter.

The ground crew call horn sounds any time a flight deck interphone call is made to the ground crew. The same horn is also used to alert the ground crew of an equipment cooling failure, or if the Inertial Reference (IR) unit is on and AC power is not supplied to the aircraft. If the IR is allowed to operate on battery power only, the battery will quickly discharge.

Interphone maintenance and troubleshooting

Maintenance and troubleshooting of the interphone system typically consists of wiring (connector) problems related to the various system's components, or removal and replacement of various LRUs. The AMU contains BITE circuitry that communicates with both Central Maintenance Computers (CMCs). The CMCs record messages related to an AMU failure, ACP failure, or ARINC 429 interface bus failure.

MICs and headsets are items that require regular maintenance. These items are constantly being moved, bounced, and dropped about the cabin. This causes wires to shake loose, wires to break, and intermittent audio transmissions. The easiest way to troubleshoot MICs and headsets is to operate the system; shake the suspect unit and/or related wiring while listening for problems. Then, swap the suspect unit and try the system a second time. Service interphone connection jacks also seem to be problem components. Many of these jacks are exposed to the outside environment and are often treated roughly by ground personnel. In most cases, the jack becomes loose causing an intermittent connection. Be sure to inspect these components carefully when dealing with interphone faults.

ADVANCED CABIN ENTERTAINMENT SERVICE SYSTEM

The **Advanced Cabin Entertainment Service System (ACESS)** is used on the B-747-400 to control five major subsystems: the passenger entertainment audio system, the cabin interphone system, cabin lighting, the passenger service system, and the passenger address system. In short, ACESS controls and distributes passenger audio and lighting. ACESS is software driven and allows for improved cabin flexibility. That is, ACESS can easily reconfigure cabin audio and video signals to compensate for changes in cabin configuration. Through ACESS software, audio and video signals can be changed for virtually any seating arrangement. Older aircraft would have required changes in system wiring and components.

ACESS is a multiplexed digital system; therefore, the majority of ACCESS information is transmitted on a serial data bus. Transmission of serial digital data provides a great reduction in wiring; however, multiplexer and demultiplexer circuits are required. As shown in figure 7-20, multiplexers/demultiplexers are required at the beginning and end of each audio transmission route. Figure 7-20 also shows that the audio input signals go through an A/D converter circuit and the audio output goes through a D/A converter circuit. Since the initial audio signals are analog, they must be converted to digital signals prior to multiplexing. (Multiplexing is the process of converting parallel digital data into serial digital data.) After multiplexing, the data is then transmitted. After transmission, the data is demultiplexed. (Demultiplexing is the process of converting serial data into parallel data.) The D/A converter is then used to change the digitized audio into an analog signal. The analog audio signal is then sent to the passenger headphones. It should be noted: multiplexers and demultiplexers, as well as A/D and D/A converters are typically circuits found within an LRU. These circuits are not stand-alone components.

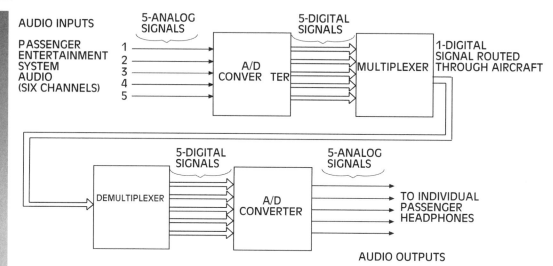

Fig. 7-20   Typical configuration for the transmission of multiplexed data.

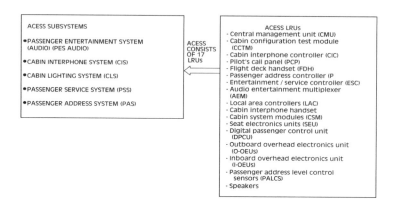

Fig. 7-21   ACESS system and subsystem interface.

Architecture

As shown in figure 7-21, ACESS is comprised of 17 different LRUs. ACESS can be thought of as a pyramid of control units which are used to receive inputs, process data, coordinate activities, and send audio and control signals to their proper destination (see figure 7-22). Please refer to this pyramid periodically during the following discussion on ACESS.

Three of the ACESS LRUs are called main controllers, they are:

    1) The **Cabin Interphone Controller (CIC)**, supervises the Cabin Interphone System (CIS) through the cabin/flight deck handsets, and the flight deck interphone system.

    2) The **Entertainment/Service Controller (ESC)**, controls the Passenger Entertainment System (PES) audio to the passenger headphones, Cabin Lighting System (CLS), and the Passenger Service System (PSS).

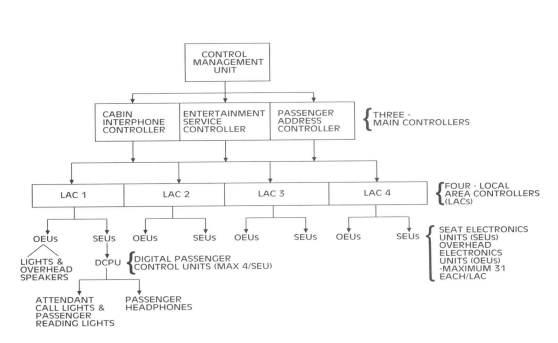

Fig. 7-22   The hierarchy of ACESS control units.

3) The **Passenger Address Controller (PAC)**, controls the Passenger Address System (PAS) audio output to the cabin speakers and passenger headphones.

Each of the three main ACESS controllers interface with four local area controllers and the control management unit (see figure 7-23). The **Control Management Unit (CMU)** is a digital interface unit used to coordinate programming, testing and monitoring of ACESS. The CMU receives inputs from the cabin configuration test module, the software data loader, and the EFIS/EICAS interface units. The **Cabin Configuration Test Module (CCTM)** is used to control portions of ACESS and to perform system tests. The CCTM will be discussed later in this chapter. The **software data loader** is used to program ACESS in the event of an aircraft configuration change. The **EFIS/EICAS Interface Units (EIUs)** are used to interface ACESS with the central maintenance computer system and the integrated display units.

Figure 7-24 shows the interface between the three main ACESS controllers and the local area controllers. (Remember the three main ACESS controllers are the CIC, ESC, and the PAC.) Each **Local Area Controller (LAC)** is comprised of three independent circuits, one dedicated to each main controller. As seen in figure 7-25, each LAC controls a given zone of the aircraft; LAC 1 controls zones A & B, LAC 2 controls zones C & D, LAC 3 controls zone E, and LAC 4 controls the upper deck.

The main controllers are located in the aircraft's main equipment rack, the LACs are located in the cabin area. LACs are typically located above various ceiling panels.

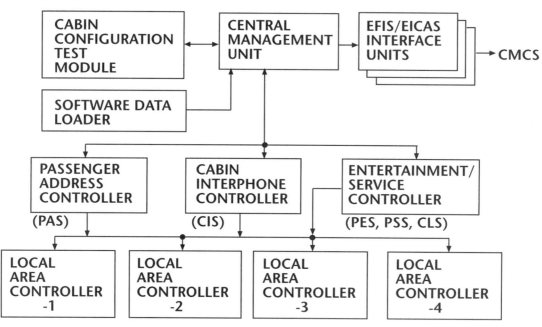

Fig. 7-23 Block diagram of the ACESS controller interface.

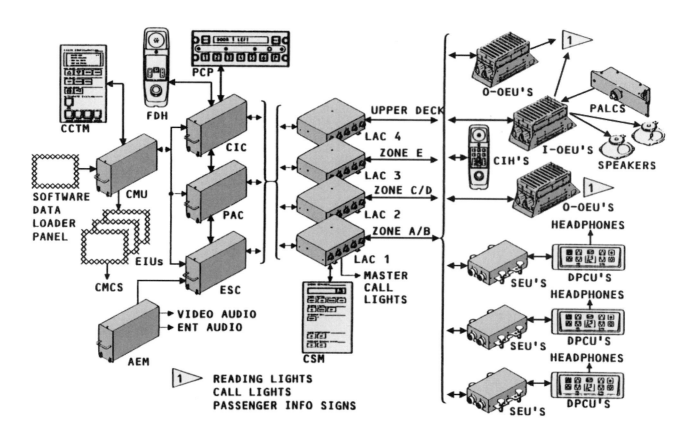

Fig. 7-24 Pictorial diagram of the Advanced Cabin Interface Service System. (Northwest Airlines, Inc.)

The LACs act as a distribution unit for each of the three main controllers and send lighting commands and digitized audio signals to the **overhead electronic units** and the **seat electronic units** for distribution to the individual passenger stations and related lighting.

The ACESS **Overhead Electronic Units (OEUs)** control the overhead speakers, cabin lighting, passenger reading lights, flight attendant call lights, and passenger information lights (*no smoking*, and *fasten seat belts*). A maximum of 31 OEUs can be connected to one LAC. The **Seat Electronic Units (SEUs)** are used to interface with each seat control for selecting various audio channels, volume, reading lights, and attendant call lights. The SEU also distributes the requested audio to each passenger headphone connection. One SEU is located under each group of seats, and can interface with up to four seats in a group. A maximum of 31 SEUs can connect to one OEU. For locations of OEUs and SEUs refer to figure 7-25.

During the preceding discussions, the major LRUs of ACESS have been introduced. The ultimate goal of ACESS is for these LRUs to interface with the aircraft crew and passengers. There are various control panels, handsets, headphones, and speakers which are all part of ACESS. These items will be discussed as the specific systems are presented in upcoming sections of this chapter.

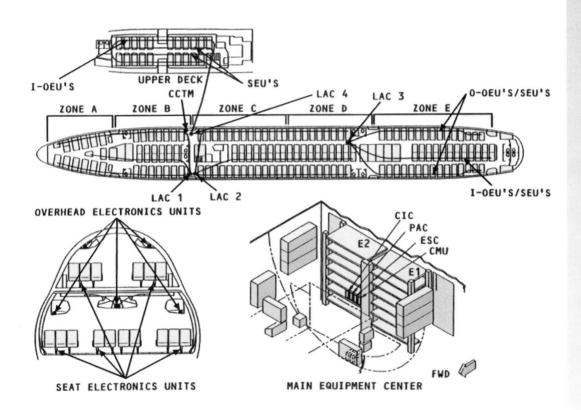

Fig. 7-25 ACESS zones and component locations. Note: LAC = local area controller, OEC = overhead electronic units, SEC = seat electronic unit, CCTM = cabin configuration test module, CIC = cabin interphone controller, PAC = passenger address controller, ESC = entertainment/service controller, CMU = control management unit. (Northwest Airlines, Inc.)

Configuration database

The **configuration database** is the software program which is used to inform ACESS of the current cabin layout. There are several LRUs within ACESS which contain the configuration database stored in a nonvolatile memory. The Central Management Unit (CMU) stores the configuration data base. The CMU also coordinates data base loading and storage for other ACESS LRUs.

The database can be used to identify items, such as seating configurations, cabin interphone dial codes, entertainment audio channels, passenger address areas, and the passenger address volume levels. This system provides flexibility to the airline for easy reconfiguration of the passenger compartment. The configuration database is typically modified by a shop technician or the engineering department using a personal computer. The database is then stored on a 3.5" floppy disk and transferred to the CMU using the aircraft's data loader. After the database has been downloaded to the CMU, the ACESS Cabin Configuration Test Module (CCTM) is used to download the database from the CMU to the various ACESS LRUs.

Cabin Interphone System

The **Cabin Interphone System (CIS)** provides communications between different flight attendant stations, and between flight attendant stations and the flight deck. A handset, which resembles a telephone receiver, is located at each flight attendants station (see figure 7-26). From the flight deck, the pilots can interface with the CIS using either the flight deck, handset or the flight interphone system.

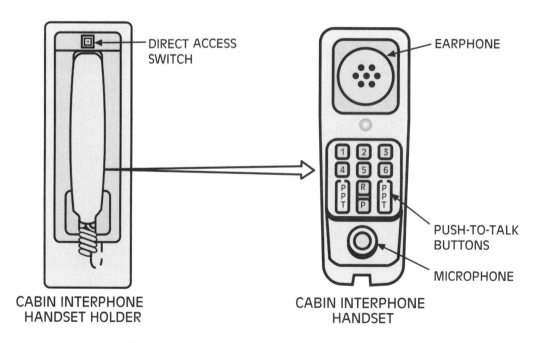

Fig. 7-26 Cabin interphone handset.

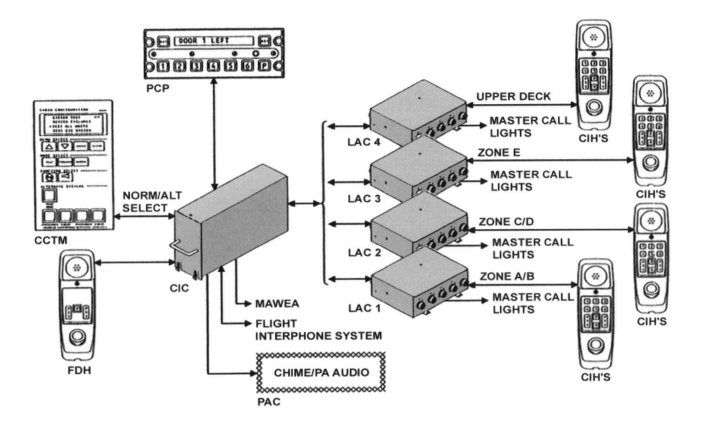

Fig. 7-27   Cabin interphone interface pictorial diagram.

Architecture

A CIS interface diagram is shown in figure 7-27. Please refer to this diagram during the following discussion on CIS architecture. Up to 25 Cabin Interphone Handsets (CIHs) can be connected to the CIS. Each cabin handset is wired to one of the four Local Area Controllers (LACs). The LACs connect to the cabin interphone controller via a digital data bus. The **Cabin Interphone Controller (CIC)** is the central multiplexer/demultiplexer for the CIS. All CIS audio goes through the CIC.

The CIC has two fully redundant circuits: normal and alternate. Each redundant circuit is capable of operating the complete CIS. If the normal circuit fails, the alternate is activated using the Cabin Configuration Test Module (CCTM) (see figure 7-28). The CIC performs all cabin interphone coordination activities and interfaces with other necessary LRUs. The CIC must be programmed with the current configuration database for proper operation. The CCTM interfaces with the CIC to down load the database from the CMU when updating the system. The various CIC interfaces and their related functions are listed in table 7-1.

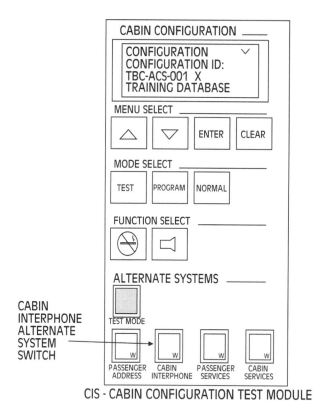

CIS - CABIN CONFIGURATION TEST MODULE

Fig. 7-28 Cabin configuration test module.

ACESS CIC INTERFACE LIST	
CIC Interface	Purpose of Interface
Cabin Configuration Test Module (CCTM)	To download configuration database files, and to activate alternate CIC circuitry
Pilot Control Panel (PCP)	Used to place an interphone call from the flight flight deck to the cabin
Modularized Avionics & Warning Electronics Assembly (MAWEA)	To sound a flight deck chime for interphone calls from the cabin to the flight crew
Flight Interphone System (FIS)	To allow flight crew personnel to communicate to CIS using the FIS
Passenger Address Controller (PAC)	To sound the appropriate cabin chime when a request is made

Table 7-1   ACESS CIC interfaces

Operation

To describe the operation of the cabin interphone system, three examples will be presented. The sequence of events described in these examples are typical of those which take place during cabin interphone use. The event sequence becomes very important when troubleshooting the system. The sequence of events can help a technician determine which LRUs are used during the various phases of an interphone call.

Example 1 - When the flight deck wishes to place a call to a flight attendant's station, the following occurs:

1) The process begins with entering the correct code into the PCP.

2) The CIC transmits a signal to the PAC to sound a chime at the attendants station. The CIC also sends a signal to the appropriate LAC which turns on the master call light at the attendant's station.

3) When the attendant picks up the handset, the CIC connects the pilot's and attendant's station and extinguishes the master call light. The pilot can use either the flight deck handset or flight interphone system for the communication. The proper selection must be made on the pilot's Audio Control Panel (ACP). The first officer's ACP is used to control the first officer's audio. The ACP was discussed under the flight interphone section of this chapter.

Example 2 - If a flight attendant initiates a call to the flight deck, the following would occur:

1) The flight attendant enters the interphone station code on the handset.

2) The station code signal is sent through the LAC to the CIC.

3) The CIC sends a signal to MAWEA to sound the flight deck chime, tells the PCP to display an incoming message, and sends a signal to the ACP to illuminate the *CAB INT* light.

4) When the pilot activates his/her handset, the CIC connects the call and turns off all call lights.

Example 3 - If a flight attendant calls another attendant station the following series of events would take place:

1) The station code is entered into the handset.

2) The station code goes to the CIC through the LAC.

3) The CIC sends a signal to the PAC that sounds a chime at the appropriate attendants station. The CIC also initiates a signal to the appropriate LAC to light the master call light at the attendant's station being called.

4) When the called station handset is removed from its holder, the CIC connects the two stations and turns off the master call light via the LAC.

Call priority

The cabin interphone system has a priority program which routes calls according to their importance. The order of CIS call priority is:

First, *Pilot Alert Call*. A *pilot alert call* can be made from any flight attendant's station using the code *PP*. *Pilot alert call* overrides all other interphone operations except another pilot alert call.

Second, *All Call*. An *All Call* can be placed from any station using code *55*. *All call* contacts all interphone stations.

Third, *Priority Line Calls*. One of the cabin interphone stations can be selected as a priority station. This station takes priority over other cabin stations.

Fourth, *Attendant's All Call*. Using code *54* any cabin station will call all other cabin stations. The flight deck is not included in the attendants all call.

Fifth, *Normal station to station calls*. In this mode, one station connects to another in the order received by the CIC. The CIC can store up to four call requests. Each call is processed in order of reception. The waiting call is processed after the preceding call is finished.

Passenger address system

The **Passenger Address System (PAS)** transmits audio through a series of cabin speakers to announce messages from: 1) the flight crew, 2) flight attendants, 3) the prerecorded tape reproducer (messages and boarding music), 4) the video tape reproducer (audio only), or 5) to sound cabin chimes. The PAS on the B-747-400 is segmented into various zones, so that different cabin sections can receive specific announcements, or so all sections of the cabin can receive the same announcement.

Architecture

Figure 7-29 shows a PAS interface diagram; please refer to this figure during the following discussion on the PAS. The **Passenger Address Controller (PAC)**, one of the three main ACESS controllers, provides the main coordinating functions for the PAS. The PAC contains two complete circuits, normal and alternate. A list of the PAC input interfaces and their basic functions is shown in table 7-2.

There are six discrete inputs shown on the interface PAS diagram (upper left of figure 7-30). The *no smoking* and *fasten seat belt* discrete are used to signal the need for a cabin chime. The cabin chime system is controlled through the PAC. The engine start, airborne, Vmo-15 KTS (maximum speed less 15 knots), and decompression discrete are used to inform the PAC of different aircraft configurations. This configuration information is used for automatic PA volume adjustments.

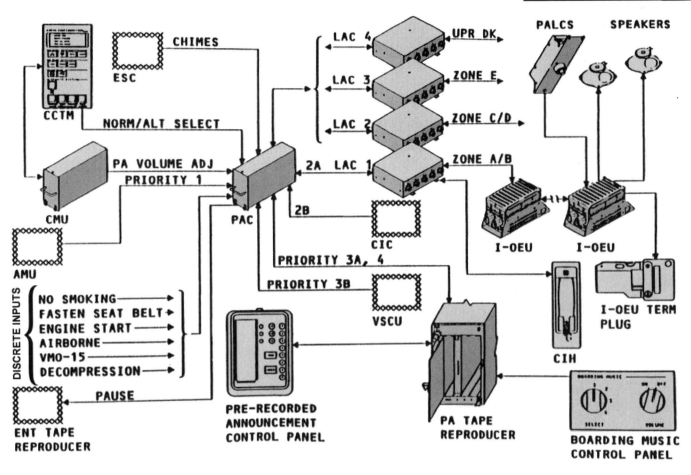

Fig. 7-29 Passenger address system interface pictorial diagram. (Northwest Airlines, Inc.)

PASSENGER ADDRESS CONTROLLER (PAC) ARINC 429 DATA INPUTS	
PAC Interfaces With	Purpose of Interface
Cabin Configuration Test Module (CCTM)	To select normal or alternate PAC circuits to control database updates from CMU, controls
Central Management Unit (CMU)	To receive configurations database information, and for fault monitoring
Audio Management Unit (AMU)	For connection to the flight interphone system
Video System Control Unit (VSCU)	For inputs of video system audio
PA Tape Reproducer	For inputs of prerecorded audio tape signals
Cabin Interphone Controller (CIC)	To coordinate calls from all cabin handsets (CIH) to the PAC for PA announcements
Local Area Controller (LAC) 1	Used for direct connection of 2 cabin handsets to the PAC
Local Area Controller (LAC) 1-4	Multiplexes and digitizes audio and control signal to connect all CIHs to the PAC for announcements from any attendants station
Entertainment/Service Controller (ESC)	Provides signals to PAC for operations of cabin chimes

Table 7-2 Passenger Address Controller ARINC 429 data inputs

The main PAC output signals include: 1) data to the CMU for system fault monitoring, 2) control signals to the entertainment, and PA tape reproducers, and 3) multiplexed and digitized audio signals to the Local Area Controllers (LACs). The LACs demultiplex the signals and send the digitized audio to the Overhead Electronics Units (OEUs). The OEUs convert the audio to an analog signal and send that signal to two cabin speakers (see figure 7-30). Each OEU contains two separate amplifiers, one for each speaker.

Operations

The PAS has six operating modes related to the following functions: 1) flight deck interphone PA announcements, 2) direct access PA messages, 3) cabin interphone PA transmissions, 4) broadcasting PAS tape reproducer music/messages, 5) broadcasting video system audio, and 6) cabin chime activation. The following three examples will describe some of the operational functions of the PAS.

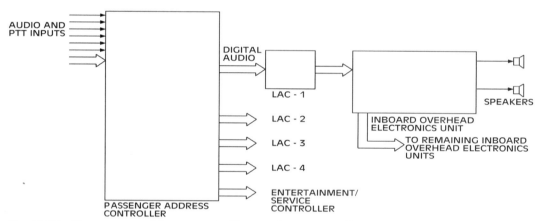

Fig. 7-30 Passenger address controller audio output diagram. (Northwest Airlines, Inc.)

Example 1 - Placing a PA announcement from the flight interphone system (see figure 7-31):

1) The pilot selects the PA system for transmission and presses the PTT switch on the audio control panel (ACP). This signal goes to the AMU.

2) The pilot announces the message into the headset or oxygen mask MIC. This signal goes to the AMU.

3) The AMU sends the press-to-talk (PTT) discrete and audio signals to the PAC.

4) The PAC multiplexes and digitizes the signals, sending them to the LACs (reference the PAS interface diagram, figure 7-30).

5) The LACs demultiplex the data and send the signals to the OEUs.

6) The OEUs convert the digital audio to analog signals, amplifies, and send the signals to the cabin speakers.

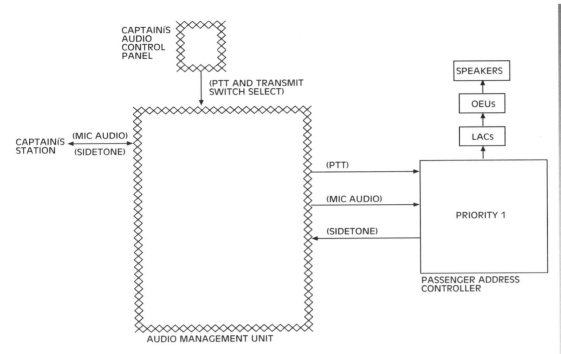

Fig. 7-31 Passenger address system interface with the flight interphone system. (Northwest Airlines, Inc.)

Example 2 - Placing a PA announcement from a direct access handset. Direct access PA stations are located at each interphone station, the number one and four doors. PA announcements made from these stations have "*direct access*" to the PAC, bypassing the LAC circuitry.

    1) Direct access is initiated by pressing the direct access switch on the handset station (see figure 7-32). This sends a direct access signal to the LACs. The PTT switch is pressed to begin PA announcement.

    2) When receiving the direct access signal, the appropriate LAC closes an internal switch and connects the handset directly to the PAC.

    3) The PAC digitizes, multiplexes, and transmits the audio/control signal to the LACs.

    4) The LACs demultiplex the data and send the audio signal to the OEUs.

    5) The OEUs convert the digital audio to an analog signal, amplify the signal, and send audio to the cabin speakers.

Example 3 - Using the cabin interphone system to place a PA announcement from any cabin interphone station (see figure 7-33):

    1) The flight attendant presses the PTT switch at any flight interphone handset. (Note: The PTT switch is used for PA announcements only.)

    2) Pressing the PTT signals the appropriate LAC that a PA announcement is to begin from that handset station. The LAC receives a PTT discrete, MIC audio, and a handset station identification signal from the handset.

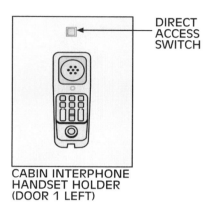

CABIN INTERPHONE
HANDSET HOLDER
(DOOR 1 LEFT)

Fig. 7-32   Handset holder showing the direct access switch.

3) The LAC digitizes and mutiplexes the input data. The LAC then transmits the data to the cabin interphone controller (CIC).

4) If more than one transmission is placed at a time, the CIC determines handset priority. The CIC then sends the appropriate multiplexed signals to the PAC.

5) The PAC then distributes the multiplexed audio and control signals to the LACs.

6) The LACs demultiplex the data and send the audio signal to the OEUs.

7) The OEUs convert the audio to analog, amplifies, and send the signals to the cabin speakers.

Understanding the sequence of events for each activity conducted by the PAS will help a technician during troubleshooting. The three preceding examples contained a brief outline of the activities. Be sure to become familiar with the system and which components are being used during a given PA announcement prior to troubleshooting.

Automatic volume control

The PAS **automatic volume adjustment** system is used to set PA volume levels according to the amount of cabin noise. The automatic volume adjustment can be done by the OEUs or PAC (see figure 7-34). The OEUs are the primary means for adjusting PA volume. Each OEU receives inputs from a Passenger Address Level Control Sensor (PALCS) which monitors local cabin noise levels. The OEUs adjust their audio output level according to the noise level signal received from the PALCSs.

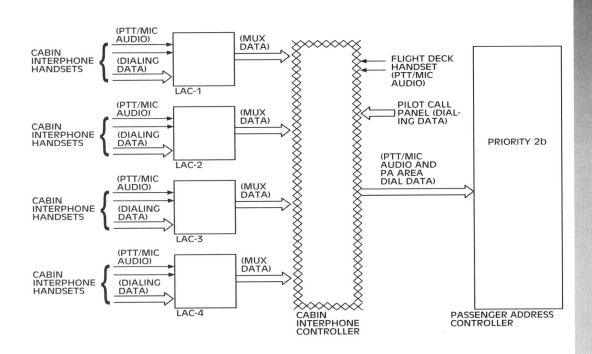

Fig. 7-33   Cabin interphone system input diagram. (Northwest Airlines, Inc.)

The PAC software contains automatic volume adjustment circuitry which is used to increase PA volume as cabin noise increases. The PAC determines cabin noise indirectly by monitoring aircraft status: engine on, airborne, Vmo - 15 KTS (maximum speed less 15 knots), and decompression. The PAC is the backup adjustment source and is used only if the PALCSs are turned off or malfunction. The PAC also receives a volume level adjustment command from the Central Management Unit (CMU). The CMU receives volume level commands from the Cabin Configuration Test Module (CCTM).

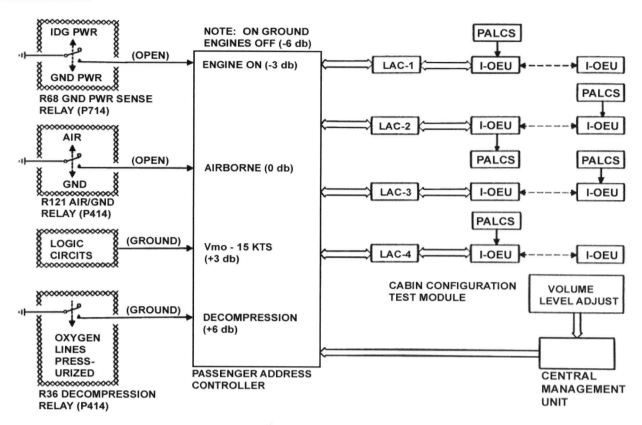

Fig. 7-34 Passenger address controller automatic audio level control diagram. (Northwest Airlines, Inc.)

Passenger entertainment audio

The **Passenger Entertainment System (PES)** audio provides prerecorded audio signals to each passenger seat interface unit. The PES typically contains several channels of music or other entertainment audio, as well as the audio portion of any in-flight movie (video) presentation. Each passenger can select a given channel for listening pleasure through the individual seat mounted control unit. The selected audio is transmitted through headphones to the passenger.

Architecture

During the following discussion on the PES audio, refer to the interface diagram shown in figure 7-35. The **Entertainment/Service Controller (ESC)** is the main ACESS controller which coordinates the distribution of the audio tape reproducer and the passenger video system audio signals. The system uses the basic ACESS philosophy, which digitizes and multiplexes the audio signals and sends them to local areas for distribution to the passenger headsets. The ESC can operate on one of two completely independent circuits, normal or alternate. The LRU input/output interfaces with the ESC. The purposes for these interfaces are listed in table 7-3.

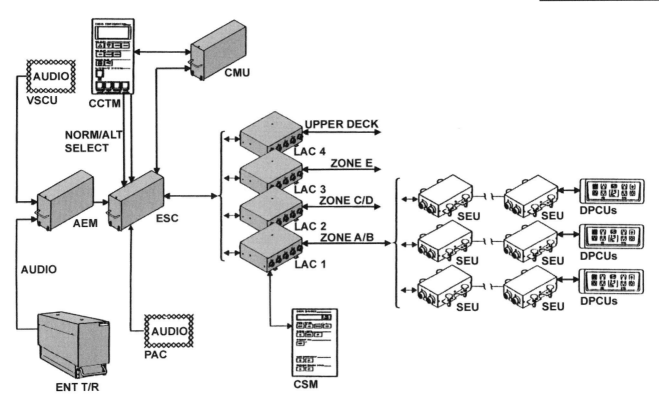

Fig. 7-35 Passenger entertainment audio system pictorial interface diagram. (Northwest Airlines, Inc.)

PASSENGER ENTERTAINMENT SYSTEM (AUDIO) MAIN CONTROLLER
INTERFACE LIST

LRU With Inputs To ESC	Purpose of Interface
Cabin Configuration Test Module (CCTM)	Controls normal/alternate circuit selection, used to install the configuration database onto the ESC, runs PES test function
Audio Entertainment Multiplexer (AEM)	Supplies the multiplexed audio signals to the ESC for control and distribution (the AEM receives the audio signals from both the audio and video tape reproducers
Passenger Address Controller (PAC)	Sends passenger address announcements to the ESC for distribution to passenger headphones during PA announcements
Local Area Controllers (LAC)	Receives multiplexed audio signals from the ESC for distribution to the seat electronic units (SEUs)

Table 7-3   Passenger Entertainment System (Audio) main controller interfaces.

The ESC can receive audio signals from three different sources: the Video System Control Unit (VSCU), the Audio Entertainment Multiplexer (AEM), or the Passenger Address Controller (PAC). Each of these three sources send multiplexed digital audio to the ESC. The ESC then controls the audio for distribution. One of the major functions of the ESC is to replace entertainment audio with passenger address audio when needed. The passenger address audio system will always have priority over any entertainment audio.

As seen in figure 7-36, the ESC sends multiplexed audio to the local area controllers (LACs). Each LAC has three output channels: left, center, and right. Each LAC output channel is connected to a maximum of 31 Seat Electronic Units (SEUs). The SEUs receive passenger audio selection and volume commands from the Digital Passenger Control Units (DPCUs) located on the armrest of each seat (see figure 7-37). The SEUs demultiplex and amplify the ESC signals. The SEUs also convert the digital audio to an analog signal. Using the DPCU command data, the SEU sends the selected audio to the passenger headphone jack.

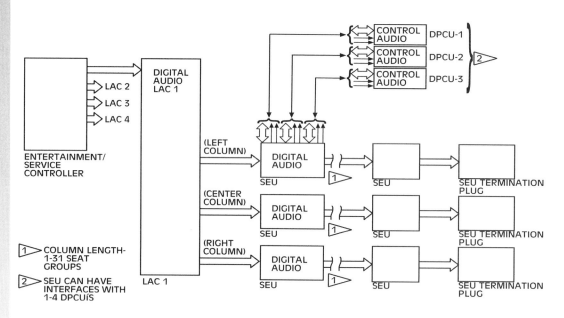

Fig. 7-36 Passenger entertainment audio interface diagram. (Northwest Airlines, Inc.)

A termination plug must be installed at the end of each series of seat electronic units (see figure 7-38). The output from the LAC connects to the J1 (Jack 1) plug of the SEU, the J2 connection of that SEU connects to the J1 plug of the next SEU in the series, and so on (see figure 7-38). On the last SEU in the series, the J2 plug must be connected to a 75 ohm termination plug. The termination plug must connect to the aircraft electrical ground. The termination plug is needed to ensure the transmission cable's impedance is correct.

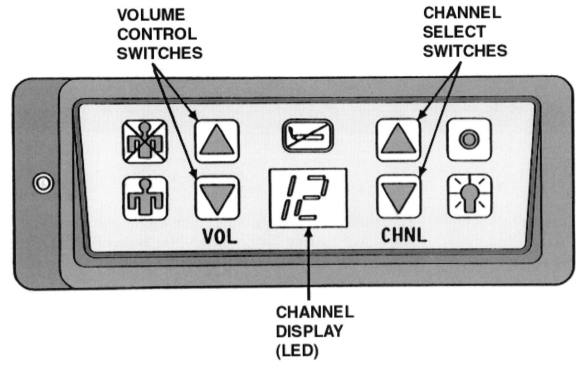

VOLUME CONTROL SWITCHES

CHANNEL SELECT SWITCHES

CHANNEL DISPLAY (LED)

VOL

CHNL

PES AUDIO - DIGITAL PASSENGER CONTROL UNIT

Fig. 7-37   Digital Passenger Entertainment Control Unit; located in the armrest of each passenger seat. (Northwest Airlines, Inc.)

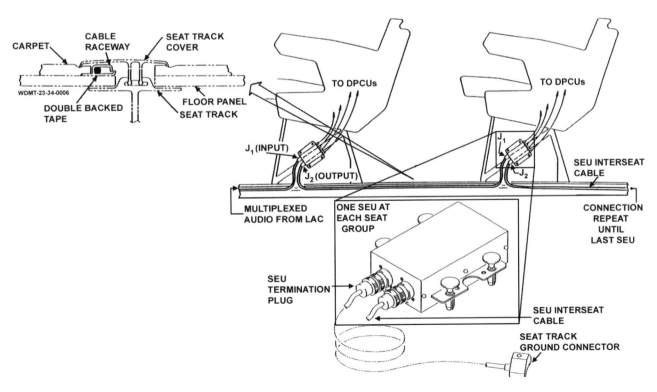

CARPET

CABLE RACEWAY

SEAT TRACK COVER

WDMT-23-34-0006

DOUBLE BACKED TAPE

FLOOR PANEL SEAT TRACK

TO DPCUs

TO DPCUs

J₁ (INPUT)

J₂ (OUTPUT)

J₁

J₂

SEU INTERSEAT CABLE

MULTIPLEXED AUDIO FROM LAC

ONE SEU AT EACH SEAT GROUP

CONNECTION REPEAT UNTIL LAST SEU

SEU TERMINATION PLUG

SEU INTERSEAT CABLE

SEAT TRACK GROUND CONNECTOR

Fig. 7-38   SEU cable interconnect diagram. (Northwest Airlines, Inc.)

Operation

The following example will cover the operation of the entertainment audio tape system as it applies to the PES. Other audio subsystems of the PES operate in a similar manner.

The following are the series of events for entertainment audio tape use:

1) The Cabin System Module (CSM) is used to turn on/off audio entertainment. The CSM is a small switch panel activated by the flight attendant. The CSM sends a digital signal to LAC 1 to turn on/off the Entertainment Tape Reproducer (ENT T/R).

2) As seen in figure 7-39, the ENT T/R sends up to 12 channels of analog audio to the Audio Entertainment Multiplexer (AEM). The AEM also receives six analog audio signals from the Video System Control Unit (VSCU).

3) The AEM digitizes and multiplexes the audio channels and transmits the signals to the Entertainment/Service Controller (ESC).

4) The ESC prioritizes entertainment and passenger address audio. The ESC sends multiplexed signals to the LACs.

5) The LACs distribute the signals to the SEUs.

6) The SEUs convert the signals to analog and distribute them to the passenger headphone jacks. The specific audio channel and volume is controlled by the individual DPCUs.

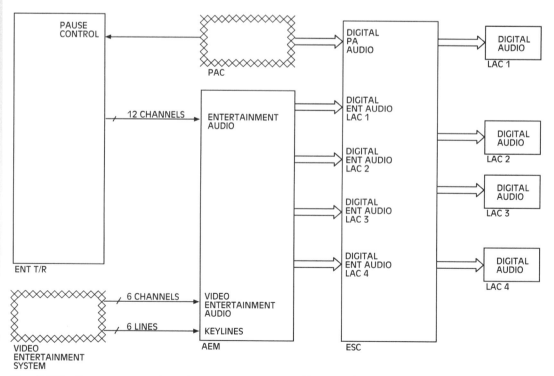

Fig. 7-39 Passenger entertainment system audio interface diagram. (Northwest Airlines, Inc.)

Passenger Service System

The **Passenger Service System (PSS)** is used by passengers to control reading lights and attendant call functions. The PSS is also used to control passenger information signs and attendant call functions from the lavatory. Passenger information signs are "no smoking", "fasten seatbelts", and "lavatory occupied" (see figure 7-40).

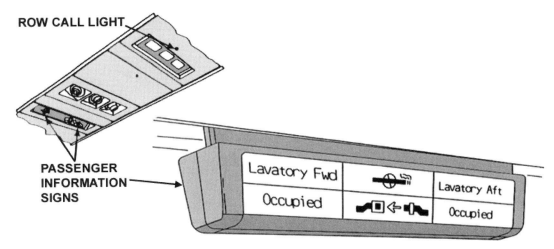

Fig. 7-40 Passenger information signs.

Architecture

During operation of the PSS most of the functions are controlled by the LACs. On the audio system just discussed, most of the functions are controlled by the system's main controller. Refer to the PSS interface diagram in figure 7-41 during the following discussions on PSS architecture. Table 7-4 gives a list of LRUs which interface with the ESC and a description of their function. The ESC is the main controller for the system. The ESC provides coordination functions and sends multiplexed control signals to the LACs for distribution.

Operation

This section will present two common functions performed by the PSS: operation of the passenger reading lights and the passenger attendant call.

Example 1 - When a passenger turns on the reading light using the seat mounted control:

1) The passenger pushes the light *ON* button on the Digital Passenger Control Unit (DPCU).

2) The DPCU sends a signal to the Seat Electronic Unit (SEU).

3) The SEU sends a digital multiplexed request signal to the Local Area Controller (LAC).

4) The LAC sends a command signal to the appropriate Overhead Electronics Unit (OEU).

5) The OEU turns on the reading light as requested by the passenger.

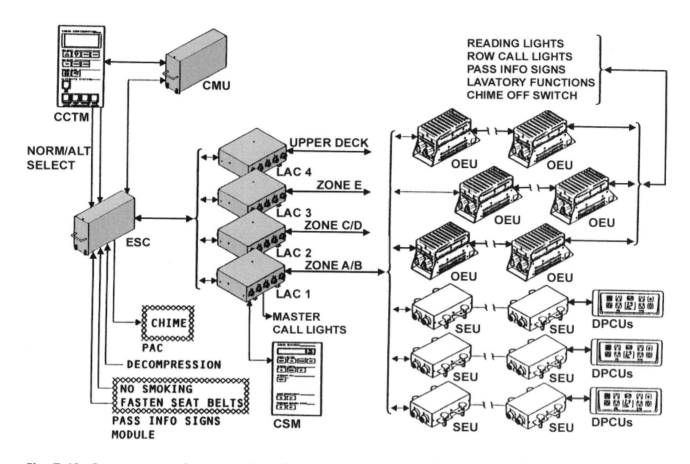

Fig. 7-41 Passenger service system interface pictorial diagram. (Northwest Airlines, Inc.)

PASSENGER SERVICE SYSTEM MAIN CONTROLLER INTERFACE LIST	
ESC Interface	Purpose of Interface
Cabin Configuration Test Module (CCTM)	Controls normal/alternate circuit selection, used to install the configuration database onto the ESC, to configure ìNo Smokingî sign locations, and to run test functions
Passenger Address Controller (PAC)	Used to sound attendant station chimes when nedded for attendant call functions
Passenger Information Sign Module	Supplies discrete inputs to the ESC for passenger information signs
Local Area Controller (LAC)	Transmits attendant call signals to ESC for sounding chime
Central Management Unit (CMU)	Controls data from CCTM to ESC

Table 7-4 Passenger Service System main controller interfaces.

Example 2 - When the passenger activates the attendant call function:

1) The passenger pushes the attendant call button on the DPCU.

2) The DPCU sends a command signal to the LAC.

3) The LAC controls three functions: a) the LAC turns on the appropriate master attendant call light, b) sends a command to the inboard OEU to turn on the call light for the appropriate seat row, and c) sends a signal to the ESC to sound the attendant call chime.

4) The ESC then sends the chime request to the PAC.

5) The PAC activates the chime.

It should be noted the attendant call chime can be turned off by a switch on the chime/light control module. The chime/light control module is located at the attendants station. The threshold and work station lights are also controlled at this module (see figure 7-42).

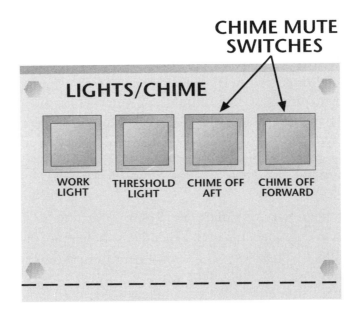

Fig. 7-42   Chime light control module. (Northwest Airlines, Inc.)

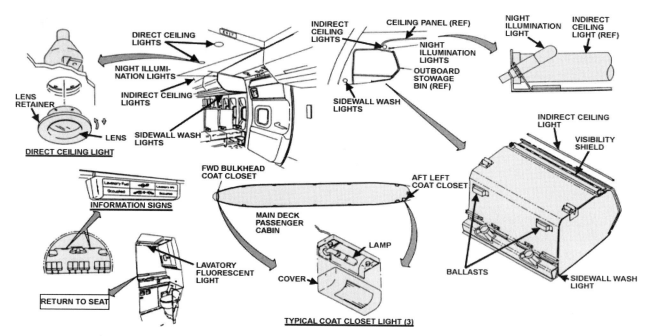

Fig. 7-43   Cabin lighting locations. (Boeing Commercial Airplane Company)

Cabin Lighting

The **Cabin Lighting System (CLS)** is part of ACESS which controls the following cabin lighting: 1) indirect ceiling lights, 2) sidewall wash lights, 3) direct ceiling lights, and 4) night lights. Figure 7-43 shows the various types of cabin lighting. Fluorescent type lights are used for wall wash, indirect ceiling lights, and some night lighting. Incandescent lamps are used for direct lighting. The cabin lights are controlled by flight attendants using the cabin system modules (see figure 7-44).

The Entertainment/Service Controller (ESC) is the main ACESS control unit used to process cabin lighting commands. The interface diagram in figure 7-44 shows the ESC connecting to the Local Area Controllers (LACs). The LACs receive lighting selection commands from the Cabin Systems Module (CSM). These commands are sent to the ESC where they are distributed to the appropriate LACs. The LACs which receive the ESC commands send a signal to the appropriate Overhead Electronics Unit (OEU). The OEUs then turn on the appropriate cabin lights.

ACESS Maintenance and Troubleshooting

The maintenance and repair of ACESS is similar to multiplexed passenger address/ entertainment/interphone systems found on other aircraft. In general, these multiplexed systems are somewhat high maintenance simply due to their complexity. Hundreds of light bulbs and switches are linked by miles of wires controlled by dozens of electronic units. With this large number of components comes a massive number of connector plugs. Connector assemblies are often a weak link in any electronic system and the large number required for ACESS makes the system vulnerable to failures. Be sure to carefully inspect all suspect connectors whenever troubleshooting the system.

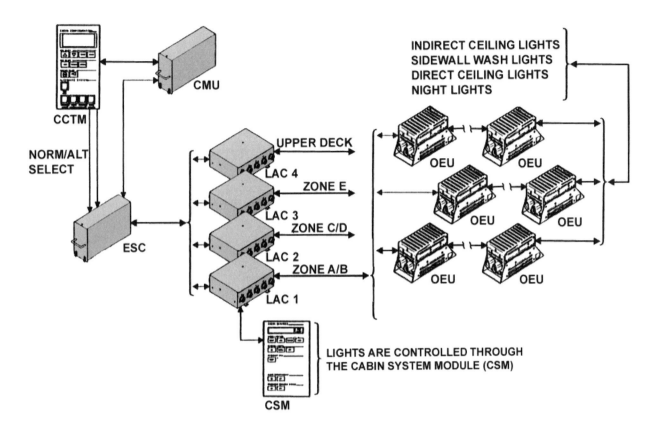

Fig. 7-44   ESC/LAC interface diagram.

Most of the ACESS LRUs are located in the cabin behind decorative panels or mounted to seat structures. For example, Overhead Electronic Units (OEUs) are located behind ceiling panels directly above the passenger seating. The OEUs are easily replaced by depressing the three latches and removing the screws which hold the unit in place (see figure 7-45). The four Local Area Controllers (LACs) are located above the cabin ceiling in three different areas. The aircraft service manual will provide diagrams that can be used to locate various ACESS components.

On the B-747-400, one electronic LRU is located under each group of passenger seats. As seen in figure 7-46, these Seat Electronic Units (SEUs) are located in the structure of the seat, which mounts to the seat track on the aircraft floor. Since passengers often rest their feet, or store baggage in this area, SEUs are often prone to wiring failures. As passengers move items (or their feet) near the SEUs, the wiring may get snagged. This causes connector problems or wiring failures due to the excess stress placed on the wiring. Whenever working on an SEU, be sure that all of the wiring is secured correctly that and all plastic covers are in place. This will help to prevent wire/connector damage.

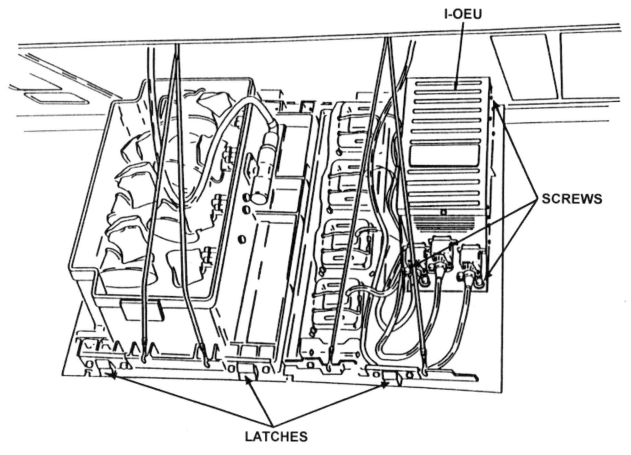

Fig. 7-45   Access to the overhead electronic unit. (Northwest Airlines, Inc.)

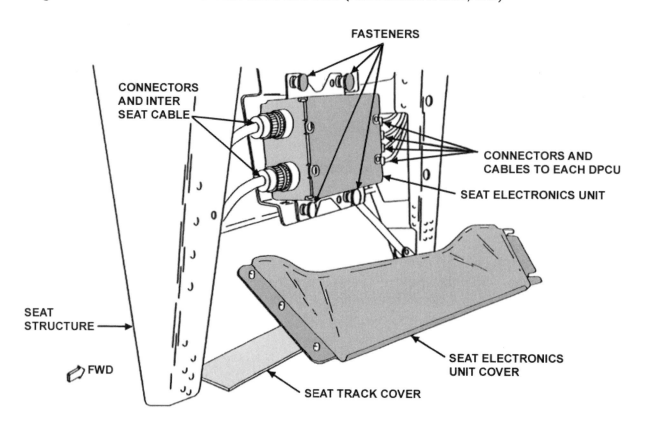

Fig. 7-46   A seat electronic unit installation. (Northwest Airlines, Inc.)

To aid in troubleshooting, ACESS employs built-in-test equipment that detects faulty LRUs, interface connections, and lighting components. The various ACESS LRUs continuously monitor system status and report any failures to the Central Management Unit (CMU). The CMU sends fault data to the EFIS/EICAS Interface Units (EIUs), which sends all fault data to the Central Maintenance Computer (CMC) and the Integrated Display System (IDS). EICAS will display seven different messages related to ACESS (see table 7-5). Since ACESS is a relatively non-flight critical system, all EICAS messages for ACESS are status messages.

ACESS MAINTENANCE MESSAGES AVAILABLE ON EICAS		
MESSAGE DISPLAYED	TYPE OF MESSSAGE	DESCRIPTION
ACESS MGT UNIT	Status	Failure of the central management unit (CMU)
PASS ADDRESS 1	Status	Failure of the passenger address controllers (PAC) normal controller circuits
PASS ADDRESS 2	Status	Failure of the passenger address controller (PAC) alternate controller circuits
CABIN INT 1	Status	Failure of the cabin interphone controllers (CICs) normal controller circuits
CABIN INT 2	Status	Failure of the cabin interphone controllers (CICs) alternate controller circuits
PASS SERVICE 1	Status	Failure of the entertainment/service controllers (ESC's) normal controller circuits
PASS SERVICE 2	Status	Failure of the entertainment/service ccontrollers (ESC's) alternate controller circuits

Table 7-5   Possible ACESS messages displayed by EICAS.

When requested by the technician, the ACESS CMU sends all fault data to the **Cabin Configuration Test Module (CCTM)**. The CCTM provides a means to monitor fault activity for the various ACESS LRUs. The CCTM is located in the aircraft cabin and can be used to perform **system status checks, system tests,** and **software configuration updates and verifications.**

System status checks

The ACESS **system status check** can be performed during flight or on the ground to retrieve failure data stored in the CMU memory. During operation, the CMU constantly monitors the main ACESS controllers for fault data. Any faults reported to the CMU are stored in memory. This memory is accessed using the CCTM located at attendant's station door #2 right. The operation of the ACESS components is unaffected by the system status check.

The following steps are needed to retrieve fault data stored in the CMU (see figure 7-47):

1) Press the normal button on the CCTM twice. Pressing the normal switch the first time activates the ACESS display. Pressing it a second time initiates the normal ACESS operating mode. While in the normal mode, the CMU will access all fault data from the reporting LRUs.

2) Press the enter button on the CCTM. This will cause the CMU to send a list of active fault data to the CCTM display. One of two messages can be displayed; *NO ERRORS TO REVIEW*, or the last reported failure. If more than one fault is present, each failure will then be displayed in succession.

3) Press the normal button to exit the system status mode.

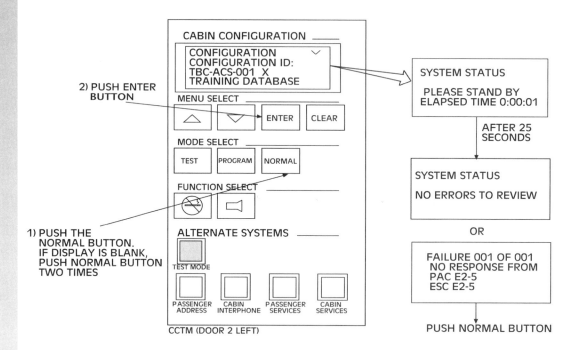

Fig. 7-47 Sequence of events to perform an ACESS status check. (Northwest Airlines, Inc.)

Any failures presented on the CCTM display will always contain five types of information: 1) failure number, 2) total number of failures, 3) description of the failure,

4) unit which detected the failure.

5) the failed component or interface. Figure 7-48 shows an example of failure data. In this case, the fault (2nd in a total of 4) is a program error, which was detected by LAC 1, and the fault lies in OEU 1-2-7 row 5. This same format is used for other types of fault data presented by the CCTM.

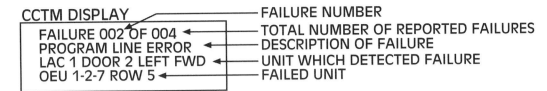

```
CCTM DISPLAY
    FAILURE 002 OF 004
    PROGRAM LINE ERROR
    LAC 1 DOOR 2 LEFT FWD
    OEU 1-2-7 ROW 5
```
FAILURE NUMBER
TOTAL NUMBER OF REPORTED FAILURES
DESCRIPTION OF FAILURE
UNIT WHICH DETECTED FAILURE
FAILED UNIT

Fig. 7-48   CCTM display of a failed ACESS status check.

System tests

The ACESS **system tests** feature allows technicians to run real time tests of the various ACESS LRUs. Since the system test function disables ACESS functions, the test can only be performed on the ground. Five ACESS subsystems are tested using the systems test feature: 1) Passenger Service System (PSS), 2) Passenger Entertainment System audio (PES audio), 3) Cabin Lighting System (CLS), 4) Cabin Interphone System (CIS), and 5) Passenger Address System (PAS). The attendant master call lights, the row call lights, and the passenger information sign lights can also be tested. A *test all units* function is included in this feature which performs a test on each of the five systems and records their faults.

The system tests are initiated using the CCTM. First, press the *test mode* button (see figure 7-49). Second, press the *test* button and the CCTM display will show the system test menu. Third, use the scroll down button to move to the desired system test. The system test menu is shown in table 7-6. After a test is complete, the CCTM display will list the number of errors found.

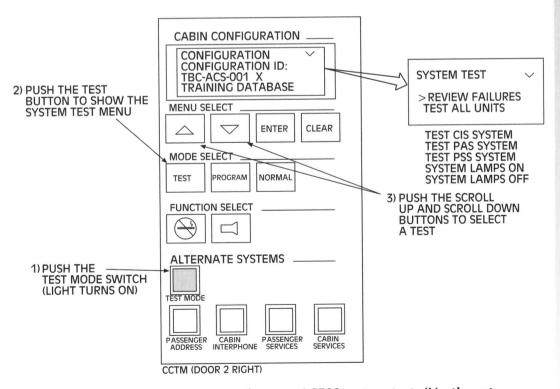

Fig. 7-49   Sequence of events to perform an ACESS system test. (Northwest Airlines, Inc.)

SYSTEM TEST MENU SELECTION	
Test Selection	Function of Test
Review failure	Used to review failures which were found during previous system tests
Test all units	Performs functional test of all ACESS subsystems (CIS, PAS. PSS, PES audio, and CLS)
Test CIS system	Used to test the cabin interphone system
Test PAS system	Used to test the passenger address system
Test PSS system	Used to test passenger service system, PES audio, and cabin lighting system
System lamps on	Used turn on the master call lights, row call lights, and passenger information signs
System lamps off	Used to turn off the master call lights, and passenger information signs

Table 7-6   System test menu selections.

If one or more errors are found, the technician can press the *test* button to return to the system test menu. From the menu, the *review failures* function can be accessed to retrieve a list of the detected faults. The CCTM will display a list of the failures in a format as described in figure 7-48. Additional failures are displayed by pressing the scroll up/down buttons on the CCTM.

Lamp test

The **lamp test** initiated through the CCTM is used to test the attendant master call lights, the row call lights, the lavatory call lights, and the passenger information sign lights (see figure 7-50). The lamp test simply turns on all call lights and information signs. The technician walks through the aircraft to identify if the lights are on. If a lamp is off the bulb is most likely defective. If a entire section of lamps are off, it is most likely caused by a defective Overhead Electronic Unit (OEU) or associated wiring.

To turn on the lights for testing, press the *test mode* switch on the CCTM. Next press the *test* button and the system test menu will appear in the CCTM display. Scroll down to the *system lamps on* selection from the menu and press enter (see figure 7-51). To turn off the lamp test function, simply scroll down the menu to *system lamps off* and press enter. The Cabin System Module (CSM) lights can also be tested through the CCTM. As mentioned earlier, there are two CSMs used to control cabin lighting: PES audio, and the passenger service system. The two CSMs are located at Door #2 left in the main cabin and in the upper deck at the attendant's station.

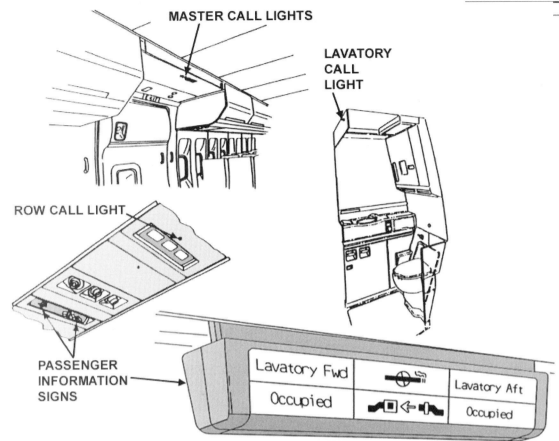

MASTER CALL LIGHTS

LAVATORY
CALL
LIGHT

ROW CALL LIGHT

PASSENGER
INFORMATION
SIGNS

Lavatory Fwd

Occupied

Lavatory Aft

Occupied

Fig. 7-50   Assorted lights tested during the lamp test. (Northwest Airlines, Inc.)

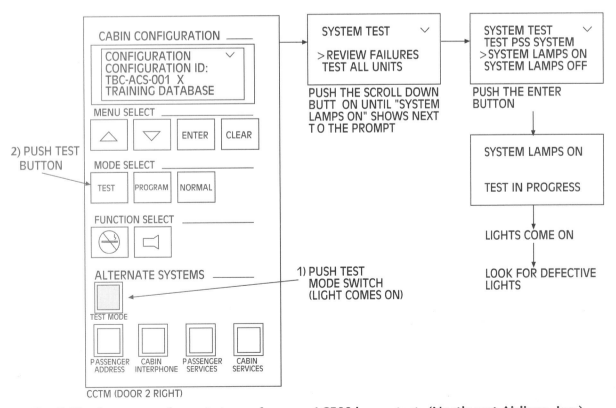

CABIN CONFIGURATION

CONFIGURATION
CONFIGURATION ID:
TBC-ACS-001 X
TRAINING DATABASE

MENU SELECT

ENTER   CLEAR

2) PUSH TEST
BUTTON

MODE SELECT

TEST   PROGRAM   NORMAL

FUNCTION SELECT

ALTERNATE SYSTEMS

1) PUSH TEST
MODE SWITCH
(LIGHT COMES ON)

TEST MODE

PASSENGER
ADDRESS

CABIN
INTERPHONE

PASSENGER
SERVICES

CABIN
SERVICES

CCTM (DOOR 2 RIGHT)

SYSTEM TEST

> REVIEW FAILURES
  TEST ALL UNITS

PUSH THE SCROLL DOWN
BUTT ON UNTIL "SYSTEM
LAMPS ON" SHOWS NEXT
T O THE PROMPT

SYSTEM TEST
TEST PSS SYSTEM
> SYSTEM LAMPS ON
  SYSTEM LAMPS OFF

PUSH THE ENTER
BUTTON

SYSTEM LAMPS ON

TEST IN PROGRESS

LIGHTS COME ON

LOOK FOR DEFECTIVE
LIGHTS

Fig. 7-51   Sequence of events to perform an ACESS lamps test. (Northwest Airlines, Inc.)

ACESS software

ACESS is a software driven system which is designed to be reprogrammed any time the cabin configuration is changed. When software modifications are made, several ACESS LRUs are reprogrammed. If an ACESS LRU is replaced for maintenance, be sure to update the software. There are two categories of software which must be loaded into ACESS: **operational software** and the **configuration database**. Although the exact procedures for down loading LRU software is beyond the scope of this text, the basic process is described in the following paragraphs.

The **operational software** is used to instruct ACESS LRUs how to function in the system. Operational software is typically installed in the shop prior to the LRU reaching the aircraft. In some cases, however, operational software can be installed with the LRU on the aircraft. Be sure to consult the manual for proper software upgrade procedures.

The **configuration database** is used by the LRUs to determine the specific layout of the aircraft's cabin. Configuration software is typically installed into an LRU's memory after the LRU has been installed in the aircraft. Using the CCTM, the configuration software is downloaded from the Central Management Unit (CMU) to other LRUs in the system. This is required only if the aircraft's cabin configuration is changed or a new LRU (requiring the database) is installed. If the aircraft's CMU is changed, both the operational and configuration software can be downloaded using the aircraft's data loader. The data loader can also be used to update the CMU.

Passenger Entertainment Video

The **Passenger Entertainment System Video (PES video)** is used to display video entertainment on various monitors and/or projectors located throughout the cabin. The system uses prerecorded VHS format tapes which contain both the video and audio tracks. The audio data is transmitted to the passenger address controller for broadcast over cabin speakers, or to the Entertainment/Service Controller (ESC) for distribution to passenger headphones.

Architecture

Reference the interface diagram in figure 7-52 during the following discussions on the PES video. The **Video System Control Unit (VSCU)** is the major control and distribution unit for the entertainment video signals. The VSCU receives video and audio inputs from the Video Tape Reproducer (VTR) and a discrete signal from the decompression relay. The decompression relay monitors for sudden cabin decompression. In the event that cabin pressure is lost, all video is automatically turned off. The VSCU sends control signals to the VTR for automatic operation of the tape unit.

The VSCU sends control and video outputs in digital format to the video distribution units. The **Video Distribution Units (VDUs)** are located throughout the cabin. Up to 24 VDUs can interface with the VSCU; although most aircraft have 10 to 12 VDUs. The VSCU also sends a *video in use* discrete signal to the cabin interphone system to inform the crew that the video is in operation. The audio signals from the video tape

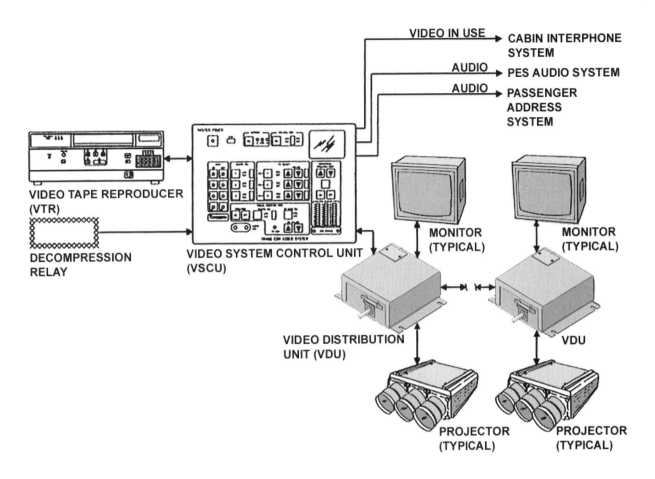

Fig. 7-52 Passenger entertainment system video interface pictorial diagram. (Northwest Airlines, Inc.)

reproducer are also sent through the VSCU to the PES (audio) and PAS. The VDUs send an *on* signal to the VSCU.

The video distribution units are each connected to the VSCU via two bidirectional data bus cables. The units are connected in series as shown in figure 7-53. A 75 ohm termination plug must be installed at the end of the data bus cables to provide proper bus impedance.

The VSCU is located under the stairway to the upper deck in the **Video Control Center (VCC)**. A maximum of two video tape reproducers will also be located in the VCC. Storage for video cassettes and other materials is also located in the VCC.

Control

All video system operating controls are located on the front panel of the VSCU (see figure 7-54). The VSCU contains controls for:

1) *MASTER POWER.*
2) *ANTENNA* selection - used for systems receiving ground based video programming (this system is typically not used on US aircraft).
3) *PREVIEW MON* (monitor) selection - to select the video source for the display monitor and audio source for VSCU monitor.

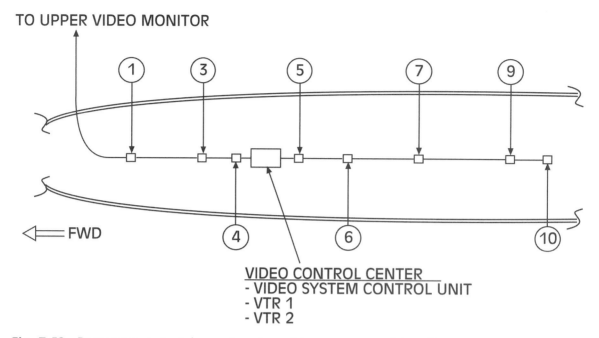

Fig. 7-53 Passenger entertainment system video component locations.

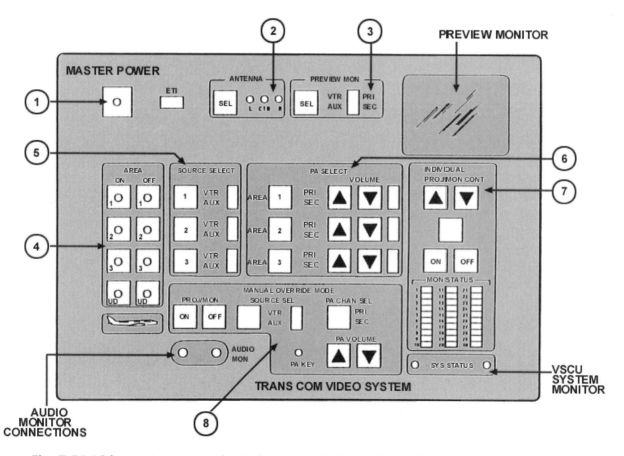

Fig. 7-54 Video system control unit front panel; Control descriptions: 1) master power, 2) antenna selection, 3) preview monitor selection, 4) area on/off controls, 5) source selection, 6) PA selection controls, 7) individual projector/monitor control, 8) manual override mode controls. (Northwest Airlines, Inc.)

4) *AREA* on/off controls - used to select the specific area where the video/audio entertainment will be distributed (area 1, 2, and 3 are in the main cabin, UD is the upper deck).

5) *SOURCE SEL* - used to select the video source (up to three input sources can feed the VSCU).

6) *PA SELECT* - used to select the primary or secondary audio source for output to the PAC, also controls volume.

7) *INDIVIDUAL PROJECTOR/MONITOR CONT* (control) - used to turn on/off individual monitors.

8) *MANUAL OVERRIDE MODE* - these controls are used to set the VSCU selections for all video zones simultaneously.

The VSCU system monitor is located in the lower right corner of the VSCU front panel. The system monitor is yellow during normal operation. The monitor illuminates orange if the VSCU BITE system detects an internal failure.

Maintenance and troubleshooting

The PES video system is a relatively simple system to maintain. Whenever troubleshooting the system, always remember which LRUs control and/or feed data to the unit in question. For example, if several monitors are inoperative, it is most likely that the area VDU which feeds those monitors is defective. Remember the VSCU controls the signals to the VDU. Therefore, if the VDU does not repair the system, the associated wiring or the VCDU should be suspected. In most cases, the LRUs are simply swapped with other units to troubleshoot the system. Connector pins are also likely fault areas. Be sure to inspect all electrical connections related to the failed system(s).

The VCSU and VDUs each have a series of **dip switches**, which are used to define the configuration of the video system (see figure 7-55). Dip switches are small two position switch that are set to binary 1 or 0 for system/component configuration. The VSCU dip switches define the number of VDUs, VTRs, video sources in the system, the audio configuration, the software configuration, the BITE enable, and the upper deck configuration. The VDU dip switches define the location of the VDU, the VDU data bus address, and the number of monitors and projectors controlled by the VDU.

Like a household VCR, the videotape reproducer periodically requires cleaning, adjustment, or belt replacement. This is typically done in the service shop and the line technician simply removes and replaces the VTR. Both the video monitors and the projection units have various adjustments to improve picture quality. Brightness, color intensity, hue, and vertical hold are typically available on each monitor and projector. The projection units also have sharpness and convergence controls to adjust the picture clarity. The appropriate service information will completely describe the adjustment of these components.

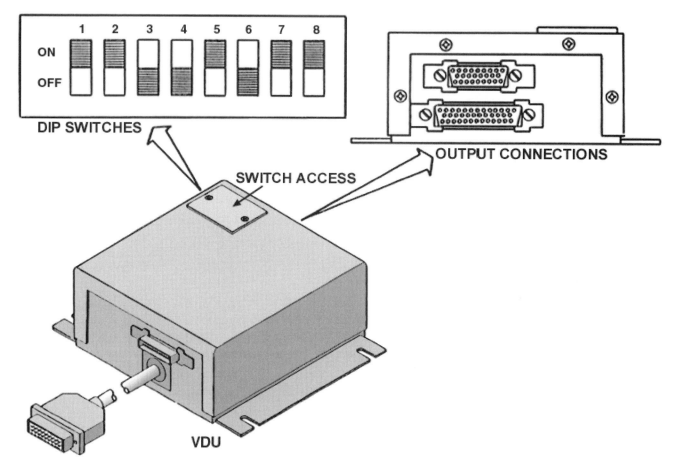

Fig. 7-55 Video distribution unit and configuration dip switches. (Northwest Airlines, Inc.)

CHAPTER 8
GLOBAL POSITIONING SYSTEM

INTRODUCTION

The **Global Positioning System (GPS)** is a satellite-based system that is capable of providing position and navigational data for ground based and airborne receivers. GPS actually determines the position of a given receiver, or receiver's antenna to be precise. The navigational capabilities of GPS are determined through multiple calculations and comparison to the **World Geodetic System (WGD)** map. The WGD map is an extremely accurate reference of the earth latitude and longitude. The global positioning system, therefore, provides location coordinates based on a latitude/longitude reference, not with reference to the known location of a ground based transmitter. Conventional forms of aircraft navigation rely on ground based transmitters that employ limited range and relatively low frequency signals. GPS employs ultra high frequency transmitters and requires only 21 transmitters for world wide navigation.

The global positioning system was first implemented by the military in the late 1970s. Through the early 80s, the system was tested and refined. During the 1980s and early 90s, the initial satellites were replaced by more accurate and powerful units. At the time that this text was written, the FAA had granted TSOs (technical standard orders) to several GPS units, which permitted installation on civilian aircraft. Today several GPS units are available and certified for enroute navigation and can be used for a non-precision IFR approach.

In the United States, the formal name for the global positioning system is the **NAVSTAR GPS**. This name was first assigned by the U.S. Deputy Secretary of Defense in 1973 and the name has stuck ever since. The United States is not the only country which has developed a usable GPS. The former Soviet Union began the implementation of their satellite navigation system called **Glanoss** several years before the country's demise. Today the Commonwealth of Independent States (made up of former Soviet Union member nations) is continuing to develop their GPS. Both nations have explored the possibility of combining efforts to reduce cost and increase worldwide coverage.

GPS SYSTEM ELEMENTS

The NAVSTAR GPS consists of three distinct elements: the **space segment**, the **control segment**, and the **user segment** (see figure 8-1). The space segment consists of a "constellation" of 24 orbiting satellites. Twenty one of the satellites are active and three are spares which can be moved into position in the event an active unit fails. The satellites are each placed in a near geosynchronous orbit approximately 10,900 nm (20,200km) above the earth. As seen in figure 8-2, the satellites are equally spaced around six different orbits to provide worldwide coverage. Each satellite completes one orbit approximately every 12 hours. The system is designed so that a minimum of five satellites should be in view of a ground based user at any given time, at any location on earth.

GPS SYSTEM

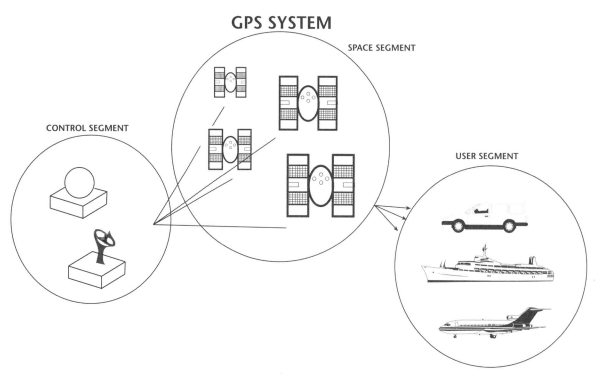

Fig. 8-1 The three segments of the Global Positioning System

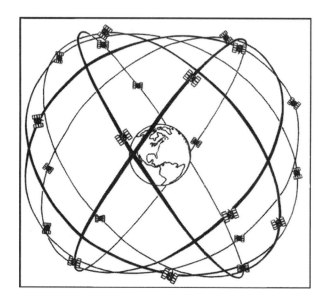

Fig. 8-2 Configuration of GPS satellites.

Each satellite in the system transmits position and precise time information on two frequencies known as L_1 and L_2. L_1 operates at **1,575.42 MHz** and L_2 has an operating frequency of **1,227.6 MHz**. The signals are digitally modulated and have a bandwidth of 20MHz or 2MHz depending on the type of information that is being transmitted. The individual satellites are identified by the information broadcast by the modulated carrier waves. It should be noted that the digital modulation of the carrier is achieved through a process known as **phase modulation (PM)**. As seen in figure 8-3 phase modulation is achieved by shifting the phase of the carrier wave to represent a change in the digital "information" signal. To improve receiver reception of faint GPS signals, each satellite employs 12 helical antennas arranged in a tight circular pattern (see figure 8-4).

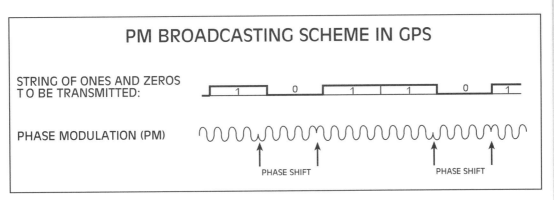

Fig. 8-3 Example of phase modulated RF signal.

Fig. 8-4 Photograph of a typical GPS satellite

The control segment of the GPS consists of five ground-based **monitor stations**, one **master control station**, and three **ground antennas** located at different sites throughout the world (see figure 8-5). The five monitor stations receive transmit time and range data from various satellites within its region. The raw data received by these unmanned stations is sent to the Master Control Station in Colorado Springs, Colorado. The Master Control Station computers analyze the data and provide correction signals as needed to the three ground antennas. The signals are periodically uplinked to the satellites and the necessary corrections are made by the system's software.

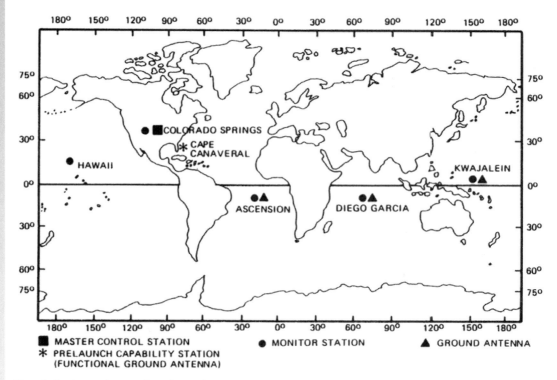

Fig. 8-5 Locations of various GPS ground facilities

When GPS was first conceived, it was intended that the entire user segment would be restricted to high accuracy military applications. Today, it would be difficult to determine if there are more civilian or military GPS receivers in operation. Once released by the Department of Defense, the civilian use of GPS has skyrocketed. GPS is currently being used (or studied for future use) by farmers, surveyors, archaeologists, miners, recreational hikers, and of course aircraft owners and pilots. Virtually anyone who needs to navigate, measure position, or determine velocity can become part of the GPS user segment.

The user segment is typically designed to receive time and position data from four or more satellites and process that data into the desired output. The specific equipment needed for these operations is a function of the equipment installation (if any), the desired accuracy, and specific output data. Output data ranges from simple display of latitude and longitude, to moving maps and displays of local airports or ground terrain.

All GPS receivers must have at least three major elements: the control/display unit, the receiver/processor circuitry, and the antenna. These elements may all be combined in one unit or consist of three individual elements. Modern hand held receivers are available for just a few hundred dollars. These units are completely self-contained however, many are not approved for aircraft use. The hand-held unit shown in figure 8-6 is a Bendix/King model KLX 100. This unit is TSOed for aircraft use and contains both a GPS and communications transceiver in one unit. Many aircraft systems are designed to be permanently installed on the aircraft and often interface with other navigational equipment. On aircraft GPS equipment, the display and receiver processor are often combined in one unit; the antenna is always mounted on the top of the aircraft fuselage to ensure proper satellite reception.

Fig. 8-6   A Bendix/King KLX 100 hand held GPS/COMM transceiver. (AlliedSignal General Aviation Avionics)

Theory of Operation

Let us start this explanation with two assumptions: 1) we know exactly where each satellite is at any given time and 2) the distance to each satellite can be calculated by the GPS receiver/processor. In reality, the exact location of a moving satellite is easy to predict. Once in orbit, each satellite will follow a consistent path. Exact satellite position data is transmitted to the receiver as part of the satellite message. Calculating distance from the satellite is also a simple matter. Since distance equals velocity multiplied by time, the receiver/processor need only measure the time it took for the GPS signal to reach the receiver. The speed at which the signal traveled to the receiver is a constant 186,000 miles/second (the speed of light). Using time and velocity to derive distance (range) is known as the **time of arrival (TOA)** ranging concept.

To keep things simple, a two-dimensional model will be presented first. Assume the GPS user segment and all satellites are located in one geometric plane. In this case, knowing the distance (range) from just two satellites would provide the location of your aircraft (see figure 8-7(a). In this example, the aircraft must be located somewhere on a circle with a radius of 30 miles from satellite A, and somewhere on a circle with a radius 40 miles from satellite B. In this two-dimensional model the aircraft can be in one of two positions. To further define the location of our aircraft in the two-dimensional model, a third GPS satellite would be added. As seen in figure 8-7(b), if the aircraft was 30 miles from satellite A, 40 miles from satellite B, and 20 miles from satellite C, the aircraft must be in position #1. As we all know, real aircraft can travel in three dimensions. To pinpoint position in each of these three dimensions, the aircraft must monitor at least four GPS satellites.

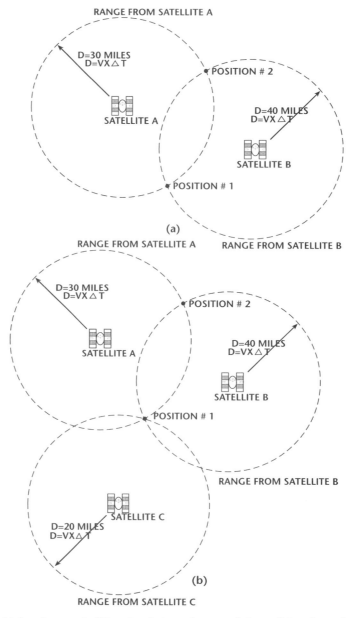

Fig. 8-7 (a) Using two satellites to determine position; (b) using three satellites to determine position.

Timing errors

In the previous example, the ability to establish distance was based on accurate time measurements. Keep in mind that the satellite radio signals travel at the speed of light; therefore, even the slightest timing error will result in significant range errors. For example, if a timing error of just 1/1000th of a second was present, the distance calculated from the satellite would be off by 186 miles. This error is obviously too large for aircraft navigation.

There are two ways to ensure accurate timing for all space and user segments of the system. One is to install very expensive atomic clocks in each GPS receiver. Each satellite in the constellation already employs four such atomic clocks. This solution, although relatively simple, would put the cost of the user segment out of practical reach for almost everyone, even for the military.

Another solution to ensure accurate timing is found through the use of mathematical formulas. Once again, let us look at a two-dimensional model of the GPS that uses three satellites to determine position. The dotted lines in figure 8-8 represent the range from all three satellites as calculated with the time bias error. This distance is known as the **pseudorange**. It should be noted that since all satellites are perfectly synchronized by their atomic clocks, the same time bias error exists between each satellite and the receiver. The solid lines in figure 8-8 represent the actual range from each satellite. Since the same range error (based on the same timing error) exists for all three satellites, the receiver/processor can simply add or subtract any error until it "finds" a common intersection from all three satellites. As before, in a three-dimensional model this is slightly more difficult; however, if four satellites are being used the GPS processor has no difficulty canceling out time bias errors.

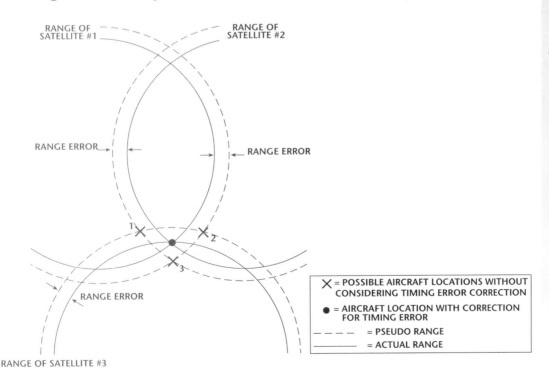

Fig. 8-8   The aircraft's exact location can only be determined after corrections for range error have been mathematically removed by the receiver/processor.

Achieving accurate time and range measurements

In most cases, the GPS receiver/processor incorporates a low cost free running crystal controlled oscillator. A **free running** oscillator will establish a given frequency without being compensated for frequency drift. That is, the accuracy of the receiver's oscillator is not extremely critical. On the other hand, the frequencies established by the oscillator within the satellites are accurate to within 0.003 seconds per one thousand years.

As mentioned earlier, the satellites transmit a digital code that repeats periodically. The transmission of this code is synchronized for each satellite in the constellation. In other words every satellite will begin and end the digital code at the same exact time. The GPS receiver also generates the same digital code; however, the receiver code is not synchronized to the satellites. Figure 8-9 shows digital codes of three satellites and one receiver. When the GPS receiver first begins to establish a position fix, the time bias between the free running clock and the satellites is calculated by the processor circuitry. In effect, the user segment now knows the correction factor to apply to the range calculations for each satellite.

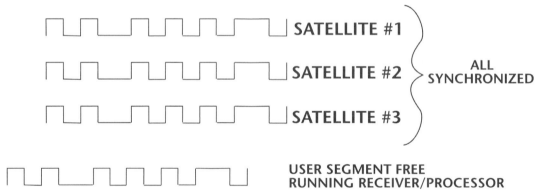

Fig. 8-9   Synchronized pseudorandom digital codes transmitted by three satellites compared to the pseudorandom code generated by the receiver/processor.

To establish the range to a given satellite, the receiver simply measures the time of arrival of the digital signal. Next, the processor calculates the time it took for the signal to travel from the satellite to the receiver ("delta" t). Lastly, the receiver multiplies the "delta" t by the speed of light.

Another way to look at the range and time corrections is shown in figure 8-10. Here it can be seen that the receiver actually measures pseudorange from four satellites. The aircraft's GPS processor mathematically determines the actual range and position. The actual range (R) is equal to the pseudorange (PR) less the range error (E). Figure 8-11 shows a simplified version of the calculations used by the receiver/processor to calculate the aircraft's position.

PSEUDORANGES

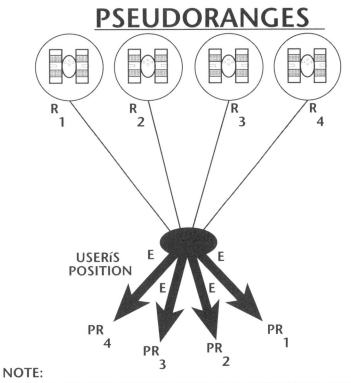

NOTE:
R=ACTUAL GEOMETRIC RANGE (FROM EACH SATELLITE TO USER)
E=RANGE ERROR CAUSED BY GPS RECEIVER CLOCK BIAS
PR=PSEUDORANGE=R+E=OBSERVED TOA VALUE x SPEED OF LIGHT

Fig. 8-10 Measurement of pseudorange from four satellites.

A. DATA PROCESSOR OBTAINS PSEUDORANGE MEASUREMENTS (PR₁, PR₂, PR₃, PR₄) FROM FOUR SATELLITES

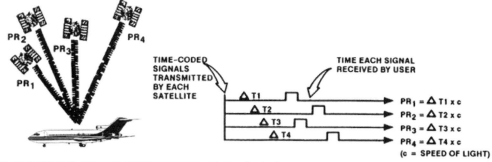

B. DATA PROCESSOR APPLIES DETERMINISTIC CORRECTIONS

PR_I = PSEUDORANGE (I = 1, 2, 3, 4)

- PSEUDORANGE INCLUDES ACTUAL DISTANCE BETWEEN SATELLITE AND USER PLUS SATELLITE CLOCK BIAS, ATMOSPHERIC DISTORTIONS, RELATIVITY EFFECTS, RECEIVER NOISE, AND RECEIVER CLOCK BIAS
- SATELLITE CLOCK BIAS, ATMOSPHERIC DISTORTIONS, RELATIVITY EFFECTS ARE COMPENSATED FOR BY INCORPORATION OF DETERMINISTIC ADJUSTMENTS TO PSEUDORANGES PRIOR TO INCLUSION INTO POSITION/TIME SOLUTION PROCESS

C. DATA PROCESSSOR PERFORMS THE POSITION/TIME SOLUTION

FOUR RANGING EQUATIONS:

$$(X_1 - U_x)^2 + (Y_1 - U_y)^2 + (Z_1 - U_z)^2 = (PR_1 - CB \times c)^2$$
$$(X_2 - U_x)^2 + (Y_2 - U_y)^2 + (Z_2 - U_z)^2 = (PR_2 - CB \times c)^2$$
$$(X_3 - U_x)^2 + (Y_3 - U_y)^2 + (Z_3 - U_z)^2 = (PR_3 - CB \times c)^2$$
$$(X_4 - U_x)^2 + (Y_4 - U_y)^2 + (Z_4 - U_z)^2 = (PR_4 - CB \times c)^2$$

$X_I \cdot Y_I \cdot Z_I$ = SATELLITE POSITION (I = 1, 2, 3, 4)
- SATELLITE POSITION BROADCAST IN 50 Hz NAVIGATION MESSAGE

DATA PROCESSOR SOLVES FOR:
- U_x, U_y, U_z = USER POSITION
- CB = GPS RECEIVER CLOCK BIAS

Fig. 8-11 Calculations used to determine an aircraft's location using the pseudorange measurements from four satellites.

Other errors

It should be noted that errors other than time bias exist in the system. The four most common errors are: 1) **satellite clock errors** which exist due to minor inaccuracies in synchronizing the satellite atomic clocks. 2) **ephemeris errors** which are caused by slight variations in the satellites position as it orbits the earth. 3) **atmospheric propagation errors** that are caused by the distortion of the transmitted signal as it travels from the satellite to the receiver. Most of these error are created as the radio signals travel through the ionosphere. 4) **Receiver errors** which are caused by local electrical noise (interference), computational errors, and errors in matching the pseudorandom digital codes.

As seen in table 8-1, the errors mentioned above are relatively minor when considering the distance from satellite to receiver. It should be noted that in some cases these errors are correctable. The control segment of the GPS monitors the satellite clock errors and ephemeris errors. Adjustments are made on a periodic basis to correct these errors. Using two signals of different frequencies and comparing the difference in time delays can minimize the amount of atmospheric error. The receiver error is often a function of equipment quality. In general, higher quality receivers incur less error.

Source of Error	Approximate Error Distance
Satellite Clock Errors	2 feet
Ephemeris Errors	2 feet
Receiver Errors	4 feet
Atmospheric Propagation Errors	2 feet

Table 8-1   Range error values

C/A- and P-codes

As mentioned earlier, the satellites transmit a pseudorandom digital code on two L-band frequencies. The digital codes transmitted by the GPS satellites are known as the **course/acquisition (C/A) code** and the **precision (P) code**. The C/A- and P-codes contain the timing and satellite position information required by the receiver to calculate navigational data. The C/A- and P-code information is often referred to as the **NAV message.**

The NAV message can be decoded using a combination of the C/A- and P-code data, which transmits at 50 bits/second and contains 1,500 bits/frame. A total of 25 data frames are transmitted before the information repeats. Since much of the data is repeated during subsequent frames, the GPS receiver can typically "lock on" to the satellite within approximately 30 seconds.

In order to ensure full reception of the CA-and P-codes, the GPS receiver must be in line of sight to four satellites at least 5° above the horizon (see figure 8-12). Satellite position is also important to achieve the proper geometry for position calculations. The ideal situation is for all four satellites to be more than 5° above the horizon and equally spaced above the aircraft. This provides the receiver with the best triangulation for position calculation.

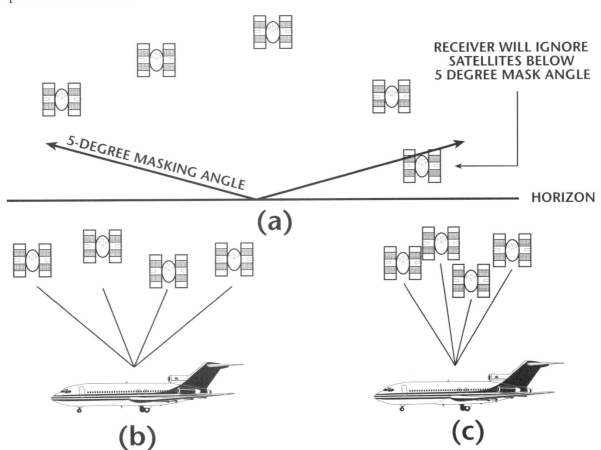

Fig. 8-12 Correct satellite location is important to accurate positioning. (a) Satellites must be at least 5° above the horizon to provide accurate data. (b) The use of four satellites with good separation provides accurate position data. (c) The use of four satellites clustered together proves less accurate position data.

GPS AND AIRCRAFT NAVIGATION

GPS provides an excellent means of general navigation; however, there are several limitations that must be addressed for civilian aircraft use. The basic GPS service fails to meet the four basic criteria: 1) *Accuracy* - the difference between the measured position and the aircraft's actual position. (It should be noted that GPS accuracy is adequate for enroute navigation, but fails to meet approach and landing requirements.) 2) *Availability* – the ability of the system to be used for navigation whenever it is needed, and the ability to provide that service throughout the entire flight. 3) *Integrity* – the ability of the system to shut itself down when it is unsuited for navigation or to provide timely warnings to the pilot of the system failure. 4) *Continuity* – the probability that GPS service will continue to be available for a period of time necessary to complete the navigation requirements of the flight.

While aircraft are flying on an IFR flight plan, their enroute separation is typically maintained at 5 miles or greater. The accuracy of GPS is sufficient for civilian enroute navigation. GPS can easily provide navigational signals capable of the five mile separation $\pm$ 4%. However, while flying an approach to land, the accuracy level must be significantly higher. The future of GPS aircraft navigation depends on improved accuracy. As for availability, the US government has stated that GPS will be consistently available for civilian use. Since systems do periodically fail, continuity may become a problem. Integrity therefore, becomes a very important issue for aircraft use of GPS. Remember, integrity is the ability to provide a timely warning in the event of a system failure.

In general, many of the GPS limitations can be solved through the use of navigation with more than four satellites. If one satellite fails, the system can simply navigate using a fix from another satellite. When using five satellites to navigate, the airborne equipment can detect a system malfunction, but cannot detect which satellite has failed. Six or more satellites are required to provide enough information to determine which satellite has failed and use the remaining satellites for accurate navigation. Modern receivers are capable of tracking several satellites at once to eliminate this problem.

The best solution is for the airborne GPS equipment to be notified immediately if a satellite fails. This is especially important for aircraft using GPS for approach and landing navigation. Each GPS satellite does broadcast an integrity message to assure the health of the system; however, one-half hour or longer may elapse from the time that a fault occurs to the time that the aircraft is notified. This time period is obviously too long.

Differential GPS

To overcome the limitation described in the previous paragraphs, the FAA has developed a concept known as **Differential GPS (DGPS)**. DGPS was designed to increase the accuracy, availability, integrity and continuity of the basic GPS to a level sufficient for complete aircraft navigation. DGPS uses one or more ground based stations to compensate for the inherent shortcomings of GPS. There are two basic subcategories of Differential GPS, the **Wide Area Augmentation System** and the **Local Area Augmentation System**.

Wide area augmentation system

The **Wide Area Augmentation System (WAAS)** is a GPS enhancement designed to improve the integrity, accuracy, availability, and continuity of the basic satellite navigation system. At the time that this text was written, WAAS was being developed to provide wide area coverage for aircraft navigation through all phases of flight, including the Category I precision approach.

Use figure 8-13 as a reference during the following discussion on WAAS. WAAS is a network of approximately 35 ground-based reference stations that receive and monitor GPS signals. Data from the reference stations is transmitted to a master station. The master station contains data on the precise location of each reference station. It can easily monitor and detect any GPS satellite errors. It can also quickly and accurately detect if a GPS satellite has failed. The master station then uplinks all correction data to one or more geostationary communications satellites. The communication satellites broadcast the correction data to the aircraft on the same frequency as GPS (L1, 1575.42 MHz). WAAS therefore, provides instantaneous correction and fault data to any aircraft in the WAAS coverage area.

WAAS will improve basic GPS accuracy to approximately seven meters vertically and horizontally. This is more than adequate for enroute navigation and for a Category I precision approach. A Category II or III approach requires even more rigid standards.

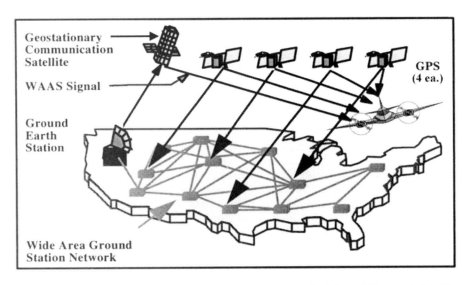

Fig. 8-13 Example of the basic system elements needed for differential GPS (DGPS).

Local area augmentation system

The Local Area Augmentation System (LAAS) is a GPS upgrade designed to be used during a Category II or Category III precision approach. This type of approach and landing requires extremely high accuracy, availability, and integrity. LAAS relies on a ground-based station to monitor the GPS satellites in a localized area. The geographic position of the ground-based station is accurately determined and loaded into the memory of the station's processor. After receiving GPS satellite data, the ground station can calculate all errors and determine the health of the GPS system. As seen in figure 8-14, the error data is then transmitted to the aircraft. LAAS will broadcast the correction message via a Very High Frequency (VHF) radio signal to the airborne receiver. The aircraft's receiver/processor then makes the necessary computations to compensate for all GPS errors. It is anticipated that LAAS will provide accuracy to less than one meter both vertically and horizontally.

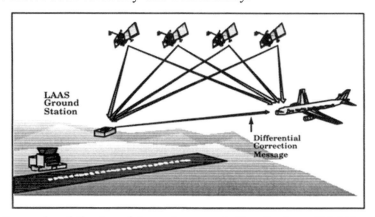

Fig. 8-14   Example of the Local Area Augmentation System (LAAS).

GPS Past, Present and Future

At the time that this text was written, both WAAS and LAAS were under development and testing. WAAS should be the first system operational, with LAAS to follow shortly thereafter. The exact outcome of Differential GPS is difficult to predict; however, some form of GPS will likely become the primary source for aircraft navigation in the near future.

All TSOed equipment is currently approved for enroute navigation. IFR approach-to-land situations are another story. In 1993, the FAA approved a nonprecision approach, commonly called the **overlay** approach. This IFR approach required relatively good visibility and high altitude minimums. The overlay approach also required that GPS equipment could only be used if the appropriate conventional ground based navigation equipment was operating onboard the aircraft during the approach. Specifically, it required that the pilot monitor the conventional systems (VOR, ILS, glide slope, and/or DME) during the approach.

As of 1996, all equipment certified under TSO-C129a allowed pilots to utilize the GPS as *supplemental area navigation equipment*. When using TSO-C129a receivers, there is no need to monitor conventional navigational systems. The equipment can only be used for nonprecision approaches and conventional navigational equipment is still required onboard the aircraft. In 1998, TSO-C145 was introduced. Equipment certified under this TSO would meet the standards for GPS navigation augmented by the Wide Area Augmentation System. As new systems are developed, additional TSOs will no doubt be introduced.

AIRCRAFT GPS EQUIPMENT

All aircraft GPS equipment must meet a minimum certification standard through a TSO. There are, however, several units currently available that are not TSO certified. Most of these GPS receivers were intended for non-aircraft uses such as hiking, marine, or automobile navigation. In some cases, pilots use non-TSOed equipment as a secondary reference during VFR flights. This is perfectly legal, but only equipment approved by a TSO may be used as a source for IFR or VFR navigation. The TSO ensures a minimum quality and accuracy standard set by the FAA; hence, it provides approval for aircraft use. In general, the major difference between an aircraft GPS receiver and one designed for marine or other use is the database. Virtually all aircraft GPS receivers contain a database of various airports and standard navigational waypoints.

As mentioned earlier, the "Global Positioning System" transmits *position* information to the user segment. The position data are **not** relative to a transmitter located on the earth's surface like other forms of aircraft navigation. The GPS user must therefore, navigate via waypoints. Waypoints can be any location (latitude and longitude) in two dimensions. The waypoints are entered into the GPS control panel using latitude and longitude coordinates or the waypoint can be selected from the airborne GPS database. It should be noted that current civilian aircraft use of GPS navigation is limited to only two dimensions; however, the GPS is capable of providing extremely accurate position data in three dimensions. Vertical navigation (altitude) is currently provided by conventional systems.

Portable GPS Equipment

There are several GPS receiver/processors currently available, which are portable or "hand-held" units. The hand-held receivers are designed for hikers, hobbyists, surveyors, boaters and even pilots. Many of the hand held units are even TSO approved. A modern portable unit is completely self-contained. The receiver/processor, control panel/display, antenna and batteries are all contained in a compact unit as seen in figure 8-15. Portable GPS units have become very popular and are being used extensively by general aviation pilots. Some advantages to the hand-held equipment include: 1) portability (the units can be moved between various aircraft, or taken to a convenient location to change waypoint or other database information), and 2) cost (there is virtually no installation cost and these units are traditionally less expensive than panel mounted systems). The major disadvantage to portable GPS receivers used for aircraft navigation is the difficulty in achieving appropriate antenna placement. Since the aircraft structure blocks the signal transmitted from the satellites, any antenna located inside the aircraft will have poor reception at best. Many portable receivers incorporate antennas that clip onto the aircraft windshield. This solution is less than optimal and may still result in temporary signal loss. One solution includes connecting the portable unit to an antenna permanently mounted on top of the aircraft.

Fig. 8-15   A typical hand-held GPS receiver. (Garmin International Inc.)

Panel-Mounted GPS Equipment

The alternative to the hand-held GPS receiver is the panel-mounted system found on many aircraft. Since increased use of GPS for enroute and approach navigation seems inevitable, many pilots and aircraft owners are selecting panel-mounted GPS receivers. These systems are permanently installed in the aircraft, similar to conventional navigation radios. A typical panel-mounted GPS receiver is shown in figure 8-16. This unit contains the receiver/processor and control/display in one unit. The antenna on all panel-mounted GPS equipment is remotely located on top of the aircraft.

Panel-mounted units are designed to fit the standard *radio rack* configuration. The GPS receiver is typically mounted in the center of the instrument panel, which provides good visibility of the display for both pilot and copilot. Many panel-mounted GPS receiver/processors have the capability to interface with other navigational displays, such as a CDI (course deviation indicator), HSI (horizontal situation indicator), RMI (radio magnetic indicator) or EFIS (electronic flight instrument system). The use of conventional displays provide better visibility of GPS data, and can improve cockpit resource management.

Fig. 8-16 A typical panel mounted GPS receiver. (Garmin International Inc.)

Features

Two of the most common airborne GPS features are the extended **database** and the **moving map display**. The extended database is found on almost all aircraft GPS receivers. It should be noted that any GPS receiver must incorporate some type of database. An "extended" database may include an encyclopedia of waypoints, airport locations, runway headings, runway lengths, and other pertinent data. Many units now include an option that allows for the entry of checklist data. Once entered, the pilot can use the GPS unit to display aircraft checklist procedures.

Since navigational data can change periodically, the GPS database must be updated at regular intervals to maintain accuracy. This update responsibility falls on the GPS user. Some units require that the database be updated manually. Some GPS systems use an advanced technology memory chip which is simply installed into the receiver/processor to update the database (figure 8-17). Another means of updating a database employs an RS-232 connection between the GPS processor and a personal computer. The update information can be loaded into the computer (typically from disc or the World Wide Web) and downloaded into the receiver/processor.

Virtually all aircraft GPS receivers will provide a digital readout of navigation data as seen in figure 8-18. This type of display is good for position data, but is difficult to interpret for continuous navigation. Many GPS receivers incorporate some type of electronic CDI (course deviation indicator). For example, the Garmin GPS 150 employs a CDI at the lower portion of the LCD display (see figure 8-19). The CDI provides a display of the aircraft's position relative to the desired course.

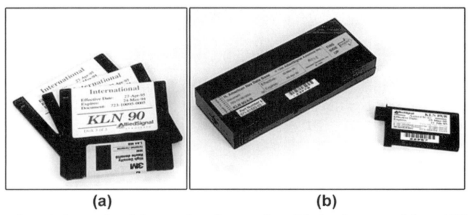

Fig. 8-17 Database information (a) stored on floppy discs (b) database modules. (AlliedSignal General Aviation Avionics)

Fig. 8-18 Representation of a typical digital display showing bearing, range, track, and estimated time of arrival. (Garmin International Inc.)

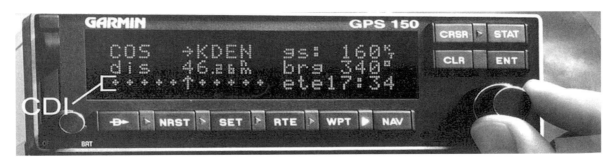

Fig. 8-19 Panel mounted GPS with CDI located at bottom of display. (Garmin International Inc.)

Moving map displays provide a horizontal "picture" of the aircraft and surrounding navigational reference points. Figure 8-20 shows a typical moving map LCD (liquid crystal display). As the name implies, the ground references move in relation to the aircraft as the flight progresses. For most pilots, the moving map is the display of choice.

Many airborne GPS receivers are now combined into one unit with a conventional NAV/COM transceiver. Figure 8-20 shows the Garmin GNS 430 and installation on a typical light aircraft panel. This unit is capable of VHF communications, VHF navigation (VOR and localizer) as well as GPS navigation. All pilot inputs are made on the GNS 430 panel and the color LCD is used to display radio tuning frequencies, VOR/LOC course deviation, and GPS moving maps. The unit is slightly taller than a typical NAV/COM radio, which allows for a larger LCD display. The unit is a standard width, which allows easy installation into a standard radio rack. This type of integrated radio system greatly simplifies the aircraft instrument panel and will most likely be the system of choice for future installations.

GPS airborne components

In general, all civilian GPS receivers/processors operate in a very similar manner. The specific circuitry, types of controls, displays and interface options may vary among various models of equipment. Each unit must contain at least four functional areas: 1) the **L-band antenna**, 2) a **spread spectrum receiver**, 3) a **data processor** (plus software), and 4) an **output interface**. Figure 8-21 shows the four functional blocks of a generic GPS receiver.

The antenna

The L-band antenna is used to receive the signal transmitted from the space segment satellites. The antenna must pick up the phase modulated signal, convert the electromagnetic energy to electrical signals, amplify those signals and send them to the receiver through the coaxial cable. This is a difficult task since the space vehicle (SV) signal is extremely weak. In fact, the signal's strength is actually less than the sky's normal background noise. This makes the reception strictly line of sight. It should be noted that although rain and clouds do not effect reception, a thin layer of ice or snow on the receiving antenna certainly will limit signal reception.

Since such a weak signal is received by the L-band antenna, almost all units contain a signal amplifier. The amplifier gets electrical power from the receiver circuitry through the coaxial cable (coax). The antenna provides amplification to overcome the anticipated signal loss through the coax. Antennas mounted less than two feet from the receiver circuitry often do not contain an amplifier.

497

Fig. 8-20   A panel mounted GPS with full color LCD moving map display.

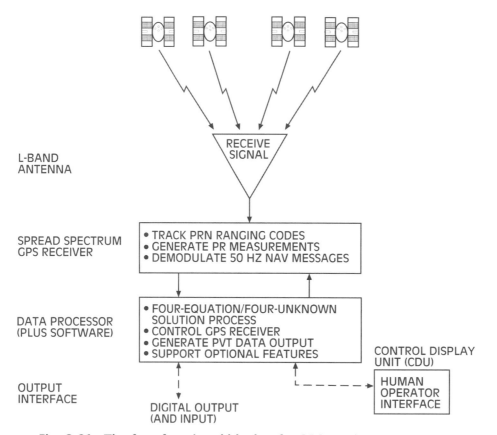

Fig. 8-21   The four functional blocks of a GPS receiver.

The receiver

There are several different types of receivers available for the GPS user segment. The type of circuitry (digital or analog) can categorize the receivers, or the number of receiving channels. Digital receivers are more popular on aircraft and multi-channel receivers provide the greatest accuracy and reliability, especially for high speed aircraft.

All receivers must perform three functions: 1) acquire and track the satellite range code information, 2) determine the pseudorange (PR) from each satellite, and 3) demodulate the 50Hz navigation message. The main input to the receiver is the antenna signal, the outputs include data to the processor unit, and in some cases to other systems which interface with the GPS. To accomplish these three functions, the generic receiver will typically include a **down converter**, a **quartz clock/frequency synthesizer**, and one or more **PM (phase modulation) signal tracking channels.**

The **down converter** filters out any extraneous signals and converts the RF (radio frequency) input from the antenna down to a given intermediate frequency (IF). The IF is a much lower frequency than the RF and easier to control while processing the transmitted information. The down converter must lower the frequencies of both L_1 and L_2 signals received by the antenna. While this occurs, the individual signals must remain completely independent to ensure signal integrity.

The **quartz clock/frequency synthesizer** provides the master time pulse and is the source of all reference frequencies within the receiver/processor. An accurate time pulse is critical to determine PR (pseudorange) measurements. The quartz clock is basically a very low noise crystal oscillator placed inside an oven to guard against frequency and time fluctuations caused by temperature change. The oven is typically a vacuum bottle containing an electric heater to keep the crystal at a constant $100°$ C. Although the quartz clock is not as accurate as the atomic clock in the satellites, it is far less expensive and is adequate for the job.

The **PM signal tracking channels** use the internal master time pulse along with the reference frequencies from the quartz clock to process the down-converted L_1 and L_2 signals at their IF. Each channel can track only one component of the transmission at any given time (the C/A- or P-code from L_1 or L_2). To simultaneously track all signal components from a satellite would require at least three channels. Most receivers contain multiple channels. Each channel operates independently under the control of a common processor circuitry. A generic five-channel receiver is shown in figure 8-22.

A **five-channel receiver** has four channels dedicated to making range measurements continuously with four different satellites. This will satisfy the processors need for a four-equation/four-unknown position solution (as shown earlier in figure 8-11). The choice of which satellites to track is made by the processor circuitry. Since only four channels are needed to solve the position equation, the fifth channel is free to handle necessary "housekeeping" functions. This includes making range measurements on the other frequency needed to calculate the errors induced by the ionosphere. Advanced planning can also be done with the fifth channel to verify data from satellites that are just rising on the horizon.

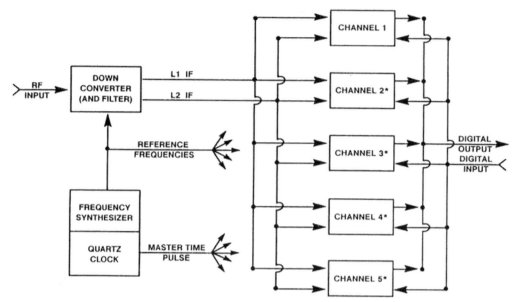

Fig. 8-22   A typical five-channel receiver

A **one-channel receiver** must still perform most of the functions of the five-channel receiver using a "time sharing" process. Since a single channel can only track one satellite at a time, the data processor uses the single channel to **perform sequential** tracking. With a one-per-second operating rate, the channel will be sequenced among the four best satellites. Using this process, the receiver can process one range measurement per second, completing the same operations of the five channel unit in five seconds. This situation works well for receivers that are relatively stationary; however, for aircraft, this update rate is typically insufficient. A **two-channel receiver** provides a low cost alternative to the five-channel unit and still provides the accuracy required by many users. The two-channel receiver can alternate between two satellites and minimize the update rate as compared to a single channel receiver. Most modern receivers contain multiple receiver channels. Multiple channels allow the GPS receiver to continuously track several satellites simultaneously, which improves system performance.

The data processor

In addition to providing the control functions for each PM signal-tracking channel in the receiver, a data processor uses the resulting range measurements and a 50 Hz NAV message to solve for position, velocity, and time (PVT). The four-equation/four-unknown position solution process described previously is useful from a conceptual point of view, but very few processors actually operate this way. Instead, the processors perform their calculations using an advanced software algorithm known as the **Kalman filter**. The mathematical theory behind Kalman filters goes beyond the scope of this text; however, with advanced software, there is little need for a flightline technician to be concerned with this process.

The output interface

Virtually all aircraft GPS receivers contain some type of display panel used to provide the pilot with the desired navigational data. A display can be found on both hand-held and panel mounted units. Many panel mounted units contain outputs which interface with other aircraft systems. This allows for GPS data to be integrated with other navigational displays, the flight data recorder, and even the autopilot system.

Whenever troubleshooting any aircraft GPS, be sure to examine all interface wiring carefully. A loose pin connection or broken wire should always be considered a possible cause of the problem.

THE KLN90

Bendix King avionic systems, owned by AlliedSignal Corporation, have been viewed as an industry standard for several decades. The KLN90 is a panel mounted GPS receiver produced by Bendix King (see figure 8-23). The KLN90 is designed for general aviation use and is often installed in corporate type aircraft, such as the Beechcraft King Air. The following discussion on the KLN90 will focus on a general description and installation of the system. It should be assumed that most of the

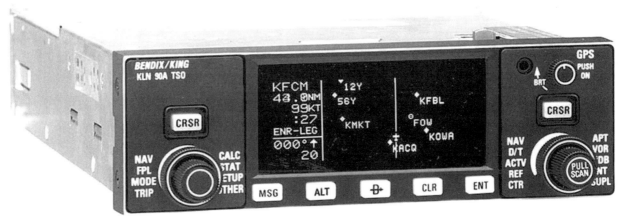

Fig. 8-23   The display panel of a Bendix/King KLN90 GPS receiver (AlliedSignal General Aviation Avionics)

information presented here is general and will apply to most panel mounted GPS receivers. It should also be noted that the following information is for training purposes only and should not be used as a guide for installation of an actual GPS receiver.

General Description

The KLN90 is designed for installation in a standard radio rack configuration with dimensions of 6.3in (16.03cm) wide, 2.0in (5.08cm) tall, and 12.5in (31.8cm) deep (see figure 8-24). The receiver processor and display/control panel are contained in one unit. The system contains an extensive database and is designed to provide real time position information with respect to a flight plan defined by the pilot.

The KLN90 can present position and course deviation information on its own LCD display or the unit can interface with a CDI, EFIS, or radar graphics unit. The internal database contains worldwide information on airports, waypoints, VORs, NDBs, intersections, and outer markers. The database memory is stored in a cartridge, which plugs into the back of the KLN90 receiver/processor. Database information can also be updated using an interface with a personal computer. Many of the input/output signals for the processor unit are formatted according to the ARINC 429 digital data specification. Gray code inputs are accepted for altitude information, and several analog output signals are used for interface message annunciators, such as the CDI and navigation flags. If needed to communicate with various equipment, an interface adaptor kit is available to convert ARINC 429 signals to analog data.

It should be noted, this system can be installed for VFR or IFR certification. If the KLN90 is to be used for IFR flights, it must meet certain installation criteria that are not required for VFR use. In other words, additional equipment or installation procedures may be required for IFR certification. For example, when the KLN 90B is used for IFR, additional annunciators must be installed, an altitude source (encoding altimeter) must be interfaced with the receiver/processor, and a specific antenna must be used. Of course, an approved pilots' guide must be accessible to the pilot(s) during flight.

Basic Installation

As shown in figure 8-24, the KLN90 is designed for a standard panel mount configuration. The receiver/processor/display unit is one component and designed to fit into a typical slide-in radio rack. The connectors for both the antenna and related wiring are mounted at the rear of the rack. A jack for the remote data loader can be installed on the aircraft panel using a 0.693-inch hole. As seen in figure 8-25, a special cable can be made to connect a PC to the data loader jack.

As with any avionics, equipment cooling is extremely important to the long-term reliability of the product. Compared to older units, modern systems are in some cases more vulnerable to heat problems since they employ a large number of electronic components "squashed" into a compact case. For this reason, most panel mounted GPS receiver/processors should be cooled by forced air blowers. Forced air-cooling is most important when several avionics units are installed, one on top of the next or side-by-side.

The KLN90 antenna is a relatively small unit that is installed on top of the aircraft. As shown in figure 8-26, the antenna is a streamline shape mounted with four # 10-32 machine screws. The antenna should be mounted to the aircraft skin using a backer plate. It is also important that the antenna be mounted in a location so it is level ±5° when the aircraft is in level flight. This will help to ensure capture of satellites that are close to the horizon. To ensure communication signals do not interfere with satellite reception, the antenna must be mounted at least three feet from any communications antenna. Some GPS antennas are directional, be sure to install the unit facing the correct direction.

Proper installation of the antenna coaxial cable is critical for system operation. The maximum allowable loss through the coaxial cable is 8.0 dB. At the time of installation, the specific coax wiring kit should be ordered. Different kits are available for aircraft requiring between 0-40, 0-80 or 0-100 ft of cable. The coax must be run using smooth curves and must not be crushed or pinched. It is recommended that a right angle coax connector (similar to figure 8-27) be used for certain installations between the cable and the antenna (or between the cable and the processor). This type connector will ensure the coax is not kinked at this location.

The system wiring for the KLN 90B is relatively simple. If the unit is installed as a stand-alone system, the only wiring required is the antenna coax, the electrical power, ground, and the display lighting wires. If the receiver/processor is to interface with other aircraft systems, such as the EFIS or an HSI, additional wiring will be required. When connecting the receiver/processor to other aircraft systems, **pin programming** may also be required. **Program pins** are used to configure the GPS to your particular

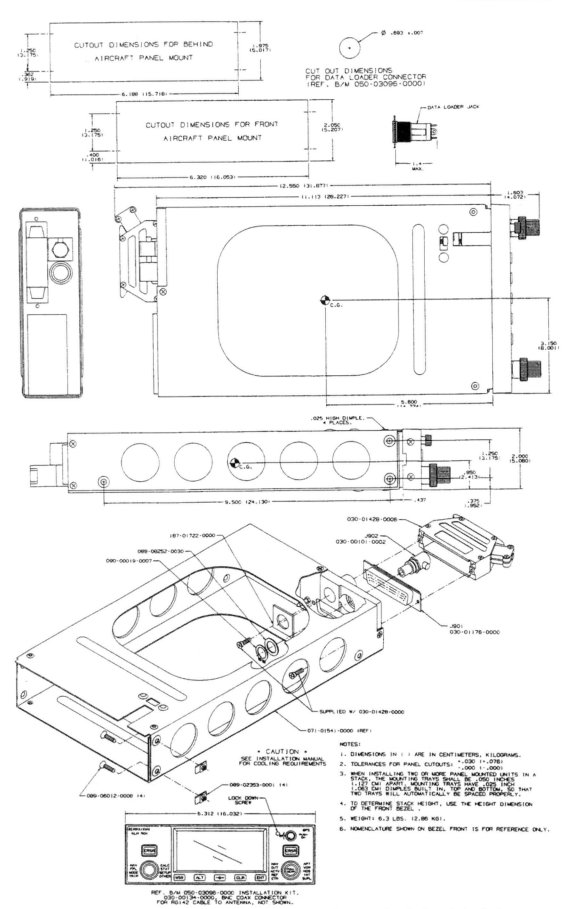

Fig. 8-24 Installation drawing for the KLN 90B (AlliedSignal General Aviation Avionics)

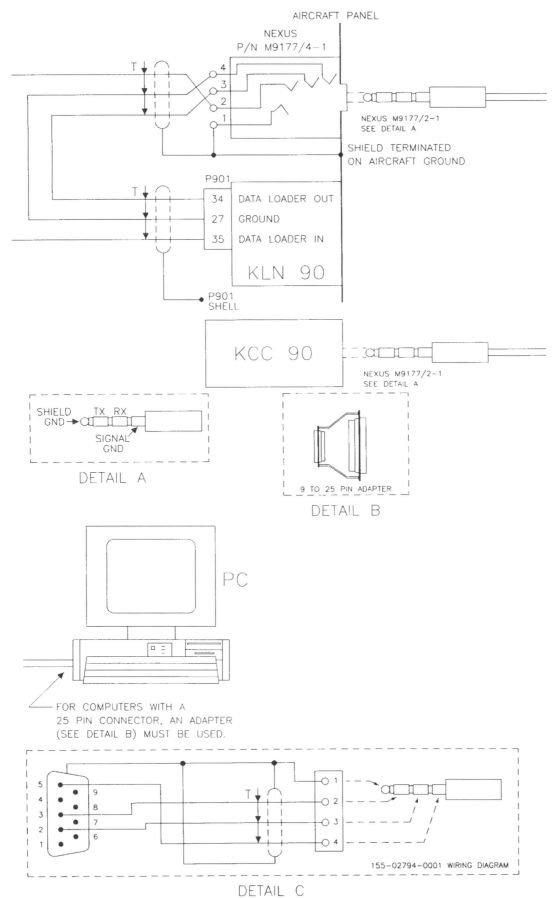

Fig. 8-25 Data loader interface cable for the KLN 90B. (Allied Signal General Aviation Avionics)

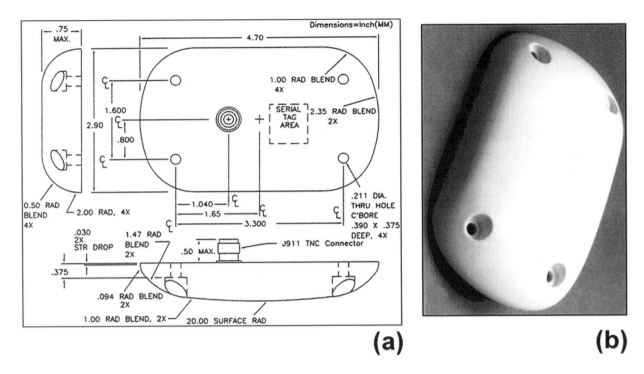

Fig. 8-26   A typical aircraft mounted GPS receiver antenna. (a) Installation drawing, (b) photograph. (AlliedSignal General Aviation Avionics)

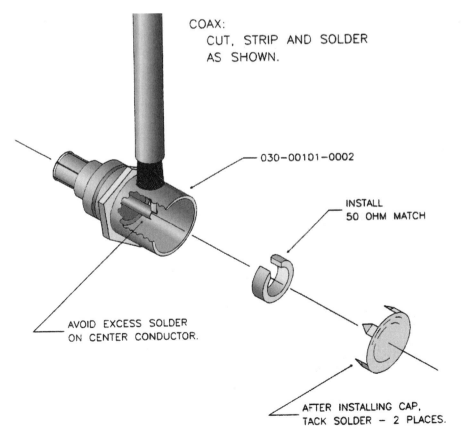

Fig. 8-27   A right angle coaxial cable connector used to eliminate cable kinks at the base of the antenna. (AlliedSignal General Aviation Avionics)

505

installation. For example, if pin #1 is connected to high (+ voltage), the KLN 90B is set to use an external analog OBS (omnibearing selector). If pin #1 is connected to ground, no external OBS will be used.

Operational tests

After installation, the system must be thoroughly tested for proper operation of the receiver/processor, the display, and any unit that interfaces with the GPS. Proper antenna (and coax) installation is verified by operation of the receiver/processor. Before applying power to the system, verify that all wiring is installed properly and that adequate equipment cooling will be supplied during the test. Move the aircraft to a location outside of the hanger away from any tall structures that may block satellites low on the horizon.

On some equipment, the initialization requires entering the current time and position of the aircraft. This is typically done through the GPS display/control unit. To test the system, simply follow the procedures outlined in the pilot's guide. Once the receiver/ processor has "locked on" to four satellites, the GPS display should provide your current location in latitude and longitude. To verify system operation, simply compare this to an accurate value of your known position. Be patient; in some locations it is difficult to get a "quick" lock on all four satellites. A receiver may require several minutes to provide accurate position data, although typically 30 seconds is adequate.

The GPS must also be tested for interference caused by other equipment on the aircraft. Two common sources of interference are the communication (comm) radio and the emergency locator transmitter (ELT). To ensure proper GPS operations, the verification check outlined in the last paragraph should be repeated while transmitting on the communication radio frequency 121.50 MHz. The test should be repeated for all comm radios installed on the aircraft and at any frequency recommended by the manufacturer. Common "interfering frequencies" include 121.175, 121.20, 131.250, 131.275, and 131.30 MHz. The 12th and 13th harmonics of these frequencies are often strong enough to interfere with GPS reception. On some comm radios, a simple in-line filter may solve the interference problem. Other situations may require that the comm and GPS antennas be moved further apart. It is also wise to test the ELT as a source of interference. The comm radio can excite a tank circuit in the ELT that can radiate GPS interference. Simply grounding the ELT antenna should verify interference from the ELT. (i.e., if grounding the antenna stops the interference, the ELT is responsible for generating the interference.) In most cases, the ELT manufacturer should be consulted for the proper repair.

Other systems that interface with the GPS must also be tested for proper operation. This includes the data loaders, CDI/HSIs, EFIS displays, encoding altimeters, and any discrete annunciators. Each of these units require a specific test which is beyond the scope of this text; however, each system must be thoroughly tested and its proper operation verified prior to completing the GPS installation.

TROUBLESHOOTING AIRCRAFT GPS EQUIPMENT

The troubleshooting of a failed GPS receiver is typically a relatively simple task. If the unit is a hand-held device, there is not much that can be done outside of returning the unit to a repair facility. A line technician may look closely at the battery, battery connections, any external power supply, the antenna and coax if applicable. If the unit is a panel-mounted unit, once again, the power supply and antenna system are the primary suspects for a failed unit.

Panel-mounted GPS receivers often interface with other aircraft systems. If a problem exists between one of these interfaces, be sure to check the system wiring. Connector pins and sockets which have become corroded or bent can often cause intermittent operation or complete system failure. If the GPS fails to operate correctly on a display device, such as the HSI, be sure to verify that the HSI is operational when driven by another source.

The antenna system of the GPS airborne equipment is often overlooked during troubleshooting. Be sure that the coaxial cable is in good condition, with no kinks or tight bends. Poor coaxial cable can cause poor or no reception of satellites low on the horizon. It should also be noted that any remotely mounted antenna can contain an amplifier used to boost signal strength. The power source for the antenna amplifier is located in the receiver circuitry. If the power source fails, the antenna will be inoperative even though it is in perfect working condition. As seen in figure 8-28, a DC voltage is sent to the antenna unit through the coaxial cable. In most cases, a voltmeter connected to the "antenna end" of the coax can be used to test the power supplied to the antenna amplifier. Of course, the receiver/processor unit must be operating for this test.

Various GPS simulators (or test units) are commercially available. One such system, produced by IFR Systems Inc., is shown in figure 8-29. The GPS-101 is a portable unit that simulates the satellite output and can be used to test the antenna, coax and connectors, and the receiver/processor. In general, testing with this unit is a simple process of elimination. For example, let us look at a system with a defective coaxial cable between the receiver and the antenna. A technician would take the following steps: 1) Place the GPS-101 transmitting antenna (the triangular shaped unit in figure 8-29) over the GPS receiver antenna on the aircraft. This will permit testing of the entire airborne system. In

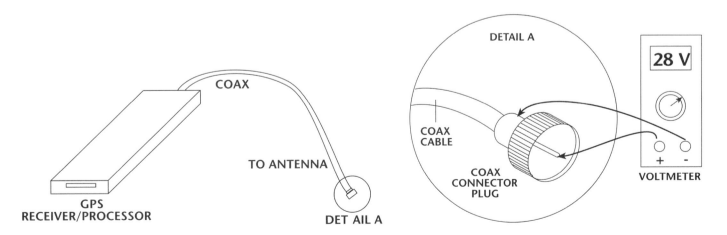

Fig. 8-28   Measuring the DC voltage sent from the GPS receiver to the antenna.

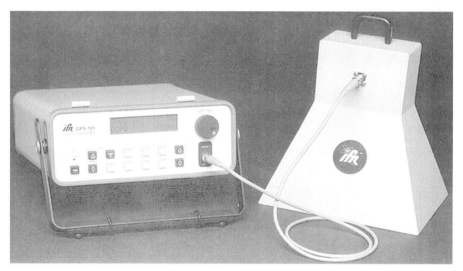

Fig. 8-29   Model GPS-101 GPS simulator. (IFR Systems, Inc.)

this case, we will assume the unit fails the test. 2) To eliminate the antenna as a possible fault, simply plug the GPS-101 simulator into the "antenna-end" of the coax and run another test. Once again, we will assume the airborne equipment fails the test. 3) Move the test unit inside the aircraft and connect the simulator output directly to the receiver/ processor (i.e., the "receiver-end" of the antenna connection). If the coaxial cable is faulty, the GPS receiver will pick up the signals transmitted by the test unit.

In the above example, we simply used the GPS-101 simulator to eliminate different sections of the airborne GPS equipment (the antenna, the coax, and the receiver). When troubleshooting, always keep in mind the power supply to the antenna amplifier. If the receiver had worked properly with the antenna bypassed (step 2 above), the fault lies in the antenna or the power sent to the antenna for the amplifier. Check for the proper voltage to the antenna before condemning the antenna assembly.

At the time that this text was written, there were only limited self-diagnostics available through the GPS receiver/processor. Typically, if the antenna and wiring have been eliminated as possible faults, the receiver/processor is defective. The receiver/ processor cannot be repaired on the flight line and must be sent to the proper repair facility.

Interference created by other avionics systems can also cause GPS problems. If the GPS is experiencing intermittent signal loss, be sure to question the pilot as to when signal loss occurs. Ground based and aircraft radio transmitters may interfere with the GPS reception. Although this is extremely limited for ground-based transmitters, airborne VHF communication can easily cause interference. Ask the pilot where and when the GPS signal was lost. If the communication radio was in use at the time, be sure to suspect interference from the communication radio.

APPENDIX A
Excerpt from ATA Specification 2200 (formally ATA 100) Revision 2001.1
(Reproduced with permission from the Air Transport Association of America, INC.)

Note 1: Only assigned numbers shown in this table.
Note 2: Sub-System/Section set to zero are all general Information (not shown below)

SYS CHAP	SUB-SYS/ SECTION	TITLE	SYS CHAP	SUB-SYS/ SECTION	TITLE
1		*Reserved for Airline Use	24		**ELECTRICAL POWER**
2		*Reserved for Airline Use		-10	Generator Drive
3		*Reserved for Airline Use		-20	AC Generation
4		*Reserved for Airline Use		-30	DC Generation
5		**TIME LIMITS & MAINTENANCE CHECKS**		-40	External Power
	-10	Time Limits		-50	AC Electrical Load Distribution
	-20	Scheduled Maintenance Checks		-60	DC Electrical Load Distribution
	-50	Unscheduled Maintenance Checks	25		**EQUIPMENT & FURNISHINGS**
6		**DIMENSIONS & AREAS**		-10	Flight Compartment
7		**LIFTING & SHORING**		-20	Passenger
	-10	Jacking		-30	Gallery
	-20	Shoring		-40	Lavatories
8		**LEVELING & WEIGHING**		-50	Additional Compartments
	-10	Weighing & Balancing		-60	Emergency
	-20	Leveling		-70	Available
9		**TOWING & TAXIING**		-80	Insulation
	-10	Towing	26		**FIRE PROTECTION**
	-20	Taxiing		-10	Detection
10		**PARKING, MOORING, STORAGE & RETURN TO SERVICE**		-20	Extinguishing
				-30	Explosion Suppression
	-10	Parking & Storage	27		**FLIGHT CONTROLS**
	-20	Mooring		-10	Aileron & Tab
	-30	Return to Service		-20	Rudder & Tab
11		**PLACARDS & MARKINGS**		-30	Elevator & Tab
	-10	Exterior Color Schemes & Markings		-40	Horizontal Stabilizer
	-20	Exterior Placards & Markings		-50	Flaps
	-30	Interior Placards		-60	Spoiler, Drag Devices & Variable Aerodynamic Farings
12		**SERVICING**			
	-10	Replenishing		-70	Gust Lock & Dampener
	-20	Scheduled Servicing		-80	Lift Augmenting
	-30	Unscheduled Servicing	28		**FUEL**
18		**VIBRATION & NOISE ANALYSIS (Helicopter Only)**		-10	Storage
				-20	Distribution
	-10	Vibration Analysis		-30	Dump
	-20	Noise Analysis		-40	Indicating
20		**STANDARD PRACTICES-AIRFARME (Ref. [Subject 3-1-5])**	29		**HYDRAULIC POWER**
				-10	Main
21		**AIR CONDITIONING**		-20	Auxiliary
	-10	Compression		-30	Indicating
	-20	Distribution	30		**ICE & RAIN PROTECTION**
	-30	Pressurization Control		-10	Airfoil
	-40	Heating		-20	Air Intakes
	-50	Cooling		-30	Pitot & Static
	-60	Temperature Control		-40	Windows, Windshields & Doors
	-70	Moisture & Air Contaminant Control		-50	Antennas & Radomes
22		**AUTOFLIGHT**		-60	Propellers & Rotors
	-10	Autopilot		-70	Water Lines
	-20	Speed & Attitude Correction		-80	Detection
	-30	Auto Throttle	31		**INDICATING & RECORDING SYSTEMS**
	-40	System Monitor		-10	Instrument & Control Panels
	-50	Aerodynamic Load Alleviating		-20	Independent Instruments
23		**COMMUNICATIONS**		-30	Recorders
	-10	Speech Communications		-40	Central Computers
	-20	Data Transmission & Automatic Calling	32		**LANDING GEAR**
	-30	Passenger Address, Entertainment & Comfort		-10	Main Gear & Doors
				-20	Nose Gear & Doors
	-40	Interphone		-30	Extension & Retraction
	-50	Audio Integrating		-40	Wheels & Brakes
	-60	Static Discharging		-50	Steering
	-70	Audio & Video Monitoring		-60	Position & Warning
	-80	Integrated Audio Tuning		-70	Supplementary Gear

509

APPENDIX B
A List of Common Acronyms Used in This Text

AAP - Altitude Awareness Panel

ACARS – Airborne Communications Addressing and Reporting System

ACESS - Advanced Cabin Entertainment Service System

ADC - Air Data Computer

AFDS - Autopilot Flight Director System

AHC - Attitude Heading Computer

AIM - Acknowledgement, Iso-alphabet, Maintenance

AIMS - Airplane Information Management System

APP - Autopilot Panel

ARP - Air Data Reference Panel

ASCB - Avionics Standard Communication Bus

ASM - Aircraft Schematic Manual

ATA - Air Transport Association

ATC - Automatic Trim Coupler

AWL - Aircraft Wiring List

AWM - Aircraft Wiring Manual

BCD - Binary Coded Decimal

BITE - Built In Test Equipment

BNR - Binary Data

CCTM - Cabin Configuration Test Module

CDC - Control Display Coupler

CDU - Control Display Unit

CFDIU - Centralized Fault Display Interface Unit

CFDS - Centralized Fault Display System

CHP - Course Heading Panel

CIC - Cabin Interphone Controller

CIS - Cabin Interphone System

CLS - Cabin Lighting System

CMC - Central Maintenance Computer

CMC - Current Mode Coupler

CMCS - Central Maintenance Computer System

CMU - Control Management Unit

CSDB - Commercial Standard Digital Bus

DAU - Data Acquisition Unit

DCP - Display Control Panel

DEMUX - Demultiplexer

DEU - Display Electronic Unit

DGPS - Differential Global Positioning System

DMC - Display Management Computer

DMU - Data Management Unit

DSDL - Dedicated Serial Data Line

DSP - Display Select Panel

EAD - Engine Alert Display

EADI - Electronic Attitude and Director Indicator

ECAM - Engine Condition and Monitoring

ECS - Environmental Control System

ED - Electronic Display

EEC - Electronic Engine Control

EFD - Electronic Flight Display

EFIS - Electronic Flight Instrument System

EHSI - Electronic Horizontal Situation Indicator

EICAS - Engine Indicating and Crew Alerting System

EIS - Electronic Instrument System

EIU - Electronic Interface Unit

ES - Elevator Station

ESC - Entertainment Service Controller

FAC - Flight Augmentation Computer

FCC - Flight Control Computer

FDE - Flight Deck Effect

FIM - Fault Isolation Manual

FIN - Functional Item Number

FMC - Flight Management Computer

FMS - Flight Management System

FPD - Flat Panel Display

FRM - Fault Reporting Manual

FS - Fuselage Station

FWC - Flight Warning Computer

GAMA - General Aviation Manufacturers Association

GMT - Greenwich Mean Time

GPS - Global Positioning System

GPWC - Ground Proximity Warning Computer

IAPS - Integrated Avionics Processor System

IDS - Integrated Display System

IDU - Integrated Display Unit

INS - Inertial Navigation Systems

IRS - Inertial Reference System

IRU - Inertial Reference Unit

LAAS - Local Area Augmentation System

LASER - Light Amplification by Stimulated Emission of Radiation

LCD - Liquid Crystal Display

LSK - Line Select Keys

LVDT - Linear Voltage Differential Transducer

MAWEA - Modularized Avionics and Warning Electronics Assembly

MCDU - Maintenance Control Display Unit

MFD - Multifunction Display

MCDU - Multipurpose Control and Display Unit

MPU - Multifunction Processor Unit

MSP - Mode Select Panel

MU - Management Unit

MUX - Multiplexer

NBPT - No Break Power Transfer

ND - Navigational Display

NRZ - Non-Return to Zero

NS - Nacelle Station

PAC - Passenger Address Controller

PAS - Passenger Address System

PES - Passenger Entertainment System

PFD - Primary Flight Display

PM - Phase Modulation

PSS - Passenger Service Systems

PWR - Power module

RAT - Ram Air Turbine

RCP - Radio Communications Panel

RLG - Ring Laser Gyro

RMM - Ramp Maintenance Manual

RVDT - Rotary Voltage Differential Transducer

RZ - Return to Zero

SAT - Static Air Temperature

SD - System Display

SDD - Sensor Display Driver

SDI - Source Destination Identifier

SDU - Sensor Display Unit

SG - Synchronization Gap

SIM - Serial Interface Module

SSM - Sign Status Matrix

TAT - True Air Temperature

TC - Terminal Controller

TG - Terminal Gap

TI - Transmit Interval

TR - Transformer Rectifier

TSM - Trouble Shooting Manual

VCC - Video Control Center

VDU - Video Distribution Unit

VSCU - Video System Control Unit

WAAS - Wide Area Augmentation System

WGD - World Geodetic System

WS - Wing Station

GLOSSARY

Accelerometer - device that senses aircraft acceleration

Acknowledgement, ISO-alphabet, Maintenance (AIM) - an ARINC 429 data word format used for systems that require large amounts of data transfer

Active Matrix LCD - a type of LCD with a transistor located at each pixel intersection

Activity Monitor - monitors the data bus signals to ensure data transmission at regular intervals

Air Data Computer (ADC) - a system that monitors pitot pressure, static pressure and air temperature to determine various parameters, such as airspeed, altitude, and vertical climb

Air Transport Association (ATA) – an organization that represents airlines, aircraft manufacturers, and various system manufactures in an effort to ensure uniformity in various facets of the aviation industry

Airborne Communications Addressing and Reporting System (ACARS) - a digital air/ground communications service designed to reduce the amount of voice communications on the increasingly crowded VHF frequencies

ARINC 629 (also called 629) - a digital data bus format that permits up to 120 receiver/transmitters to share a bi-directional serial data bus

ARINC Incorporated - a global corporation made up of various U.S. and international airlines, aircraft operators and their subsidiaries. The company provides services related to a variety of aviation communication and navigation systems

Attenuation - occurs when a digital data signal becomes weak or the transmitted voltage becomes too low

Availability - the capability of the GPS system to be used for navigation whenever it is needed, and the ability to provide that service throughout the entire flight

Avionics Standard Communication Bus (ASCB) - a bi-directional data bus operating at 0.667 MHz

Binary Coded Decimal (BCD) - a specific ARINC 429 data word format

Bipolar - a digital data format that reverses polarity (two-polarity) when it changes from binary 1 to binary 0

BITS Mode - used to access real time data from the RAM memory of the DPU and/or MPU on the Collins EFIS 85/86

Block Diagram - defines electrical circuits by separating a system or subsystem into functional blocks

Built In Test Equipment (BITE) - systems used in conjunction with many digital circuits to aid in system troubleshooting

Bus Hierarchy - a design concept to ensure that the most critical electrical systems are the least likely to fail

Butt Line - a vertical reference plane, which divides the aircraft front to rear through the center of the fuselage

Cabin Interphone System (CIS) - provides communications between different flight attendant stations, and between flight attendant stations and the flight deck on the B-747-400

Caution - a notation that will call attention to any methods, materials, or procedures, which must be followed to avoid damage to equipment on the aircraft

Central Maintenance Computer System (CMCS) - the advanced built-in troubleshooting systems found on Boeing aircraft

Centralized Fault Display Interface Unit (CFDIU) - the main computer that manages the CFDS information

Centralized Fault Display System (CFDS) - is the advance diagnostic system used on Airbus Industries transport category aircraft

Character Generator - processor circuitry used to produce letters, numbers, and symbols

Cold Cathode Fluorescent Tube - a highly efficient fluorescent tube often used to illuminate a transmissive LCD

Commercial Standard Digital Bus (CSDB) - a one way data bus system between one transmitter and a maximum of 10 receivers

Compensator Unit - used to correct for magnetic errors and flux detector misalignment

Configuration Strapping - (also known as pin programming) specific electrical connections used to determine various display, input/output signal formats and other system parameters

Continuity - the probability that a GPS service will continue to be available for a period of time necessary to complete the navigation requirements of the flight

Control Display Unit (CDU) - an LRU containing a display and alphanumeric keyboard typically used to control the auto-flight and other system functions

Current Mode Coupler (CMC) - an inductive coupling device, which connects the LRU to the 629 data bus

Data Bus Analyzer - test equipment used to troubleshoot digital systems

Datum - a vertical reference plane for aircraft, typically located toward the front of the fuselage

Dedicated Serial Data Line (DSDL) - a digital bus structure unique to Airbus aircraft

Demultiplexer (DEMUX) - a circuit that converts serial data into parallel data

Differential GPS (DGPS) - a design concept to increase the accuracy, availability, integrity and continuity of the basic GPS to a level sufficient for complete aircraft navigation

display and processor circuitry

Differential Transducer - a solid-state device used to generate an electrical signal, typically used to identify the position of moving components

Discrete Word Format - an ARINC 429 data word format used to transmit the status of several individual components

Display Electronic Unit (DEU) - used on the B-747-400 to provide the processing power for the six display units and other aircraft systems

Display Management Computer (DMC) - the main processor for the CRT displays found on the Airbus A-320

Doublet Signal - a short positive and negative pulse (spike) on the bus whenever the data value changes from binary one to binary zero or back

Down Converter - a circuit found in GPS receivers used to filter out any extraneous signals and converts the RF input to a given intermediate frequency (IF)

Dutch Roll - a slow oscillation of the aircraft about its longitudinal axis

Dynamic Pressure - the difference between pitot pressure and static pressure

Effectivity - used to determine if the aircraft is covered by the information stated in that section (page) of a manual

Electric Servo - utilizes an electric motor and clutch assembly to move the aircraft s control surface according to autopilot commands

Electronic Flight Display (EFD) - an LRU used to provide primary flight and navigation data on a LCD or CRT display

Electronic Flight Instrument System (EFIS) - employs 2 or more CRTs or LCDs to present alphanumeric data and graphical representations of aircraft flight instruments

Electronic Instrument System (EIS) - found on the B-747-400 to display flight and navigational data, as well as engine parameters and warning information

Electrostatic Discharge - the discharge (movement of electrons) created when any material containing a static charge is exposed to a material containing a different or neutral charge

EICAS - Engine Indicating and Crew Alerting System

Ephemeris Errors - caused by slight variations in the GPS satellites positions as they orbit the earth

Equipment Identifier - part of the ARINC 429 data word used to further define the word label

Fault Flag - any indication made on one or more displays related to a complete or partial system failure

Fault Isolation Manual (FIM) - a part of the aircraft maintenance manuals, which contains various repair strategies for given faults

Flag Note - used to bring attention to a specific effectivity on a diagram or schematic

Flat Panel Display (FPD) - a solid-state device used to display various formats of video information in modern aircraft flight displays

Flight Deck Effects - any EFIS or EICAS display, or discrete annunciator used to inform the flight crew of a system fault

Flight Director - a visual aid to the pilot, present on the ADI, EADI or PFD used for reference or manual control of the aircraft

Flight Interphone - a system used to provide communications between flight crewmembers and/or other aircraft operations personnel

Flight Management System (FMS) - computer-based systems that reduce pilot workload by providing: automatic radio tuning, lateral and vertical navigation, thrust management, and the display of flight plan maps

Flight Phase — a time period of a given flight often recorded by the CMC to help the technician determine the airplanes configuration at the time of a system fault

Flight phase Screening - a software function used to help eliminate nuisance messages from entering the central maintenance system

Flux Detector - remote sensor used in a Flux Gate Compass system

Flux Gate Compass - a system that employs one or more remote sensors to produce an electric signal that can be used to determine the aircraft s position relative to magnetic North

Free Running Oscillator - a circuit to establish a given frequency

Functional Item Number (FIN) - a unique number given to each line replaceable unit on Airbus aircraft

Fuselage Station (FS) - used to indicate locations longitudinally along the aircraft fuselage

Glanoss - the former Soviet Union s satellite navigation system

Global Positioning System (GPS) - a satellite-based system that is capable of providing position and navigational data for ground based and airborne receivers

GLOBALink - the name of the ACARS service provided by ARINC Incorporated

Home Diagram - the diagram / schematic where the component is shown in full detail

Hydraulic Servo - the most powerful type of servo actuator; hence, these units are typically used on transport category aircraft

Inertial Navigation Systems (INS) - an aircraft and navigation system that does not rely on external radio signals; laser gyros and accelerometers provide three dimensional navigation capabilities

Inertial Reference System (IRS) - a combination of laser gyros and accelerometers used to sense the aircraft s angular rates and accelerations

Inertial Reference Unit (IRU) - an LRU containing laser gyros and accelerometers used for aircraft navigation

Integrated Avionics Processor System (IAPS) - provides integration functions for various avionics systems

Integrity - the ability of the GPS system to shut itself down when it is unsuited for navigation or to provide timely warnings to the pilot of the system failure

Interphone System - found on transport category aircraft to provide communications between flight crew, ground crew, flight attendants, and maintenance personnel

Ionization Air Blower - a special air blower that will safely delete any static charge formed on nonconductors

Kalman Filter - an advanced software algorithm used in GPS receivers to process the complex position calculations necessary to determine aircraft position

Line Select Keys (LSK) - push button switches found on the control display units used to select items for display or activate functions available on the CDU

Link Test - used to test ACARS link between airborne equipment and ground based equipment

Liquid Crystal Display (LCD) - a type of flat panel display that employs the use of liquid crystals to control light

Local Area Augmentation System (LAAS) - a GPS upgrade designed to be used during a Category II or Category III precision approach

Maintenance Control Display Unit (MCDU) - used on some Boeing aircraft to monitor and test the flight control computers, flight management computers, and the thrust management computers

Maintenance Manual - provides specific information for flight line and hangar maintenance activities

Manchester II Code - (often simply referred to as Manchester) is a serial digital data format that incorporates a voltage change in each data bit

Microfiche - a 4 by 6 inch filmcard, which is capable of storing up to 288 pages of data

Microfilm Strips - referred to as film, and used mainly by airlines to store maintenance and parts

Multifunction Display (MFD) - an EFIS display used to display weather radar data, course and flight plan information, system checklist, and provide back-up functions in the event of a partial system failure

Multifunction Processor Unit (MPU) - a processor used with the Collins EFIS 85/86 that provides signal processing and switching to the multifunction display

Multiplexer (MUX) - converts parallel data into serial data

Multipurpose Control and Display Unit (MCDU) - LRU containing an alpha numeric keyboard and CRT display found on Airbus aircraft used to input data to the CMCS and the CFDS

Nacelle Station (NS) or Elevator Station (ES) - used on larger aircraft to help technicians find components in specific areas of the aircraft

NAVSTAR GPS - the formal name for the global positioning system in the United States

No Break Power Transfer (NBPT) - an automated system that allows the aircraft to switch AC power supplies without a momentary interruption of electrical power

Non-Return to Zero (NRZ) - a self-clocking data bus format that does not return to zero at the end of each data bit

Note – found in various service manuals, used to draw your attention to a particular procedure, that will make the task easier to perform

Nuisance Message - any FDE or CMC message that is not caused by an actual fault

Pad Bits - "fill in" any portion of the data field not used for the transmission of data

Page Block - used to categorize a given chapter-section-subject of the maintenance manual

Parity Bit - a binary bit set by the transmitter that permits error checks by each receiver connected to the bus

Passenger Address System (PAS) - system for transmitting audio through a series of cabin speakers to announce messages

Passive Matrix LCD - employs a grid of conductors with pixels located at each intersection of the grid

Phase Modulation (PM) - digital modulation of the carrier for GPS and satellite

Pin Programming - (also known as configuration strapping) specific electrical connections used to determine various display, input/output signal formats and other system parameters

Pitot Pressure - the absolute pressure of the air that is "pressed" into the front of the aircraft

Pneumatic Servo - vacuum actuated autopilot component used to move control surfaces; typically found on light aircraft

Power Distribution System - one or more electrical distribution points used to connect the main power supply for the aircraft

Ram Air Turbine (RAT) - a propeller driven system which is deployed from the fuselage to provide emergency power in the event of a catastrophic hydraulic or electrical system failure

Ramp Maintenance Manual (RMM) – an abbreviated maintenance manual designed specifically for line maintenance and minor troubleshooting

Resolution - the smallest unit of data that can be measured or transmitted by a given system

Return to Zero (RZ) - a self-clocking data bus format that returns to zero during the second part of each bit

Ring Laser Gyro (RLG) - an angular rate sensor that employs a laser to detect aircraft motion.

Rotating Mass Gyro and Ring Laser Gyro - used by many autopilots to detect movement of the aircraft, gyro outputs are also used for reference on certain navigation systems

Sensitive Wires - considered critical to flight safety and must not be modified without specific manufacturer approval

Serial - transmission of data one binary digit or bit at a time

Serial Interface Module (SIM) - an ARINC 629 subsystem that changes the Manchester current signal from the LRU into an analog voltage doublet signal

Servo - a device used to move the flight control surface in accordance with autopilot command signals

Sign Status Matrix (SSM) - a portion of the ARINC 429 data word provides information, which might be common to several peripherals

Simplified Schematic - wiring illustration with intermediate depth and scope

Single Wire Electrical System - uses the airframe to distribute negative voltage

Split Parallel - power distribution systems used on some modern four-engine aircraft such as the Boeing 747-400

Standard Practices - the procedures and practices used repeatedly during aircraft maintenance, troubleshooting, and repair

Static Air Temperature (SAT) - the temperature of the undisturbed air surrounding the aircraft

Structural Modal Oscillations - an undesired effect created by turbulence, which causes bending of the fuselage around the wing area

Stub Cable - is a 4-wire cable used to connect the TC to the current mode coupler on an ARINC 629 subsystem

Synchronization Gap (SG) - a time period common to all transmitters on the ARINC 629 data bus

Tach Generator - used in electric servo systems as rate sensors

Terminal Controller (TC) - an ARINC 629 subsystem that moves data to and from the LRU memory

Terminal Gap (TG) - period unique for each transmitter connected to the ARINC 629 data bus

Timing Tolerance Fault - occurs when the rise or fall of a digital signal responds too slowly to be within specifications of the data bus standard

Transformer Rectifier (TR) - A device to change 115 VAC to 26 VDC

Transmissive Display - LCD containing a dedicated light source mounted to the rear of the display

Transmit Interval (TI) - a common period for all transmitters on the ARINC 629 data bus

Trouble Shooting Manual (TSM) - designed specifically for system troubleshooting

True Air Temperature (TAT) - a measure of the air temperature as it is compressed by the moving aircraft

Vector Generator - processor circuitry used to draw lines on a CRT display

Vertical Navigation - an autopilot mode used to fly the selected altitude or glide path

VU Number — specific number used to designate all panels and racks on Airbus aircraft

Warning - call attention to any methods, materials, or procedures, which must be followed to avoid injury or death

Water Line - a horizontal reference plane, which runs the length of the aircraft from nose to tail

Wide Area Augmentation System (WAAS) - a GPS enhancement designed to improve the integrity, accuracy, availability, and continuity of the basic satellite navigation system over a large area of coverage

Wing Leveler - a simple autopilot system used to provide guidance along only the longitudinal axis of the aircraft. These systems, often found on light single engine aircraft, were called because they were used to keep the wings level

Wing Station (WS) - used to indicate locations longitudinally along the wings of the aircraft

Wire Identification Number — a label placed on all wires three inches or longer

Wire Routing Chart - used to locate wire bundles that run through the aircraft

Wiring Diagram - very specific diagrams with details on wires, connectors, and pin numbers for a given system

Wiring Manual - contains various diagrams, charts, lists and schematics needed to maintain the various electrical/electronic systems on the aircraft

Yaw Damper - system designed to control rudder and eliminate dutch roll

Zoning — a reference system designed by the Air Transport Association (ATA) to further identify the location of components on large aircraft

INDEX